Microbiology of the phyllosphere

# Microbiology of the phyllosphere

Edited by

N. J. FOKKEMA
and
J. VAN DEN HEUVEL

Willie Commelin Scholten
Phytopathological Laboratory
Baarn
The Netherlands

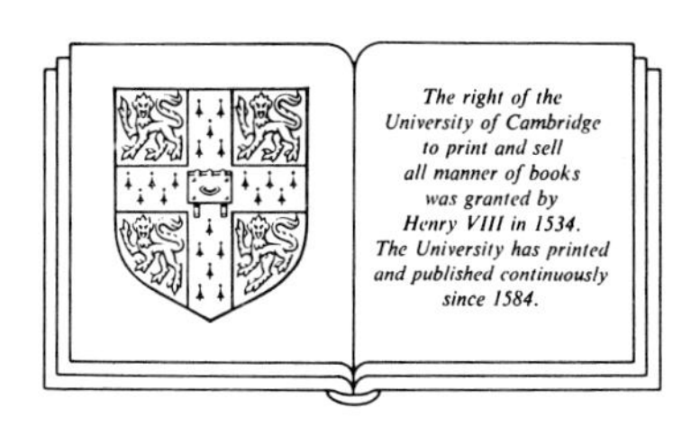


CAMBRIDGE UNIVERSITY PRESS
Cambridge
London New York New Rochelle
Melbourne Sydney

Published by the Press Syndicate of the University of Cambridge
The Pitt Building, Trumpington Street, Cambridge CB2 1RP
32 East 57th Street, New York, NY 10022, USA
10 Stamford Road, Oakleigh, Melbourne 3166, Australia



First published 1986

Printed in Great Britain at the University Press, Cambridge

*British Library cataloguing in publication data*

Microbiology of the phyllosphere

1. Micro-organisms, Phytopathogenic
2. Leaves – Microbiology

I. Fokkema, N.J. II. Heuvel, J. van den

581.2′3 QR351

*Library of Congress cataloguing in publication data available*

ISBN 0 521 32344 4

CONTENTS

SECTION III: ENDOPHYTIC LEAF FUNGI



SECTION IV: PLANT-PATHOGENIC AND SAPROPHYTIC PROKARYOTES



SECTION V: BIOLOGICAL CONTROL ON AERIAL PLANT SURFACES

# PREFACE

The microbiology of the phyllosphere, the environment created by the leaf and colonized by bacteria, yeasts and moulds, attracts scientists from different disciplines such as plant pathology, mycology and microbial ecology. The phyllosphere has been an inspiring and unifying theme since 1970 when the first symposium on this subject was organized by T.F. Preece and C.H. Dickinson. The present volume contains the review-type papers presented at the Fourth International Symposium on the Microbiology of the Phyllosphere, which was held from 2 to 6 September 1985 at Wageningen, the Netherlands.

The programme selected by the organizing committee included many new developments in techniques on monitoring micro-organisms and in theoretical concepts on colonization, as well as several new subjects not treated in previous symposium volumes.

Phyllosphere research is starting to benefit from genetic manipulation: new techniques that greatly improve the possibilities of tracing micro-organisms, elucidating the role of particular traits of micro-organisms and preparing 'ideal' biocontrol agents are now becoming available. The term phyllosphere implies a three-dimensional habitat; going upwards from the leaf surface we may meet micro-organisms not directly contacting the surface, such as sooty moulds and hyperparasites, going downwards we may encounter non-pathogenic endophytic fungi which exert a newly discovered fungal mutualism with potentially great implications for the host population as far as resistance to insects and plant pathogens is concerned. The difference between inside and outside the leaf, and also between pathogen and saprophyte is sometimes vague and irrelevant, particularly in the case of bacteria. All these different micro-organisms may interact with each other and, although not being dealt with as a particular group, plant pathogens are controlled directly or indirectly by coexisting micro-organisms and this can be exploited by practical biological control in some instances. This area will expand provided that basic research on colonization, survival and climate adaptation of micro-organisms increases as well.

We are very grateful for the assistance we received during our editorial work from Ms M.A. Williamson correcting the English and from P.A.H.M. Bakker, G.T.N. de Leeuw and R.J. Scheffer for specialist advice. D. Dutscher is acknowledged for his participation in the preparation of some of the illustrations, A.J. Schreuder for checking bibliographic data and

A. Gerrits van de Ende for his help with the compilation of the index. The production of this camera-ready book would have been impossible without our flexible and cooperative departmental secretaries Ms. C. Chornelis-Sinai and Ms W.G.M. Schellingerhout-Huyg. Their assistance and that of Ms L.M. den Bak, D.T. Fokkema, Ms W.A.J. Jansen and J.H. van Pelt in the preparation of the typescript is gratefully acknowledged.

The assembly of international authorities on different aspects of phyllosphere research and the organization of the symposium would not have been possible without the generous support of the following sponsors, who are gratefully acknowledged:
State University Utrecht
Royal Netherlands Academy of Arts and Sciences
Heineken N.V.
Shell Nederland B.V.
BASF Nederland B.V.
Bayer AG, Geschäftsbereich Pflanzenschutz
Schering AAgrunol B.V.
Aseptafabriek B.V.

Finally, we thank Cambridge University Press for the trust in the success of this book and the pleasant cooperation. We expect that this volume will become a major reference text for all scientists involved in phyllosphere research and related fields.

Baarn, June 1986

N.J. Fokkema
J. van den Heuvel

Organizing committee:

J.H. Andrews
J.P. Blakeman
G.C. Carroll
N.J. Fokkema
J. van den Heuvel
T.F. Preece

# SECTION I

# NEW TECHNIQUES

# MICROBIOLOGICAL AND SAMPLING CONSIDERATIONS FOR QUANTIFICATION OF EPIPHYTIC MICROBIAL COMMUNITY STRUCTURE

C.E. Morris and D.I. Rouse

*Department of Plant Pathology, University of Wisconsin-Madison, 1630 Linden Drive, Madison, WI 53706, USA*

## *INTRODUCTION*

When thinking about the structure of epiphytic microbial communities two questions come to mind. Firstly, what are the number and relative proportions of the various species present? Secondly, what physical and biological interactions exist between the species present and their environment? An answer to the first question must necessarily precede an answer to the second question since one must know what organisms are present and in what proportion before their interactions with each other and the environment can be studied. The number of species and/or their relative proportions within a community characterize what ecologists call the 'diversity' of a community.

Ecologists have exerted a massive effort to elucidate the factors which govern the number and relative proportions of species in plant and animal communities. The result of their efforts is a body of theory, reviewed by Pianka (1966), explaining the effect of time, resource limitation, climate, and species interactions on the diversity of plant or animal species in a community. This theory is based almost entirely on observations which include broad generalizations about environmental conditions (Buzas 1972), and is supported by very little experimental evidence. Consequently, the validity and universality of the proposed relationships of diversity to the physical and biological environment have been questioned (Good 1975; Houston 1979).

Recently, microbial ecologists have devoted much attention to the measurement of diversity in microbial ecosystems (Gamble *et al.* 1977; Kaneko *et al.* 1977; Griffiths & Lovitt 1980; Martin & Bianchi 1980; Mills & Wassel 1980; Hauxhurst *et al.* 1981; Troussellier & Legendre 1981; Atlas *et al.* 1982; Bianchi & Bianchi 1982; Horowitz *et al.* 1983; Atlas 1984). Only a few of these studies have been of epiphytic microbial communities. The microbial diversity studies to date have only partially addressed the following basic questions:

i) How does one express numerically and concisely the number and relative proportions of species present in a microbial community?
ii) What is the minimum number of isolates that should be sampled to estimate diversity and to compare diversity between different microbial communities?
iii) Should isolates be classified according to the currently accepted taxonomy or should they be classified based on their function or behaviour in the environment?

iv) What is the relationship between the level of diversity and biological or physical environmental factors?

The microbial diversity literature to date has assumed answers to some of these questions and obtained preliminary answers to others.

Thus microbial ecologists have perpetuated the efforts of plant and animal ecologists to measure diversity and to find relationships between the diversity of organisms in a community and parameters of the physical or biological environment. Frequently, results of their studies are consistent with the existing body of theory, but they are also based on observations and broad generalizations rather than experimental evidence. Microbial ecosystems can be manipulated more readily than plant and animal ecosystems. However, with the exception of one study of bacteria and protozoa (Hairston *et al.* 1968), microbial ecologists have not exploited the opportunity to provide experimental evidence for the theories generated by plant and animal ecologists. They have directly applied the sampling strategies used by plant and animal ecologists to estimate diversity, and have not come to grips with the conceptual or experimental aspects of the question of how to estimate diversity in microbial communities where there are often $10^6$ or more individuals per sampling unit.

In this chapter we will review some of the literature on diversity and indicate some of the problems that need to be addressed to advance our understanding of epiphytic microbial diversity. Because of the paucity of work on diversity of epiphytic microbial diversity much of the following literature review is drawn from research reported on other microbiotic systems or macrobiotic systems.

## *DEFINITION OF DIVERSITY*

There is no single generally accepted definition of diversity despite the considerable effort devoted by plant and animal ecologists to explain the patterns of diversity in ecological systems (Peet 1974). The multiple definitions that exist fall into one of three categories:

i) the number of species in a community (cf. Pielou 1977; McGuinness 1984),
ii) the number of species and their relative proportions (Margalef 1958),
iii) parameters of the distribution which describes the frequency of species containing 1, 2, 3,...n individuals (Fisher *et al.* 1943; Preston 1948; Edden 1971).

Each of these three definitions was developed to address specific observations about the structure of plant and animal communities. These definitions of diversity fall into Whittaker's category of 'α diversity' (Whittaker 1972), that is, the variety of organisms in a defined region. Whittaker (1972) has noted that there is also 'β diversity', defined as the rate of change of species composition across a gradient. This discussion will be confined to α diversity.

### *The number of species in a community*

The number of species in a community is the simplest of all the definitions of diversity. This definition was suggested to be ecologically meaningful because it addressed the concept that the maximum number of species that can be supported in a given region depends on the number of habitable sites in that region. It is often observed that number of species is positively correlated with the area of a habitat. The resulting hypothesis of habitat diversity states that the number of species increases with increasing area due to the addition of habitable sites. The development of this hypothesis is extensively reviewed from a historical perspective by McGuinness (1984).

There are a number of studies of epiphytic populations which provide at least a partial list of the species found on leaf surfaces. A review of the papers presented at the past three meetings in this series (Preece & Dickinson 1971; Dickinson & Preece 1976; Blakeman 1981) suffices to show that much effort has been expended to obtain lists of micro-organisms that can be found on plant surfaces. Most of the studies conducted do not address the question of diversity directly because the lists of species are usually not obtained on a quantitative basis (Parbery *et al.* 1981). In other words, there is usually no statistical estimation given of the precision or the bias associated with the sampling methods used.

### *The number of species and their relative proportions*

Many contributors to the ecological literature were uncomfortable with defining diversity in terms of number of species and felt that a better definition would include both the number and relative proportions of species (Peet 1974). They believed that, for example, a community with 10 equally abundant species should have a different level of diversity than a community with 91% of the individuals in one species and 1% in each of the other nine species. Incorporation of the components of species richness and evenness into the concept of diversity was first formalized by Simpson (1949). Speculation about the relevance of this definition of diversity to the actual biology or ecology of organisms came shortly thereafter. Hurlbert (1971) proposed that measures of diversity which incorporate species richness and evenness estimate the probability of interspecific encounters. This assumes that each individual has the potential to come into contact with all other individuals in the community. These encounters are important for predation, competition, and other types of antagonism between organisms. Such measures of diversity purportedly also estimate the 'information content' of communities, but the nature of this 'information' has been defined vaguely (Margalef 1958; Atlas 1984).

The most widely used measure of diversity in the microbial ecology literature is the Shannon index (Shannon & Weaver 1949). There are few examples of the use of a diversity index accounting for species richness applied to a microbial epiphytic community. Morris (1985) studied the diversity of bacterial epiphytic communities on bean leaflets. She examined the statistical properties of the Shannon index as applied to those communities. Calculated values of this index were 1.71, 1.81, 3.72

and 2.77 for samples of isolates taken from each of four randomly selected leaflets indicating considerable variation in diversity from leaflet to leaflet.

*The frequency distribution of species containing 1, 2, ... n individuals*

The basis for the third definition of diversity was the observation of a pattern of relative abundances of species common to many communities. Plant (Williams 1950), bird (MacArthur 1957), insect (Fisher *et al.* 1943; Preston 1948; Williams 1953; Van Emden & Williams 1974), marine snail (Kohn 1959), and diatom (Patrick 1968) communities are all characterized by a high frequency of species represented by only one or two individuals and a low frequency of species represented by many individuals in a sample. Based on goodness of fit, Preston (1948) characterized the frequency distribution describing the relative abundances of species as lognormal. Alternatively, Fisher *et al.* (1943) reasoned that this distribution could be described by the logarithmic series. Their reasoning was as follows. They noted that the probability of sampling 'x' individuals from a population was a Poisson random variable with some mean 'm' that varied with the efficiency of the sampling technique and the size of the population. For a heterogeneous mixture of populations, there would be an 'm' for each component population. They then assumed an underlying distribution of 'm''s such that the probability of observing 'n' individuals of any component population was a random variable distributed as a negative binomial. Finally they noted that the logarithmic series was the best approximation of the negative binomial distribution for censored data.

Ecological and biological explanations for the skewed distribution of species containing 1, 2, 3, ... n individuals in a community (i.e. species abundance distribution) have been proposed. Odum *et al.* (1960) suggested that in a community of organisms there are a few producer classes which provide 'services' needed by all members of the community, and many small classes of organisms that provide 'services' needed by only a few members. They developed this model based on an analogy of the abundance of occupations in human societies, where many individuals are involved with food production, for example, but few individuals occupy themselves with each of a variety of tasks that provide goods to only a small proportion of the community. This analogy may be useful for explaining the skewed abundance distribution of species from many trophic levels, but does not offer an explanation for a skewed distribution of species on the same trophic level. Kendall (1948) demonstrated mathematically that species abundance will be distributed according to a log series if the birth, death and immigration rates of the individuals in each species are distributed according to a negative exponential function of time.

Estimation of diversity indices that are parameters of the frequency distribution of the abundance of groups or species is problematic for epiphytic communities, and may be problematic in the studies of other types of microbial communities as well. The distribution of numbers of groups containing 1, 2, ... n individuals may not be adequately described by a logarithmic series distribution. There is no indication that the distribution of species abundance in any microbial community is symmetrical and efforts to fit such distributions as the lognormal may be inappropriate. Morris (1985) constructed frequency distributions of group (species) abundance for epiphytic bacterial communities on bean leaflets and found that the logarithmic series distribution was inadequate to describe her data.

## *ESTIMATION OF SPECIES DIVERSITY*

Numerous indices have been proposed in accordance with the definitions of diversity listed above, and are described by Peet (1974). Although the relevance of the existing definitions of diversity to the biology and ecology of organisms is unclear, the statistical aspects of diversity indices have been thoroughly investigated. The major statistical problems associated with measuring diversity, as reflected by the number of publications devoted to them, are i) estimation of population parameters from a sample, and ii) comparison of the levels of diversity among different samples. Many of the other difficulties associated with measuring diversity, such as determining the appropriate sample size, have been addressed in the attempt to resolve these two basic problems.

### *Estimation of population parameters*

Measures of diversity of organisms in a community are usually based on a sample of individuals and are rarely based on a complete census of the community. Many of the diversity indices which incorporate information about the number of species in a sample, e.g. the Shannon index (Shannon & Weaver 1949), increase with increasing sample size. Therefore, methods for estimating the total number of species in a community, $s^*$, have been sought.

Fisher *et al.* (1943) and Preston (1948) were the first to introduce methods of estimating $s^*$. According to either method, the first step in estimating $s^*$ involves fitting the data for the number of species containing 1, 2, ... n individuals to either the log series (Fisher *et al.* 1943) or lognormal distribution (Preston 1948). The total number of species in the community can then be estimated from the parameters of the respective distribution. To use the model of Fisher *et al.* (1943) accurately, one must know that the number of species is related to the logarithm of the sample size (Fisher *et al.* 1943). This can be determined from virtually any sample of reasonable size. In contrast, Preston's model assumes knowledge of the mean number of individuals per species in the population (Edden 1971). This information can only be obtained from samples that are a large proportion, or a complete census, of the total population. Small samples may lead to inaccurate estimates of these parameters, and to an inappropriate assumption that the lognormal actually describes the species abundance in the population.

Many of the models to estimate $s^*$ proposed, subsequent to those of Fisher *et al.* and Preston, have involved intricate probability theory and assumptions about the population that are difficult to verify from a sample (Efron & Thisted 1976; Starr 1979). Additionally, these methods, as well as those of Fisher *et al.* and Preston, assume that the relationship between sample size and number of species is constant over all sample sizes. Given this assumption, direct extrapolation of the relationship between the number of species and any appropriate transformation of sample size (Margalef 1958; Odum *et al.* 1960; Menhinick 1964) may offer the simplest solution to estimating the total number of species in a community.

The proportion of a species in the total population, $p_i$*, also can be estimated inaccurately from a sample as noted by Good (1953). To increase the accuracy of estimates of $p_i$*, he proposed 'smoothing' the sample species abundance frequency distribution and then estimating the $p_i$* from the smoothed curve. He assumed that species abundance should conform to a smooth distribution (e.g. lognormal or negative binomial), and that the sample distribution deviated from a smooth distribution due to sampling error. Pielou (1966) recommended modified formulae for indices incorporating measures of $p_i$*, when $p_i$* is believed to be inaccurately estimated from a sample.

### *Comparison of diversity among different samples*

The techniques for estimating population parameters have rarely been used successfully. Therefore, the diversity of total populations is unknown and comparisons of the levels of diversity for different communities must be based on samples. As mentioned previously, the value of many diversity indices increases with increasing sample size. Furthermore, the relationship between sample size and the value of an index is a function of the species abundance distribution. Consequently, community 'A' may be more diverse than 'B' for sample sizes greater than N, but 'B' may be more diverse than 'A' for sample sizes less than N. This dilemma was discussed in detail by Peet (1974).

Sanders (1968) was one of the first to address the problem of determining the appropriate sample size for comparing levels of diversity among communities. He introduced the technique of rarefaction whereby the mean and standard deviation of an index are estimated over a range of sample sizes less than the total number of individuals collected. This permitted comparison of diversity values which had been estimated with different sample sizes, and also revealed communities that were relatively less diverse at some sample sizes than at others. The technique has since been modified by Simberloff (1978).

A statistical approach to the comparison of values of diversity has been proposed by Hutcheson (1970). Using the Shannon index (Shannon & Weaver 1949) (H') as an example, he noted that it is asymptotically normally distributed for samples from a single population, as confirmed by Bowman *et al.* (1971) and later by Tong (1983). Given this distribution and an estimate of the variance associated with estimates of H', the significance of differences between H' values could be evaluated based on a simple Student's t-test.

The extremely large population size of many microbial communities, compared to plant or animal communities, necessitates examination of the statistical aspects of diversity measurement in these communities as well. As indicated above there already exists a large literature on the statistical aspects of diversity measurement. Because of the well defined sampling unit (leaves) available to epiphytic microbial ecologists, and the abundance of these sampling units it should be possible to address experimentally several of the statistical problems raised above.

### *METHODOLOGICAL PROBLEMS*

Problems more basic than sample size complicate the measurement of diversity of epiphytic communities and need to be addressed in more detail than accomplished by research to date. The first of these is the operational definition of an epiphytic community. Although Pielou (1975) noted that the definition and scope of a community is arbitrary and based on the interests of the researcher, the definition of an epiphytic community is, in reality, limited by isolation procedures. One may be interested in the diversity of bacteria on leaf surfaces, for example, but may actually be measuring the diversity of bacteria that can be removed from leaf surfaces, grow on a general plating medium, and survive some period of time in storage before being characterized. The difference between the diversity of the community one desires to examine and that of the community actually examined may or may not affect the conclusions of the study. How to determine the bias of our samples in this sense is unresolved.

The second basic problem of measuring epiphytic microbial diversity is the ability to sample randomly. This is a major assumption of all diversity indices (Pielou 1975). Although a random aliquot of washing buffer may be sampled, some types of micro-organisms in the sample may be clumped or they may inhibit the growth of other types of microbes on media. The estimated proportion of the inhibited group of bacteria is therefore influenced by a factor other than sample size. Parbery *et al.* (1981) compared several indirect plating techniques including leaf printing and dilution plating onto various general media. They demonstrated considerable variability between techniques in recovery of epiphytic fungi. Morris (1985) compared several techniques for removal of bacterial epiphytes and found that sonication of bean leaflets for 21 min removed more bacteria than did washing leaflets in flasks for 2 h on a rotary shaker. It is not known what effect these different techniques for removing bacteria had on the composition of observed bacteria species although some trends appeared depending on the general growth medium selected for plating (Morris 1985).

There is also a within leaf sampling problem associated with dilution plating techniques. The proportions of certain groups in the bacterial community from one leaflet can vary drastically among aliquots of the leaf washing. This may be true for fungi as well. These results suggest that small samples should be composed of isolates randomly chosen from platings of multiple aliquots.

The third question which must be addressed to determine the diversity of epiphytic microbial community is how to characterize epiphytic micro-organisms. This is particularly true for epiphytic bacteria. Standard procedures are well documented for the characterization of isolates according to currently accepted taxonomy. Although the concept of species is meaningful for mammals, birds, insects and plants, the meaning of bacterial species, as distinguished by the currently accepted taxonomy, has been debated (e.g. Trüper & Krämer 1981). Furthermore, the species or even generic identity of many bacterial isolates cannot be readily determined in ecological studies as in the case of the study, by Griffiths & Lovitt (1980), of bacteria isolated from oil storage tanks. It may be desirable to characterize isolates based on traits that presumably relate to the function or behaviour of these isolates in their environment, but the effect on diversity of the nature and number of traits selected should be examined. Factors regulating the diversity of bacteria may not be operating at the level of currently defined species but, rather, at the level of a few traits associated with growth and survival. Among individuals in a community of epiphytic bacteria there may be a diversity of nutrient affinities, mechanisms for surviving starvation and desiccation, and levels of tolerance to environmental pollutants.

Of the many environmental factors influencing epiphytic bacteria, the quality and quantity of nutrients available on leaf surfaces certainly play a role in the growth of these bacteria. Several researchers have adapted a numerical taxonomic system based on nutrient utilization to characterize epiphytic bacteria (Goodfellow *et al.* 1976; Austin *et al.* 1978; Morris 1985). For example, Morris (1985) focused on the nutritional aspect of the environment by characterizing isolates of epiphytic bacteria on the basis of their ability to utilize single carbon and nitrogen sources in a defined minimal medium. She selected nutrients for this study that had been found in leachate from bean leaves or were commonly found in leachate of a wide range of plants. Many of the nutrients selected had also been used in taxonomic studies of the fluorescent pseudomonads as well as in studies of epiphytic bacteria from other plants (Stanier *et al.* 1966; Sands *et al.* 1970). Morris & Rouse (1985) studied the effect of nutrients on the bacterial community on bean leaflets. They found that application of low concentrations affected the composition of the epiphytic community by increasing the proportion of those bacteria able to utilize the nutrient applied. This result was used as the basis for suggesting the application of bacterial nutrients as a means of achieving biological control.

The final problem in measuring bacterial diversity, to be discussed here, is how to account for the composition of communities with the existing indices. For example, the Shannon index measures the probability of correctly guessing the identity of a randomly sampled individual (Piclou 1975). If $H'=0$, then this probability equals 1 because there is only one species or group of organisms present. Of course, this is only true if the composition of the community is known. Of considerable importance is the fact that many epiphytic micro-organisms are not found on every leaf. For example, several epiphytic bacterial species have been found to be lognormally distributed on leaves of several plants (Hirano & Upper 1983, this volume). This means that many leaves may have undetectable levels of these bacteria present or they may not be present at all, while other individual leaves in close proximity may have very large numbers of these bacteria present. What, then, is the probability of correctly guessing the identity of a randomly sampled isolate from a randomly sampled leaflet? As a simplistic example of the importance of acknowledging composition of communities in measuring diversity, assume that for epiphytic communities of all leaflets in a large field, H'=0. Clearly the conclusions about the ecology of these epiphytes would be different if all leaflets harboured the same species of epiphyte than if each leaf harboured a different species. A question immediately arises here: should one consider the diversity of epiphytic micro-organisms on populations of leaves or within the individual leaf?

The definitions of community and of diversity of epiphytic micro-organisms need to be carefully considered, and measurement of diversity needs to incorporate information about the identity of species or groups as well as their numbers and relative abundances. The diversity of epiphytic micro-organisms may be readily measured, but leaves, themselves, constitute a population characterized by variability in exposure to certain environmental conditions and to sources of epiphytic microbial inoculum. Modified methods of quantifying diversity need to be explored for microbial communities for which there is a population of sampling or colonizable units.

## *REFERENCES*

Atlas, R.M. (1984). Use of diversity measurements to assess environmental stress. *In* Current Prospectives in Microbial Ecology, eds M.J. Klug & C.A. Reddy, pp. 540-5. Washington, DC: Am. Soc. Microbiol.

Atlas, R.M., Busdosh, M., Krichevsky, E.J. & Kaneko, T. (1982). Bacterial populations associated with the Arctic amphipod *Boeckosimus affinis*. Canadian Journal of Microbiology, *28*, 92-9.

Austin, B., Goodfellow, M. & Dickinson, C.H. (1978). Numerical taxonomy of phylloplane bacteria isolated from *Lolium perenne*. Journal of General Microbiology, *104*, 139-55.

Bianchi, M.A.G. & Bianchi, A.J.M. (1982). Statistical sampling of bacterial strains and its use in bacterial diversity measurement. Microbial Ecology, *8*, 61-9.

Blakeman, J.P., ed. (1981). Microbial Ecology of the Phylloplane. London: Academic Press.

Bowman, K.O., Hutcheson, K., Odum, E.P. & Shenton, L.R. (1971). Comments on the distribution of indices of diversity. *In* Statistical Ecology, III: Many Species Populations, Ecosystems, and Systems Analysis, eds G.P. Patil, E.C. Pielou & W.E. Waters, pp. 315-66. University Park: Penn. State Univ. Press.

Buzas, M.A. (1972). Patterns of species diversity and their explanation. Taxon, *21*, 275-86.

Dickinson, C.H. & Preece, T.F., eds (1976). Microbiology of Aerial Plant Surfaces. London: Academic Press.

Edden, A.C. (1971). A measurement of species diversity related to the lognormal distribution of individuals among species. Journal of Experimental Marine Biology and Ecology, *6*, 199-209.

Efron, B. & Thisted, R. (1976). Estimating the number of unseen species: how many words did Shakespeare know? Biometrika, *63*, 435-47.

Emden, H.F. van & Williams, G.F. (1974). Insect stability and diversity in agro-ecosystems. Annual Review of Entomology, *19*, 455-75.

Fisher, R.A., Corbet, A.S. & Williams, C.B. (1943). The relationship between the number of species and the number of individuals in a random sample from an animal population. Journal of Animal Ecology, *12*, 42-58.

Gamble, T.N., Betlach, M.R. & Tiedje, J.M. (1977). Numerically dominant denitrifying bacteria from world soils. Applied and Environmental Microbiology, *33*, 926-39.

Good, D. (1975). The theory of diversity-stability relationships in ecology. Quarterly Review of Biology, *50*, 237-66.

Good, I.J. (1953). The population frequency of species and the estimation of population parameters. Biometrika, *40*, 237-64.

Goodfellow, M., Austin, B. & Dickinson, C.H. (1976). Numerical taxonomy of some yellow pigmented bacteria isolated from plants. Journal of General Microbiology, *97*, 219-33.

Griffiths, A.J. & Lovitt, R. (1980). Use of numerical profiles for studying bacterial diversity. Microbial Ecology, *6*, 35-43.

Hairston, N.G., Allan, J.D., Colewell, R.K., Futuyama, D.J., Holwell, J., Lubin, M.D., Mathias, J. & Vandermeer, J.H. (1968). The relationship between species diversity and stability: an experimental approach with protozoa and bacteria. Ecology, *49*, 1091-101.

Hauxhurst, J.D., Kaneko, T. & Atlas, R.M. (1981). Characteristics of bacterial communities in the Gulf of Alaska. Microbial Ecology, *7*, 167-82.

Hirano, S.S. & Upper, C.D. (1983). Ecology and epidemiology of foliar bacterial plant pathogens. Annual Review of Phytopathology, *21*, 243-69.

Horowitz, A., Krichevsky, M.I. & Atlas, R.M. (1983). Characteristics and diversity of subarctic marine oligotrophic, stenoheterotrophic, and euryheterotrophic bacterial populations. Canadian Journal of Microbiology, *29*, 527-35.

Houston, M. (1979). A general hypothesis of species diversity. American Naturalist, *113*, 81-101.

Hurlbert, S.H. (1971). The nonconcept of species diversity: a critique and alternative parameters. Ecology, *52*, 577-86.

Hutcheson, K. (1970). A test for comparing diversities based on the Shannon formula. Journal of Theoretical Biology, *29*, 151-4.

Kaneko, T., Atlas, R.M. & Krichevsky, M. (1977). Diversity of bacterial populations in the Beaufort Sea. Nature (London), *270*, 596-9.

Kendall, D.G. (1948). On some models of population growth leading to R.A. Fisher's logarithmic series distribution. Biometrika, *35*, 6-15.

Kohn, A. (1959). The ecology of *Conus* in Hawaii. Ecological Monographs, *29*, 47-90.

MacArthur, R.H. (1957). On the relative abundance of bird species. Proceedings of the National Academy of Sciences of the U.S.A., *43*, 293-5.

McGuinness, K.A. (1984). Equation and explanation in the study of species-area curves. Biological Reviews of the Cambridge Philosophical Society, *59*, 423-40.

Margalef, D.R. (1958). Information theory in ecology. General Systems, *3*, 36-71.

Martin, Y.P. & Bianchi, M.A. (1980). Structure, diversity, and catabolic potentialities of aerobic heterotrophic bacterial populations associated with continuous cultures of natural marine phytoplankton. Microbial Ecology, *5*, 265-79.

Menhinick, E.F. (1964). A comparison of some species-individuals diversity indices applied to samples of field insects. Ecology, *45*, 859-61.

Mills, A.L. & Wassel, R.A. (1980). Aspects of diversity measurement for microbial communities. Applied and Environmental Microbiology, *40*, 578-86.

Morris, C.E. (1985). Diversity of epiphytic bacteria on snap bean leaflets based on nutrient utilization abilities: Biological and statistical considerations. Ph.D. thesis, University of Wisconsin, Madison.

Morris, C.E. & Rouse, D.I. (1985). Role of nutrients in regulating epiphytic bacterial populations. *In* Biological Control on the Phylloplane, eds C.E. Windels & S.E. Lindow, pp. 63-82. St. Paul: Am. Phytopath. Soc.

Odum, H.T., Cantoln, J.E. & Kornicker, L.S. (1960). An organizational hierarchy postulate for the interpretation of species-individual distributions, species entropy, ecosystem evolution, and the meaning of a species-variety index. Ecology, *41*, 395-9.

Parbery, I.H., Brown, J.F. & Bofinger, V.J. (1981). Statistical methods in the analysis of phylloplane populations. *In* Microbial Ecology of the Phylloplane, ed. J.P. Blakeman, pp. 47-65. London: Academic Press.

Patrick, R. (1968). The structure of diatom communities under varying ecological conditions. Annals of the New York Academy of Sciences, *108*, 359-65.

Peet, R.K. (1974). The measurement of species diversity. Annual Review of Ecology and Systematics, *5*, 285-307.

Pianka, E.R. (1966). Latitudinal gradients in species diversity: a review of concepts. American Naturalist, *100*, 33-46.

Pielou, E.C. (1966). The measurement of diversity in different types of biological collections. Journal of Theoretical Biology, *13*, 131-44.

Pielou, E.C. (1975). Ecological Diversity. New York: John Wiley and Sons, Inc.

Pielou, E.C. (1977). Mathematical Ecology. New York: John Wiley and Sons, Inc.

Preece, T.F. & Dickinson, C.H. (1971). Ecology of Leaf Surface Micro-organisms. London: Academic Press.

Preston, F.W. (1948). The commonness and rarity of species. Ecology, *29*, 254-83.

Sanders, H.L. (1968). Marine benthic diversity: a comparative study. American Naturalist, *102*, 243-82.

Sands, D.C., Schroth, M.N. & Hildebrand, D.C. (1970). Taxonomy of phytopathogenic pseudomonads. Journal of Bacteriology, *101*, 9-23.

Shannon, C.E. & Weaver, W. (1949). The Mathematical Theory of Communications. Urbana: Univ. of Illinois Press.

Simberloff, D. (1978). Use of rarefaction and related methods in ecology. *In* Biological Data in Water Pollution Assessment: Quantitative and Statistical Analysis, ASTM STP652, eds K.L. Dickson, J. Cairns, Jr. & R.J. Livingstone, pp. 150-65. Philadelphia: Am. Soc. for Test. and Mat.

Simpson, E.H. (1949). Measurement of diversity. Nature (London), *163*, 688.

Stanier, R.Y., Palleroni, N.J. & Doudoroff, M. (1966). The aerobic pseudomonads: a taxonomic study. Journal of General Microbiology, *43*, 159-271.

Starr, N. (1979). Linear estimation of the probability of discovering a new species. Annals of Statistics, *7*, 644-52.

Tong, Y.L. (1983). Some distribution properties of the sample species-diversity indices and their applications. Biometrics, *39*, 999-1008.

Troussellier, M. & Legendre, P. (1981). A functional evenness index for microbial ecology. Microbial Ecology, *7*, 283-96.

Trüper, H.G. & Krämer, J. (1981). Principles of characterization and identification of prokaryotes. *In* The Prokaryotes. A Handbook on Habits, Isolation, and Identification of Bacteria. Vol. I, eds M.P. Starr, H. Stolp, H.G. Trüper, A. Balows & H.G. Schlegel, pp. 176-93. Berlin: Springer-Verlag.

Whittaker, R.H. (1972). Evolution and measurement of species diversity. Taxon, *21*, 213-51.

Williams, C.B. (1950). The applications of the logarithmic series to the frequency of occurrence of plant species in quadrats. Journal of Ecology, *38*, 107-38.

Williams, C.B. (1953). The relative abundance of different species in a wild animal population. Journal of Animal Ecology, *22*, 14-31.

# HOW TO TRACK A MICROBE

J.H. Andrews

*Department of Plant Pathology, University of Wisconsin-Madison, 1630 Linden Drive, Madison, WI 53706, USA*

## *INTRODUCTION*

What factors influence the distribution and abundance of organisms? This is a central issue in ecology. Distribution tells us something of the biotic and abiotic constraints a species can tolerate. Abundance provides a measure of dominance and a point of departure for assessing the organism's actual role in community energy flow or nutrient cycling. Meaningful answers to the foregoing question hinge on having a reliable way to first determine distribution and abundance.

Measurement of distribution and abundance obviously requires that the organisms involved be identifiable. Occasionally a species may be naturally absent from a particular habitat. It is then relatively easy to experimentally release organisms and track them and their progeny over time directly (by microscopy if the microbe is morphologically distinct) or indirectly (by bioassay or as plate counts on a selective medium). Usually, however, microbes are either indistinguishable microscopically, or the system is subject to influx of other members of the same species, or both. In these situations, either a unique attribute of the population of interest must be exploited, or a specific tag introduced as a marker. Thus, markers can serve to assess both the extent of dispersal and the density of labelled populations.

In this chapter I attempt to provide an overview of the key considerations associated with using markers in microbial ecology. The thrust is on the development and limitations of markers, rather than on the results of studies in which they have been employed. Antimicrobial resistance markers are emphasized becaused recently they have become prominent in ecological studies.

## *MARKERS AND THE KINETICS OF EPIPHYTIC MICROBES*

Of the eight types of populations discussed by Brock (1971) with respect to growth kinetics, the 'dividing transit' category probably best typifies that found on a leaf surface, i.e. cells are entering and leaving the system as well as dividing within it. In some microbial habitats immigration and emigration can be ignored for all practical purposes. For example, in flowing water, Brock (1971) contends that microbes adhere so tightly to substrata that emigration is effectively inconsequential. To distinguish growth of periphytic bacteria in hot springs from immigration, slides were immersed, raised, irradiated with ultraviolet (UV) light at intervals about equal to the generation time of the population (Bott & Brock 1970), and then lowered again. Since newly attached arrivals were killed before they divided, the rate of population increase on non-irradiated slides represented both growth and immigration whereas that on the irradiated slides reflected only immigration. They concluded that influx was quantitatively unimportant. In a related study of the thermophilic blue-green alga, *Synechococcus* sp., immigration of cells could be excluded because high temperatures (*ca* 70°C) of the water upstream from the sampling stations precluded algal growth (Brock and Brock 1968).

For leaf-microbe systems, unlike the above examples, influx is significant (J.H. Andrews and L.L. Kinkel, this volume) and outflow may also be appreciable because plant canopies can constitute a major source of airborne bacteria (Lindemann *et al.* 1982). Thus, population density assayed at any time will reflect the net result of four variables: immigration (I) or influx, emigration (E) or outflow, growth (G) and death (D). It is difficult to isolate the variables and this complicates efforts to assess growth or the contribution of each parameter to actual densities observed. If a natural population declines, this does not necessarily mean that it is not growing, but merely that $I + G < E + D$. Similarly, an increase might simply reflect influx without growth. However, if a marked population can be introduced, and if it increases, then this must be due to growth alone which is sufficient to overcome losses from both emigration and death.

An interesting perspective is that the use of marked organisms by microbial ecologists is analogous to the mark, release, and recapture approach which is standard practice in animal ecology (Andrewartha & Birch, 1984). Animals are captured, marked, released and the population density estimated from the proportion of tagged individuals caught subsequently. Although the approaches and interpretations differ in detail, the important point is that wherever markers are used, the fundamental premise is that marked and unmarked organisms behave similarly. Not only are such assumptions rarely fulfilled, but typically they are not addressed critically and often not even acknowledged. Just as some animals are more likely to be trapped than others (some seek the trap, others avoid it), marking microbes often results in associated genetic changes which influence competitiveness or some other ecologically important trait. These and related limitations are examined in some detail later.

## *OPTIONS FOR TRACKING*

The common types of markers and their associated assay procedures are summarized in Table 1 (see also Brockwell *et al.* (1978) and Schwinghamer & Dudman (1980)). Although theoretically all could be used in ecological studies, the method(s) of choice is limited by logistics and the microbial system to be studied. Direct enumeration based on distinctive morphology of the organism (as in *Chromatium* and *Leucothrix* spp.; Brock 1971), colour variants (albino or pigmented mutants), or fluorescent antibody (FA) procedures overcomes the limitations of indirect counts but it is not commonly used. One problem is that, even with FA, due to lack of specificity, an introduced biotype generally cannot be discriminated from background members of the same species. Recent introduction of monoclonal antibody procedures may remedy this shortcoming. Although the new molecular techniques such as gene probing (Maniatis *et al.* 1982; Kosuge & Nester 1984) are highly sensitive and specific (Grunstein & Hogness 1975; Southern 1975; Broome & Gilbert 1978; Clarke *et al.* 1979), they are presently too laborious for large-scale use. Non-radioactive (biotinylated) variants of the procedure, based on formation of a coloured product, are now available in

Table 1. Some types of markers.

| Marker | Assay |
|---|---|
| Antigenic or cell wall structure | serotyping, fluorescent antibodies or modification thereof; phage sensitivity (plaques) |
| Cryptic host DNA | autoradiography (isotope dilution); Southern blot; in situ colony hybridization or modification thereof |
| Introduced, non-replicating DNA (superinfecting phage) | loss of phage marker from dividing lysogenic bacteria by UV induction of lytic cycle |
| 'Morphological' | colour, colony morphology, ice nucleation acitivity |
| Mating type | compatibility, formation of heterokaryons or sexual state |
| Metabolic or biochemical | ability to grow on minimal medium with appropriate marker (e.g. lactose) as sole carbon source; ability to tolerate metabolite (amino acid) analogs; ability to form coloured colonies on media with appropriate precursors |
| Auxotrophic | inability to grow on media lacking the specific nutrient |
| Drug resistance | ability to grow on media containing the antimicrobial agent (antibiotic or fungicide) |

medicine as ready-to-use, diagnostic kits for identification of selected pathogens in fixed microscopic samples.

The radiographic (Brock 1971) and superinfecting phage (Meynell 1959) methods (Table 1) are interpreted similarly. For the former, cellular DNA is labelled (e.g. with $^{3}$H-thymidine) and the microbe then introduced to the environment of interest. Because DNA replicates in semi-conservative fashion, the label will be reduced by half at each generation. The marked population can be tracked (and the growth rate estimated by the time taken for the grain count to halve) until the dilution effects reduce the grain count to background levels. Likewise, the nonreplicating genetic marker will decrease by half at each generation and the loss assessed by plating procedures. However, in general, neither method would be suitable for phylloplane studies or other 'open' systems where appreciable immigration of non-labelled individuals would preclude prolonged tracking and result in gross overestimates of growth.

In short, markers of the drug resistance type (e.g. to antibiotics, phage, metals, various organic and inorganic chemicals, and numerous physical stressors) have attained widespread use because they can be easily obtained, and accompanied by direct or positive selection procedures (see below). Since an activity lacking in the wild-type parental line is gained by the mutant, relatively rare mutants can be detected within a dense milieu of non-mutant, sensitive cells. Thus the technique has great resolving power. Enrichment procedures, commonly needed for other types of markers are not needed.

### *CHOICE OF ANTIMICROBIAL AGENT AS MARKER*

Some of the most commonly used antibiotics and fungicides, their mode of action, and mechanisms of resistance are summarized in Table 2. Close attention should be paid to the correct preparation and storage of drug solutions, the details of which are beyond the scope of this chapter (see Anhalt & Washington 1980).

What is the best choice of marker? Should more than one marker be used? These questions are difficult to answer. Authors rarely say why a particular inhibitor was selected. Many choices are undoubtedly made empirically or based on precedent or convenience. Ideally, a marker should be clearly distinctive, easily obtained, stable, possess minimal pleiotropic effects, and the organism within which it resides be fully characterized (discussed later). Indications of probable performance can be gleaned from the literature, but evidence as it pertains to a specific system can only be obtained *post facto* from preliminary experiments.

Occasionally, especially with antibiotics, multiple markers such as rifampicin-nalidixic acid resistance are used. This has been done to check the persistence of one of the markers, in which case it should be shown in advance that they operate independently, i.e., that there is no cross-resistance. Better yet, a complementary approach, such as use of antigenic markers (Brockwell *et al.* 1977) should be taken. However, in general, the rationale for multiple markers is rarely explained or

Table 2. Some common antimicrobial agents: mode of action and mechanism of resistance.

| Agent | Mode of action | Mechanism of resistance conferred by | |
|---|---|---|---|
| | | R-plasmids | Chromosomal mutation |
| Ampicillin | kills growing bacteria by interfering with a terminal step in cell wall synthesis[1] | resistance gene specifies β-lactamase which cleaves β-lactam ring of antibiotic[1] | peptidoglycan synthesis altered[2] |
| Kanamycin | bactericidal agent; binds to 70S ribosomes; causes misreading of m-RNA[1] | enzyme specified modifying antibiotic, preventing interaction with ribosomes[1] | membrane transport protein altered[2] |
| Rifampicin | bactericide, binds to DNA-dependent RNA polymerase blocking transcription[4] | - | subunit of RNA polymerase modified[4] |
| Strepto-mycin | bactericide; binds to 30S ribosome subunit; causes misreading of m-RNA[1] | enzyme specified modifying antibiotic, preventing interaction with ribosomes[1] | ribosomal subunit altered[2] |
| Nalidixic acid | bactericide; inhibits DNA gyrase activity[2] | NR[4 7] | gyrase sub-unit altered[3] |
| Benomyl[5] | fungistat or fungicide; prevents microtubule assembly thereby interfering with mitosis | - | change in tubulin structure |

Table 2 (continued)

| Agent | Mode of action | Mechanism of resistance conferred by | |
|---|---|---|---|
| | | R-plasmids | Chromosomal mutation |
| Nystatin[4 6] | fungistat or fungicide; complex with membrane sterols causing disruption of permeability | - | alteration of sterols in cell membranes |

[1] Maniatis *et al.* (1982).
[2] Lin *et al.* (1984).
[3] Bainbridge (1980).
[4] McEvoy (1985).
[5] Davidse & De Waard (1984).
[6] Molzahn & Woods (1972).
[7] NR: not reported.

justified. They are apparently installed to minimize the likelihood that tolerant strains might exist indigenously.

### *RESISTANCE PATTERNS AND GENETICS*

Resistance to antimicrobics by bacteria in nature is commonly carried on plasmids (R-factors), rather than resulting from spontaneous mutation of the bacterial chromosome (e.g. Cohen *et al.* 1972; Lin *et al.* 1984). The distinction between these two types is important because the resistance used in tracking has been obtained by mutation and is commonly associated with pleiotropic and other effects discussed later. Although plasmids have recently been found in fungi (Beringer & Hirsch 1984; Mishra 1985), genetic analysis of resistant strains indicates that generally the mutation of a single chromosomal gene is responsible for drug resistance. More than one locus may be involved (Davidse & De Waard 1984; Dekker 1984).

For most antibiotics, typified by penicillin, highly resistant mutants cannot be isolated in a single step but resistance increases stepwise in approximately geometric fashion ('multi-step' mutants). For the streptomycin category (aminoglycosides), different *individual* mutations confer a wide range in resistance. Thus, high resistance can be found as a result of a single mutation to only certain antimicrobials and, other things being equal, these are the compounds of choice as markers (Hopwood 1970).

### *MUTAGENESIS AND ADDITION OF FOREIGN GENES*

Occasionally the microbe population of interest will be sufficiently heterogeneous with respect to the marker and level of resistance desired that spontaneous mutants can be selected without resorting to mutagenesis. Historically, spontaneous mutants have been preferred, where possible, for ease of isolation and to avoid extraneous 'cryptic' mutations, but this situation is changing as markers can now be routinely inserted biologically into an otherwise unaltered organism (see below).

Mutagenesis procedures have been reviewed at length elsewhere (e.g. Hopwood 1970; Carlton & Brown 1981; Rowlands 1983) and only some general considerations can be summarized here.

#### *Type of mutation*

Broadly speaking, mutations can be of the point, deletion, or insertion type (Table 3; Carlton & Brown 1981). Depending on the research objective, a particular sort is usually preferable. For example, in biochemical studies, various of the point mutations (e.g. frameshift) are useful because behaviour of the mutant can be compared with functional revertants to the wild-type parent. On the other hand, for ecological studies involving a marker, ideally one should seek a stable, nonreverting, mutant of the deletion or insertion type. Unfortunately, deletion mutations are unlikely to result in antibiotic resistance and chemical mutagenesis, which is often used, does not produce deletions with high efficiency.

#### *Mutagen*

The foregoing suggests that choice of mutagen depends logically on the type of mutation desired, and how effectively the mutagenic agent induces it (Carlton & Brown 1981). In practice, however, agents seem to be chosen more subjectively, based on convenience or empirical success (Hopwood 1970). Occasionally more than one type of mutation and mutagen may be desirable (Carlton & Brown 1981). Mutagens may be classified as physical (e.g. UV light), chemical (e.g. nitrosoguanidine) or biological (transposons). With physical or chemical mutagens, dose is important and varies with circumstances. When a positive selection screen (i.e. for characteristic(s) gained by the mutant) is available, as it is for antimicrobial agents, one can strive for a high absolute number of viable mutants per culture, rather than proportion among survivors (Hopwood 1970). The limitation on this is that as dose increases, the likelihood of accompanying undesired mutational change also increases (Hopwood 1970; Rowlands 1983).

UV irradiation is a simple, convenient way to obtain a wide variety of mutants of the point and deletion type. It can be used most effectively for non-pigmented rather than coloured organisms, including bacteria (Carlton & Brown 1981) and fungi (Leben *et al.* 1955; Papavizas *et al.* 1982).

Nitrosoguanidine, which is highly mutagenic at low killing dosages, causes primarily point mutations of the transition type. The main problems associated with its use are that closely-linked multiple mutations tend to be induced and the chemical is a carcinogen. It has been used to induce benomyl-resistant markers (Cullen & Andrews 1985) and auxotrophic mutants (Churchill & Mills 1984) in fungi, as well as extensively in industrial strain improvement programs (Rowlands 1983) and in bacteriology (Carlton & Brown 1981).

Changing microbes by inserting foreign DNA is the most novel and exciting approach to obtaining desired phenotypes. Although there are numerous modes of bacterial gene transfer (Low & Porter 1978), the most common involve conjugation (cell-to-cell contact), transformation (introduction of naked DNA), or transduction (phage infection). One important use of conjugation is transposon mutagenesis. Transposons are mobile sequences of DNA which insert themselves into host chromosome or plasmid DNA (Kleckner *et al.* 1977; Kleckner 1981; Beringer *et al.* 1984;Kosuge & Nester 1984). They are usually delivered to recipient bacteria in non-replicating vectors known as 'suicide plasmids' (which disappear because they cannot replicate in their new host) within donor cells. The transposon carries its own transposase functions and remains only in the cells in which integration into the host genome has occurred. Among other things, bacterial transposons encode drug resistance markers, for which screens can be developed; they also cause insertional inactivation

Table 3. Origins and mechanisms of antibiotic resistance.

| Type of genetic change | Nature of host chromosomal alteration | Induced or conferred by | Effect on host gene function | Reversions |
|---|---|---|---|---|
| Point mutation | generally base substitutions | spontaneous; physical; chemical | variable to complete loss | usual |
| Deletion mutation | loss of DNA | physical; chemical | complete loss | extremely low frequency |
| Insertion mutation | addition of DNA | phage; transposon; insertional sequence | complete loss | usually low frequency |
| Introduction of independent replicons | none | plasmids | addition | variable frequency |

of genes which can lead to mutant phenotypes for which selection is possible. Alternatively, cells can be transformed by extrachromosomal DNA that replicates as an independent replicon (Kado & Lurquin 1982). The foreign DNA introduced may also determine selectable phenotypes such as antibiotic resistance.

Thus, both transposons and independent replicons can be used to biologically 'label' an organism (Table 3). For instance, *Escherichia coli* has been transformed to multiple drug resistance by purified R-factor DNA (Cohen *et al.* 1972), and Beringer *et al.* (1978) have inserted Tn*5*, a transposon coding for kanamycin resistance, into the chromosome of *Rhizobium leguminosarum*. Tn*1*, a marker for ampicillin resistance has been transposed in *R. meliloti* (Casadesus *et al.* 1980), and Tn*7* (trimethoprim and streptomycin resistance) has been inserted into the Ti plasmid of *Agrobacterium tumefaciens* (Hernalsteens *et al.* 1978). Bioluminescence genes (*lux*) of *Vibrio fischeri* have been inserted into cloning vectors that can introduce any gene set into any Gram-negative bacteria so far tested to monitor phytopathogenic bacteria in plants (C.I. Kado, personal communication; Shaw *et al.* 1985).

Transformation and transpositional events are less well characterized in fungi (Mishra 1985). Transformation systems involving complementation of auxotrophic markers or the utilization of unusual carbon or nitrogen sources have been developed (Hinnen *et al.* 1978; Case *et al.* 1979; Tilburn *et al.* 1983). More recently, systems based on dominant selectable markers such as G418 (a 2-deoxystreptamine antibiotic), methotrexate, oligomycin, and benomyl resistance have been developed for yeasts (Zhu *et al.* 1985) and filamentous fungi (e.g. Bull & Wootton 1984; D.C. Cullen, personal communication).

To conclude, the use of movable drug-resistance elements offers a precise and highly controlled means of manipulating microbes for genetic, biochemical and ecological investigations. Currently the greatest application is in molecular biology, where these marker elements are useful in isolating mutants, strain construction, localized mutagenesis, chromosome mapping, and so forth. Drug resistance is a preferred phenotype, because positive selection can be used to isolate and identify genes of interest (Kleckner *et al.* 1977). In ecology, transposon and related technologies are research breakthroughs because investigators can now demonstrate single site mutations by mapping the insertion and because the genes transferred code for enzymes that alter the drug rather than the target site of the drug in the host. Regardless of how attractive prospective ecological applications may be, studies in nature will be deferred until the legal restrictions on use of genetically engineered organisms are removed (see conclusion).

### *Mutant expression*

Devising logical, efficient procedures for detecting the desired mutants is probably the most important step of the protocol. It is better to invest time developing a suitable screen than wasting time by taking numerous isolates of the mutagenized line (Hopwood 1970).

Fortunately, screening for resistance to antimicrobial agents is done simply and directly by 'rational' screens (Rowlands 1983), without the need for involved enrichment methods, because the mutants have gained a selectable phenotype and hence are easy to detect. Where a property is lost, as is the case for auxotrophs, detection must be indirect or essentially random, although facilitated by replica-plating procedures. The main questions with antimicrobial markers concern what concentration of agent to use and when to select mutants after plating. The first question is best resolved by using a range of concentrations, starting at *ca* 3x that tolerated by the wild-type. This is accomplished most conveniently using gradient plates, described later. Since higher cell concentrations may make levels of resistance appear higher (Schwinghamer 1967), different cell densities should be used in pilot trials when determining drug sensitivities of parent and mutant strains. The test medium may also influence susceptibility, depending on mode of action of the drug (Schwinghamer 1967).

Delayed selection in choosing mutants is advised since pure clones may not become established immediately because of the nuclear condition of the cell or phenotypic lag (delay between homozygous mutation in the genome and manifestation of the new character) (Hopwood 1970). A uninucleate, haploid stage for mutagenesis and cloning is preferable because mutant genes are typically recessive to the wild-type allele. If the mutation occurs in a diploid or heterokaryotic cell, it will not be expressed until segregation occurs. In studies with auxotrophic mutants of *Tilletia caries*, Churchill & Mills (1984) found that no mutants could be detected unless a period of growth in complete medium after mutagen treatment was permitted before assay. This allowed time for replication and separate packaging of the mutated DNA strand. Likewise, since bacterial cells in culture usually do not contain a single genome, growth in broth for a few generations after mutagenesis is recommended before plating single colonies (Hopwood 1970).

Whether mutants are spontaneous or induced, they are isolated by direct plating of a large population ($10^{10}$ to $10^{11}$ cells), occasionally following growth in liquid medium. Most approaches are similar to the one summarized in Fig. 1 (modified from Meynell & Meynell 1970, p. 265; Park 1978). An outline of a typical experimental protocol after obtaining the mutant is presented for the *Rhizobium* sp.-root nodule system by Schwinghamer & Dudman (1980, p. 364).

Hopwood (1970) has emphasized the 'one culture-one mutant' rule to avoid isolating strains carrying descendants of the same mutant gene. Hence, in the above protocol, one would start with, e.g., 10 clones of the species and follow the flow sheet separately for each, rather than choosing 10 of the desired phenotype at the end from the same original culture.

The ingenious gradient plate technique (Bryson & Szybalski 1952) provides an efficient, convenient way to expose populations to a range of concentrations. Briefly, the plates contain two 10- to 20-ml layers of agar (see Carlton & Brown (1981) for details). The bottom layer of nutrient agar is poured and hardens with the plate sloping so that agar extends sufficiently to just cover the entire bottom. The plate is then placed horizontally and an overlay, containing the antibiotic, is added. The drug diffuses downward, becoming diluted proportionally to the ratio of the thickness of the agar layers, thereby establishing a uniform concentration gradient during incubation. Plates are prepared 24 h before use and the microbes are spread or streaked on the agar surface. Several variants of the procedure for drugs and other screens exist (Hartman 1968, p. 155). Additionally, the technique facilitates stepwise increase of resistance in situ and permits visual distinction of different patterns of inhibition. We have used it effectively with bacteria and yeasts; presumably filamentous fungi could be screened, with detergents (Hopwood 1970) being added if necessary to restrict colony size.

If more than one marker is to be installed, the usual procedure is to screen for resistance to each successively (e.g. Kuykendall & Weber 1978; Mendez-Castro & Alexander 1983). Alternatively a double gradient plate or modifications thereof could be used, with one compound added to each layer (reviewed in Hartman (1968)).

### *Strain characterization*

Whether of spontaneous or induced origin, marker strains for ecological studies have been very poorly characterized as a rule. Authors have apparently either overlooked the fundamental premise on which all studies with such mutants are based, or have considered that to report such documentation is trivial and superfluous (e.g. Dupler & Baker (1984)). Where attempts at comparison are made, they are generally restricted to a few obvious attributes such as gross morphology, or physiology (e.g. Bedford *et al.* 1984), and a selected biological function (which in mutants is often impaired) such as ability to nodulate (Hartel & Alexander 1983), to cause disease (Hsu & Dickey 1972; Bennett & Billing 1975), or to perform antagonistically (Cullen & Andrews 1985). The size of the experiment influences sensitivity of detection of undesirable changes, which may be apparent on a field scale, but not in the

Inoculate broth with wild-type or mutagenized cells

↓

Incubate until culture is turbid

↓

Add antimicrobial compound (*ca* 10 to 1,000 $\mu g\ ml^{-1}$, depending on drug; typically ≧ 3x tolerance of wild-type)

↓

Incubate; spread onto solid medium amended with the drug

↓

Incubate; choose an isolated colony to increase resistance where feasible or to initiate stock cultures

Fig. 1. Procedure for isolation of mutants.

growth chamber (Fry *et al.* 1984).

Two types of genetic alteration can occur. The mutant strain could be different from the wild-type at a locus or loci other than for antimicrobial resistance. As noted above, multiple, 'cryptic' mutations are common with certain mutagens such as nitrosoguanidine. However, differences are usually casually ascribed to pleiotropy, wherein a gene (in this case the one governing drug resistance) has more than one phenotypic effect. Multiple phenotypic effects associated with resistance might arise from (i) deletion of a vulnerable step in synthesis of a metabolite, perhaps also essential for biological activity (e.g. nodulation, pathogenicity), or (ii) enhanced synthesis of a compound responsible both for drug tolerance and biological activity by feedback or repressor action (Schwinghamer 1964). A more general effect might be an overall metabolic change that affects biological activity. If revertants to drug susceptibility have restored wild-type phenotypes, probably the effect is pleiotropic. One way to positively separate cryptic mutations from pleiotropy is to use nucleic acid sequencing and genetic mapping procedures. Otherwise, site-specific insertion mutants can be developed or multiply isolated independent mutants can be used. If all have the same phenotype, pleiotropy is probably involved as it would be unlikely to have the same cryptic mutation in more than one mutant. However, arguments can be made that the act of adding or removing specific genes might cause other, unrelated genes to be turned on or off.

Rather than dealing with extraneous genetic changes, Josey *et al.* (1979) used intrinsic resistance to multiple antibiotics as a tag. This avoids the potential detractions of highly resistant mutants, but how similar overall such fortuitous isolates from nature may be to the accepted or type strain is questionable. Also, strains similar but not identical to that being tracked may be isolated incidentally. Use of spontaneous mutants is a similar but more controlled approach.

Among the few detailed characterizations are the studies by Schwinghamer and colleagues on marker strains of *Rhizobium* spp. In tests of 15 antibiotics, parental strains exhibited as much intraspecific as interspecific variation in sensitivity (Schwinghamer 1964, 1967). In a related study involving eight strains representing four species of *Rhizobium* (Schwinghamer & Dudman 1973), about one fifth of the spectinomycin-resistant mutants exhibited partial or full loss of nodulating ability, but significantly, two-thirds of these originated from one parental line. Both these points emphasize the importance of selecting several prospective clones for experimentation. Some strains were highly resistant to all antibiotics; conversely some drugs such as tetracycline were consistently toxic at relatively low levels to all strains (Schwinghamer 1967). Mutants varied in their cross-resistance to antibiotics, some being strongly so (e.g. neomycin$^r$) and others (streptomycin$^r$) to a much lesser degree. Mutants also varied in loss of nodulating ability when tested on seedlings under sterile conditions, from showing little or no change (streptomycin$^r$) to becoming ineffective (viomycin$^r$). Level of resistance seems correlated with pleiotropy because mutants resistant to low concentrations of novobiocin generally nodulated well, whereas clones reselected for higher tolerance were markedly less effective. There was no apparent correlation between chemical structure (as opposed to mode of action) of the drug and symbiotic behaviour of resistant mutants.

In overview, there is a fundamental difference between resistance conferred in bacteria by R-plasmids and that by mutation of the bacterial chromosome (Table 2; Lin *et al.* 1984). As noted before, R-plasmid-type resistance usually involves inactivation of the drug, whereas mutations involve changes in the target site in the host. These latter alterations are usually somewhat deleterious. This illustrates the problems inherent in assuming that marked and unmarked organisms behave similarly and emphasizes the great advantage in using either independent replicons to incorporate resistance or transposon rather than physical or chemical mutagenesis. As a minimum prerequisite, the plating efficiency and ecological competence of the marker population should be assessed and reported.

### *SUBCULTURE AND STORAGE*

Since the two problems most commonly encountered in cloning are cross-contamination of strains and gain or loss of markers (Maniatis *et al.* 1982) it is ironic that maintenance of marker strains (and the corresponding parental lines) is evidently treated as casually as is strain characterization. Most authors do not state their storage conditions including, importantly, whether the marked microbe was subcultured on media containing or without the antimicrobic. Correct preservation to preserve viability, function, and genetic stability is essential. For long-term storage the preferred methods are lyophilization, the use of liquid nitrogen, ultra-low temperature maintenance at $\leq$-70°C with cryoprotectants, or for some fungi, refrigeration under oil (Kirsop & Snell 1984). These guidelines should be followed as a point of departure in studies with markers. However, marked strains may behave differently from parental lines, e.g. be refractory to cold (Karunakaran & Johnston 1974).

Wherever maintenance conditions permit growth to occur, the question is whether or not the medium should contain the drug. In the few cases that such information is reported, opinions evidently differ: marker strains are subcultured on amended (e.g. Hartel & Alexander 1983, 1984; Churchill & Mills 1984) or unamended (e.g. Schwinghamer 1964; Brockwell *et al.* 1977; Fry *et al.* 1984) media. The only argument that can be made for storage on amended media is to maintain selection pressure for the marker. To do so presumes either that the antimicrobic compound is stable or that subcultures are made frequently to new media. Some antibiotics (e.g. ampicillin), are highly unstable, some are very light-sensitive (e.g. rifampicin), and all can adsorb to the walls of glass containers (Anhalt & Washington 1900). The counter-argument is that drug dependence may develop (Park 1978), or that continued presence of the drug may decrease viability or foster additional mutations (M.N. Schroth, personal communication). In this context it is noteworthy that organisms stressed on selective media may be debilitated and difficult to recover

(Andrew & Russell 1984).

I feel that mutant strains should be stored without the antimicrobic (and checked periodically for persistence of the marker). If the mutant is so unstable that the culture must be maintained on the drug to prevent occurrence of revertants, then it seems pointless or at best risky to use it in nature.

*STABILITY*

To be useful a marker must persist in the labelled cell and its progeny in nature for at least the duration of the experiment. That is, it cannot be lost by exchange with the native population (Schwinghamer & Dudman 1973; Bennett & Billing 1975; Lai *et al.* 1977 b) or by reversion to wild-type. Where resistance is plasmid-borne, non-conjugative plasmids which lack transfer functions might be used (Gross & Vidaver 1981; Lindemann 1985). As with strain characterization and storage, stability has been documented insufficiently. Frequently no information is given, or an isolate is said to be 'stable' without criteria or evidence being presented (Dupler & Baker 1984). In practice, marker strains are arbitrarily considered 'stable' if they can be subcultured ≧3 times on unamended media (e.g. Liang *et al.* 1982; Cullen & Andrews 1985) or survive one round of host passage (e.g. Schwinghamer & Dudman 1973; Bedford *et al.* 1984).

Disappearance of a tagged microbe could result from either instability of the label or death of the cell that carried it. To distinguish these alternatives in environments subject to influx, a second, independent marker is needed. In one of the few thorough examinations of stability, Brockwell *et al.* (1977) used serology and streptomycin resistance ($strep^r$) to track strains of *Rhizobium trifolii*. At five harvests involving >600 nodules over the 41-month study, there was at least 95% correspondence between data based on either marker. Some anomalies suggested that either inocula lost antigenic determinants, and/or streptomycin resistance, or that naturally occurring strains found in the nodules acquired these traits. They proposed use of a third marker (a drug) to help explain these inconsistencies. Most importantly, however, the observations showed convincingly that, in their system, either tag was stable and could be used reliably. These results are supported by those of Renwick & Jones (1985) who compared the fluorescent ELISA procedures with antibiotic resistance.

To summarize, stability assays in vitro should always be followed by (rather than extrapolated to) tests in nature. Interestingly, markers on plasmids may be more stable when the microbe is maintained in vivo rather than in culture (Lai *et al.* 1977 a). Pilot trials ideally should be based upon complementary types of markers. Finally, data on stability have a rightful and important place in a research report!

*OVERVIEW AND PROSPECTS*

Historically, markers of the drug resistance type have been

used to some extent in medical epidemiology and environmental microbiology (Greenberg 1969; Danso *et al.* 1973; Park 1978; Liang *et al.* 1982), but largely by soil microbiologists interested in population dynamics of *Rhizobium* spp. (e.g. Obaton 1971; Hossain & Alexander 1984). In parallel but independent studies, plant pathologists have tracked the survival of bactericide- (e.g. Schroth *et al.* 1974) or fungicide- (e.g. McGee & Zuck 1981) resistant mutants, because of the need to know whether a chemical control programme selected for such strains, and how long they might persist after the compound was withdrawn from use.

Only comparatively recently have markers been exploited in phylloplane microbiology. Lindow (1983, 1985) studied antagonists marked with streptomycin or rifampicin resistance to track antagonists to ice-nucleating bacteria on pear leaves and flowers. Lindemann (1985) tagged near-isogenic lines of *Pseudomonas syringae* pv. *tomato* with kanamycin or rifampicin resistance and subjected them to reciprocal competition studies on tomato leaves in a greenhouse. Neither strain increased over a 3-day period when applied as challenge inoculum 48 h after application of its counterpart. Since antibiosis is apparently not involved (Lindemann 1985), the organism arriving first is able to pre-empt and retain the niche, at least temporarily, by some other mechanism. Cullen & Andrews (1985) tagged the apple scab antagonist *Chaetomium globosum* with benomyl resistance (Ben$^{r}$) and studied its ability to persist on apple leaves, with and without the chemical, over 3 weeks in an orchard. Relative to the wild-type parents, Ben$^{r}$ populations declined more rapidly, but the rate of decline was reduced by applying benomyl with the inoculum. Fokkema (1983) used a benomyl-resistant strain of *Sporobolomyces roseus* in attempts to demonstrate natural biological control by yeasts on wheat leaves. Under agricultural conditions, however, naturally occurring fungicide resistance may limit the use of benomyl resistance as a marker of introduced yeasts or hyphal fungi in dispersal studies. In an unsprayed wheat crop about 5% of all isolates of *S. roseus* appeared to be resistant to benomyl due to benomyl applications in previous seasons. In addition, benomyl sprayings did not reduce the density of *S. roseus*, and the percentage resistant isolates amounted to more than 60% of the total number of isolates (N.J. Fokkema, personal communication).

If a marked strain, whether tagged 'naturally' (e.g. bacteriocin, siderophore; Schwinghamer & Dudman 1980) and sufficiently distinctive, or 'artificially' (drugs, fungicides) lends itself to disease control, then a dual role is possible in integrated control with the fungicide. Plasmids encoding resistance to arsenate and antimony (Hendrick *et al.* 1984) and copper (Stall *et al.* 1984) have been reported, which suggest a similar potential for chemical and biological control. Although the Ben$^{r}$ mutant of Cullen & Andrews (1985) did not reduce scab appreciably under the prevailing test conditions, disease was significantly less with the combined fungicide and antagonist than with benomyl alone. Here and elsewhere (Odeyemi & Alexander 1977; Mendez-Castro & Alexander 1983; Hossain & Alexander 1984) the chemical apparently acts in part by promoting colonization of the mutant. Prospective synergism has been extended by mutagenizing antagonists and searching for improved and fungicide-tolerant biotypes (Abd-El Moity *et al.* 1982; Papavizas *et al.* 1982; Papavizas & Lewis 1983). The microbes thus altered cannot strictly be used as markers because they are intentionally unlike the parental strain, but they have interesting potential for biological and integrated control strategies.

Unquestionably the exciting developments in genetic engineering will contribute substantially to marker technology and subsidiary roles, such as described above, that marked organisms might play. This is already the case for bacteria and will soon be realized with the filamentous fungi. As one example, well-characterized bacteria such as *Escherichia coli* can be used as 'factories' in which to modify the DNA of other bacteria introduced as plasmids. For instance, DNA of a *Rhizobium* species can be cloned, mutated in *E. coli*, returned to the same *Rhizobium* species and then recombined with host DNA, replacing the wild-type genes (Ruvkun & Ausubel 1981). Hence, it is now a realistic goal to install multiple genes of interest sequentially in phylloplane microbes.

Two ethical and public policy issues attend the use of antibiotic resistance markers. The first involves the indiscriminate deployment of genes for resistance to medically important antibiotics which is to be avoided. This is especially important in organisms having poorly understood conjugational properties where drug resistance could be spread in native bacterial populations. In response to such concerns, current National Institutes of Health (U.S.A.) guidelines for recombinant DNA methods discourage the use of *E. coli* strains carrying conjugation-proficient plasmids, even for laboratory use. The problem is of less concern in tracking studies where mutated strains are used, since mutated chromosomal genes do not have the potential to spread through a population to the degree that R-factors on plasmids do. Secondly, whatever potential molecular biology holds for developing markers, legal constraints on genetically engineered organisms will preclude their use in nature for the foreseeable future. Furthermore, as Lindemann (1985) points out, deletion mutants will probably be acceptable before mutants involving gene additions (such as the drug resistance type). As in all such cases where there is a threat, even remote, of an undesirable outcome, it is logistically impossible to ever prove absence of an effect. Ultimately, public policy decisions on use will fall within the subjective domain of risk assessment. The current scientific debate and public uproar on this issue (Cohen 1977; Brill 1985) could be largely appeased if there were an established, reliable reference point in microbial ecology on which to confidently base predictions. As things stand, that theoretical framework does not exist and it will not exist for some time.

*ACKNOWLEDGEMENTS*

A contribution from the College of Agricultural and Life Sciences, University of Wisconsin, Madison, U.S.A. I thank the USDA (grant no. 81-CRCR-10707), and the NSF (grant no. DEB 8110199) for support and the following colleagues for discussions or comments on the manuscript: D.C. Cullen, J. Handelsman, R.F. Harris, L.L. Kinkel, H.C. Kistler, J. Lindemann and S.E. Lindow.

*REFERENCES*

Abd-El Moity, T.H., Papavizas, G.C. & Shatla, M.N. (1982). Induction of new isolates of *Trichoderma harzianum* tolerant to fungicides and their experimental use for control of white rot of onion. Phytopathology, *72*, 396-400.

Andrew, M.H.E. & Russel, A.D. (eds) (1984). The Revival of Injured Microbes. London: Academic Press.

Andrewartha, H.G. & Birch, L.C. (1984). The Ecological Web: More on the Distribution and Abundance of Animals. Chicago: University of Chicago Press.

Anhalt, J.P. & Washington, J.A., II (1980). Preparation and storage of antimicrobic solutions. *In* Manual of Clinical Microbiology, 3rd ed., ed.-in-chief E.H. Lennette, pp. 495-6. Washington: Am. Soc. Microbiol.

Bainbridge, B.W. (1980). The Genetics of Microbes. New York: John Wiley and Sons.

Bedford, K.E., MacNeill, B.H. & Bonn, W.G. (1984). Survival of a genetically marked strain of the blister spot pathogen *Pseudomonas syringae* pv. *papulans* in leaf scars and buds of apple. Canadian Journal of Plant Pathology, *6*, 17-20.

Bennet, R.A. & Billing, E. (1975). Development and properties of streptomycin-resistant cultures of *Erwinia amylovora* derived from English isolates. Journal of Applied Bacteriology, *39*, 307-15.

Beringer, J.E., Beynon, J.L., Buchanan-Wollaston, A.V. & Johnston, A.W.B. (1978). Transfer of the drug resistance transposon Tn*5* to *Rhizobium*. Nature (London), *276*, 633-4.

Beringer, J.E. & Hirsch, P.R. (1984). The role of plasmids in microbial ecology. *In* Current Perspectives in Microbial Ecology, eds M.J. Klug & C.A. Reddy, pp. 63-70. Washington: Am. Soc. Microbiol.

Beringer, J.E., Ruiz Sainz, J.E. & Johnston, A.W.B. (1984). Methods for the genetic manipulation of *Rhizobium*. *In* Microbiological Methods for Environmental Biotechnology, eds J.M. Grainger & J.M. Lynch, pp. 79-94. New York: Academic Press.

Bott, T.L. & Brock T.D. (1970). Growth and metabolism of periphytic bacteria: methodology. Limnology and Oceanography, *15*, 333-42.

Brill, W.J. (1985). Safety concerns and genetic engineering in agriculture. Science, *227*, 381-4.

Brock, T.D. (1971). Microbial growth rates in nature. Bacteriological Reviews, *35*, 39-58.

Brock, T.D. & Brock, M.L. (1968). Measurement of steady state growth rates of a thermophilic alga directly in nature. Journal of Bacteriology, *95*, 811-5.

Brockwell, J., Diatloff, A. & Schwinghamer, E.A. (1978). An appraisal of methods for distinguishing between strains of *Rhizobium* spp. and their application to the identification of isolates from field environments. *In* Microbial Ecology, eds M.W. Loutit & J.A.R. Miles, pp. 390-7. New York: Springer-Verlag.

Brockwell, J., Schwinghamer, E.A. & Gault, R.R. (1977). Ecological studies of root-nodule bacteria introduced into field environments--V. A critical examination of the stability of antigenic and streptomycin-resistance markers for identification of strains of *Rhizobium trifolii*. Soil Biology & Biochemistry, *9*, 19-24.

Broome, S. & Gilbert, W. (1978). Immunological screening method to detect specific translation products. Proceedings of the National Academy of Sciences of the U.S.A., *75*, 2746-9.

Bryson, V. & Szybalski, W. (1952). Microbial selection. Science, *110*, 45-51.

Bull, J.H. & Wootton, J.C. (1984). Heavily methylated amplified DNA in transformants of Neurospora crassa. Nature, *310*, 701-4.

Carlton, B.C. & Brown, B.J. (1981). Gene mutation. *In* Manual of Methods for General Bacteriology, ed.-in-chief P. Gerhardt, pp. 222-42. Washington: Am.

Soc. Microbiol.

Casadesus, J., Ianez, E. & Olivares, J. (1980). Transposition of Tn1 to the *Rhizobium meliloti* genome. Molecular & General Genetics, *180*, 405-10.

Case, M.E., Schweizer, M., Kushner, S.R. & Giles, N.H. (1979). Efficient transformation of *Neurospora crassa* by utilizing hybrid plasmid DNA. Proceedings of the National Academy of Sciences of the U.S.A., *76*, 5259-63.

Churchill, A.C.L. & Mills, D. (1984). Selection and culture of auxotrophic and drug-resistant mutants of *Tilletia caries*. Phytopathology, *74*, 354-7.

Clarke, L., Hitzeman, R. & Carbon, J. (1979). Selection of specific clones from colony banks by screening with radioactive antibody. Methods in Enzymology, *68*, 436-42.

Cohen, S.N. (1977). Recombinant DNA: fact and fiction. Science, *195*, 654-7.

Cohen, S.N., Chang, A.C.Y. & Hsu, L. (1972). Non chromosomal antibiotic resistance in bacteria: genetic transformation of *Escherichia coli* by R-factor DNA. Proceedings of the National Academy of Sciences of the U.S.A., *69*, 2110-4.

Cullen, D. & Andrews, J.H. (1985). Benomyl-marked populations of *Chaetomium globosum*: survival on apple leaves with and without benomyl and antagonism to the apple scab pathogen, *Venturia inaequalis*. Canadian Journal of Microbiology, *31*, 251-5.

Danso, S.K.A., Habte, M. & Alexander, M. (1973). Estimating the density of individual bacterial populations introduced into natural ecosystems. Canadian Journal of Microbiology, *19*, 1450-1.

Davidse, L.C. & Waard, M.A. de (1984). Systemic fungicides. Advances in Plant Pathology, *2*, 191-257.

Dekker, J. (1984). Development of resistance to antifungal agents. *In* Mode of Action of Antifungal Agents, eds A.P.J. Trinci & J.F. Ryley, pp. 89-111. Cambridge: Cambridge University Press.

Dupler, M. & Baker, R. (1984). Survival of *Pseudomonas putida*, a biological control agent, in soil. Phytopathology, *74*, 195-200.

Fokkema, N.J. (1983). Naturally-occurring biological control in the phyllosphere. Les Colloques de l'INRA, *18*, 71-7.

Fry, W.E., Yoder, O.C. & Apple, A.E. (1984). Influence of naturally occurring marker genes on the ability of *Cochliobolus heterostrophus* to induce field epidemics of southern corn leaf blight. Phytopathology, *74*, 175-8.

Greenberg, B. (1969). *Salmonella* suppression by known populations of bacteria in flies. Journal of Bacteriology, *99*, 629-35.

Gross, D.C. & Vidaver, A.K. (1981). Transformation of *Pseudomonas syringae* with nonconjugative R plasmids. Canadian Journal of Microbiology, *27*, 759-65.

Grunstein, M. & Hogness, D. (1975). Colony hybridization: a method for the isolation of cloned DNAs that contain a specific gene. Proceedings of the National Academy of Sciences of the U.S.A., *72*, 3961-5.

Hartel, P.G. & Alexander, M. (1983). Decline of cowpea rhizobia in acid soils after gamma-irradiation. Soil Biology & Biochemistry, *15*, 489-90.

Hartel, P.G. & Alexander, M. (1984). Temperature and desiccation tolerance of cowpea rhizobia. Canadian Journal of Microbiology, *30*, 820-3.

Hartman, P.A. (1968). Miniaturized Microbiological Methods. New York: Academic Press.

Hendrick, C.A., Haskins, W.P. & Vidaver, A.K. (1984). Conjugative plasmid in *Corynebacterium flaccumfaciens* subsp. *oortii* that confers resistance to arsenite, arsenate and antimony (III). Applied and Environmental Microbiology, *48*, 56-60.

Hernalsteens, J.P., DeGreve, H., Montagu, M. van & Schell, J. (1978). Mutagenesis by insertion of the drug resistance transposon Tn*7* applied to the Ti plasmid of *Agrobacterium tumefaciens*. Plasmid, *1*, 218-25.

Hinnen, A., Hicks, J.B. & Fink, G.R. (1978). Transformation of yeast. Proceedings of the National Academy of Sciences of the U.S.A., *75*, 1929-33.

Hopwood, D.A. (1970). The isolation of mutants. *In* Methods in Microbiology. Vol. 3A, eds J.R. Norris & D.W. Ribbons, pp. 363-433. New York: Academic Press.

Hossain, A.K.M. & Alexander, M. (1984). Enhancing soybean rhizosphere colonization by *Rhizobium japonicum*. Applied and Environmental Microbiology, *48*, 468-72.

Hsu, S.T. & Dickey, R.S. (1972). Interaction between *Xanthomonas phaseoli*, *Xanthomonas vesicatoria*, *Xanthomonas campestris* and *Pseudomonas fluorescens* in bean and tomato leaves. Phytopathology, *62*, 1120-8.

Josey, D.P., Beynon, J.L., Johnston, A.W.B. & Beringer, J.E. (1979). Strain identification in *Rhizobium* using intrinsic antibiotic resistance. Journal of Applied Bacteriology, *46*, 343-50.

Kado, C.I. & Lurquin, P.F. (1982). Prospectus for genetic engineering in agriculture. *In* Phytopathogenic Prokaryotes. Vol. 2, eds M.S. Mount & G.H. Lacy, pp. 303-25. New York: Academic Press.

Karunakaran, V. & Johnston, J.R. (1974). Death of nystatin-resistant mutants of *Saccharomyces cerevisiae* during refrigeration. Journal of General Microbiology, *81*, 255-6.

Kirsop, B.E. & Snell, J.J.S. (1984). Maintenance of Microorganisms. A Manual of Laboratory Methods. New York: Academic Press.

Kleckner, N. (1981). Transposable elements in prokaryotes. Annual Review of Genetics, *15*, 341-404.

Kleckner, N., Roth, J. & Botstein, D. (1977). Genetic engineering *in vivo* using translocatable drug-resistance elements. Journal of Molecular Biology, *116*, 125-59.

Kosuge, T. & Nester, E.W. (1984). Plant-Microbe Interactions. Molecular and Genetic Perspectives. Vol. 1. New York: Macmillan Publishing Co.

Kuykendall, L.D. & Weber, D.F. (1978). Genetically marked *Rhizobium* identifiable as inoculum strain in nodules of soybean plants grown in fields populated with *Rhizobium japonicum*. Applied and Environmental Microbiology, *36*, 915-9.

Lai, M., Panopoulos, N.J. & Shaffer, S. (1977 a). Transmission of R plasmids among *Xanthomonas* spp. and other plant pathogenic bacteria. Phytopathology, *67*, 1044-50.

Lai, M., Shaffer, S. & Panopoulos, N.J. (1977 b). Stability of plasmid-borne antibiotic resistance in *Xanthomonas vesicatoria* in infected tomato leaves. Phytopathology, *67*, 1527-30.

Leben, C., Boone, D.M. & Keitt, G.W. (1955). *Venturia inaequalis* (Cke.) Wint. IX. Search for mutants resistant to fungicides. Phytopathology, *45*, 467-72.

Liang, L.N., Sinclair, J.L., Mallory, L.M. & Alexander, M. (1982). Fate in model ecosystems of microbial species of potential use in genetic engineering. Applied and Environmental Microbiology, *44*, 708-14.

Lin, E.C.C., Goldstein, R. & Syvanen, M. (1984). Bacteria, Plasmids, and Phages. An Introduction to Molecular Biology. Cambridge, MA: Harvard University Press.

Lindemann, J. (1985). Genetic manipulation of microorganisms for biological control. *In* Biological Control on the Phylloplane, eds C.E. Windels & S.E. Lindow, pp. 116-30. St. Paul, MN: Am. Phytopath. Soc.

Lindemann, J., Constantinidou, H.A., Barchet, W.R. & Upper, C.D. (1982). Plants as sources of airborne bacteria, including ice nucleation-active bacteria. Applied and Environmental Microbiology, *44*, 1059-63.

Lindow, S.E. (1983). Methods of preventing frost injury caused by epiphytic ice-nucleation-active bacteria. Plant Disease, *67*, 327-33.

Lindow, S.E. (1985). Integrated control and role of antibiosis in biological control of fireblight and frost injury. *In* Biological Control on the Phylloplane, eds C.E. Windels & S.E. Lindow, pp. 83-115. St. Paul, MN: Am. Phytopath. Soc.

Low, K.B. & Porter, D.D. (1978). Modes of gene transfer and recombination in bacte-

ria. Annual Review of Genetics, *12*, 249-87.
Maniatis, T., Fritsch, E.F. & Sambrook, J. (1982). Molecular Cloning: A Laboratory Manual. Cold Spring Harbor, NY: Cold Spring Harbor Laboratory.
McEvoy, G.K. (1985). American Hospital Formulary Service Drug Information 85. Bethesda, MD: Am. Soc. Hosp. Pharm.
McGee, D.C. & Zuck, M.G. (1981). Competition between benomyl-resistant and sensitive strains of *Venturia inaequalis* on apple seedlings. Phytopathology, *71*, 529-32.
Mendez-Castro, F.A. & Alexander, M. (1983). Method for establishing a bacterial inoculum on corn roots. Applied and Environmental Microbiology, *45*, 248-54.
Meynell, G.G. (1959). Use of superinfecting phage for estimating the division rate of lysogenic bacteria in infected animals. Journal of General Microbiology, *21*, 421-37.
Meynell, G.G. & Meynell, E. (1970). Theory and Practice in Experimental Bacteriology, 2nd ed. Cambridge: Cambridge University Press.
Mishra, N.C. (1985). Gene transfer in fungi. Advances in Genetics, *23*, 73-178.
Molzahn, S.W. & Woods, R.A. (1972). Polyene resistance and the isolation of sterol mutants in *Saccharomyces cerevisiae*. Journal of General Microbiology, *72*, 339-48.
Obaton, M.M. (1971). Utilisation de mutants spontanés résistants aux antibiotiques pour l'étude écologique des *Rhizobium*. Comptes Rendus Hebdomadaires des Séances de l'Académie des Sciences Paris, Série D, *272*, 2630-3.
Odeyemi, O. & Alexander, M. (1977). Use of fungicide-resistant rhizobia for legume inoculation. Soil Biology & Biochemistry, *9*, 247-51.
Papavizas, G.C. & Lewis, J.A. (1983). Physiological and biocontrol characteristics of stable mutants of *Trichoderma viride* resistant to MBC fungicides. Phytopathology, *73*, 407-11.
Papavizas, G.C., Lewis, J.A. & Abd-El Moity, T.H. (1982). Evaluation of new biotypes of *Trichoderma harzianum* for tolerance to benomyl and enhanced biocontrol capabilities. Phytopathology, *72*, 126-32.
Park, R.W.A. (1978). The isolation and use of streptomycin-resistant mutants for following development of bacteria in mixed culture. *In* Techniques for the Study of Mixed Populations, eds D.W. Lovelock & R. Davies, Soc. Appl. Bact. Tech. Ser. no. 11, pp. 107-12. New York: Academic Press.
Renwick, A. & Jones, D.G. (1985). A comparison of the fluorescent ELISA and antibiotic resistance identification techniques for use in ecological experiments with *Rhizobium trifolii*. Journal of Applied Bacteriology, *58*, 199-206.
Rowlands, R.T. (1983). Industrial fungal genetic and strain selection. *In* The Filamentous Fungi. Vol. 4, eds. J.E. Smith, D.R. Berry & B. Kristiansen, pp. 346-90. London: Edward Arnold.
Ruvkun, G.B. & Ausubel, F.M. (1981). A general method for site-directed mutagenesis in prokaryotes. Nature (London), *289*, 85-8.
Schroth, M.N., Thomson, S.V., Hildebrand, D.C. & Moller, W.J. (1974). Epidemiology and control of fire blight. Annual Review of Phytopathology, *12*, 389-412.
Schwinghamer, E.A. (1964). Association between antibiotic resistance and ineffectiveness in mutant strains of *Rhizobium* spp. Canadian Journal of Microbiology, *10*, 221-33.
Schwinghamer, E.A. (1967). Effectiveness of *Rhizobium* as modified by mutation for resistance to antibiotics. Antonie van Leeuwenhoek, *33*, 121-36.
Schwinghamer, E.A. & Dudman, W.F. (1973). Evaluation of spectinomycin resistance as a marker for ecological studies with *Rhizobium* spp. Journal of Applied Bacteriology, *36*, 236-72.
Schwinghamer, E.A. & Dudman, W.F. (1980). Methods for identifying strains of diazotrophs. *In* Methods for Evaluating Biological Nitrogen Fixation, ed. F.J. Bergersen, pp. 337-65. New York: John Wiley and Sons.
Shaw, J.J., Close, T.J., Engebrecht, J. & Kado, C.I. (1985). Use of bioluminescence

to monitor *Agrobacterium*, *Erwinia*, *Pseudomonas* and *Xanthomonas* in plants. Phytopathology, *75*, 1288 (Abstr.).

Southern, E. (1975). Detection of specific sequences among DNA fragments separated by gel electrophoresis. Journal of Molecular Biology, *98*, 503-17.

Stall, R.E., Loschke, D.C. & Rice, R.W. (1984). Conjugational transfer of copper resistance and avirulence to pepper within strains of *Xanthomonas campestris* pv. *vesicatoria*. Phytopathology, *74*, 797 (Abstr.).

Tilburn, J., Scazzocchio, C., Taylor, G.G., Zabicky-Zissman, J.H., Lockington, R.A. & Davis, R.W. (1983). Transformation by integration in *Aspergillus nidulans*. Gene, *26*, 205-21.

Zhu, J., Contreras, R., Gheysen, D., Ernst, J. & Fiers, W. (1985). A system for dominant transformation and plasmid amplification in *Saccharomyces cerevisiae*. Bio/Technology, *3*, 451-6.

# THE USE OF AUTORADIOGRAPHY AND OTHER ISOTOPE TECHNIQUES IN PHYLLOSPHERE STUDIES

M.C. Edwards[1] and J.P. Blakeman[2]

[1]*Department of Plant Science, University of Aberdeen, St. Machar Drive, Aberdeen AB9 2UD, UK*
[2]*Department of Mycology & Plant Pathology, Queen's University of Belfast, Newforge Lane, Belfast BT9 5PX, Northern Ireland*

## *INTRODUCTION*

Water droplets on leaf surfaces become enriched with organic and inorganic material leached from the leaf. A great variety of substances has been recorded in leached material from leaves, including simple sugars, amino acids, organic acids, growth regulators, vitamins, alkaloids, phenols and many inorganic constituents, including all the essential minerals and many trace elements (Blakeman 1973). Intensive competition for these nutrients, particularly amino acids and sugars, is a characteristic of the phyllosphere, and hence a major factor in the development of populations of micro-organisms and in their potential as biological disease control agents.

Various techniques have been devised to study competition for nutrients on leaf surfaces. The potential of the isotope techniques now available will be illustrated and possible future developments of these techniques will be discussed.

### *Objectives and scope of isotope techniques*

Most techniques in biology are basically analytical, allowing the separation of a heterogeneous mixture of individuals into groups on the basis of a common similarity between the members of each group. However, living systems undergo many and varied transformations, and analytical procedures are often very cumbersome and unreliable as a means of studying such changes. Radioactive tracer techniques enable these changes to be followed with considerable accuracy. After the addition of radioactively labelled cells of one species of micro-organism to a mixed natural population of micro-organisms, possible transformations between the introduced organisms and the resident natural organisms can be determined by looking for the distribution of radio-activity in the analyzed population. Hence the main purpose of using radioactive isotopes is to provide dynamic information to supplement that provided by analytical techniques. In certain cases, stable (non-radioactive) isotopes can also be used in a similar way, and their distribution determined by means of mass spectrometry.

Since the leaf surface is such a rapidly changing ecosystem, tracer methods are ideal for studying the uptake and ultimate fate of particular molecules important in phyllosphere ecology, and to find out which micro-organisms are most metabolically active. Such studies monitoring competition between micro-organisms, by studying the effect of various external factors, changes with time of day and season, and the ultimate fate of the molecules and/or atoms being competed for. Studies can be made in terms of either individual cells of micro-organisms or of bulk populations of micro-organisms, depending on the techniques employed.

## *CHOICE OF BASIC TECHNIQUES*

The techniques available for the use of isotopes on leaf surfaces fall into three main groups, depending on the way in which the distribution of the tracer is measured.

### *Scintillation counting*

Scintillation counting is an indirect, gross method of examining isotope uptake by whole populations of micro-organisms. Cells are incubated with the labelled nutrient solution after which they are separated from the incubation medium and the radioactivity taken up is measured using a scintillation counter. The radioactivity remaining in the incubation medium is also measured and respired. $^{14}CO_2$ can be trapped and counted if the label used is $^{14}C$. These methods have the advantages of a wide range of sensitivity and the relative ease with which they can be carried out. In addition it may be possible, using suitable methods of extraction, to separate out various chemical forms of the isotope after the experiment. However, experiments can only provide information on a whole population basis. Since it has proved impossible to separate mixed cells of different species after incubation, isotope uptake by the component species of mixed populations must be extrapolated from results obtained with single-species cultures.

### *Mass spectrometry*

Stable isotopes such as $^{15}N$ cannot be detected by scintillation counting or other techniques which rely upon radiation, because by their nature they are not radioactive. They can, however, be detected by mass spectrometry, and hence experiments can be carried out with $^{15}N$ in a manner similar to scintillation counting experiments, the isotope being measured at the end of the experiment by mass spectrometry.

### *Autoradiography*

Autoradiography is a direct microscopical method which will give information on nutrient relationships between individual cells or groups of cells. Cells are incubated with labelled nutrient solutions as before, but at the end of the incubation period the cells are covered with a layer of emulsion containing silver bromide crystals in a gelatine matrix. β-Particles from decaying radioactive atoms in the cells act upon the emulsion in a manner similar to that of light, producing a 'latent image' of reduced silver, which is then developed using a photographic developer to produce an autoradiograph in which silver grains are superimposed upon a background of the cells. The distribution of silver grains over the cells indicates the relative amount of tracer taken up by each cell. Autoradiography allows a detailed study of interrelationships between individual cells or groups of cells, but is generally complex to carry out and should not be considered if there are other methods of obtaining the required information.

## *SCINTILLATION COUNTING*

Scintillation counting methods are appropriate for studies of nutrient uptake by whole populations of micro-organisms, particularly for time-course studies and for investigations aimed at identifying the various final forms of the isotope.

Three main techniques have been developed, two for use on surfaces and one for use with suspension cultures. Brodie & Blakeman (1975) described a technique for use in studies of carbon compound uptake on surfaces on which droplets of up to 0.04 ml containing micro-organisms and labelled nutrients were placed on a coverslip in a perspex chamber (Fig. 1). Humidified air was drawn through the chamber and respired $^{14}CO_2$ collected in sodium hydroxide traps placed beyond the exit of the chamber. At the end of the experiment micro-organisms were removed from the glass surface by ultrasonic cleaning and separated from the nutrient solution using a membrane filter. Distribution of $^{14}C$ label amongst the various fractions such as micro-organisms, glass, nutrient solution and contents of sodium hydroxide traps was determined by scintillation counting and almost 100% of applied label was recovered. Later a similar technique was described in which the coverslip was replaced by a leaf (Blakeman & Brodie 1977). The method was also modified to study the effect of leaching on nutrient leakage, respiration and germination of *Botrytis cinerea* conidia (Brodie & Blakeman 1977); labelled conidia on a membrane filter (Nuclepore) were laid on a bed of sterile acid-washed sand in the

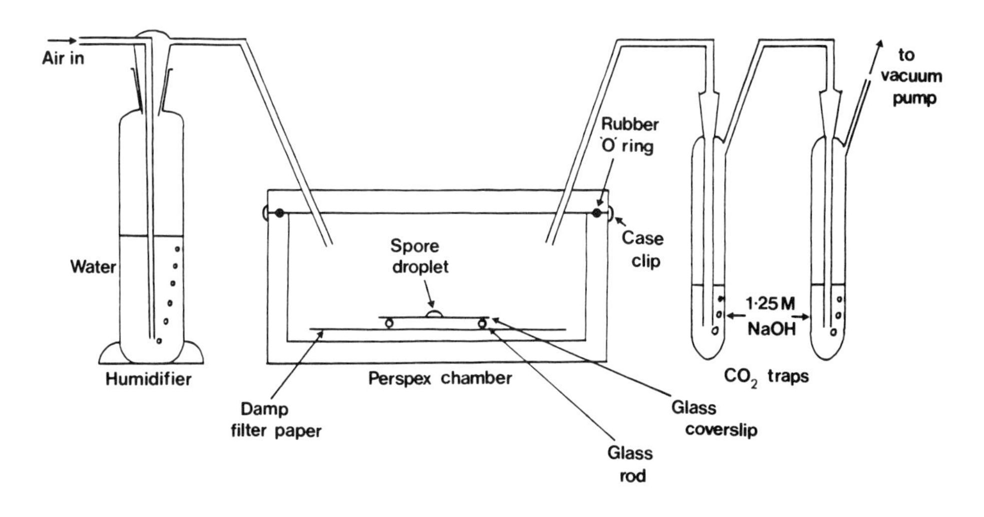


Fig. 1. Perspex chamber technique of Brodie & Blakeman (1975).

perspex chamber, and water or an appropriate leaching solution was dripped onto the sand and collected from an outlet low on the side of the chamber. The conidia could then be leached by the diffusion gradient brought about by the flow of water, and label lost from the conidia could thus be measured in the leaching solution (Fig. 2).

The failure of attempts to completely separate out different species of micro-organisms at the end of mixed-species experiments led Rodger (1981) to develop a technique for suspension cultures, in which two species of micro-organism suspended in a labelled nutrient solution were placed in the two halves of a perspex cell, separated by two polycarbonate membrane filters, allowing free passage of labelled nutrients but not of cells (Fig. 3). At the end of the experiment the microbial cells were again separated from the nutrient solution and the various components of the system subjected to scintillation counting. This permitted the separation of individual species of micro-organisms after the experiment enabling competition for the nutrient supply to be studied. Suspension culture is unlike the spatial arrangement of cells in two dimensions on the leaf surface, and the separation of cells throughout the experiment means that they compete for nutrients as a population rather than as individual cells.

In addition to these techniques developed specifically for phyllosphere studies, Mount & Ellingboe (1969) used a Parlodion stripping method to remove powdery mildew mycelium from $^{35}S$ or $^{32}P$-labelled leaves to study uptake of nutrients from the host leaf by the pathogen. Radioactivity in the surface mycelium and in the leaf was then measured by scintillation counting.

### *Results illustrating uses of techniques*

Using their original perspex chamber method Brodie & Blakeman (1976) showed that the ability of leaf surface bacteria and yeasts

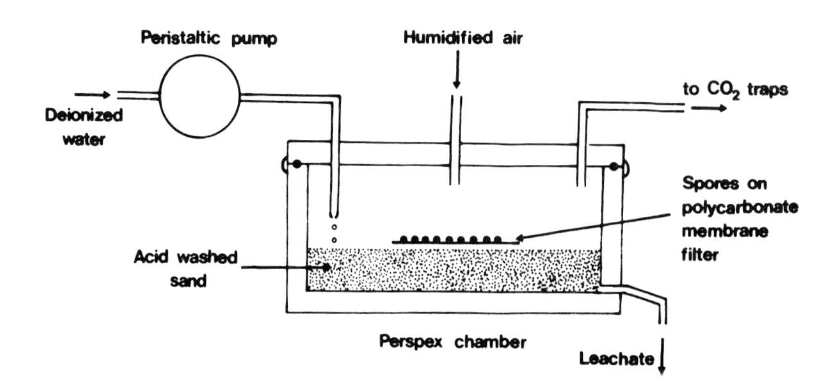


Fig. 2. Apparatus for leaching spores (Brodie & Blakeman 1977).

to inhibit germination of *B. cinerea* conidia on glass was closely correlated with their ability to remove amino acids from mixed amino acids/glucose nutrient solutions, but that no such relationship existed for sugars. This attempt to mimic on glass nutrient conditions on leaf surfaces therefore suggested that competition for amino acids was a major factor in inhibition of fungal germination by phyllosphere micro-organisms. Experiments on leaves supported this conclusion (Blakeman & Brodie 1977) and showed that on the leaf surface inhibition could be achieved with populations of micro-organisms of the order of one-tenth the size of those on glass (Blakeman 1978). By incubating water droplets on sterile $^{14}$C-labelled leaves, Blakeman & Brodie (1977) showed that a varying amount of label was leached from the leaf, which was therefore available to the phyllosphere microflora. Using the leaching technique Blakeman & Brodie (1976) found that on wetting, $^{14}$C-labelled *B. cinerea* conidia lost from 2.5 to 20% of the incorporated label. In the absence of leaching or micro-organisms the products of initial leakage were rapidly reabsorbed by the conidia, but bacteria, or leaching which mimicked the effect of bacteria, deprived the conidia of these lost endogenous reserves, and inhibited germination and germ tube growth. Parker & Blakeman (1984) used the original perspex chamber technique to investigate nutrient uptake by uredospores of the obligate parasite *Uromyces viciae-fabae* (broad bean rust) and found that these spores took up very little exogenous nutrient compared with cells of a leaf surface yeast, *Cryptococcus* sp. The latter could consume over 90% of the glucose in a solution of glucose and amino acids (at concentrations similar to those in leaf leachates) before germinating *U. viciae-fabae* uredospores had taken up 10% of the label. Rodger & Blakeman (1984) used the leaf method in a study of the development of the phyllosphere flora on sycamore leaves during the growing season, and found intense competition for carbon compounds, particularly for amino acids during midsummer, although at other periods availability of water was the main limiting factor on

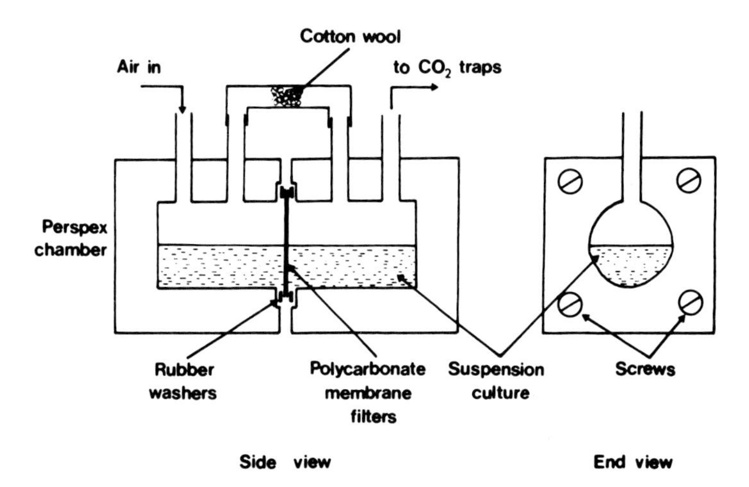


Fig. 3. Mixed pure culture apparatus (Rodger 1981).

growth of the phyllosphere microflora. Data of Frossard *et al.* (1983), obtained using a technique in which wheat leaves were labelled with $^{14}CO_2$ and subsequently washed to leach labelled assimilates, suggested that phyllosphere saprophytes do not increase leaching of assimilates from leaves by acting as a nutrient sink.

Also important in the phyllosphere are ectoparasites such as the powdery mildews. Mount & Ellingboe (1969) demonstrated the very rapid establishment of a nutritional relationship between wheat powdery mildew and its host after the initial infection.

### *MASS SPECTROMETRY*

Nitrogen fixation by micro-organisms in the phyllosphere has been measured by a technique modified by Jones (1970) from a method of Stewart (1967). Leaf or shoot samples were placed in an exposure flask in which the air was replaced by a gas mixture containing the isotope $^{15}N$. The system was then sealed and incubated for a known period. After digestion, the $^{15}N$ enrichment of the samples was determined using a mass spectrometer, allowing the rate of nitrogen fixation to be calculated.

#### *Results illustrating uses of the technique*

The activity of nitrogen-fixing micro-organisms in the phyllosphere (Ruinen 1974) has been investigated using $^{15}N$ mass spectrometry. Initial results (Jones 1970; Jones *et al.* 1974) suggested that rates of nitrogen fixation in the phyllosphere were rather high, but later results indicated that these early data may have exaggerated the importance of phyllosphere nitrogen fixation, and suggested that nitrogen fixation is ephemeral or only takes place under suitable conditions (Jones 1982). Nitrogen-fixing bacteria isolated from leaves may not be adapted to the leaf environment but may be merely casual inhabitants (Ruinen 1974; Jones 1976). However, experiments with $^{15}N$-labelled glycine indicated that plants may be able to utilize atmospheric nitrogen-fixed compounds in the form of amino acids produced by micro-organisms in the phyllosphere (Jones 1976).

### *AUTORADIOGRAPHY*

Autoradiography is ideal for detailed studies of nutrient uptake by individual cells or groups of cells in the phyllosphere. It has previously been quite widely used in aquatic ecology to study uptake of nutrients, usually by free-living micro-organisms which are subsequently collected onto a membrane filter for autoradiography (Brock & Brock 1968; Faust & Correll 1977; Fuhs & Canelli 1970). Various isotopes have been used in this work including $^{3}H$, $^{14}C$ and $^{32}P$. Although these studies enabled distinctions to be made between nutrient uptake by different types of cells in a mixed population, they did not exploit one of the other major advantages of autoradiography, the ability to study nutrient relationships in the context of spatial relationships. However, Sol (1969) used autoradiography in a study of uptake of leaf exudates by fungal conidia and Waid *et al.* (1973) used the method in a study of decomposer fungi on leaf litter surfaces. Autoradiography has been used in studies of nutritional relationships between host leaves and fungal pathogens (e.g. Andrews 1975; Mendgen & Heitefuss 1975; Manners & Gay 1980) and a technique for use on both leaf and artificial surfaces was developed by Edwards & Blakeman (1984 a).

Autoradiography requires skill and practice, and time must be allowed to develop a technique for a new situation even if an established method is to be used. A full discussion of the various approaches and methods is given by Rogers (1979). Autoradiography requires the use of a dark-room, and it can be time-consuming, due to the delay during exposure of the autoradiographs, which may extend to several weeks. Examination of the finished autoradiographs often involves much microscopic recording since sample sizes are necessarily small compared with scintillation counting and mass spectrometry methods.

Edwards & Blakeman (1984 a) described an autoradiographic technique for use on artificial and leaf surfaces. The important requirements for such a technique are good resolution and sensitivity, an even thickness of emulsion over the specimen to give reliable quantitative results and a system which does not permit any loss of soluble label by leaching during preparation of the autoradiograph. These requirements were met by

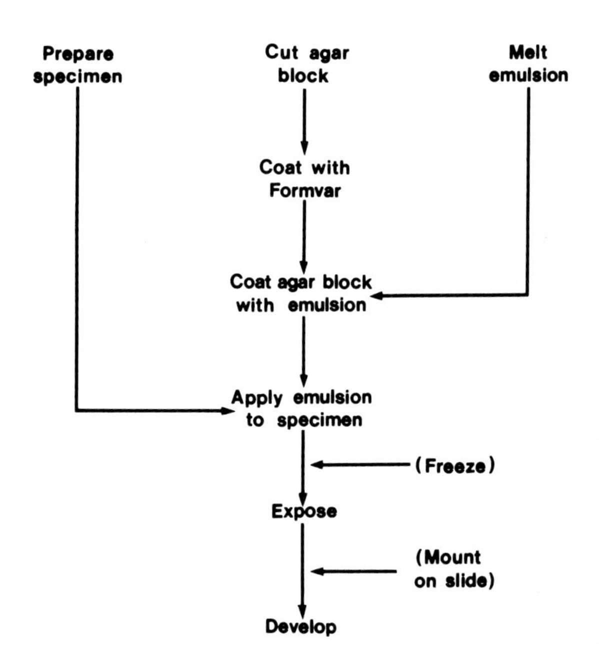


Fig. 4. Flow chart illustrating the steps in producing an autoradiograph. Steps in parentheses are included only when producing autoradiographs on leaf surfaces. Reproduced from Edwards & Blakeman (1984 a) by courtesy of the Journal of Microscopy.

the application of a thin, even pre-formed layer of gelled and dried emulsion to the specimen (Fig. 4). Micro-organisms were incubated with $^{14}C$-labelled nutrient solution in a 40-µl droplet on a polycarbonate membrane filter on a coverslip inside a petri dish moist chamber. After incubation, the membrane filter was briefly dried on filter paper, stuck to a microscope slide, and transferred to a dark room for coating with emulsion. The emulsion was prepared as a thin layer on an agar block to give a smooth even layer. Blocks of agar were cut from agar plates, and flooded with a solution of Formvar in 1,2-dichloroethane, drained and dried. The emulsion was melted and diluted 2:1 with distilled water, and a platinum wire loop was used to deposit a film of emulsion on each Formvar-coated agar block. When dry the emulsion was applied to the specimen by pressing down gently but firmly. The agar block was then carefully removed, leaving the specimen coated with emulsion and Formvar. After exposure in a light-tight box at -20°C the slides were thawed and developed in photographic developer, fixed and rinsed, then air-dried and stained with neutral red. The dry autoradiographs were cleared for microscopic examination using immersion oil. Silver grains superimposed over the cells showed the distribution of radioactivity (Fig. 5). Results can be expressed as relative grain counts or as visual assess-

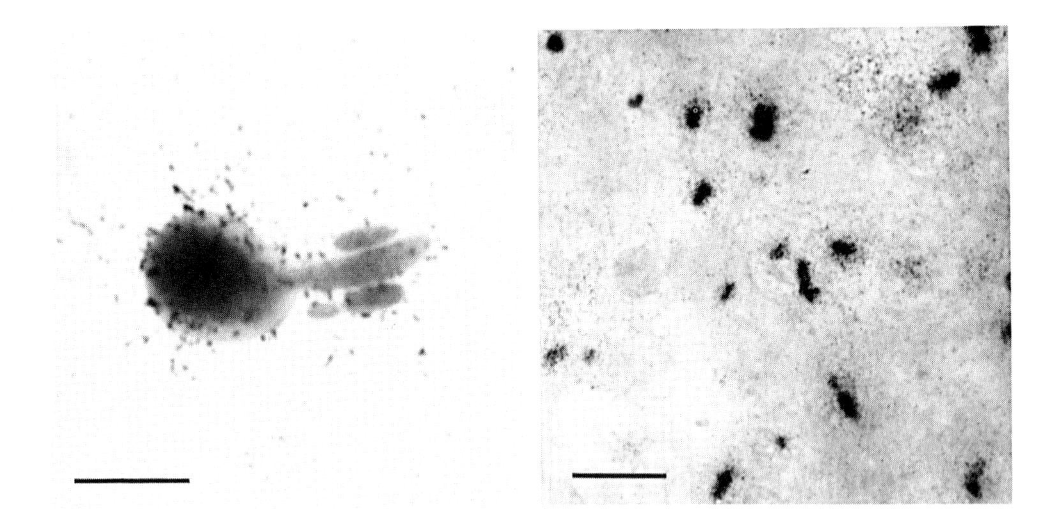

Fig. 5. Autoradiograph under transmitted light of a germinated conidium of *Botrytis cinerea* and three yeast cells (*Sporobolomyces* sp.) incubated for 5 h in a nutrient solution containing $^{14}C$-labelled mixed amino acids. The bar marker represents 10 µm.

Fig. 6. Autoradiograph of yeast cells (*Sporobolomyces* sp.) on the abaxial leaf surface of *Phaseolus vulgaris*, incubated for 5 h in a tracer solution containing $^{14}C$-labelled glucose. Note the very heavy labelling of the yeast cells. The bar marker represents 10 µm.

ments of grain density.

It is important that unlabelled controls are included in each experiment to check for background or spurious silver grains. The finding of silver grains over germ tubes of *B. cinerea* in control autoradiographs led to experiments indicating apparent silver uptake by the fungus (Edwards & Blakeman 1984 b).

A similar technique was used for studies on leaf surfaces. After incubation of micro-organisms on the leaf a disc was removed, and stuck to a large glass coverslip in place of the polycarbonate membrane filter used in the in vitro method. After coating with emulsion, the specimen is quickly frozen in 2-methylbutane and transferred to liquid nitrogen before exposure. For development, thawed coverslips are stuck to microscope slides and placed in special horizontal racks to prevent displacement of the emulsion during processing. Specimens can then be mounted directly in glycerol or dehydrated for mounting in DPX mountant.

An epifluorescent microscope is the ideal means for viewing autoradiographs of leaf discs as the staining is quick and gives a clear view of

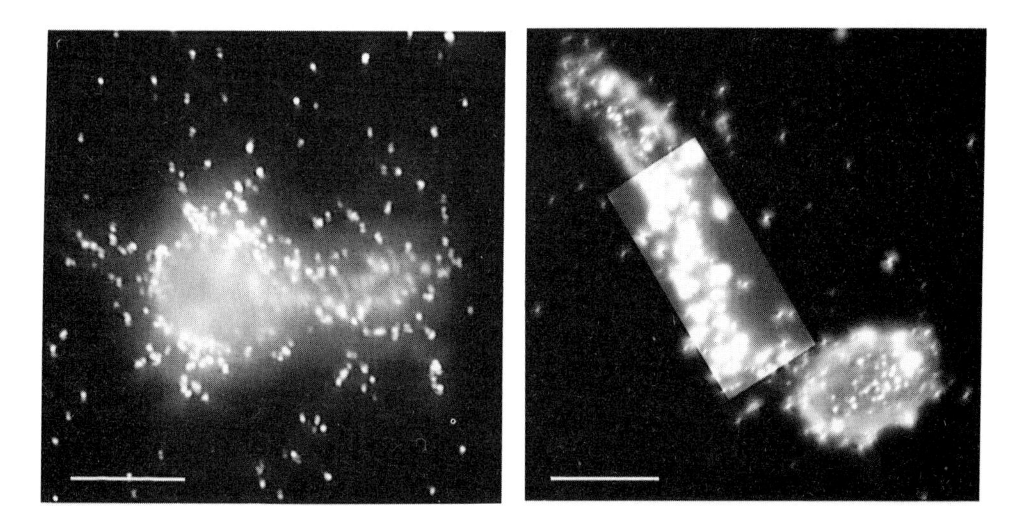

Fig. 7. The same autoradiograph as in Fig. 5, seen under incident polarized light. The light reflected by the silver grains can be measured to give an estimate of the quantity of silver present. The bar marker represents 10 µm.

Fig. 8. Autoradiograph of a germinated conidium of *B. cinerea* seen under incident polarized light. Conditions of incubation were the same as in Fig. 5. The view of the autoradiograph is masked down using the diaphragm built into the photometer. Only the light from the silver grains in the delimited area is measured. The bar marker represents 10 µm.

the micro-organisms on the leaf surface. If necessary, transmitted white light can be used to view the silver grains (Fig. 6). A mixture of acridine orange and Tinopal AN optical brightener (Paton & Jones 1973) was found to give good results (Edwards & Blakeman, unpublished results). After one minute in the stain (one drop saturated aqueous solution of Tinopal AN in 10 ml 0.1% aq. acridine orange solution) autoradiographs were rinsed in tap water, mounted in water with a coverslip and examined under ultraviolet (UV) light using a Leitz Ortholux II microscope fitted with a Ploemopak epifluorescence illuminator, using an exciter/barrier filter block H2.

Counting silver grains by eye is a laborious and time-consuming process and so a quicker means of gathering data is often needed. Edwards & Blakeman (1984 a) used a polarized light microscope in which the polarized light is directed onto the specimen through the microscope objective lens, and the light reflected by the silver grains (Fig. 7) is measured with a photometer mounted on top of the microscope. The amount of light reflected is proportional to the number of silver grains in the field (Rogers 1972). For this work, the polycarbonate filter was omitted because it produced a green birefringence colour in polarized light, and the micro-organisms were instead incubated on a slide from which the nutrient solution was carefully drawn off with a micropipette before preparation of the autoradiograph. A unit length or area of field can be used for assessment of nutrient uptake using diaphragms built into the photometer (Fig. 8). Where different micro-organisms are close together, it can often be impossible to assign silver grains to specific cells. In such cases, reflectance measurements can be made over a standard area, and the numbers of yeasts, bacteria and standard lengths of germ tube noted. Using the data from a series of such measurements, mean numbers of grains due to each class of organism can be estimated by multiple regression analysis (Edwards & Blakeman 1984 a).

### *Results illustrating uses of techniques*

The main purpose of using autoradiography is to examine nutrient competition in the individual micro-sites found on the leaf surface, and to study spatial influences on interactions between micro-organisms. Edwards & Blakeman (1984 a) showed that *B. cinerea* conidia in competition with bacteria and yeasts took up fewer exogenous $^{14}C$ amino acids than conidia without microbial competitors. Competition for endogenous nutrients leached from labelled conidia was almost entirely local, with yeasts or bacteria only a short distance from the conidia taking up some of the leached nutrients (Edwards & Blakeman, unpublished results).

Sol (1969) showed that $^{14}C$-labelled leaf surface exudates of *Vicia faba* were adsorbed by conidia of *B. fabae*. Uptake of both glucose and amino acids by *B. cinerea* and *Colletotrichum lindemuthianum* (French bean anthracnose) conidia was reduced by competition from epiphytes, and more *C. lindemuthianum* conidia produced appressoria (Edwards & Blakeman, unpublished results). Yeasts and bacteria were evenly distributed over the leaf surface when first inoculated, but after one or two days colonization became concentrated in particular areas. The presence of a second epiphyte modified the pattern of colonization, and more nutrients were taken up by the micro-organisms in areas which appeared to favour their growth. Differences in nutrient uptake between treatments or between different micro-sites were much greater on leaves than on artificial surfaces. Differences of a similar order of magnitude were found by scintillation counting (Blakeman 1978).

Light and electron microscope autoradiography techniques have been used by various workers to study the uptake of nutrients from host leaves by pathogenic fungi such as the downy mildews (Andrews 1975) or the rusts (Mendgen & Heitefuss 1975). These studies employed vertical sections through leaf surfaces, the sections then being coated with emulsion.

Autoradiography in combination with scanning electron microscopy has been used in tissue culture work (Hodges & Muir 1974) and for whole cell smears (Junger & Bachmann 1980). This method was examined for use on leaf surfaces, but appeared to have no particular advantage over high magnification light microscope autoradiography, and yielded quantitative data much more slowly (Edwards & Blakeman, unpublished results).

Autoradiography has a narrower range of sensitivity than scintillation counting, leading to problems in comparing numbers of silver grains from relatively large and small cells in the same autoradiograph. This is because there is a finite number of potential silver grains in a given area of emulsion, and as the number of β-particles from a cell striking the emulsion increases, so does the chance of two β-particles hitting the same point in the emulsion, giving rise to an underestimate of radioactivity. Thus an exposure time long enough to produce silver grains above a small, lightly labelled cell may be too long to give an accurate estimate of radioactivity in a large, heavily labelled cell.

General statistical problems in phyllosphere studies have been reviewed previously (Baker 1981; Parbery *et al.* 1981), but autoradiography brings its own statistical difficulties (Rogers 1979). These are mainly due to 'crossfire' effects: β-particles will emerge from a cell in an autoradiograph at all angles, and so may not strike the emulsion and produce a silver grain directly above that cell, but slightly to one side of it. These crossfire problems can be overcome by the use of multiple regression analysis, as discussed earlier.

The choice of isotope for autoradiography is governed by several factors. The aim of the work involved will often leave little choice, for if competition for carbon is being followed, $^{14}C$ is the only isotope available. However, if the initial uptake of amino acids is the only point of interest, then $^{3}H$ would be an alternative. Secondly, only isotopes with relatively low energies such as $^{3}H$, $^{14}C$ and $^{125}I$ are really suitable for autoradiography if resolution is an important factor, since isotopes with β-particles of higher energy will produce a wide scatter of silver grains around the labelled cell due to the greater path length of the β-particle. The great value of $^{3}H$ in autoradiography lies in its very low maximum energy and consequent short path length (typically 1-2 μm) (Falk & King 1963; Rogers 1979), but this has the disadvantage that the path may start and finish within a cell of only moderate size and so will never reach the emulsion, leading to an underestimate of radioactivity. This is particularly important where cells of varying size are being studied. This same factor led to the use of $^{14}C$ rather than $^{3}H$ by Brodie & Blakeman (1975) and subsequent scintillation counting work. Tritium is, however, ideal for autoradiography using sections (Andrews 1975; Mendgen & Heitefuss 1975) provided that the sections themselves are sufficiently thin and in close enough contact with the emulsion.

### *FUTURE DEVELOPMENTS*

Isotope techniques must always be set in their proper context of supplementing the information provided by analytical techniques. However, analytical techniques have provided a great deal of information about the biology and chemistry of the phyllosphere, and in so doing have posed many questions which existing isotope techniques can help to answer. Examples of such studies which could be extended by isotope techniques include metabolic activity in micro-habitats on the leaf surface (Dickinson & Wallace 1976), and effects on leaf cuticle by phylloplane micro-organisms (MacNamara & Dickinson 1981).

The autoradiography method described by Edwards & Blakeman (1984 a) is suitable for soluble tracers and hence will localize labelled amino acids and sugars which have only just been taken up by cells, in addition to insoluble incorporated label. If insoluble incorporated label alone is the subject of study, simpler, more conventional autoradiographic techniques could be used (after rinsing soluble material from the specimen) such as stripping film autoradiography (Rogers 1979). Filtering to separate cells from nutrient solutions, which is used in scintillation counting techniques, may leach a variable amount of soluble label from the cells back into the solution and this should be taken into account (Edwards & Blakeman 1984 a).

There is scope for greater development of both scintillation counting methods and autoradiography to obtain further information about nutrient relationships amongst phyllosphere micro-organisms. As an example of this, the dual-culture cell technique could be extended to a series of single-culture chambers, arranged in a ring surrounding a central stirred sterile chamber from which each is separated by a polycarbonate filter. This would enable competition between a number of different species to be studied at one time. However, the finding of Bright & Fletcher (1983) that marine bacterial cells attached to a surface assimilated more amino acids than free-living cells of the same species suggests that culturing cells in suspension, which in nature are attached to surfaces, may lead to anomalous results.

Preece (1971) and Baker (1981) have reviewed the use of fluorescence microscopy on leaf surfaces, and Frankland *et al.* (1981) have successfully used immunofluorescence as a means of identifying species on surfaces. This technique should be combined with autoradiography to enable metabolically active species to be identified in autoradiographs of leaf surfaces. The 2-(*p*-iodophenyl)-3-(*p*-nitrophenyl)-5-phenyl tetrazolium chloride (INT) test for respiration (Zimmerman *et al.* 1978) has been combined with autoradiography of nutrient uptake by aquatic micro-organisms (Bright & Fletcher 1983); this should also prove to be a useful test to determine whether micro-organisms, which do not appear to take up nutrients in leaf surface autoradiographs, are still alive.

The various results indicating the importance of competition for amino acids in the phyllosphere suggest that further research should be carried out using the $^{15}N$ mass spectrometry technique, which avoids the problems of $^{14}CO_2$ loss by respiration in $^{14}C$ techniques, to determine the true importance of nitrogen on the leaf surface.

*ACKNOWLEDGEMENTS*

Part of this work was supported by the Natural Environment Research Council and The Royal Society.

*REFERENCES*

Andrews, J.H. (1975). Distribution of label from $^3H$-glucose and $^3H$-leucine in lettuce cotyledons during the early stages of infection with *Bremia lactucae*. Canadian Journal op Botany, *53*, 1103-15.

Baker, J.H. (1981). Direct observation and enumeration of microbes on plant surfaces by light microscopy. *In* Microbial Ecology of the Phylloplane, ed. J.P. Blakeman, pp. 3-14. London: Academic Press.

Blakeman, J.P. (1973). The chemical environment of leaf surfaces with special reference to spore germination of pathogenic fungi. Pesticide Science, *4*, 575-88.

Blakeman, J.P. (1978). Microbial competition for nutrients and germination of fungal spores. Annals of Applied Biology, *89*, 151-5.

Blakeman, J.P. & Brodie, I.D.S. (1976). Inhibition of pathogens by epiphytic bacteria on aerial plant surfaces. *In* Microbiology of Aerial Plant Surfaces, eds C.H. Dickinson & T.F. Preece, pp. 529-57. London: Academic Press.

Blakeman, J.P. & Brodie, I.D.S. (1977). Competition for nutrients between epiphytic micro-organisms and germination of spores of plant pathogens on beetroot leaves. Physiological Plant Pathology, *10*, 29-42.

Bright, J.J. & Fletcher, M. (1983). Amino acid assimilation and electron transport system activity in attached and free-living marine bacteria. Applied and Environmental Microbiology, *45*, 818-25.

Brock, M.L. & Brock, T.D. (1968). The application of micro-autoradiographic techniques to ecological studies. Mitteilungen Internationale Vereinigung für Theoretische und Angewandte Limnologie, *15*, 1-29.

Brodie, I.D.S. & Blakeman, J.P. (1975). Competition for carbon compounds by a leaf surface bacterium and conidia of *Botrytis cinerea*. Physiological Plant Pathology, *6*, 125-35.

Brodie, I.D.S. & Blakeman, J.P. (1976). Competition for exogenous substrates *in vitro* by leaf surface micro-organisms and germination of conidia of *Botrytis cinerea*. Physiological Plant Pathology, *9*, 227-39.

Brodie, I.D.S. & Blakeman, J.P. (1977). Effect on nutrient leakage, respiration and germination of *Botrytis cinerea* conidia caused by leaching with water. Transactions of the British Mycological Society, *68*, 445-7.

Dickinson, C.H. & Wallace, B. (1976). Effects of late applications of foliar fungicides on activity of micro-organisms on winter wheat flag leaves.

Transactions of the British Mycological Society, *76*, 103-12.

Edwards, M.C. & Blakeman, J.P. (1984 a). An autoradiographic method for determining nutrient competition between leaf epiphytes and plant pathogens. Journal of Microscopy, *133*, 205-12.

Edwards, M.C. & Blakeman, J.P. (1984 b). Apparent silver accumulation by germinating conidia of *Botrytis cinerea* in autoradiographs. Transactions of the British Mycological Society, *82*, 161-4.

Falk, G.J. & King, R.C. (1963). Radioautographic efficiency for tritium as a function of section thickness. Radiation Research, *20*, 466-70.

Faust, M.A. & Correll, D.L. (1977). Autoradiographic study to detect metabolically active phytoplankton and bacteria in the Rhode River estuary. Marine Biology, *41*, 293-305.

Frankland, J.C., Bailey, A.D., Gray, T.R.G. & Holland, A.A. (1981). Development of an immunological technique for estimating mycelial biomass of *Mycena galopus* in leaf litter. Soil Biology & Biochemistry, *13*, 87-92.

Frossard, R., Fokkema, N.J. & Tietema, T. (1983). Influence of *Sporobolomyces roseus* and *Cladosporium cladosporioides* on leaching of $^{14}$C-labelled assimilates from wheat leaves. Transactions of the British Mycological Society, *80*, 289-96.

Fuhs, W.G. & Canelli, E. (1970). Phosphorus-33 autoradiography used to measure phosphate uptake by individual algae. Limnology and Oceanography, *15*, 962-7.

Hodges, G.M. & Muir, M.D. (1974). Autoradiography of biological tissues in the scanning electron microscope. Nature (London), *247*, 383-5.

Jones, K. (1970). Nitrogen fixation in the phyllosphere of the Douglas Fir, *Pseudotsuga douglasii*. Annals of Botany, *34*, 239-44.

Jones, K. (1976). Nitrogen fixing bacteria in the canopy of conifers in a temperate forest. *In* Microbiology of Aerial Plant Surfaces, eds C.H. Dickinson & T.F. Preece, pp. 451-63. London: Academic Press.

Jones, K. (1982). Nitrogen fixation in the canopy of temperate forest trees: a re-examination. Annals of Botany, *50*, 329-34.

Jones, K., King, E. & Eastlick, M. (1974). Nitrogen fixation by free living bacteria in the soil and in the canopy of Douglas Fir. Annals of Botany, *38*, 765-72.

Junger, E. & Bachmann, L. (1980). Methodological basis for an autoradiographic demonstration of insulin receptor sites on the surface of whole cells: a study using light and scanning electron microscopy. Journal of Microscopy, *119*, 199-211.

MacNamara, O.C. & Dickinson, C.H. (1981). Microbial degradation of plant cuticle. *In* Microbial Ecology of the Phylloplane, ed. J.P. Blakeman, pp. 455-73. London: Academic Press.

Manners, J.M. & Gay, J.L. (1980). Autoradiography of haustoria of *Erysiphe pisi*. Journal of General Microbiology, *116*, 529-33.

Mendgen, K. & Heitefuss, R. (1975). Micro-autoradiographic studies on host-parasite interactions. 1. The infection of *Phaseolus vulgaris* with tritium labelled uredospores of *Uromyces phaseoli*. Archives of Microbiology, *105*, 193-9.

Mount, M.S. & Ellingboe, A.H. (1969). $^{32}$P and $^{35}$S transfer from susceptible wheat to *Erysiphe graminis* f. sp. *tritici* during primary infection. Phytopathology, *59*, 235.

Parbery, I.H., Brown, J.F. & Bofinger, V.J. (1981). Statistical methods in the analysis of phylloplane populations. *In* Microbial Ecology of the Phylloplane, ed. J.P. Blakeman, pp. 47-65. London: Academic Press.

Parker, A. & Blakeman, J.P. (1984). Nutritional factors affecting the behaviour of *Uromyces viciae-fabae* uredospores on broad bean leaves. Plant Pathology, *33*, 71-80.

Paton, A.M. & Jones, S.M. (1973). The observation of micro-organisms on surfaces by incident fluorescence microscopy. Journal of Applied Bacteriology, *36*, 441-3.

Preece, T.F. (1971). Fluorescent techniques in mycology. *In* Methods in Microbiology. Vol. 4, ed. C. Booth, pp. 509-16. London: Academic Press.
Rodger, G. (1981). Microbial competition for nutrients on leaves of sycamore and lime. Ph.D. thesis, University of Aberdeen.
Rodger, G. & Blakeman, J.P. (1984). Microbial colonization and uptake of $^{14}C$ label on leaves of sycamore. Transactions of the British Mycological Society, *82*, 45-51.
Rogers, A.W. (1972). Photometric measurements of grain density in autoradiographs. Journal of Microscopy, *96*, 141-53.
Rogers, A.W. (1979). Techniques of Autoradiography, 3rd edition. Amsterdam: Elsevier/North-Holland Biomedical Press.
Ruinen, J. (1974). Nitrogen fixation in the phyllosphere. *In* The Biology of Nitrogen Fixation, ed. A. Quispel, pp. 121-67. Amsterdam: North Holland Publishing Co.
Sol, H.H. (1969). Adsorption of $^{14}C$-labelled exudates by conidia of *Botrytis fabae* on *Vicia faba* leaves. Netherlands Journal of Plant Pathology, *75*, 227-8.
Stewart, W.D.P. (1967). Nitrogen turnover in marine and brackish habitats. II. Use of $^{15}N$ in measuring nitrogen fixation in the field. Annals of Botany, *31*, 385-407.
Waid, J.S., Preston, K.J. & Harris, P.J. (1973). Autoradiographic techniques to detect active microbial cells in natural habitats. Bulletin Ecological Research Committee Swedish Natural Science Research Council, *17*, 317-22.
Zimmermann, R., Iturriaga, R. & Becker-Birck, J. (1978). Simultaneous determination of the total number of aquatic bacteria and the number thereof involved in respiration. Applied and Environmental Microbiology, *36*, 926-35.

# QUANTITATIVE SEROLOGICAL ESTIMATIONS OF FUNGAL COLONIZATION

K. Mendgen

*Universität Konstanz, Fakultät für Biologie, D-7750 Konstanz, Fed. Rep. Germany*

## *INTRODUCTION*

A variety of methods has been used in the past to quantify pathogenic fungi present in or on host plants. Observations on degree of colonization range from verbal descriptions (e.g. very slight to abundant) and number of plants infected to more elaborate light-microscopical estimations of amounts of hyphae in sectioned tissue (Toth & Toth 1982). Phase-contrast lenses (Frankland 1974) and, in my experience, optics for differential interference contrast, help to recognize very thin and even empty hyphae. Specific fluorescent stains for hyphal walls make it easy to discriminate between fungal and plant structures (Patton & Johnson 1970; Rohringer *et al.* 1977).

In morphometric studies, the relative proportions of different fungal components or fungal structures are determined in tissue sections or on host surfaces. Instrumentation is now available that makes such an analysis easy (Moesta *et al.* 1983).

Direct methods also include isolations from freshly collected tissue or air-dried material followed by dilution plating on a selective medium (e.g. Davis *et al.* 1983). This technique needs specific adaptations for each fungus assayed. It may indicate the amount of propagules in an infected tissue, but not the amount of fungal biomass. The value of these assays is discussed in the proceedings of the previous meeting in this series (Blakeman 1981).

Ride & Drysdale (1972) recommended the chemical determination of the chitin content as a quantitative measure for fungal invasion. This method is rapid and simple and has been used for different fungi: *Botrytis cinerea* (Harding & Heale 1978), *Erysiphe cichoracearum* (Onogur & Schlösser 1976), *Fusarium* spp. (Zak 1976; Raghu Kumar & Subramanian 1977; Schönbeck *et al.* 1977; Dehne & Schönbeck 1979), *Puccinia* spp. (Lösel & Lewis 1974; Pearce & Strange 1977; Whipps & Lewis 1980), *Verticillium* spp. (Pegg 1978) and some mycorrhizal fungi (Becker & Gerdemann 1977; Hepper 1977; Haselwandter 1979). This method has its pitfalls since, in contrast to hyphae, spores often have much higher chitin contents and this results in an inaccurate estimate of fungal growth. Furthermore, the chitin content is not directly correlated with hyphal growth (Mayama *et al.* 1975; Kaminsky & Heath 1982), probably because of changes in the fungal wall composition during growth in the host (Mendgen *et al.* 1985).

Similary, ergosterol has been recommended as a measure of fungal growth (Seitz *et al.* 1979). This sterol is the predominant sterol component of most fungi and is either absent or a minor constituent in most higher plants. This method seems to be more sensitive than the chitin assay. However, like the chitin determination, the ergosterol assay cannot discriminate between different fungi although different fungi seem to have different ergosterol contents. Therefore, this assay is only a rough method for estimating fungal biomass in or on a solid substrate.

As an alternative to the methods mentioned above, the total amount of fungal material present in the infected tissue may be measured serologically (e.g. with the enzyme-linked immunosorbent assay, ELISA). In ELISA, the antibody binding reaction is combined with the adsorption of the antibodies to a solid plastic matrix; this complex selectively retains antibodies linked with an enzyme. Alternatively, the enzyme-linked complex may be replaced by a radioactive-labelled antibody (radioimmunosorbent assay, RISA), but this test is more complicated to use (Savage & Sall 1981; Richardson & Warnock 1983) and the isotopes may be difficult to handle under normal laboratory conditions.

## *SEROLOGICAL ESTIMATIONS WITH EILISA*

The ELISA method of Clark & Adams (1977) is used very often for virus detection in plants: non-labelled antibodies are adsorbed to the wells of microtiter plates. A test solution with the antigen is added, and then enzyme-labelled antibodies are added. The latter antibodies also bind to the antigen. Subsequently, an enzyme substrate is added and the colour is measured. Alkaline phosphatase and horse-radish peroxidase are mainly used as enzyme label. Assuming that the rate of colour formation is proportional to the amount of conjugate reacted, the assay will indicate the amount of antigen or antibody present. The value of this technique in plant pathology has been reviewed recently (Clark 1981; Gugerli 1983). In the present review, new aspects for the estimations of fungal structures are summarized. Table 1 lists the fungal species for the detection of which ELISA has been used.

## *SPECIFICITY OF THE ANTIGEN AND SENSITIVITY OF ELISA*

In most cases whole cells are used as antigen. These may consist of unbroken thalli, e.g. of *Botrytis cinerea* (Savage & Sall 1981), obtained from cultures in the growth phase, or the complete fungal mycelium including spores (Casper & Mendgen 1979; Mendgen 1981; Johnson *et al.* 1982). Also an acetone precipitate of culture fluid, e.g. of *Phoma tracheiphila* (Nachmias *et al.* 1979), may be used. More specific antigens are enzymes from fungi, e.g. invertase from *Phytophthora megasperma* f. sp. *glycinea* (Moesta *et al.* 1983), $NADP^+$-glutamate dehydrogenase from *Sphaerostilbe repens* (Martin *et al.* 1983) or polygalacturonases from *Fusarium oxysporum* (Suresh *et al.* 1984). Scheffer & Elgersma (1981) used a phytotoxic glycopeptide from *Ophiostoma ulmi*, the causal organism of Dutch elm disease, as an antigen. Banowetz *et al.* (1984) used membranes and proteins, carbohydrates and complex polysaccharides from spore surfaces of *Tilletia controversa* and *T. caries*. In this latter study, monoclonal antibodies were used. Although this new technique seems to become very promising because of its high specificity (Benhamou & Ouellette 1985), it was not possible to differentiate with monoclonal antibodies between *T. controversa* and *T. caries* (Banowetz *et al.* 1984).

When polyclonal antibodies against *Epichloë typhina* were used, only three (*Thelephora*, *Rhizoctonia* and *Claviceps*) of 14 different genera of plant-pathogenic fungi tested, reacted to any extent in the ELISA system. A high concentration (50 mg $ml^{-1}$) of sclerotia of *Claviceps* sp. gave an absorbance value of 0.043, roughly equivalent to that of mycelium of *E. typhina* at 100 ng $ml^{-1}$ (Johnson *et al.* 1982). Obviously, the ELISA technique may also be useful in studying taxonomic relationships between fungi. Testing different Endogonaceae, Aldwell *et al.* (1985) were able to differentiate between these mycorrhizal fungi serologically. The data obtained corresponded well with the current classification based on morphological features and demonstrated the high specificity of ELISA.

Table 1. Fungi detected in plant tissues with ELISA.

| Fungal species | Reference |
|---|---|
| *Acremonium coenophialum* | Welty *et al.* (1984) |
| *Alternaria solani* | Vargo & Baumer (1984) |
| *Armillaria* sp. | Lung-Escarmant & Dunez (1979) |
| *Aspergillus* sp. | Hommell *et al.* (1976) |
| *Bipolaris* sp. | Vargo & Baumer (1984) |
| *Clitocybe* sp. | Lung-Escarmant & Dunez (1979) |
| Endomycorrhizal fungi (*Gigaspora margarita*, *G. calospora*, *Glomus mosseae*, *G. clarum*, *Acaulospora laevis*, *Sclerocystis dussii*) | Aldwell *et al.* (1983, 1985) |
| Endophytic fungi | Funk *et al.* (1983); Johnson *et al.* (1983); Musgrave (1984) |
| *Epichloë typhina* | Funk *et al.* (1983); Halisky *et al.* (1983); Johnson *et al.* (1983, 1985); Musgrave (1984) |
| *Gremmeniella abietina* | Liese *et al.* (1982) |
| *Phytophthora syringae* | Kimishima *et al.* (1984) |
| *Phoma tracheiphila* | Nachmias *et al.* (1979) |
| *P. exigua* | Aguelon & Dunez (1984) |
| *Plasmopara halstedii* | Liese *et al.* (1982) |
| *Verticillium albo-atrum* | Leach & Swinburne (1984) |
| *V. lecanii* | Casper & Mendgen (1979); Mendgen (1981) |

The results mentioned above show that it may be possible to differentiate between two fungi in or on a leaf. For example, the hyperparasite, *Verticillium lecanii*, could easily be distinguished in pustules of various rust fungi (Casper & Mendgen 1979; Mendgen 1981). In that system, there were no cross-reactions between extracts of the plants, the rust fungus and the hyperparasite. The specificity of the antibody could easily be checked with a microscope equipped for epifluorescence: fluorescein isothiocyanate-labelled antibodies against *V. lecanii* were restricted to this fungus in the cross-sections through a leaf of *Phaseolus vulgaris* infected with *Uromyces appendiculatus* and *V. lecanii* (Mendgen & Casper 1980). Moreover, the specificity of the antiserum may be enhanced by pretreatment with antigens (proteins) from similar fungi and discarding the precipitate formed (Frankland *et al.* 1981; Leach & Swinburne 1984).

The sensitivity of ELISA was at least $10^3$ to $10^5$ times greater than that of the double diffusion test (Nachmias *et al.* 1979). *E. typhina* was detected in concentrations down to 100 ng $ml^{-1}$ (Johnson *et al.* 1982). This value is similar to that possible with RISA for the detection of

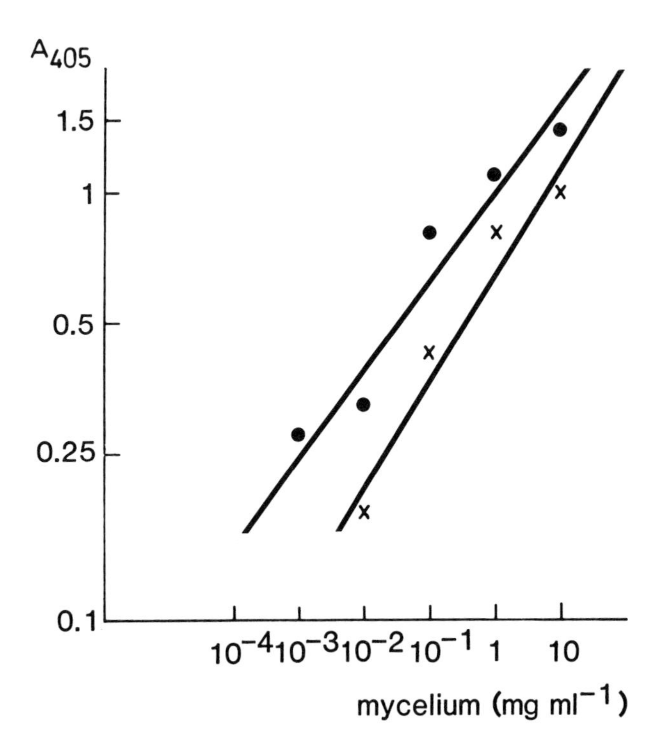


Fig. 1. Enzyme-linked immunosorbent assay (ELISA) for a dilution series of mycelium of two *E. typhina* isolates. Absorbance at 405 nm after incubation of *E. typhina* isolated from tall fescue (•) and bentgrass (x). The absorbance of the control buffer (PBS) was 0.007. Drawn after data from Johnson *et al.* (1982).

*B. cinerea* (Savage & Sall 1981). Johnson *et al.* (1983) detected antigens in samples consisting of only one endophyte-infected seed extracted together with 19 endophyte-free seeds.

### *QUANTITATIVE ESTIMATIONS*

It may be possible to correlate spectrophotometric ELISA readings with the total amount of fungal material in infected host plants. An example of such a correlation is shown for a dilution series of mycelium of two *E. typhina* isolates (Fig. 1). Quantitative measurements have also been made with *V. lecanii* in pustules of stripe rust (*Puccinia striiformis*) under different growth conditions (Fig. 2). A good correlation was obtained between the amount of the hyperparasitic fungus growing in the rust pustules and the ELISA values (Casper & Mendgen 1979; Mendgen 1981).

There are some principal restrictions in using the ELISA technique for the quantitative evaluation of fungal biomass or fungal growth. Comparable to the problems with quantitative chitin or ergosterol measurements, the antigen chosen for the detection of a fungus within plant tissue may not be evenly distributed within all fungal structures. As a consequence, the increase or decrease of that antigen, rather than the growth of the fungus will be measured. Changes in protein patterns during the development of a fungal pathogen have been reported (Huang & Staples

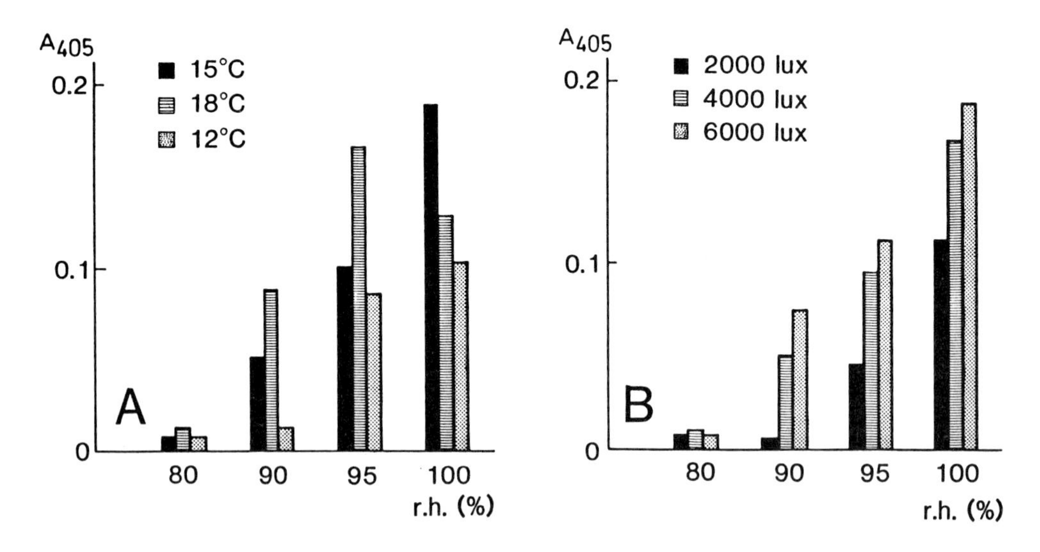


Fig. 2. The amount of *V. lecanii* in rust pustules on wheat, estimated by ELISA and expressed as absorbance at 405 nm (adapted from Mendgen 1981). A. Influence of relative humidity and temperature on the development of *V. lecanii* grown under 4000 lux. $A_{405}$ values of the control (rusted leaves without *V. lecanii*) were <0.01. B. Influence of relative humidity and light on the development of *V. lecanii* grown at 15°C. $A_{405}$ values of the control were <0.01.

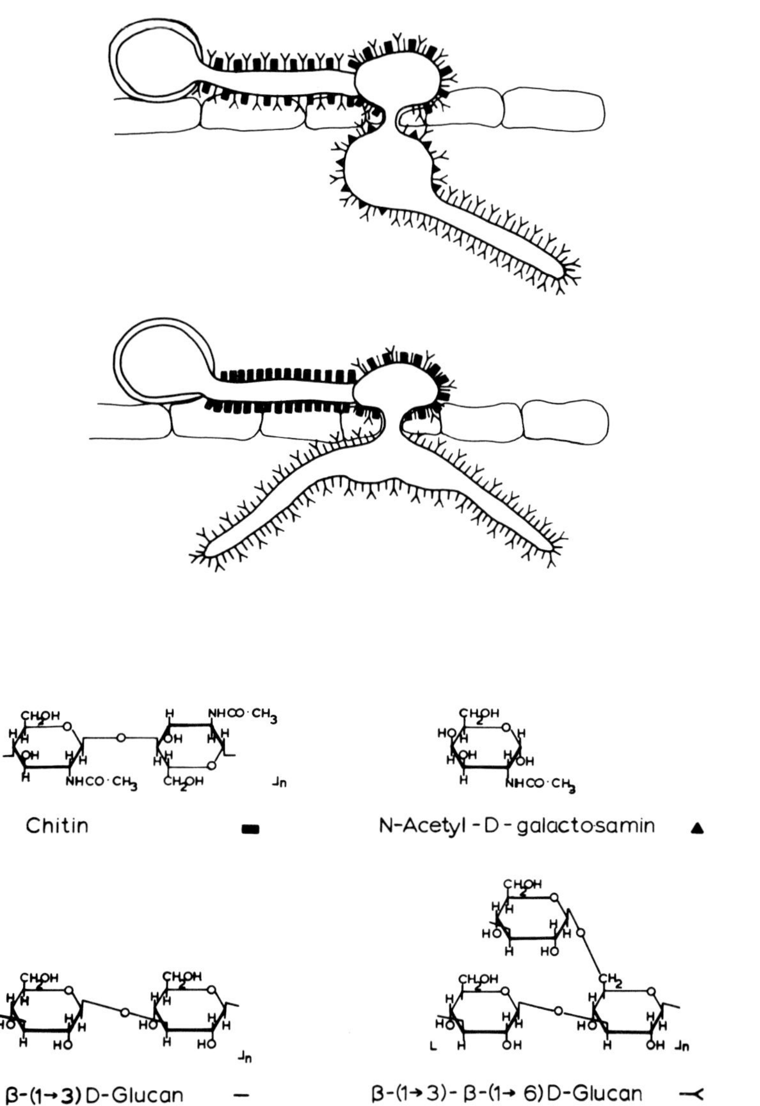


Fig. 3. A, B. Surface carbohydrate pattern on outer and inner infection structures of two rust fungi: *Uromyces appendiculatus* (A) and *Puccinia coronata* (B). C. Structures of the carbohydrates involved (Mendgen 1984).

1982; Kim *et al.* 1982). Also the wall carbohydrates (Bartnicki-Garcia 1968) and especially the surface carbohydrates (Fig. 3) of the wall change drastically during the course of infection of a rust fungus (Mendgen *et al.* 1985). Such carbohydrates (e.g. polyglucans or polygalactans) may be effective antigenic determinants. To avoid some of these problems, antibodies against a mixture of antigens derived from all parasitic structures of a fungus are recommended for quantitative studies of fungal development. This technique would overcome inaccuracies of other methods in which only one component of a fungus is measured, which may change with age, growth rate and environmental conditions.

## *ACKNOWLEDGEMENTS*

I thank Dr R.E. Gold, Konstanz, for reading the manuscript and the Deutsche Forschungsgemeinschaft for supporting the research.

## *REFERENCES*

Aguelon, M. & Dunez, J. (1984). Immunoenzymatic techniques for the detection of *Phoma exigua* in infected potato tissues. Annals of Applied Biology, *105*, 463-9.

Aldwell, F.E.B., Hall, I.R. & Smith, J.M.B. (1983). Enzyme-linked immunosorbent assay (ELISA) to identify endomycorrhizal fungi. Soil Biology & Biochemistry, *15*, 377-8.

Aldwell, F.E.B., Hall, I.R. & Smith, J.M.B. (1985). Enzyme-linked immunosorbent assay as an aid to taxonomy of the Endogonaceae. Transactions of the British Mycological Society, *84*, 399-402.

Banowetz, G.M., Trione, E.J. & Krygier, B.B. (1984). Immunological comparisons of teliospores of two wheat bunt fungi, *Tilletia* species, using monoclonal antibodies and antisera. Mycologia, *76*, 51-62.

Bartnicki-Garcia, S. (1968). Cell wall chemistry, morphogenesis and taxonomy of fungi. Annual Review of Microbiology, *22*, 87-108.

Becker, W.N. & Gerdemann, J.W. (1977). Colorimetric quantification of vesicular-arbuscular mycorrhizal infection in onion. New Phytologist, *78*, 289-95.

Benhamou, N. & Ouellette, G.B. (1985). Les anticorps monoclonaux: une technologie de pointe en phytopathologie. Phytoprotection, *66*, 5-15.

Blakeman, J.P. ed. (1981). Microbial Ecology of the Phylloplane. London: Academic Press.

Casper, R. & Mendgen, K. (1979). Quantitative serological estimation of a hyperparasite: detection of *Verticillium lecanii* in yellow rust infected wheat leaves by ELISA. Phytopathologische Zeitschrift, *94*, 89-91.

Clark, M.F. (1981). Immunosorbent assays in plant pathology. Annual Review of Phytopathology, *19*, 83-106.

Clark, M.F. & Adams, A.N. (1977). Characteristics of the microplate method of enzyme-linked immunosorbent assay for the detection of plant viruses. Journal of General Virology, *34*, 475-83.

Davis, J.R., Pavek, J.J. & Corsini, D.L. (1983). A sensitive method for quantifying *Verticillium dahliae* colonization in plant tissue and evaluating resistance among potato genotypes. Phytopathology, *73*, 1009-14.

Dehne, H.-W. & Schönbeck, F. (1979). Untersuchungen zum Einfluss der endotrophen Mycorrhiza auf Pflanzenkrankheiten. I. Ausbreitung von *Fusarium oxysporum* f. sp. *lycopersici* in Tomaten. Phytopathologische Zeitschrift, *95*, 105-10.

Frankland, J.C. (1974). Importance of phase-contrast microscopy for estimation of total fungal biomass by the agar-film technique. Soil Biology & Biochemistry, *6*, 409-10.

Frankland, J.C., Bailey, A.D., Gray, T.R.G. & Holland, A.A. (1981). Development of an immunological technique for estimating mycelial biomass of *Mycena galopus* in leaf litter. Soil Biology & Biochemistry, *13*, 87-92.

Funk, C.R., Halisky, P.M., Johnson, M.C., Siegel, M.R. Stewart, A.V., Ahmad, S., Hurley, R.H & Harvey, I.C. (1983). An endophytic fungus and resistance to sod webworms: association in *Lolium perenne* L. Bio/Technoloy, *1*, 189-91.

Gugerli, P. (1983). Use of enzyme immunoessay in phytopathology. *In* Immunoenzymatic Techniques, eds S. Avrameas, B.V.P. Druet, R. Masseyeff & G. Feldmann, pp. 369-84. Amsterdam: Elsevier Science Publishers.

Halisky, P.M., Funk, C.R. & Vincelli, P.C. (1983). A fungal endophyte in seeds of turf-type perennial ryegrass. Phytopathology, *73*, 1343 (Abstr.).

Harding, V. & Heale, J.B. (1978). Post-formed inhibitors in carrot root tissue treated with heat-killed and live conidia of *Botrytis cinerea*. Annals of Applied Biology, *89*, 348-51.

Haselwandter, K. (1979). Mycorrhizal status of ericaceous plants in alpine and subalpine areas. New Phytologist, *83*, 427-31.

Hepper, C.M. (1977). A colorimetric method for estimating vesicular-arbuscular mycorrhizal infection in roots. Soil Biology & Biochemistry, *9*, 15-8.

Hommell, M., Truong, T.K. & Bidwell, D.E. (1976). Technique immunoenzymatique (ELISA) appliquée au diagnostic sérologique des candidoses et aspergilloses humaines. Nouvelle Presse Médicale, *5*, 2789-91.

Huang, B.-F. & Staples, R.C. (1982). Synthesis of proteins during differentiation of the bean rust fungus. Experimental Mycology, *6*, 7-14.

Johnson, M.C., Anderson, R.L., Kryscio, R.J. & Siegel, M.R. (1983). Sampling procedures for determining endophyte content in tall fescue *Festuca arundinacea* seed lots by ELISA. Phytopathology, *73*, 1406-9.

Johnson, M.C., Pirone, T.P., Siegel, M.R. & Varney, D.R. (1982). Detection of *Epichloë typhina* in tall fescue by means of enzyme-linked immunosorbent assay. Phytopathology, *72*, 647-50.

Johnson, M.C., Siegel, M.R. & Schmidt, B.A. (1985). Serological reactivities of endophytic fungi from tall fescue and perennial ryegrass and of *Epichloë typhina*. Plant Disease, *69*, 200-2.

Kaminskyj, S.G.W. & Heath, M.C. (1982). An evaluation of the nitrous acid - 3-methyl-2-benzothiazolinone hydrazone hydrochloride - ferric chloride assay for chitin in rust fungi and rust-infected tissue. Canadian Journal of Botany, *60*, 2575-80.

Kim, W.K., Howes, N.K. & Rohringer, R. (1982). Detergent-soluble polypeptides in germinated uredospores and differentiated uredosporelings of wheat stem rust. Canadian Journal of Plant Pathology, *4*, 328-33.

Kimishima, E., Nishio, T., Takayama, M. & Nagao, N. (1984). Studies on serological detection and identification methods for species of *Phytophthora*. 3. Detection of *Phytophthora syringae* in plant tissues by means of enzyme-linked immunosorbent assay. Research Bulletin of the Plant Protection Service, *20*, 1-6.

Leach, J.E. & Swinburne, T.R. (1984). An indirect ELISA for quantitative estimation of *Verticillium albo-atrum* in hops. Phytopathology, *74*, 845 (Abstr.).

Liese, A.L., Gotlieb, A.R. & Sakston, W.E. (1982). Utilization of the ELISA technique for the diagnosis of two fungal diseases: Scleroderris canker of conifers and downy mildew of sunflower. Phytopathology, *72*, 263 (Abstr.).

Lõsel, D.M. & Lewis, D.H. (1974). Lipid metabolism in leaves of *Tussilago farfara* during infection by *Puccinia poarum*. New Phytologist, *73*, 1157-69.

Lung-Escarmant, B. & Dunez, J. (1979). Differentiation of *Armillaria* and *Clitocybe* species by the use of the immunoenzymatic ELISA procedure. Annales de Phytopathologie, *11*, 515-8.

Martin, F., Botton, B. & Msatef, Y. (1983). Enzyme-linked immunosorbent assay of

fungal NADP$^+$-glutamate dehydrogenase. Plant Physiology, *72*, 398-401.
Mayama, S., Rehfeld, D.W. & Daly, J.M. (1975). A comparison of the development of *Puccinia graminis tritici* in resistant and susceptible wheat based on glucosamine content. Physiological Plant Pathology, *7*, 243-57.
Mendgen, K. (1981). Growth of *Verticillium lecanii* in pustules of stripe rust (*Puccinia striiformis*). Phytopathologische Zeitschrift, *102*, 301-9.
Mendgen, K. (1984). Wirtsfindung, Wirtserkennung biotropher Pilze und Abwehrreaktionen der Wirtspflanze. Mitteilungen aus der Biologischen Bundesanstalt für Land- und Forstwirtschaft Berlin-Dahlem, *223*, 6-16.
Mendgen, K. & Casper, R. (1980). Detection of *Verticillium lecanii* in pustules of bean rust (*Uromyces phaseoli*) by immunofluorescence. Phytopathologische Zeitschrift, *99*, 362-4.
Mendgen, K., Lange, M. & Bretschneider, K. (1985). Quantitative estimation of the surface carbohydrates on the infection structures of rust fungi with enzymes and lectins. Archives of Microbiology, *140*, 307-11.
Moesta, P., Grisebach, H. & Ziegler, E. (1983). Immunohistochemical detection of *Phytophthora megasperma* f. sp. *glycinea* in situ. European Journal of Cell Biology, *31*, 167-9.
Musgrave, D.R. (1984). Detection of an endophytic fungus of *Lolium perenne* using enzyme-linked immunosorbent assay (ELISA). New Zealand Journal of Agricultural Research, *27*, 283-8.
Nachmias, A., Bar-Joseph, M., Solel, Z. & Barash, I. (1979). Diagnosis of mal secco disease in lemon by enzyme-linked immunosorbent assay. Phytopathology, *69*, 559-61.
Onogur, E. & Schlösser, E. (1976). Quantitative Bestimmung der Pilzmasse echter Mehltaupilze in infiziertem Gewebe. Phytopathologische Zeitschrift, *87*, 91-3.
Patton, R.F. & Johnson, D.W. (1970). Mode of penetration of needles of eastern white pine by *Cronartium ribicola*. Phytopathology, *60*, 977-82.
Pearce, R.B. & Strange, R.N. (1977). Glycinebetaine and choline in wheat in relation to growth of stem rust. Physiological Plant Pathology, *11*, 143-8.
Pegg, G.F. (1978). Effect of host substrate on germination and growth of *Verticillium albo-atrum* and *V. dahliae* conidia and mycelia. Transactions of the British Mycological Society, *71*, 483-9.
Raghu Kumar, C. & Subramanian, D. (1977). Studies on *Fusarium* wilt of cotton. I. Host colonization. Phytopathologische Zeitschrift, *90*, 223-35.
Richardson, M.D. & Warnock, D.W. (1983). Enzyme-linked immunosorbent assay and its application to the serological diagnosis of fungal infection. Sabouraudia, *21*, 1-14.
Ride, J.P. & Drysdale, R.B. (1972). A rapid method for the chemical estimation of filamentous fungi in plant tissue. Physiological Plant Pathology, *2*, 7-15.
Rohringer, R., Kim, W.K., Samborski, D.J. & Howes, N.K. (1977). Calcofluor: an optical brightener for fluorescence microscopy of fungal plant parasites in leaves. Phytopathology, *67*, 808-10.
Savage, S.D. & Sall, M.A. (1981). Radioimmunosorbent assay for *Botrytis cinerea*. Phytopathology, *71*, 411-5.
Scheffer, R.J. & Elgersma, D.M. (1981). Detection of a phytotoxic glycopeptide produced by *Ophiostoma ulmi* in elm by enzyme-linked immunospecific assay (ELISA). Physiological Plant Pathology, *18*, 27-32.
Schönbeck, F., Dehne, H.-W. & Zimmermann, J. (1977). Untersuchungen über den Einfluss von Diallat auf den Befall von Weizen mit *Fusarium culmorum* und *F. avenaceum*. Phytopathologische Zeitschrift, *90*, 77-86.
Seitz, L.M., Sauer, D.B., Burroughs, R., Mohr, H.E. & Hubbard, J.D. (1979). Ergosterol as a measure of fungal growth. Phytopathology, *69*, 1202-3.
Suresh, G., Balasubramanian, R. & Kalyanasundaram, R. (1984). Enzyme-linked immunosorbent assay of polygalacturonases in cotton plants infected by *Fusarium oxysporum* f. sp. *vasinfectum*. Zeitschrift für Pflanzenkrankheiten und Pflanzenschutz, *91*, 122-30.

Toth, R. & Toth, D. (1982). Quantifying vesicular-arbuscular mycorrhizae using a morphometric technique. Mycologia, *74*, 182-7.

Vargo, R.H. & Baumer, J.S. (1984). Soaking as a method of preparing samples for an enzyme-linked immunosorbent assay (ELISA) for *Bipolaris sorokiniana*. Phytopathology, *74*, 884 (Abstr.).

Welty, R.E., Milbrath, G.M., Faulkenberry, D., Azevedo, M.D. & Meek, L. (1984). Detecting *Acremonium coenophialum* in seeds of tall fescue. Phytopathology, *74*, 1142 (Abstr.).

Whipps, J.M. & Lewis, D.H. (1980). Methodology of a chitin assay. Transactions of the British Mycological Society, *74*, 416-8.

Zak, J.C. (1976). Pathogenicity of a gibberellin-producing and a non-producing strain of *Fusarium moniliforme* in oats as determined by a colorimetric assay for N-acetyl glucosamine. Mycologia, *68*, 151-8.

# SECTION II

# ECOLOGY OF EPIPHYTIC FUNGI

# COLONIZATION DYNAMICS: THE ISLAND THEORY

J.H. Andrews and L.L. Kinkel

*Department of Plant Pathology, University of Wisconsin-Madison, 1630 Linden Drive, Madison, WI 53706, USA*

## *INTRODUCTION*

Community ecologists generally search for repetitions, rather than isolated events. The processes underlying such patterns are then sought and, by induction, the results are generalized so that predictions can be made about the system. Although theory-free data gathering has a (limited) role, especially in a young science, we are at the point in phylloplane microbiology where guiding principles are badly needed. This is particularly evident for the species dynamics of epiphytic microbes.

As the title implies, this chapter presents a theoretical approach to colonization; more specifically, for the fluctuation in numbers of fungal species on individual leaves. Our idea is that microbial colonization of the phylloplane is roughly analogous to colonization of islands by macro-organisms, expressed in the theory of island biogeography. Briefly, this theory postulates that the number of species on an island results from a dynamic equilibrium between species immigrating onto the island and extinction of established species within its boundary. The theory predicts that the equilibrial number will be higher on large islands close to the mainland species pool than on small, more remote islands.

The chapter is developed as follows. We first define colonization and other key terminology within the semantics of island biogeography. We then describe the island model as developed by macro-ecologists and its potential applicability to phylloplane communities. A synopsis of the major results from fungal colonization studies to date is presented. Finally, we assess our evidence for and against the concept based on a 5-year study that began in 1980, and suggest directions for future research.

## *COLONIZATION AND RELATED TERMINOLOGY*

The relationship of colonization to the key terms, immigration, extinction, and propagule, must first be clarified. *Immigration* involves the arrival of a new species 'propagule' (see below) on an island unoccupied by the species (MacArthur & Wilson 1967; Simberloff 1969). *Extinction* is the disappearance (by death or emigration) of a species from an island. It does not preclude recolonization and, most importantly, is inferred if the species is not tallied in a subsequent sampling or census. The inherent limitation of detectability threshold in all island studies is discussed later. A *propagule* represents the minimum number of individuals of a species capable of breeding and population increase under ideal conditions for that species (Simberloff 1969; Simberloff & Wilson 1969). Excluded from censuses are animals of zero reproductive value such as a male ant landing after a nuptial flight (Simberloff & Wilson 1969). By analogy, in microbiology, a propagule represents a biological unit capable of reproducing itself under ideal conditions. Operationally it is a colony-forming unit under standard laboratory conditions. In our phylloplane context, a species immigrated when it was first isolated from leaf surfaces and it became extinct when it was no longer cultured from samples.

*Colonization* has been defined variously, both within and between disciplines. In island semantics, it implies the existence of at least one propagule of a species, and each species for which at least one propagule exists is designated a colonist (Simberloff & Wilson 1969). It is also an immigrant if not tallied at the previous census; every immigrant is also by definition a colonist. This concept of colonist and colonization implies nothing about establishment, unlike the original definition proposed by MacArthur & Wilson (1967; p. 188) which denoted relatively lengthy persistence, especially where breeding and population increase occur. Thus, according to Simberloff & Wilson (1969), a species whose propagule lands on an island is a colonist, "even if it is doomed to quick extinction for purely physical reasons". We use their interpretation.

Put another way, the foregoing interpretation means that a colonist may be either a *transient* or a *resident*. Hence, in the following discussion a colonist is any fungal propagule that was isolated from the leaf surface; nothing should be assumed about its life form, growth status, or ultimate fate, other than that it was obviously viable. The implication of this assumption and its place in criticisms of the model are discussed later.

## *THE ISLAND THEORY*

### *Origin and development*

The fascination of biologists with island communities dates at least since Darwin's collections of the Galapagos finches in 1835. The contribution of the theory was in providing a common framework for interpreting many of the biological observations which hitherto were unconnected or unexplained. Among these were two general, empirical relationships. The first was a formal association between number of species and area of an island, developed by Preston (1948, 1960), who observed that communities often contain many more rare than common species. Graphically expressed in a typical species-abundance curve form, this corresponds roughly to a lognormal distribution. Based on this distribution, Preston (1962 a,b) postulated a direct relationship between the number of individuals and the number of species present in a community. A key assumption in the development of the species-area model was that for relatively uniform islands, the number of individuals present is related to island area. Thus, area, through its influence on numbers of individuals, affects the number of species an island supports. Studies of plants and animals, including birds, ants and beetles, show an empirical species number(S)-area(A) relationship, expressed as $S = cA^z$ where c and z are constants (Preston, 1960; Diamond & May, 1976); e.g. species number decreases by about half for each tenfold decrease in area. Second, for islands of similar area, species number decreases with increasing distance from the mainland or colonizing source (MacArthur & Wilson 1967; Diamond & May 1976).

MacArthur & Wilson (1963, 1967) and less explicitly Preston (1962 a,b) proposed that the number of species (N) on an island tended towards a dynamic equilibrium, resulting from a balance between the rate of immigration (I) of new species onto the island and the rate of extinction (E) of species from it. In a plot of I and E versus N, I falls because as more species immigrate, fewer arrivals belong to *new* species, and reaches zero when all mainland species have arrived (Fig. 1). In contrast, E rises for the purely mathematical reason that the more species there are present, the more there are to go extinct. If one assumes that all species have the same, constant rates of I and E, then the plots for both functions are linear. However, as usually depicted, I is a concave function because rapidly dispersing species arrive first, followed by slower colonists, dropping I to an ever diminishing extent. Biologically, the theory predicts that a greater number of species will result in more species at smaller population sizes and thus greater extinction probabilities "due to the samller average population size acting through both ecological and genetical accident" (MacArthur & Wilson 1967, p. 22). These biological effects account for the concave depiction of E. Equilibrium in species number, $\hat{S}$, occurs where the falling I curve intersects the rising E curve. The theory also addresses other issues such as the nature of the selection pressure on colonists (r- and K-selection), islands as 'stepping stones' for species dispersal, and niche shifts, which are beyond the scope of this discussion. Numerous elaborations on this simple model exist (e.g. Wilson 1969; Pielou 1979; Simberloff 1983 a).

The theory predicts that $\hat{S}$ is a function of the island's area and proximity to the species pool (MacArthur & Wilson 1967, p. 22; Simberloff 1976 b), which influence I and E (Fig. 2). I is lower for more distant islands and E is higher for smaller islands. MacArthur & Wilson (1967) did not develop the reasons for this. However, it can be inferred from their 1963 paper and subsequent elaborations (Wilson 1969; MacArthur 1972, p. 79-126) that the main reasons are inaccessibility of remote islands to colonists, and increased competition plus smaller population sizes on smaller islands, respectively.

The elegant simplicity of the equilibrium postulate led to an enthusiastic and, in retrospect, generally uncritical reception. As Simberloff (1976 a) said, the theory reached the status of a paradigm. It was immediately obvious that islands could be not only 'real' in the oceanic context, but 'virtual' as is any habitat patch surrounded by dissimilar terrain. Among the many so-called 'tests' of the theory have been

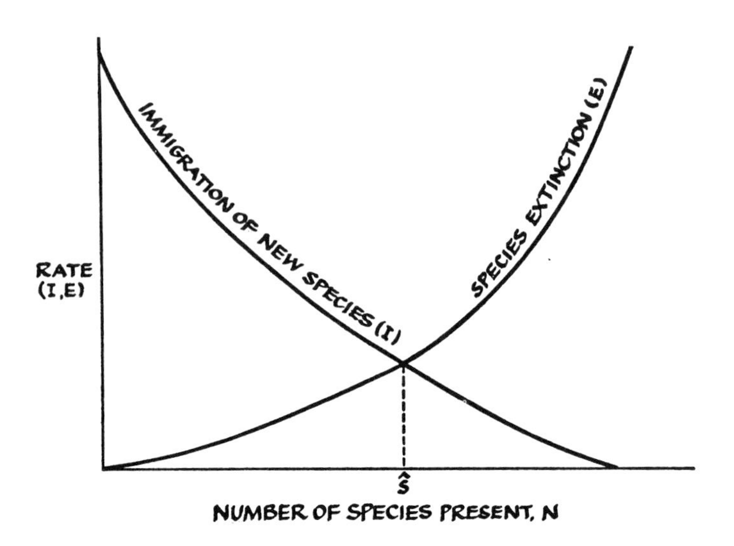


Fig. 1. Equilibrium model of biota on a single island. Equilibrium in species number occurs at $\hat{S}$, where the plot for immigration of species new to the island intersects that of extinction of species from the island. From MacArthur & Wilson (1967); reprinted with minor changes by permission of Princeton University Press.

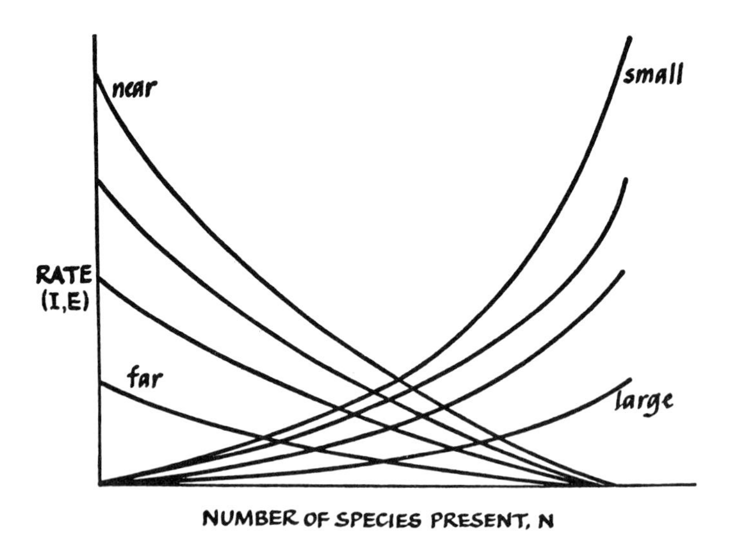


Fig. 2. Equilibrium models of biotas of several islands varying in size and proximity to the mainland source area. Immigration is lowered as distance increases; extinction is lowered as island area increases. From MacArthur & Wilson (1967); reprinted with minor changes by permission of Princeton University Press.

several studies of birds (e.g. Diamond 1969; Johnson 1972); insects on real (e.g. Simberloff & Wilson 1969) or virtual (e.g. Seifert 1975) islands; small mammals on montane islands (Brown 1971); mites on rodents (Dritschilo *et al.* 1975); and arthropods in caves (Culver 1970). Janzen (1968) proposed the imaginative idea that, for insects, host plant species in evolutionary time, and individual plants in contemporary time, act as islands. The theory was embraced among conservationists as a guide for refuge design (Diamond & May 1976).

Two examples will serve to illustrate the applications of the theory. The first, typical of most, is based solely on 'natural experiment' or observational data. Diamond (1969) surveyed the birds known to breed on each of the Channel Islands 12 to 98 km off the southern California coast and compared his results to those of a census made about 50 years earlier. The major findings, consistent with the theory, are as follows (Diamond 1969; MacArthur 1972, p. 79-81): (1) no island had as many species as a comparable mainland area; (2) many species had gone extinct, but (3) these were counterbalanced by immigrants, which resulted in (4) the islands having about the same number of species overall at the two censuses.

The second approach involves an experimental manipulation of the variables and remains today the most elegant and rigorous test of the theory. The entire arthropod faunas of six small mangrove islands (10 to 20 m diameter; 2 to 533 m from nearest source) in the Florida Keys were killed by methyl bromide fumigation; the process of recolonization was then monitored over a 2-year period and communities were compared to those censused prior to 'defaunation' (Simberloff 1969; Wilson & Simberloff 1969; Simberloff & Wilson 1969, 1970). Within 250 days after defaunation the original species number was restored on all but the most remote island and species composition was similar to that on untreated islands, although population densities were lower. Colonization curves had the logarithmic form predicted by the theory. Records taken two years after defaunation provided further evidence for a dynamic equilibrium on all but the furthest island whose species were considered to be still climbing toward equilibrium. The influence of area on species number was confirmed in a subsequent study (Simberloff 1976 b) with a different set of mangrove islands, by censusing, then removing sections of the habitat, and re-censusing.

### *Applications in microbial ecology*

There has been some speculation but little experimental study on the role that island theory might play in microbial ecology. At the third International Symposium on the Microbiology of Leaf Surfaces (1980), the MacArthur-Wilson model was used to postulate how pesticides might affect phylloplane species diversity (Andrews 1981); independently, H.G. Wildman and J.C. Zak presented a poster on leaves as ecological islands. Elsewhere, Wicklow (1981) discussed the theory briefly and concluded that fungal systems may provide a good means to test it. Patrick (1967) examined the effect of 'island' area and invasion rate on diatom species colonizing glass slides. Species number increased with size of area exposed. When the invasion rate was lowered (to simulate immigration rates onto distant islands) by reducing flow rate of the water, species number declined. Cairns *et al.* (1969) followed the colonization by protozoans of artificial substrates immersed in a lake. The colonization curve approximated that predicted, but equilibrium was not reached and there was a fairly wide scatter of points for the extinction rate.

In short, there have been insufficient studies with microbes to make any unequivocal general statements as to applicability of the theory.

### *Criticisms and limitations of the theory*

Gilbert (1980) stated that three criteria must be fulfilled to substantiate the model for any particular situation: (i) a close relationship must exist between species number and insular area (Simberloff 1983 a); (ii) species number must remain constant; and (iii) turnover or species replacement must occur. To these, Simberloff (1983 a) added a fourth point, that immigration and extinction, not simply 'transient intrapopulation movement', has to account for the observed population processes, i.e. the extent to which population dynamics are accounted for by breeding on the island as opposed to invasion must be known. This raises two issues. First, the species tallied must be truly capable of population increase (i.e. they are 'propagules'). Second, the system under study must not be so swamped by repeated invasions of prolific colonists that the population immigration and extinction dynamics represented by the model are inapplicable. By the above standards, the only convincing test of the theory is the study of arthropods on mangrove islands in the Florida Keys, discussed previously.

The criticisms, most of which have been put forward recently, are reviewed in depth elsewhere (Pielou 1979; Abbott 1980; Gilbert 1980; Simberloff 1983 a,b,c; McGuinness 1984) and can be addressed only briefly here. Critics have rarely made a distinction between criticisms of the theory per se and the studies which purport to substantiate it. Strikingly, many of the critical assumptions on which the theory depends remain unvalidated. Among these are the lognormal distribution of species, and the relationships between individuals and numbers of species, and numbers of individuals and area. Most 'support' for the concept has come from field observations where variables are not controlled. Correlations (not causation) can be established, but even some of these are suspect because they have been based on selected data or few data points (Gilbert 1980). In many studies, colonization is not actually observed, but inferred from species-area curves (reviewed by Simberloff 1974). Historical data, often inadequately documented, and collected originally for reasons other than tests of the theory, are frequently used. There is no consensus on definition of equilibrium (Simberloff 1983 c), which has been interpreted graphically and subjectively, rather than quantitatively. Species turnover calculations are complicated by the artifacts of sampling, inter-island movement, or environmental change (e.g. human activity) (Simberloff 1983 a). For example, Diamond's (1969) study discussed earlier, of birds on the Channel Islands, was challenged on those grounds by Lynch & Johnson (1974) who believed, contrary to the model's predictions, that extinction is not as common an event for birds as artifactual data due to 'pseudoturnover' suggest. Other sources of experimental error include incomplete surveys, taxonomic errors, habitat differences among islands, and incorrect determination of the actual mainland source (Diamond 1969; Simberloff 1969). Jones & Diamond (1976) refuted most of the criticisms. Sampling, especially in microbiology, can alsways be criticized as inexhaustive. Presence can be enumerated reliably (although in macro-ecology evidence for breeding is frequently questionable); absence is only the failure to detect, not necessarily extinction. Finally, a detected correlation between species number and area accords with predictions, but is not unique to the model. In view of these criticisms, the model remains unconfirmed (Simberloff 1983 a,b) and, as such, should not be broadly applied.

Finally, it should be noted that the equilibrium theory is limited because it only treats species qualitatively, i.e. as present or absent. To know actual population numbers and the activity of the organisms might be more ecologically interesting and informative.

### *Phylloplane colonization studies*

The colonization of leaf surfaces by fungi has been reviewed comprehensively elsewhere (e.g. Hudson 1971; Dickinson 1976). Only some generalizations and their implications for a conceptual approach will be considered here. Although relative proportions vary over time, the dominant mycoflora isolated from deciduous leaves are usually members of the following taxa: *Alternaria* spp., *Cladosporium* spp., *Epicoccum purpurascens*, *Aureobasidium pullulans*, *Cryptococcus* spp. and *Sporobolomyces* spp. (e.g. Hudson 1971; Dickinson 1976). The remainder of any single leaf surface community generally consists of a group of rare species, represented in the sample by one or a few individuals. In our study, approximately 27% of the 120 species detected were isolated only once out of a total of 14,000 individuals. This pattern of species abundance, with a few common and many rare species, extends also to the bacterial epiflora (C.E. Morris, personal communication), and is reported in most macro-communities (Preston 1948).

Successional trends in both presence and relative abundances over time have been reported on numerous hosts (e.g. Hudson 1962; Sherwood & Carroll 1974; Wildman & Parkinson 1979). In general, proportions of the total community comprised of filamentous fungi increase towards the latter part of the growing season, while yeasts and bacteria are relatively more numerous earlier in community development (Dickinson 1976). These trends may be due to the changing biology or micro-environment of the leaves, to microbial interactions on the leaf surface, or simply to a changing immigrant pool (air spora). A pattern of increasing densities of individuals as the season progresses seems to be a general epiphytic phenomenon (Andrews *et al.* 1980; Rodger & Blakeman 1984).

### *Leaves as islands*

Leaf-microbe systems present advantages and complications

for tests of the theory. Both conceptually and physically a leaf is an island for microbe colonists from its relatively depauperate or uncolonized state at bud break until abscission. Leaves are accessible and easily replicated natural islands. They possess relatively simple microbial communities compared to other habitats such as soil, and can be easily manipulated. Leaves support communities typified by few common and many rare species, corresponding to a premise of the theory. The leaf as a micro-environment presents a more uniform and better understood system than do large habitat patches or real islands. Dispersal of microbes is largely passive, which precludes the complication of behavioural responses which may play a role in animal dispersal (Simberloff 1969).

On the other hand, complications include the fact that a non-destructive survey of microbial populations is not feasible; hence, the same island cannot be followed through time. This introduces the problem of leaf-to-leaf variability. Not all microbes can be physically removed for enumeration and not all that are will grow because conditions are not 'ideal' or the propagule, unlike a bird or mammal, may be constitutively dormant. Among those showing growth, it is logistically impossible to identify every colony of the thousands or millions isolated from each leaf at each sampling date. The increasing numbers of individuals per leaf over time complicates sampling. To facilitate taxonomy, certain assumptions must be made and only a 'representative' subset of the population actually identified. Leaves grow, so island area changes, as do other habitat characteristics. The 'mainland' source is not discrete, so it is impossible without contriving sources to test the distance effect. Dispersal patterns may have a strong seasonal (or diurnal) component, and may be affected by inter-island rain wash or vectors. Since leaves are ephemeral it might be expected that no equilibrium would be reached. They are also subject to ecological 'catastrophes' such as wind or rain scouring and drought; biotas with these characteristics typically have not been well described by the theory (Strong 1979).

### *The apple leaf-fungal community system*

*Experimental.* There are three ways to examine the theory with a leaf-microbe system. First, leaves can be sampled sequentially to detect colonization patterns during the evolution of foliar islands from bud break until senescence or abscission. This is comparable to the 'unmanipulated variable' type experiment common in macro-ecology. Second, 'new' islands, similar to natural leaves, can be introduced. Third, pre-existing leaf islands can be sterilized analogously to the 'defaunation' of real islands in the Florida Keys. The first approach approximates most closely what is happening in nature, but is complicated by uncontrolled variation (e.g. in leaf ontogeny and phenology) and because even buds and unfolding leaves are colonized to some extent (Andrews & Kenerley 1980). The latter approaches are more exciting conceptually and largely overcome the shortcomings of the first method. We used all three procedures in a study of fungi (filamentous fungi or moulds, plus yeasts) on apple leaves, which is outlined very briefly below.

To document natural patterns, at 9 dates in 1981 and at 11 dates in 1982 between April and October, three large and three small leaves were sampled from the same position in a McIntosh apple tree in an unsprayed orchard. 'New' islands, represented by sterile seedling leaves, were produced from apple seeds sterilized in hydrogen peroxide. The resulting sterile, potted seedlings were placed in an orchard at three intervals in 1982 and 1983 and assayed for microflora at about biweekly intervals for 8 weeks. For the 'defaunation' approach, mature leaves were sterilized in situ with hydrogen peroxide and colonization was followed by sampling daily, and subsequently weekly, for about 8 weeks.

For each of the three approaches, moulds and yeasts were assayed sequentially by: (i) washing each leaf separately and plating washings onto potato dextrose agar (PDA); (ii) imprinting the leaf onto PDA, and then (iii) incubating the leaf intact on PDA. To estimate relative species abundance, 200 colonies were isolated randomly from dilution plates of leaf washings. Species richness (numbers) was estimated for each leaf using the 200 colony set, plus colonies visually different from these 200, from all assays (dilution plating, leaf imprinting, leaf incubation). Our sampling and sorting procedures were validated by several checks which will be described elsewhere along with other details of the study.

*Results*. The hypotheses tested were as follows: (1) that fungal colonization of apple leaves reaches an equilibrium; (2) that species comprising the mycoflora community continually change (i.e. that immigration and extinction occur); and (3) that variation in leaf island size is reflected in species richness. To date these hypotheses have been examined only for the mould component of the fungal community. More formally, each hypothesis was posed in the standard null format, and the evidence examined to determine whether there were sufficient grounds to reject it. Testing these hypotheses required the development of formal definitions of I and E for data collected from different leaves over time. Actual numbers were not considered significant because the theory focuses instead on the relationships of the parameters to each other over time. We used estimated I and E values rather than I and E rates because rate calculations are highly influenced by the sampling interval in relation to the generation time of the organism.

The null hypothesis of non-stabilization was rejected. There was strong evidence for an equilibrium condition later in the season with no statistically significant changes in the number of species per leaf over time (Fig. 3). As predicted by the model and consistent with other microbial studies (Patrick 1967; Cairns *et al*. 1969), the early time course of colonization was approximately exponential. In fact, the most rapid change in numbers occurred during the first few hours and days of exposure and leaves rapidly reached an equilibrium.

The null hypothesis of static species composition at equilibrium was rejected. There was strong evidence for turnover in species or a dynamic equilibrium (Fig. 4). Both immigration, the arrival of new species, and

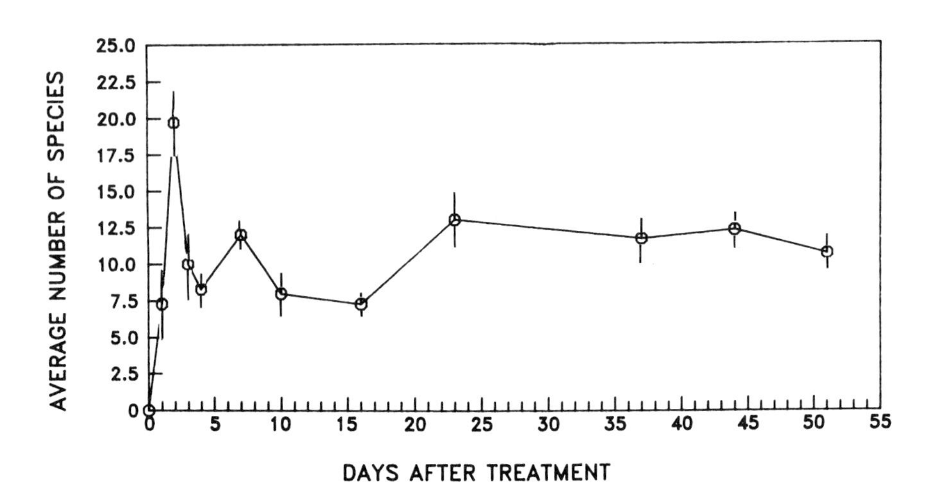


Fig. 3. Colonization of mature, surface-sterilized apple leaves by species of filamentous fungi over time in an apple orchard. Points represent the number of filamentous fungal species per leaf, averaged over three leaves on each date, with standard error. Equilibrium occurred at about 12 species per leaf.

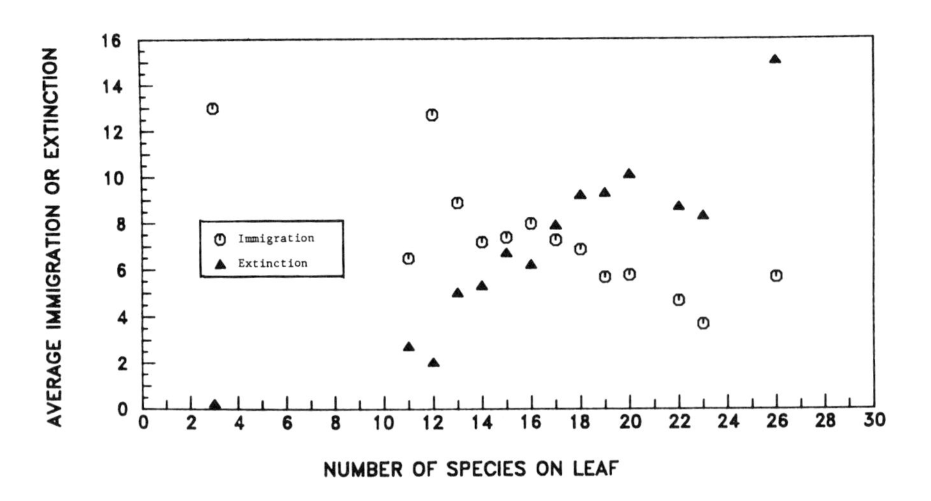


Fig. 4. Immigration and extinction of mould species as a function of species present on apple leaves from bud-break until leaf senescence, equilibrating at about 16 species per leaf. A similar pattern was obtained during recolonization of surface-sterilized mature leaves (Fig. 3).

extinction, the death or loss of established species, occurred at equilibrium. These values (which are equivalent at equilibrium) stabilized at about 7 species per leaf at an equilibrium of about 16 species per leaf.

The null hypothesis of absence of an area effect on species number could not be rejected. There was no statistically significant difference in numbers of species on large versus small leaves. At any sampling date, the relative abundance of dominant species was very similar on large and small leaves.

## *OVERVIEW AND PROSPECTS*

Three questions are central to our theoretical treatment. First, is the equilibrium model valid as a general ecological theory? Second, was our system and experimental design sufficiently rigorous to test it? Third, does the theory have any application in phylloplane microbial ecology?

With respect to general applicability, the concept has lost much of its appeal in macro-ecology. It is noteworthy, however, that MacArthur & Wilson (1967) on the whole put the theory forward cautiously and foresaw many of its shortcomings, despite whatever might be said about their use of selected or incomplete data bases (e.g. Slud 1976). As noted above, the present controversy has developed largely because of uncritical acceptance, inappropriate tests, failure to validate assumptions of the model, and unwarranted conclusions by many subsequent researchers. Thus, the concept cannot be said to be substantiated. However, this should not be interpreted to mean that it does not or cannot apply to specific situations; rather, at this point, there is insufficient evidence to justify its acceptance. The theory is eminently useful conceptually as an exciting point of departure in experimental studies and to formulate testable hypotheses.

We believe our study does provide a good test of the theory with a microbial system. The main shortcoming involved sampling and inability to test for the predicted distance (isolation) effects on species number. Sampling is a controversial, perennial issue in microbial ecology. We have approached it by using three cultural methods rather than just one. There is considerable leaf-to-leaf variability and our level of detection of species should be improved. Nevertheless, variation in species composition among leaves at any given date was less than that among leaves between dates, so a meaningful 'signal' is detectable above the 'noise'. A distance effect could probably be documented by using suitably marked organisms (J.H. Andrews, this volume) from a designated mainland species pool (e.g. the orchard floor). Interestingly, decrease in inoculum potential of a single (pathogen) species with distance from a source is well documented in plant pathology (e.g. Gregory 1977) and it is intuitively reasonable that the same principle would apply to the simultaneous dispersal of multiple species. The problem is that the sources are also multiple and diffuse.

Regarding the third question, if the tenets of Gilbert (1980) and Simberloff (1983 a) are strictly invoked, then the model is not corroborated for fungi on apple leaf surfaces. We have demonstrated a dynamic equilibrium, but the prediction of a species-area effect is not substantiated. Some factor other than area is probably most important in limiting filamentous fungal species on the phylloplane. There was also no relationship between leaf area and the number of individuals per leaf in this study. Previous studies on the density of the leaf epiflora indicate that area is not limiting for microbes on the phylloplane (e.g. Macauley & Waid 1981). Possibly the overriding factor is abiotic, such that as populations move towards an area limitation, the leaf is 'purged' of microbes by drought, or scouring by wind and rain.

Given that the island theory was not wholly confirmed for fungi on apple leaves, does it have any application in phylloplane microbial ecology? We believe firmly that the model has a role and that it should be submitted to other tests. The concept could probably be modified to account for the specifics of leaf systems and, if so, it will have made an even greater contribution. It can serve as a bridge between macro- and microbial ecology; tests in one system or the other can be complementary, highlighting problems and potential applicability in each. For example, it is exciting to see an equilibrial pattern in moulds on leaf surfaces, similar to that documented for insular communities of macro-organisms. Does this vary depending on environmental conditions, and to what extent does equilibrium number depend on the individual history of a particular leaf island? Likewise, the finding of many rare and few abundant species accords with community patterns typical of macro-organisms. What are the biological implications?

Most importantly, the essence of this and all other theories is that they provide for a conceptual synthesis of otherwise disparate, sterile bits of data. Novalis has said "Theories are nets: only he who casts will catch". It is appropriate to conclude with the words MacArthur & Wilson (1967, p. 5) used when introducing their concept:

> "A theory attempts to identify the factors that determine a class of phenomena and to state the permissible relationships among the factors as a set of verifiable propositions. A purpose is to simplify our education by substituting one theory for many facts. A good theory points to possible factors and relationships in the real world that would otherwise remain hidden and thus stimulates new forms of empirical research. Even a first, crude theory can have these virtues. If it can also account for, say, 85% of the variation in some phenomenon of interest, it will have served its purpose well".

*ACKNOWLEDGEMENTS*

A contribution from the College of Agricultural and Life Sciences, University of Wisconsin, Madison, U.S.A. This research was supported by NSF grant no. DEB 8110199. We thank our colleagues E.V. Nordheim and F. Berbee who have contributed throughout the study.

## *REFERENCES*

Abbott, I. (1980). Theories dealing with the ecology of landbirds on islands. Advances in Ecological Research, *11*, 329-71.

Andrews, J.H. (1981). Effects of pesticides on non-target microorganisms on leaves. *In* Microbial Ecology of the Phylloplane, ed. J.P. Blakeman, pp. 283-304. London: Academic Press.

Andrews, J.H. & Kenerley, C.M. (1980). Microbial populations associated with buds and young leaves of apple. Canadian Journal of Botany, *58*, 847-55.

Andrews, J.H., Kenerley, C.M. & Nordheim, E.V. (1980). Positional variation in phylloplane microbial populations within an apple tree canopy. Microbial Ecology, *6*, 71-84.

Brown, J.H. (1981). Mammals on mountaintops: nonequilibrium insular biogeography. American Naturalist, *105*, 467-78.

Cairns, J., Jr., Dahlberg, M.L., Dickson, K.L., Smith, N. & Waller, W.T. (1969). The relationship of fresh-water protozoan communities to the MacArthur-Wilson equilibrium model. American Naturalist, *103*, 439-54.

Culver, D.C. (1970). Analysis of simple cave communities. I. Caves as islands. Evolution, *24*, 463-74.

Diamond, J.M. (1969). Avifaunal equilibria and species turnover rates on the Channel Islands of California. Proceedings of the National Academy of Sciences of the U.S.A., *64*, 57-63.

Diamond, J.M. & May, R.M. (1976). Island biogeography and the design of natural reserves. *In* Theoretical Ecology. Principles and Applications, ed. R.M. May, pp. 163-86. Philadelphia: Saunders.

Dickinson, C.H. (1976). Fungi on the aerial surfaces of plants. *In* Microbiology of Aerial Plant Surfaces, eds C.H. Dickinson & T.F. Preece, pp. 293-324. London: Academic Press.

Dritschilo, W., Cornell, H., Nafus, D. & O'connor, B. (1975). Insular biogeography: of mice and mites. Science, *190*, 467-9.

Gilbert, F.S. (1980). The equilibrium theory of island biogeography: fact or fiction? Journal of Biogeography, *7*, 209-35.

Gregory, P.H. (1977). Spores in air. Annual Review of Phytopathology, *15*, 1-11.

Hudson, H.J. (1962). Succession of micro-fungi on aging leaves of *Saccharum officinarum*. Transactions of the British Mycological Society, *45*, 395-423.

Hudson, H.J. (1971). The development of the saprophytic fungal flora as leaves senesce and fall. *In* Ecology of Leaf Surface Micro-organisms, eds T.F. Preece & C.H. Dickinson, pp. 447-55. London: Academic Press.

Janzen, D.H. (1968). Host plants as islands in evolutionary and contemporary time. American Naturalist, *102*, 592-5.

Johnson, N.K. (1972). Origin and differentiation of the avifauna of the Channel Islands, California. Condor, *74*, 295-315.

Jones, H.L. & Diamond, J.M. (1976). Short-time-base studies of turnover in breeding bird populations on the California Channel Islands. Condor, *78*, 526-49.

Lynch, J.F. & Johnson, N.K. (1974). Turnover and equilibria in insular avifaunas, with special reference to the California Channel Islands. Condor, *76*, 370-84.

MacArthur, R.H. (1972). Geographical Ecology. Patterns in the Distribution of Species. Princeton: Princeton University Press.

MacArthur, R.H. & Wilson, E.O. (1963). An equilibrium theory of insular zoogeography. Evolution, *17*, 373-87.

MacArthur, R.H. & Wilson, E.O. (1967). The Theory of Island Biogeography. Princeton: Princeton University Press.

Macauley, B.J. & Waid, J.S. (1981). Fungal production on leaf surfaces. *In* The Fungal Community: Its Organization and Role in the Ecosystem, eds D.T. Wicklow & G.C. Carroll, pp. 501-31. New York: Marcel Dekker.

McGuinness, K.A. (1984). Equations and explanations in the study of species-area curves. Biological Reviews of the Cambridge Philosophical Society, *59*,

423-40.
Patrick, R. (1967). The effect of invasion rate, species pool, and size of area on the structure of the diatom community. Proceedings of the National Academy of Sciences of the U.S.A., *58*, 1335-42.
Pielou, E.C. (1979). Biogeography. New York: John Wiley & Sons.
Preston, F.W. (1948). The commonness, and rarity, of species. Ecology, *29*, 254-83.
Preston, F.W. (1960). Time and space and the variation of species. Ecology, *41*, 611-27.
Preston, F.W. (1962 a). The canonical distribution of commonness and rarity: Part I. Ecology, *43*, 185-215.
Preston, F.W. (1962 b). The canonical distribution of commonness and rarity: Part II. Ecology, *43*, 410-32.
Rodger, G. & Blakeman, J.P. (1984). Microbial colonization and uptake of $^{14}C$ label on leaves of sycamore. Transactions of the British Mycological Society, *82*, 45-51.
Seifert, R.P. (1975). Clumps of *Heliconia* inflorescences as ecological islands. Ecology, *56*, 1416-22.
Sherwood, M. & Carroll, G.C. (1974). Fungal succession on needles and young twigs of old-growth Douglas fir. Mycologia, *66*, 499-506.
Simberloff, D.S. (1969). Experimental zoogeography of islands: a model for insular colonization. Ecology, *50*, 296-314.
Simberloff, D.S. (1974). Equilibrium theory of island biogeography and ecology. Annual Review of Ecology and Systematics, *5*, 161-82.
Simberloff, D.S. (1976 a). Species turnover and equilibrium island biogeography. Science, *194*, 572-8.
Simberloff, D.S. (1976 b). Experimental zoogeography of islands: effects of island size. Ecology, *57*, 629-48.
Simberloff, D.S. (1983 a). Biogeography: the unification and maturation of a science. *In* Perspectives in Ornithology, eds A. Brush & G. Clark, pp. 411-55. Cambridge: Cambridge University Press.
Simberloff, D.S. (1983 b). Biogeographic models, species' distributions and community organization. *In* Evolution, Time and Space: The Emergence of the Biosphere, eds R.W. Sims, J.H. Price & P.E.S. Whaeley, pp. 57-83. New York: Academic Press.
Simberloff, D.S. (1983 c). When is an island community in equilibrium? Science, *220*, 1275-7.
Simberloff, D.S. & Wilson, E.O. (1969). Experimental zoogeography of islands. The colonization of empty islands. Ecology, *50*, 278-96.
Simberloff, D.S. & Wilson, E.O. (1970). Experimental zoogeography of islands. A two-year record of colonization. Ecology, *51*, 934-7.
Slud, P. (1976). Geographic and Climatic Relationships of Avifaunas with Special Reference to Comparative Distribution in the Neotropics. Smithsonian Contributions to Zoology, no. 212. Washington, DC: Smithsonian Institution Press.
Strong, D.R., Jr. (1979). Biogeographic dynamics of insect-host plant communities. Annual Review of Entomology, *24*, 89-119.
Wicklow, D.T. (1981). Biogeography and conidial fungi. *In* Biology of Conidial Fungi. Vol. 1, eds G.T. Cole & B. Kendrick, pp. 417-47. New York: Academic Press.
Wildman, H.G. & Parkinson, D. (1979). Microfungal succession on living leaves of *Populus tremuloides*. Canadian Journal of Botany, *57*, 2800-11.
Wilson, E.O. (1969). The species equilibrium. *In* Diversity and Stability in Ecological Systems, eds G.M. Woodwell & H.H. Smith, Brookhaven Symposia in Biology no. 22, pp. 38-47. Springfield, VA: National Technical Information Service.
Wilson, E.O. & Simberloff, D.S. (1969). Experimental zoogeography of islands: defaunation and monitoring techniques. Ecology, *50*, 267-78.

# ADAPTATIONS OF MICRO-ORGANISMS TO CLIMATIC CONDITIONS AFFECTING AERIAL PLANT SURFACES

C.H. Dickinson

*Plant Biology Department, University of Newcastle upon Tyne, Ridley Building, Newcastle upon Tyne NE1 7RU, UK*

## *INTRODUCTION*

The discovery of extensive populations of microbial epiphytes on aerial, living plant surfaces has inevitably led to speculations about their ability to withstand the various climatic extremes to which they are exposed. These discussions have, in part, been motivated by comparisons between the environmental conditions which determine the growth of micro-organisms in the soil and on roots as against those affecting above-ground populations. Thus, in addition to the usual problems of nutrient supply, competition and antagonism, the organisms living on stems, leaves, petals and fruits must be able to withstand drastic, and repeated, fluctuations in radiation levels, of temperature and in the water regime. These hazards occur during even the most favourable season for microbial growth; at other times the microbes not only have to withstand prolonged periods of cold, heat or drought, but in addition their host may shed many of its tissues or even die. It is, however, notable that despite repeated references to these likely hardships there has been relatively little research directed towards an understanding of their importance for the microbial epiphytes on aerial plant surfaces.

Micro-organisms arrive at aerial plant surfaces by various routes and the primary problem for many must be to adhere to the surface so that they are not immediately dislodged by the rain or wind which brought them there in the first place. This preliminary attachment mechanism must, later, be reinforced as the spore or cell develops to form a colony. Such development requires favourable conditions, but these will almost certainly not persist and changes in one or other environmental factor will result in growth ceasing. Whether this cessation of activity is temporary or permanent will depend on the biology of the organism and on the severity and duration of the adverse conditions. Further 'stop-go' development may be expected in all but the most favourable habitats, until the supporting tissues eventually die, fall and become incorporated into the litter layer. This pattern of microbial colonization may be modified further by the scarcity of nutrient supplies on most plant surfaces, and by the competition and antagonism which results when large numbers of organisms attempt to become established in these relatively impoverished habitats.

An understanding of the biology of these microbial epiphytes thus requires some analysis of their ability to grow or survive under a wide range of conditions. Data from cultural and direct examination techniques suggest that single-celled organisms are especially well adapted to life on these surfaces. Many of these organisms produce copious quantities of extracellular (or exocellular) polysaccharides (and some proteins) which confer a number of advantages on the ensheathed cells, they grow quickly with a rapid increase in cell number which enables them to colonize newly formed surfaces and to recover from catastrophic environmental conditions and they can, in many instances, flourish on relatively small quantities of metabolites. More complex organisms, e.g. the filamentous fungi, some algae and animals, rarely colonize leaves to the same extent as the single-celled organisms. This may, in part, be due to the longer time scale needed for their development, which is often incompatible with the life span of individual host organs, and it may also be due to their inability to thrive in, or even tolerate, the environmental conditions which prevail around these habitats. It is, for example, well established that several filamentous phylloplane fungi, e.g. *Alternaria* and *Cladosporium* species, frequently complete their life cycles by sporulating on fallen leaves in the litter layer.

## *ARRIVAL AND ESTABLISHMENT*

### *Forces of attraction*

The initial attachment of microbial propagules to plant surfaces may be accomplished through the agency of one or more of a wide variety of forces (Table 1). Little is known about which of these forces

Table 1. Forces of attraction between microbial cells and absorbent surfaces (reproduced by permission from Daniels 1980).

| Type of forces |
|---|
| Chemical bonding (hydrogen, thio, amide, and ester bonds) |
| Ion-pair formation ($-\overset{+}{N}_3 \ldots {}^{-}OOC-$) |
| Ion-triplet formation ($-COO^{-} \ldots Ca^{2+} \ldots {}^{-}OOC-$) |
| Interparticle bridging (polyelectrolytes) |
| Charge fluctuations |
| Charge mosaics |
| Charge attraction of opposite signs |
| Electrostatic attraction between surfaces of similar charge |
| Electrostatic attraction due to image forces |
| Surface tension |
| Van der Waals forces of attraction |
| Electromagnetic forces |
| Hydrodynamic forces |
| Diffusional forces |
| Gravitational forces |
| Positive chemotaxis (cellular mobility) |

holds propagules on aerial plant surfaces, but it has been suggested that electrical charges play an important role in the retention of spores and vegetative cells on leaves. As noted later, the presence of a capsule can confer an electrical charge on microbial cells.

Whatever the mechanisms involved it is clear that these 'Stage I' adhesion forces (Fletcher 1979) are insufficient, in most instances, to hold an organism once it begins to grow. Thus at this time further bridges are developed between the organism and the underlying surface. These 'Stage II' systems, which are probably more substantial than the initial forces, may take the form of capsules and slime, of fibrils or of specific organs specialized for attachment.

*Capsules and slime*

The production of extracellular polymers, which form cell capsules or accumulate around the colony as slime, is a recognized characteristic of many bacteria and yeasts. The distinction between a slime layer, which can be removed by simple centrifugation, and a micro- or macrocapsule, which cannot be so removed, is, however, blurred by the fact that the production of these materials is affected by the conditions under which the organism is growing (Ward & Berkeley 1980). Much is known about the chemistry of the polymers involved, which are predominantly polysaccharides with the capacity to absorb considerable quantities of water. These polysaccharides may be either homopolysaccharides (e.g. levans produced by some pseudomonads and xanthomonads) or heteropolysaccharides (e.g. xanthan produced by *Xanthomonas campestris* and an alginate-like polymer produced by *Pseudomonas aeruginosa* (Ward & Berkeley 1980)). In a few instances, notably in species of *Bacillus*, the capsule is composed of polyglutamic acid (Troy 1979). Species of *Erwinia*, *Pseudomonas* and *Xanthomonas*, are known to produce slimes which may function as non-specific adhesives or which may facilitate specific interactions with the surfaces of green plants (Corpe 1980).

Similar materials are produced by many yeasts (Phaff 1971) and it is notable that several of the genera which inhabit aerial plant surfaces are particularly prolific producers of such polysaccharides (Spencer & Gorin 1965). Indeed, scanning electron micrographs of bacterial and yeast colonies on leaf surfaces frequently show them to be covered with a film of material which may well be slime which has become desiccated during the preparative processes (Figs 1 A,B; 2 A-C; Bashi & Fokkema 1976).

Some investigations of the dynamics of mucilage production have been carried out by Bashi & Fokkema (1976), who studied *Sporobolomyces roseus* on wheat leaves. They found that when the yeast was supplied with adequate nutrients and the relative humidity was 95% or above it formed extensive, mucilage-covered colonies in 4 days at 18°C. At this time the colony appeared as a coherent sheet within which individual cells could hardly be distinguished. If no nutrients were added the yeast produced only small groups of cells which though roughened with what appeared to be dried mucilage were easily distinguished as separate individuals. The

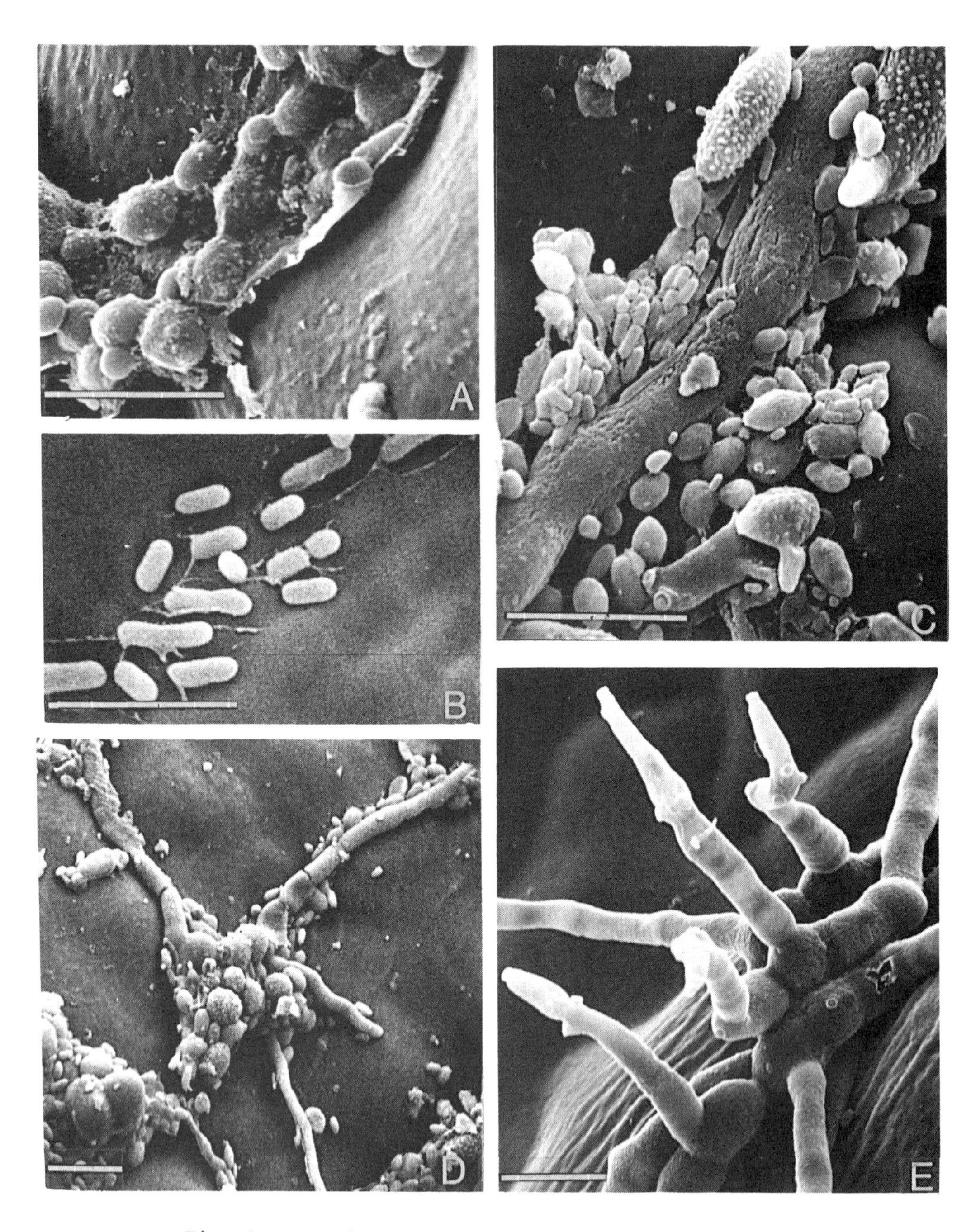


Fig. 1. Scanning electron micrographs of critical point-dried leaves of *Sambucus nigra* (A,B,E), *Ribes nigrum* (C) and *Crataegus monogyna* (D) collected in Northumberland, U.K. in July, 1985. A. Yeast colony embedded within an extracellular mucilage which has partly peeled off the leaf. B. Bacterial colony showing the outlines of capsules which have been

chemistry of the mucilage of *S. roseus* has not been extensively studied but extracellular phosphogalactans have been detected in cultures of an unknown species of *Sporobolomyces* (Slodki 1966).

Another phylloplane fungus which is especially noted for its production of slime is *Aureobasidium pullulans*. This produces a polysaccharide called pullulan, which is an α-glucan comprising mostly α-maltotriose linked endwise by 1,6 bonds (Catley 1979). It is produced in culture when the growth rate starts to decline, due to the exhaustion of the available carbohydrate and a developing shortage of nitrogen. Its formation is linked to a morphogenetic change from the development of hyphae to the production of yeast-like cells. This highly viscous, water-soluble polymer may thus be important in ensuring that these more vulnerable, single cells are firmly attached to leaf surfaces.

By contrast to these single-celled or polymorphic organisms, far less is known about the adhesion of hyphae to leaf surfaces. Electron microscopy has, however, revealed the presence of sheathing material around the hyphae of some pathogens growing on leaves, and, despite earlier suggestions that this coating was an artefact caused by disruption of the cuticle (Wheeler 1975), it is now accepted that many filamentous fungi produce hyphal sheaths (Fleet & Phaff 1981; Wessels & Sietsma 1981). These are thought to be likely to assist in the adhesion of hyphae to leaves (Wheeler & Gantz 1979), and particularly striking visual evidence that such coatings are produced when the fungi grow on plants has been provided by Hau & Rush (1982). They showed that the hyphae of *Helminthosporium oryzae* are attached to papillae on rice leaves by lateral extensions of the sheath. The strength of the attachment in this host-pathogen complex may be gauged by the fact that when the hyphae are pulled off the leaves they take with them the underlying epicuticular wax crystals. This leaves a clearly visible track, though scanning microscopy shows that the underlying cuticle does not seem to be damaged by their removal. Confirmation that the sheathing material was produced solely by the fungus was obtained by demonstrating that it was also formed when the hyphae were grown on inert leaf replicas. The importance of extracellular mucilage in ensuring good adhesion between pathogens and their hosts is emphasized by the evidence that such material is produced around the appressoria of many fungi, notably *Colletotrichum* spp. (Leach 1923; Marks *et al.* 1965; Akai *et al.* 1967; Lapp & Skoropad 1978) and *Pellicularia* spp. (Flentje 1957).

Several non-pathogenic fungi have also been shown to form hyphal sheaths. These include marine ascomycetes, such as species of *Lulworthia* (Davidson 1973), *Ceriosporopsis* and *Halosphaeria* (Hyde *et al.* 1986), and a number of wood-rotting fungi, notably *Poria placenta* (Sutter *et al.*

Fig. 1 (continued)
dried. C,D. Fungal hyphae attracting and or enhancing the growth of various single-celled organisms. E. Microcyclic conidiation in an unknown fungus. Scale bars equal 10 μm except in B (5 μm).

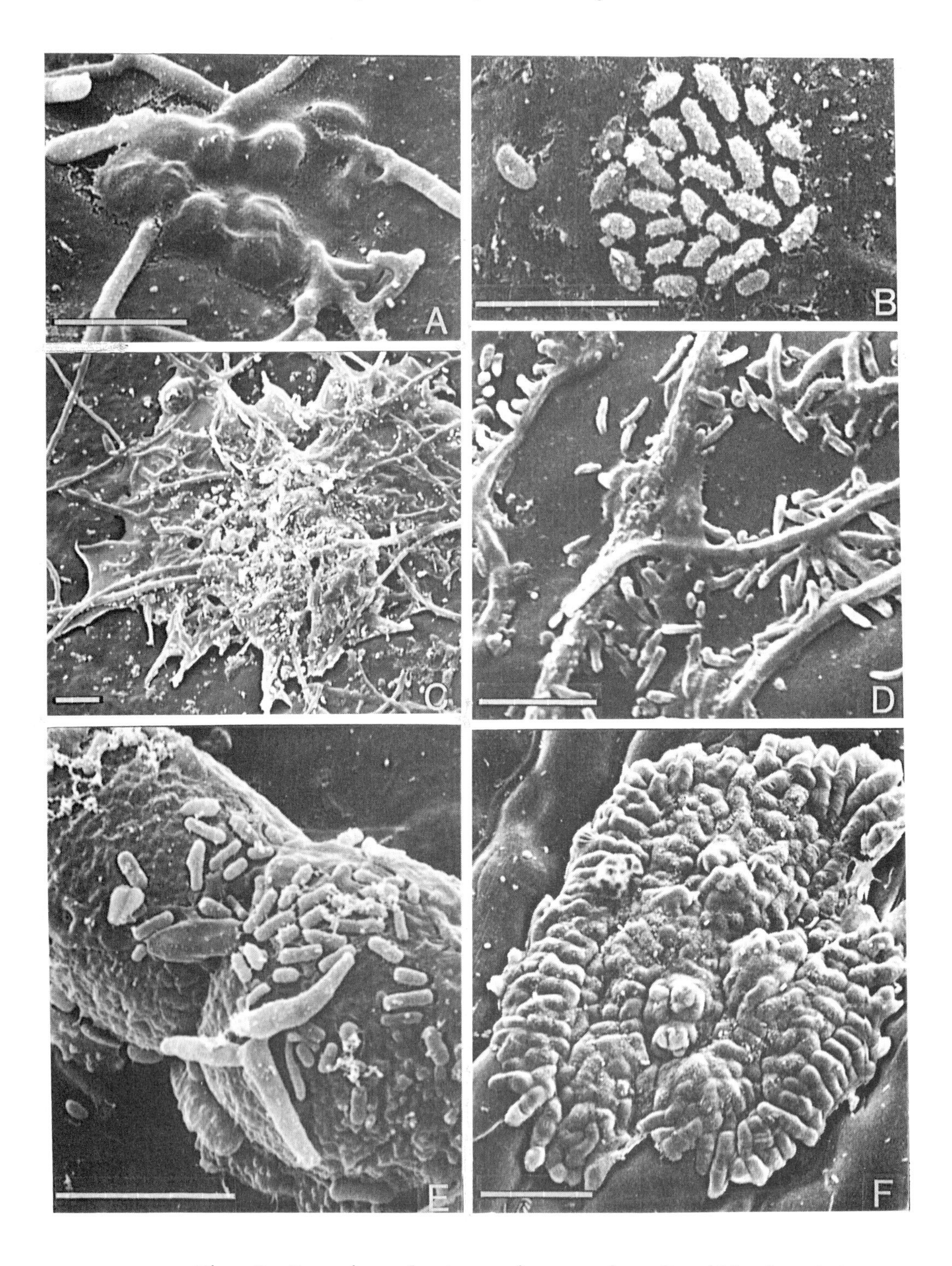


Fig. 2. Scanning electron micrographs of critical point-dried leaves of *Bosquiera phoberos* collected from an evergreen rainforest at about 2000 m, near Zomba, Malawi, March

1984). In this latter fungus the sheath mucilage contains microcrystals of calcium oxalate which give it a definite structure, such that it has also been referred to as a hyphal tube. Speculation as to the functions of these sheathing materials has suggested that, in addition to adhesion to surfaces, they also modify the hyphal ionic environment, facilitate the adhesion of bacteria and other microbes to the hyphae, retard desiccation and act as a matrix for storing or concentrating extracellular enzymes (Rees & Jones 1984; Sutter *et al.* 1984; Hyde *et al.* 1986).

Apart from *Epicoccum nigrum* (Duncan & Herald 1974), it is not known if the common filamentous fungi produce such sheathing material around their hyphae. There is, however, an increasingly large amount of evidence that these sheaths are produced by many fungi and with more critical electron microscope techniques it seems likely that they will be found to be as common on filamentous fungi as capsules and slime are on the yeasts.

In so far as capsules or slime assist in the adherence of micro-organisms to plants it is unclear to what extent this is due simply to their stickiness or to chemical interactions between the capsular material and one or more components of the plant's surface (Dazzo 1980). There is no doubt that they are composed of exceptionally sticky materials, so much so that several have been investigated to determine their potential as industrial gums (Jeanes 1977). For example, a strain of the common phylloplane yeast *Cryptococcus laurentii* has been shown to produce a highly viscous, dense, cohesive, thixotropic gum which rapidly regains its viscosity after shearing force has been applied (Jeanes *et al.* 1964).

Interactions between bacterial exopolysaccharides and components of the host's surface have been shown to occur with the xanthan produced by *Xanthomonas campestris* and several wall polysaccharides, including cellulose, galactomannans and glucomannans (Morris *et al.* 1977). Whether such interactions can occur on an intact cuticle with its covering of epicuticular wax is not clear but the possibility that such forces may exist should interest microbial ecologists studying aerial plant surfaces.

More complex interactions between the surface components of micro-organisms and green plants are being revealed by autoradiographic, histochemical and enzymological techniques. These studies have shown that the

Fig. 2 (continued)
1984. A-C. Extracellular polymers forming coatings on cells and colonies of various single-celled and filamentous organisms. D. As before with bacteria growing in the vicinity of and adhering to fungal hyphae. E. Fungal spores with several types of adhering bacteria. F. Compact and dense bacterial colony exhibiting a growth form which may enhance survival on aerial plant surfaces. Scale bars equal 5 μm except in C and F (10 μm).

zoospores of *Enteromorpha* and *Phytophthora* spp. adhere to surfaces by secreting glycoprotein from peripheral vesicles (Callow & Evans 1974; Sing & Bartnicki-Garcia 1975). Extracellular proteins have also been shown to enhance the adhesion of rust germlings to cereal leaves and it has been suggested that they also assist in stomatal recognition which is an important step in the entry of these pathogens into their host (Staples 1985).

### *Fibrils*

Detailed electron-microscopic studies have revealed that both *Agrobacterium* and *Rhizobium* species consolidate their attachment to their respective hosts by forming strands or fibrils which extend from the bacterial cell to the plant's surface (Dazzo 1980). Cultural studies have shown that the fibrils of *Rhizobium* spp. are composed of cellulose (Deinema & Zevenhuizen 1971; Napoli *et al.* 1975), and it may be that these strengthen the contact between the bacterium and the root hair so that subsequent infection events proceed efficiently (Dazzo 1980). The fibrils produced by *Agrobacterium tumefaciens* are also composed of cellulose (Sutherland 1982) and it is thought that they fulfil a dual role of ensuring adhesion and of trapping more *A. tumefaciens* cells within the mesh which is formed. Similar fibrils were observed holding bacteria onto wheat roots, though in that instance it was thought that the fibrillar network was a combined product of both the bacteria and the root (Rovira & Campbell 1974).

There have been some reports of fibrillar material being involved in the attachment of hyphae to leaf surfaces. The germ tubes and appressoria of *Cochliobolus carbonum* are bridged to the underlying leaf surface by fibrillar or granular material (Murray & Maxwell 1974) and the appressoria of *Erysiphe graminis* f. sp. *hordei* are similarly attached to barley leaves (Edwards & Allen 1970). It is not clear, however, if this was the only means of attachment in either instance. *Helminthosporium maydis* germ tubes have been shown to become attached to maize leaves by a sheath-fibril complex, which is of unknown composition but which was shown to be produced by the pathogen itself (Potter *et al.* 1980).

### *Specialized organs of attachment*

Fimbriae may be important in facilitating the attachment of bacteria to leaves but they have not, to date, been the subject of any systematic study. Likewise none of the groups of bacteria which possess these distinct holdfasts, i.e. the stalked bacteria, the sheathed bacteria, the budding bacteria and the gliding bacteria (Corpe 1980), have yet been recorded on living leaves. We should, however, be careful not to confuse ignorance with truth as there may well be many more epiphytic bacteria on aerial plant surfaces than have been discovered to date.

### *Community development*

Fletcher (1979) has noted that the attachment of individual bacterial cells to submerged surfaces may have profound ecological consequences in terms of the subsequent adhesion and development of larger organisms including algae and invertebrates. By analogy it would seem likely that the colonization of aerial plant surfaces is also determined, at least in part, by the relative abilities of different groups of micro-organisms to adhere to these surfaces and perhaps to each other. This may be one explanation for the cascade effect which is observed, whereby some leaves remain sparsely colonized even up to senescence whereas others become heavily colonized by mixed communities of bacteria, yeasts, filamentous fungi and algae (Figs 1 C,D; 2 D). There is no doubt that the provision of copious nutrient supplies can bring about the same sort of effect (e.g. the production of aphid honeydew on leaves) but dense incrustations of micro-organisms can be regularly observed on leaves, which are virtually free of aphids, pollen or any of the other recognized sources of exogenous nutrients. Further study of the development and the composition of such communities would be of interest as they may offer enhanced protection against leaf pathogens (Blakeman & Fokkema 1982).

It might be expected that all these reports concerning the attachment of microbes to surfaces would have stimulated some systematic work on the devices adopted by microbial epiphytes to stick to aerial plant surfaces. Unfortunately no such work has yet been done. Hence we must continue to extrapolate from a relatively few studies of specific pathogens to guess at the likely mechanisms which are employed by the saprophytic microbiota to cling to these exposed surfaces.

## *CLIMATIC INFLUENCES ON MICROBIAL GROWTH*

It is axiomatic that any treatment of the effects of extremes of humidity, radiation or temperature must take account of the interrelationships between them. These are further highlighted by the multiple roles which can be assigned to many of the adaptations exhibited by epiphytic organisms. In addition it must be noted that a number of biological factors can affect the responses of organisms, growing singly or in mixed communities. Experimental analysis of the influence of, for example, low and high temperatures is thus extremely difficult and the comparison of results obtained under only slightly different conditions may be virtually impossible. Hence, the intention here is to survey the likely climatic hazards facing microbial epiphytes on aerial plant surfaces and then to discuss some of the characteristics which may ensure their survival in these habitats.

### *Temperature*

The tacit assumption of many microbiologists who have cultured leaf surface micro-organisms is that they are dealing with mesophiles, which may be defined as organisms which flourish in an environment which fluctuates between 10 and 45°C (though it should be noted that there is no agreement between authors as to the exact definition of this term). As far as aerial plant surfaces are concerned the upper end of this range may well cover the maximum temperatures which are experienced by phylloplane organisms. It is, however, hardly appropriate at the lower end as epiphytic organisms on conifer needles and other evergreen leaves and stems have to tolerate long periods when the temperatures are below 10°C, and more importantly they must withstand repeated freezing and thawing for several months each year. Hence we should be looking for psychrophiles on some plants and we should be considering the abilities of various microbial structures to withstand the damaging effects of ice formation. Psychrophiles may be defined in a conventional way as those organisms with a maximum temperature for growth of about 20°C and an optimum of 15°C or less (Morita 1976), or one may adopt the more practical approach advocated by Ingraham & Stokes (1959) whose criterion was good growth (i.e. visible colony formation on a solid medium) at 0°C within two weeks. Rose (1968) adopted this latter criterion and he emphasized that growth at low temperatures may be both quantitatively and qualitatively different from that in warmer conditions. For example, there is a tendency for organisms to produce increased amounts of pigment and extracellular polysaccharides when grown at temperatures well below their optima.

In an excellent review of the problems which are created by freezing and thawing, Mazur (1968) discussed the main causes of cell damage; these he identified as the physical damage caused by ice crystal formation, the physiological disruption which follows the progressive rise in the concentration of solute in the cell and the problems which result from desiccation when all the water in the cell has been converted into ice. Mazur (1968) demonstrated that in most instances, and especially in nature, fungi survive freezing because they can withstand desiccation relatively well and because the rates of cooling which are experienced are such that intracellular ice is rarely formed. This observation does, however, represent the most optimistic situation and it is clear from many published studies that 100% survival rates are rare, especially where vegetative cells are being manipulated. A complete analysis of the situation is further complicated by the large number of factors which are known to affect the survival of microbes subjected to freezing and thawing (Table 2). Hence repeated freeze/thaw cycles are potentially hazardous, even if only a small proportion of the population succumbs at any one time.

### *Free water and humidity on aerial plant surfaces*

The water relations of plant surfaces can vary substantially and rapidly with the conditions alternating between a surfeit of free water, which may be important in redistributing cells, and drought, which is made worse by a low atmospheric humidity. These fluctuations are reduced in a dense canopy, which can create microclimatic conditions significantly different from those which prevail on a regional basis (Burrage 1971, 1976). The duration of leaf surface wetness is, of course, an important parameter as regards the performance of many plant pathogens but, surprisingly, it has not been seriously considered in studies of microbial epiphytes in the phylloplane.

Variations in humidity can affect the performance, or even the survival, of micro-organisms but the effect is not straightforward in that survival is often best at both high and low humidities (Sussman 1968). The most drastic effects of humidity have often been recorded at intermediate levels. Much of this information has come from studies of the survival of bacteria in aerosols (Strange & Cox 1976), from work on the persistence of pathogenic bacteria on contaminated surfaces in hospitals and from experiments to improve the efficiency of freeze-drying as a means of preserving microbes (Mazur 1968). However, as with the other factors relatively little work has been done on the impact of alternating wet and dry conditions such as affect both vegetative cells and spores on leaf surfaces.

*Radiation*

The effects of visible and near-visible radiation on the growth and survival of micro-organisms have been the subject of reviews by Leach (1971), Bridges (1976), Krinsky (1976) and others. Much of the work on this subject has, not unnaturally, focussed on the ultraviolet (UV) end of the spectrum which is known to have damaging or even lethal effects on many organisms. However, despite many casual suggestions to the contrary it is now widely accepted that very little of the most lethal, far-UV, radiation (<300 nm) reaches the earth's surface. Thus we must consider the likely significance of near-UV (300-380 nm) and visible radiations (380-750 nm) for the micro-organisms living on aerial plant surfaces, which as far as this factor is concerned, must constitute some of the most exposed habitats on earth.

Light has direct effects on organisms and it is also important in that it is transformed into heat, which in the tropics can raise the tempera-

Table 2. Factors known to affect survival of microbes from freezing and thawing (reproduced by permission from MacLeod & Calcott 1976).

| Factor |
|---|
| Type and strain of organism |
| Population density |
| Nutritional status |
| Growth phase and rate of growth |
| Composition of the freezing menstruum |
| Cooling rate to the freezing point of the suspension |
| Cooling rate from the freezing point of the suspension |
| Time held at low temperature |
| Holding temperature |
| Rate of warming to freezing point |
| Diluting environment prior to viability determination |
| Method of viability determination |
| Medium used to assess viability |

ture sufficiently to cause sun scorch on the sensitive young leaves and fruits. Little is known about the survival of micro-organisms under such extreme conditions.

### *ADAPTATIONS OF MICROBIAL EPIPHYTES*

#### *Protection against desiccation*

As noted earlier, extracellular polysaccharides and proteins may ensure adhesion to plant surfaces, but this may be only one of several roles which these materials fulfil in the life of microbial epiphytes (Table 3). It must be noted that there is, as yet, little evidence for some of these suggested functions of extracellular polymers, but even if only a few of these advantages accrue to any one organism the production of such material would seem to be worthwile.

The regulation of cell water relations may be a particularly important function of the capsules around microbial epiphytes, especially as it is known that many bacteria and yeasts are badly affected by desiccation. For example, when aliquots of $10^3$ cells of several Gram-negative bacte-

Table 3. Possible functions of capsules and other extracellular polymers (excluding enzymes).

| Type of function | |
|---|---|
| A. | Ensure adhesion to host's surface |
| | i. Non-specific adhesion |
| | ii. Adhesion based on specific interactions |
| B. | Regulate microbial cell water relations |
| | i. Prevent water loss leading to desiccation |
| | ii. Promote rehydration by water uptake |
| C. | Protect cells from UV radiation |
| D. | Protect cells against parasites etc. |
| | i. Block bacteriophage attachment |
| | ii. Deter animals which might ingest cells |
| E. | Assist in growth and development |
| | i. Provide surface receptors for enzymes that synthesize polymers |
| | ii. Increase uptake of nutrients from the green plant |
| | iii. Aid penetration and colonization of host |
| F. | Increase possibility of cell dispersal due to their hydrophilic properties and negative charge |

ria, including *Escherichia coli*, *Pseudomonas aeruginosa* and *Proteus vulgaris*, were exposed on 1 $cm^2$ inert surfaces at 25°C they survived for less than 2 days at 85 and 53% r.h. and for less than 7 days at 11% r.h. (McDade & Hall 1964). Studies of plant-pathogenic bacteria have shown that many die within minutes of being dried and others can only live for such limited periods that season-to-season survival as dried cells would seem to be highly unlikely (Leben 1974). This rapid decline in viability following desiccation has been attributed to the breakdown of cell membranes, with a consequent loss in differentiation between adjacent areas in the cell (Webb 1960). Survival in dry conditions is affected by a number of factors including temperature, with cells generally remaining viable for longer periods under cooler conditions (Strange & Cox 1976), and the stage of the growth cycle when cells are dried, with most organisms showing the greatest resistance to drying at the end of the logarithmic phase or in the stationary phase (Fry 1966). This can be related to the production of extracellular polysaccharides, some of which first appear when organisms enter the storage and maintenance phase of the growth cycle (Catley 1979).

It is also notable that bacteria suspended in aerosols and those which are freeze-dried survive better if various substances are added to the initial bacterial suspensions (Strange & Cox 1976). These protective additives include sugars, proteins, polyethylene glycol and complex mixtures, such as milk or broth. The effects of these materials may well parallel those of naturally-produced extracellular polysaccharides, which have also been shown to protect cells from the effects of desiccation. An example of such a protection is seen in studies of *E. coli* in which a mucoid strain survived longer in dry conditions than did a non-mucoid strain (Morgan & Beckwith 1939). This factor alone may well explain why so many phylloplane yeasts and bacteria produce capsules or slime (Davenport 1980), which are important in survival on plant surfaces and in dissemination between hosts (Leben 1974).

Not all the evidence points to mucilage having a protectant role for the cells embedded within it. Bashi & Fokkema (1977) demonstrated that *Sporobolomyces roseus* populations on wheat leaves declined quickly when the relative humidity was reduced from 95 to 65%. This decline occurred to the same extent on leaves with added nutrients, which had been shown to stimulate mucilage production, as it did on leaves with no nutrient additions, on which the yeast populations were smaller and the cells were not covered with mucilage. This result suggests that more experimental analysis is needed to determine the influence of fluctuating environmental conditions on phylloplane organisms growing under different conditions.

Little is known about the ability of fungal hyphae to withstand alternating wet and dry conditions, though there is information which shows that many fungal spores can withstand such fluctuating conditions (Sussman 1968). The survival of vegetative hyphae was tested by Good & Zathureczky (1967) who discovered that short germ tubes of *Botrytis cinerea*, *Cercospora musae*, *Cladosporium fulvum* and *Monilinia fructicola* were remarkably resistant to drying over anhydrous calcium sulphate. Their survival was apparently unaffected by their stage of development, i.e. short germ tubes were compared with longer ones, and there were only slight increases in mortality when they were dried for a second time. In a very limited test it was also shown that *M. fructicola* germ tubes could survive to approximately the same extent when the fungus was exposed outdoors as when it was dried in a desiccator. Diem (1971) followed this work with experiments on a range of filamentous fungi such as might be found in the phylloplane of temperate plants. He found that the germ tubes of three hyaline fungi, viz. a species of *Aspergillus*, a species of *Penicillium* and *Colletotrichum graminicola* were killed by exposing them to a relative humidity of 73% or less for 4 hours at 27°C. By contrast, five fungi which produce dark-pigmented hyphae were generally able to survive at 40% r.h. and even exposure to anhydrous calcium chloride for 8 hours. Interestingly, *C. graminicola* produces on leaves immediately after germination pigmented appressoria which may function as survival structures (Parbery & Emmett 1977). The humidity values examined were selected on the basis that the ambient r.h. in bright sunlight is 40 to 65% and after making allowances for transpiration the microclimate at the leaf surface can be 15 to 20% above these ambient values (Schnathorst 1965; Diem 1971). The survival rates at 27°C were also compared with those at 20°C and, by contrast with the survival of bacteria, it was found that most of the fungi tolerated adverse humidity better when grown at high rather than low temperatures.

More precise evidence concerning the ability of phylloplane fungi to tolerate fluctuating wet and dry conditions has been produced by Park (1982). He showed that a group of filamentous fungi which are commonly isolated from green leaves were capable of recommencing growth almost immediately after a dry period during which time they were dormant. By contrast, a number of other filamentous storage and soil fungi had a long lag period before they recommenced growth. Park (1982) noted that the new growth in the phylloplane fungi involved their marginal, hyphal tips and he emphasized the likely importance of this brief loss of growing time and minimal waste of previously synthesized biomass.

### *Protection against radiation*

Micro-organisms, growing in a habitat where they are exposed to all incident electromagnetic radiation are subject to a number of possible hazards. The two principal problems are caused by far-UV (<300 nm) and by near-UV and visible wavelengths (300-750 nm).

UV radiation, in common with ionizing radiation, some chemicals and high temperatures, damages DNA by causing adjacent pyrimidine bases to join up as dimers and by causing a number of other more subtle changes (Bridges 1976). These alterations to the DNA molecule are generally reversible, especially under the influence of near-UV or visible light which photoactivates enzymes which can split the dimers in situ. However, despite the existence of such repair mechanisms it is well known that UV and ionizing radiations can cause considerable mortality amongst micro-organisms.

Microbial cells vary considerably in their intrinsic tolerance of such irradiation, and their sensitivity is further affected by the presence of oxygen, by their water content, the ambient temperature and certain chemicals (Bridges 1976). They also differ in their ability to recover from radiation damage and a number of complex repair mechanisms have been discovered, in addition to the light-mediated reaction mentioned above (Bridges 1976).

As noted before, very little far-UV radiation reaches the earth's surface but any that does will impinge on the aerial plant surface habitats more effectively than on almost any other situation where micro-organisms grow. Hence, it would be of interest to know if the epiphytes which inhabit these exposed habitats are particularly tolerant of such radiation and if they have especially well developed damage repair mechanisms.

Although it has been known for many years that visible light can also cause lethal damage to micro-organisms the discovery that carotenoid pigments protect cells against such photodynamic effects has stimulated renewed interest in this phenomenon (Sistrom *et al.* 1956; Krinsky 1976). The photodynamic effect is brought about by the action of light on a photosensitizing compound (which can be any organic molecule able to absorb light in the range 320-900 nm). Once activated this compound can either generate singlet excited oxygen or it may itself participate in redox reactions. Such photosensitized reactions can result in damage to the cell's nucleic acids and consequently produce mutations or even death. They may also adversely affect the cell's membrane system. Carotenoids appear to act by quenching or reacting with the singlet excited oxygen or even by preventing its formation (Krinsky 1976). Hence these pigments are of great biological importance and it is not surprising that they are produced by many leaf surface microbes.

Notable leaf-inhabiting organisms which produce carotenoids include the yeasts of the genera *Cryptococcus*, *Rhodotorula*, *Sporidiobolus* and *Sporobolomyces* (Simpson *et al.* 1971; Valadon 1976), in which it has been shown that light stimulates the formation of the pigment, and the Uredinales, whose common name is based on the rusty orange colour of their urediniospore pustules. Austin *et al.* (1978) found that 81% of the bacteria isolated from the leaves of *Lolium perenne* were pigmented, with large numbers of strains being either yellow or pink. These examples suggest that such pigments are probably common in microbial epiphytes, and further studies would be valuable to elucidate their importance in the long-term survival of microbial cells on plant surfaces. Their relationship with the production of melanin and its role in cell protection also requires more investigation.

### *Melanin*

The deposition of melanin in the hyphal walls of some fungi is almost certainly of great importance in terms of their ability to colonize aerial plant surfaces. Melanin is not generally formed by either the Mastigomycotina or the Zygomycotina, it occurs in a number of

Basidiomycotina, but it is most frequently encountered amongst the Ascomycotina. Its formation appears to be correlated with the occupation of terrestrial habitats and especially with life cycles which involve some activity on above-ground plant surfaces. Hence it is produced by a very large number of Hyphomycetes which colonize flowers and fruits (Ellis 1971, 1976), and it is found in almost all of those remarkable epiphytes, the sooty moulds (Hughes 1976). In view of this pattern of occurrence it is perhaps surprising that it is produced by very few yeasts, and those that do produce it are atypical in other respects (De Hoog & Hermanides-Nijhof 1977) and that it is not found in the walls of bacteria, except for some streptomycetes.

Melanin may be synthesized by different fungi using at least two pathways (Table 4). Cultural studies have indicated that it is formed as a secondary metabolite when the carbohydrate source is becoming exhausted (Cooper & Gadd 1984). At this time some hyphal compartments may differentiate to form chlamydospores and this process may be closely linked with the formation of melanin.

Melanin is also found in the cells forming the outer layers of sclerotia and rhizomorphs. Granules of melanin are deposited in the outer layers of the hyphal wall and may even form an encrusting layer on the outside of the hypha (Durrell 1968; Ellis & Griffiths 1974). The chemical nature of fungal melanin is, as yet, not well understood but it appears to be related to eumelanin which is formed by various animals (Ellis & Griffiths 1974).

Table 4. Biosynthesis of melanin in fungi.

| Characteristics of melanin formation |
| --- |
| HOW FORMED? |
| - oxidation of tyrosine (e.g. *Cryptococcus neoformans* (Nurudeen & Ahearn 1979)) |
| - via pentaketide pathway (e.g. *Pyricularia oryzae* (Woloshuk *et al.* 1981), *Thielaviopsis basicola* (Wheeler & Stipanovic 1979)) |
| WHERE FORMED? |
| - in outer layer of hyphal wall and on surface of the hypha |
| - in differentiated hyphae and resting structures |
| WHY FORMED? |
| - protection from radiation |
| - protection against low water potentials |
| - prevention of lysis by bacteria |

As regards its significance, the most popular supposition is that it acts as a barrier or screen to protect the cytoplasm from the damaging effects of visible and UV light. This suggestion has been supported by the discovery of large populations of dark-pigmented fungi in habitats which receive high levels of radiation and by a number of experimental studies, especially with UV light (Durrell 1968; Sussman 1968). The interrelationship between melanin and light is further confirmed by evidence that in some fungi light is involved in its formation (Lingappa *et al.* 1963; Sussman *et al.* 1963).

A second, most important function which has been found for melanin is that of protecting hyphae and spores against lytic enzymes, and especially those released by bacteria (Gray 1976). This effect is probably due in part to the inhibition of polysaccharases by melanin (Bull 1970) and although this protection has in the past been considered in relation to soil organisms it is probable that it is also important on leaves.

The possibility that melanin also protects fungi against the effects of desiccation does not yet appear to have been widely considered. It is envisaged that the melanin would operate in this respect in much the same way as many of the additives which have been shown to improve the survival of freeze-dried microbes (Mazur 1968). Most of the evidence for this suggestion is circumstantial and more precise studies are required which will detail the time course of melanization and the relative responses of hyaline and melanized hyphae to drying conditions. Amongst the studies which provide support for this idea is that of Diem (1971) who found that several dematiaceous Hyphomycetes resisted desiccation more efficiently than a group of hyaline species. The exact significance of these data is, however, unclear as it was not stated whether the short germ tubes which were exposed to drying conditions contained melanin at that stage of development. Several studies on the effects of radiation (Sussman 1968) may also be interpreted as providing evidence for this theory, especially as high levels of radiation almost invariably have a drying effect. It may also be argued that the presence of melanin in the hyphae and the spores of many soil-inhabiting micro-organisms would be more valuable in protecting them from the effects of drought than against high levels of radiation.

### *Morphological responses to environmental stress*

Several of the organisms which inhabit leaf surfaces are noted for their ability to produce two or more different growth forms. The supreme example of such a polymorphic fungus is *Aureobasidium pullulans*, which is capable of producing five different vegetative forms (Table 5). The development of chlamydospores whilst not being considered to be reproduction nevertheless implies the onset of a period of inactivity before renewed vegetative growth. The developmental changes can be brought about by a number of environmental and biological factors which may operate separately or in combination. It is probable that polymorphism will enable this species to colonize aerial plant surfaces and to continue to thrive if the conditions alter in almost any direction. In this respect it is notable that the formation of yeast-like cells is encouraged by low levels of carbohydrate, such as might be found on plant surfaces.

Another organism which is able to alter its growth form is *Cladosporium cladosporioides* which has been shown to behave almost like a yeast when its conidia were germinated on glass at high humidity and 20 to 30°C (Dickinson & Bottomley 1980). This growth form may be regarded as an instance of microcyclic conidiation, which sort of development has also been observed by scanning electron microscopy on leaves collected from the field (Fig. 1 E). Some yeasts can produce pseudomycelium which will increase the range of conditions which they can tolerate.

Development of dormant resistant structures occurs in several organisms, though these are rare amongst the single-celled epiphytes. Chlamydospores are formed by some yeasts (Bandoni *et al.* 1971) but the majority of these organisms appear to be unable to respond to the onset of severe weather or biological antagonism. By contrast, several phylloplane filamentous fungi are able to form chlamydospores (e.g. *Aureobasidium pullulans*) and sclerotia (e.g. *Aureobasidium* and *Cladosporium* spp.). Pugh & Buckley (1971) have shown that these structures are formed when organisms are grown in paired cultures with other fungi, which suggests that there is a biological factor involved in their initiation, and they also demonstrated that they were formed on leaves and in litter. It has also been shown that chlamydospores develop when nitrogen is supplied as nitrate or when it becomes limiting in *Aureobasidium pullulans* cultures (Brown *et al.* 1973; Dominguez *et al.* 1978). As yet it is not known if they are formed following the onset of unfavourable environmental conditions, though these would create secondary changes which might encourage their development.

## *SURVIVAL THROUGH UNFAVOURABLE SEASONS*

Epiphytic micro-organisms probably grow most luxuriantly in tropical, evergreen rainforest (Fig. 2 A-F) where there are only slight seasonal variations in the climate. There the usual temperate region

Table 5. Polymorphism in *Aureobasidium pullulans*.

| Growth forms | | Factors known to affect development |
|---|---|---|
| Hyaline hyphae | | $pCO_2$ |
| Melanin-pigmented hyphae | | $pO_2$ |
| Yeast-like budding cells | multiple | Carbon source |
| Swollen hyaline cells | interactions | Nitrogen source |
| Melanin-pigmented chlamydospores | | Divalent cations |
| | | Temperature |
| | | 'Yeast extract' |

phylloplane populations of bacteria, yeasts and relatively few filamentous fungi, are joined by numerous algae, lichens, liverworts, mosses and ferns (Ruinen 1961). Elsewhere in the tropics the climate is less favourable, with a period of summer drought, creating a considerable seasonal hazard for microbial cells and propagules. In most temperate climates, there are significant fluctuations in the environmental conditions during the growing season, but these are as nothing when compared with the prolonged and extreme adverse weather conditions which characterize the winter months.

There have been few studies on the year-round fluctuations in phylloplane populations, with most investigations commencing when leaves unfold and finishing when they senesce. As a result we know very little about the ways in which phylloplane organisms survive through either hot or cold seasons. However, several possible strategies may be envisaged for such survival (Table 6).

Survival on actively growing evergreen leaves may be important in the tropics where local plant canopy microclimates permit growth which is not possible in more exposed situations. In temperate regions survival on evergreen leaves has been studied by Mishra & Dickinson (1981) who found that there were relatively few yeast cells and fungal hyphae on green *Ilex aquifolium* leaves in January and April. The populations increased in summer and autumn, but it is not known if the inoculum for this increase was the residual population which survived the winter on green leaves or the population of phylloplane organisms which were shown to survive through the winter on the leaf litter. Survival on evergreen leaves could be of great importance in the re-establishment of phylloplane populations as it appears that many of these organisms have a notably catholic ability to colonize a wide variety of green plants.

Evidence for the survival of micro-organisms in hydathodes, buds, bark crevices, beneath leaf sheaths and in other protected sites on biennial

Table 6. Strategies for survival of microbial epiphytes between growing seasons.

| Types of survival | |
|---|---|
| A. | Active growth on persistent, evergreen leaves |
| B. | Passive survival within sheltered habitats on biennial or perennial hosts |
| C. | Survival on leaf litter or in soil<br>i. Active growth and sporulation<br>ii. Passive survival as spores, chlamydospores, sclerotia or other dormant structures |

and perennial plants comes mainly from studies of plant pathogens (Leben 1974). Organisms occupying these sites may be relatively inactive, or hypobiotic, in which state they survive for relatively long periods.

A number of investigators have demonstrated that organisms which usually grow on living leaves can also be found on or in leaf litter or in the soil itself. Much of the evidence concerns the isolation of leaf and fruit yeasts from litter or soil. For example, Di Menna (1960) identified 54% of the yeasts isolated from the 2 to 4 cm soil horizon as being leaf inhabitants. These included *Cryptococcus laurentii*, *Rhodotorula glutinis*, *R. graminis*, *Sporobolomyces roseus* and *Torulopsis ingeniosa*. As regards the filamentous fungi it may be that development on senescent and fallen plant tissues is a normal part of their life cycle, during which they obtain sufficient nutrients to reproduce either sexually or asexually (Kendrick & Burges 1962; Mishra & Dickinson 1981). In all these instances there is, however, one unanswered question, concerning the method of re-inoculation of newly expanded leaves and other surfaces from these reservoirs. Rain splash and wind may play a part in this, but it is difficult to envisage large numbers of yeasts and bacteria being resuspended in the atmosphere by these agencies. Thus many of the organisms which accumulate in litter and soil may remain there in an inactive state until they eventually die. This would imply that the recolonization of large and rapidly formed expanses of aerial plant surface depends on a relatively sparse initial inoculum, which is quickly compensated for by the exceptionally rapid growth and multiplication of many single-celled microbial epiphytes.

It should also be noted that relatively little is known about the survival of saprophytic phylloplane bacteria (see also Y. Henis, this volume). This is in part due to the general paucity of studies of such bacteria on plant surfaces and also to the cumbersome and tedious identification processes which hamper any large-scale ecological study of this group.

*ACKNOWLEDGEMENT*

I would like to thank Dr G.W. Beakes for his very helpful comments during the preparation of this review.

*REFERENCES*

Akai, S., Fukutomi, M., Ishida, N. & Kunoh, H. (1967). An anatomical approach to the mechanism of fungal infections in plants. *In* The Dynamic Role of Molecular Constituents in Plant-Parasite Interaction, eds C.J. Mirocha & I. Uritani, pp. 1-20. St. Paul: Am. Phytopath. Soc.

Austin, B., Goodfellow, M. & Dickinson, C.H. (1978). Numerical taxonomy of phylloplane bacteria isolated from *Lolium perenne*. Journal of General Microbiology, *104*, 139-55.

Bandoni, R.J., Lobo, K.J. & Brezden, S.A. (1971). Conjugation and chlamydospores in *Sporobolomyces odorus*. Canadian Journal of Botany, *49*, 683-6.

Bashi, E. & Fokkema, N.J. (1976). Scanning electron microscopy of *Sporobolomyces roseus* on wheat leaves. Transactions of the British Mycological Soci-

ety, *67*, 500-5.
Bashi, E. & Fokkema, N.J. (1977). Environmental factors limiting growth of *Sporobolomyces roseus*, an antagonist of *Cochliobolus sativus*, on wheat leaves. Transactions of the British Mycological Society, *68*, 17-25.
Blakeman, J.P. & Fokkema, N.J. (1982). Potential for biological control of plant diseases on the phylloplane. Annual Review of Phytopathology, *20*, 167-92.
Bridges, B.A. (1976). Survival of bacteria following exposure to ultraviolet and ionizing radiations. *In* The Survival of Vegetative Microbes, eds T.R.G. Gray & J.R. Postgate, pp. 183-208. Cambridge: Cambridge University Press.
Brown, R.G., Hanic, L.A. & Hsiao, M. (1973). Structure and chemical composition of yeast chlamydospores of *Aureobasidium pullulans*. Canadian Journal of Microbiology, *19*, 163-8.
Bull, A.T. (1970). Inhibition of polysaccharases by melanin: enzyme inhibition in relation to mycolysis. Archives of Biochemistry and Biophysics, *137*, 345-56.
Burrage, S.W. (1971). The microclimate at the leaf surface. *In* Ecology of Leaf Surface Micro-organisms, eds T.F. Preece & C.H. Dickinson, pp. 91-101. London: Academic Press.
Burrage, S.W. (1976). Aerial microclimate around plant surfaces. *In* Microbiology of Aerial Plant Surfaces, eds C.H. Dickinson & T.F. Preece, pp. 173-84. London: Academic Press.
Callow, M.E. & Evans, L.V. (1974). Studies on the ship-fouling alga *Enteromorpha*. III. Cytochemistry and autoradiography of adhesive production. Protoplasma, *80*, 15-27.
Catley, B.J. (1979). Pullulan synthesis by *Aureobasidium pullulans*. *In* Microbial Polysaccharides and Polysaccharidases, eds R.C.W. Berkeley, G.W. Gooday & D.C. Ellwood, pp. 69-84. London: Academic Press.
Cooper, L.A. & Gadd, G.M. (1984). Differentiation and melanin production in hyaline and pigmented strains of *Microdochium bolleyi*. Antonie van Leeuwenhoek, *50*, 53-62.
Corpe, W.A. (1980). Microbial surface components involved in adsorption of microorganisms onto surfaces. *In* Adsorption of Microorganisms to Surfaces, eds G. Bitton & K.C. Marshall, pp. 105-44. New York: John Wiley & Sons.
Daniels, S.L. (1980). Mechanisms involved in sorption of microorganisms to solid surfaces. *In* Adsorption of Microorganisms to Surfaces, eds G. Bitton & K.C. Marshall, pp. 7-58. New York: John Wiley & Sons.
Davenport, R.R. (1980). Cold-tolerant yeasts and yeast-like organisms. *In* Biology and Activities of Yeasts, eds F.A. Skinner, S.M. Passmore & R.R. Davenport, pp. 215-30. London: Academic Press.
Davidson, D.E. (1973). Mucoid sheath of *Lulworthia medusa*. Transactions of the British Mycological Society, *60*, 577-9.
Dazzo, F.B. (1980). Microbial adhesion to plant surfaces. *In* Microbial Adhesion to Surfaces, eds R.C.W. Berkeley, J.M. Lynch, J. Melling, P.R. Rutter & B. Vincent, pp. 311-28. Chichester: Ellis Horwood.
Deinema, M.H. & Zevenhuizen, L.P.T. (1971). Formation of cellulose microfibrils by Gram-negative bacteria and their role in bacterial flocculation. Archiv für Mikrobiologie, *78*, 42-57.
Dickinson, C.H. & Bottomley, D. (1980). Germination and growth of *Alternaria* and *Cladosporium* in relation to their activity in the phylloplane. Transactions of the British Mycological Society, *74*, 309-19.
Diem, H.G. (1971). Effect of low humidity on the survival of germinated spores commonly found in the phyllosphere. *In* Ecology of Leaf Surface Microorganisms, eds T.F. Preece & C.H. Dickinson, pp. 211-9. London: Academic Press.
Dominguez, J.B., Goñi, F.M. & Uruburu, F. (1978). The transition from yeast-like to chlamydospore cells in *Pullularia pullulans*. Journal of General Micro-

biology, *108*, 111-7.

Duncan, B. & Herald, A.C. (1974). Some observations on the ultrastructure of *Epicoccum nigrum*. Mycologia, *66*, 1022-9.

Durrell, L.W. (1968). Studies of *Aureobasidium pullulans* (De Bary) Arnaud. Mycopathologia et Mycologia Applicata, *35*, 113-20.

Edwards, H.H. & Allen, P.J. (1970). A fine-structure study of the primary infection process during infection of barley by *Erysiphe graminis* f. sp. *hordei*. Phytopathology, *60*, 1504-9.

Ellis, M.B. (1971). Dematiaceous Hyphomycetes. Kew: Commonwealth Mycological Institute.

Ellis, M.B. (1976). More Dematiaceous Hyphomycetes. Kew: Commonwealth Mycological Institute.

Ellis, D.H. & Griffiths, D.A. (1974). The location and analysis of melanins in the cell walls of some fungi. Canadian Journal of Microbiology, *20*, 1379-86.

Fleet, G.H. & Phaff, H.J. (1981). Fungal glucans - structure and metabolism. *In* Plant Carbohydrates. II. Extracellular Carbohydrates, eds W. Tanner & F.A. Loewus, pp. 416-40. Berlin: Springer-Verlag.

Flentje, N.T. (1957). Studies on *Pellicularia filamentosa* (Pat.) Rogers. III. Host penetration and resistance, and strain specialization. Transactions of the British Mycological Society, *40*, 322-36.

Fletcher, M. (1979). The attachment of bacteria to surfaces in aquatic environments. *In* Adhesion of Microorganisms to Surfaces, eds D.C. Ellwood, J. Melling & P. Rutter, pp. 87-108. London: Academic Press.

Fry, R.M. (1966). Freezing and drying of bacteria. *In* Cryobiology, ed. H.T. Meryman, pp. 665-96. London: Academic Press.

Good, H.M. & Zathureczky, P.G.M. (1967). Effects of drying on the viability of germinated spores of *Botrytis cinerea*, *Cercospora musae*, and *Monilinia fructicola*. Phytopathology, *57*, 719-22.

Gray, T.R.G. (1976). Survival of vegetative microbes in soil. *In* The Survival of Vegetative Microbes, eds T.R.G. Gray & J.R. Postgate, pp. 327-64. Cambridge: Cambridge University Press.

Hau, F.C. & Rush, M.C. (1982). Preinfectional interactions between *Helminthosporium oryzae* and resistant and susceptible rice plants. Phytopathology, *72*, 285-92.

Hoog, G.S. de & Hermanides-Nijhof, E.J. (1977). The black yeasts and allied hyphomycetes. Studies in Mycology, *15*, 1-222.

Hughes, S.J. (1976). Sooty moulds. Mycologia, *68*, 693-820.

Hyde, K., Jones, E.B.G. & Moss, S.T. (1986). How do fungal spores attach to surfaces? Proceedings of the International Biodeteriation Symposium (in press).

Ingraham, J.L. & Stokes, J.L. (1959). Psychrophilic bacteria. Bacteriological Reviews, *23*, 97-108.

Jeanes, A. (1977). Dextrans and pullulans: industrially significant α-D-glucans. *In* American Chemical Society Symposium, 45, eds P.A. Sandford & A. Laskin, pp. 284-97.

Jeanes, A., Pittsley, J.E. & Watson, P.R. (1964). Extracellular polysaccharide produced from glucose by *Cryptococcus laurentii* var. *flavescens* NRRL Y-1401: chemical and physical characteristics. Journal of Applied Polymer Science, *8*, 2775-87.

Kendrick, W.B. & Burges, A. (1962). Biological aspects of the decay of *Pinus sylvestris* leaf litter. Nova Hedwigia, *4*, 313-42.

Krinsky, N.I. (1976). Cellular damage initiated by visible light. *In* The Survival of Vegetative Microbes, eds T.R.G. Gray & J.R. Postgate, pp. 209-39. Cambridge: Cambridge University Press.

Lapp, M.S. & Skoropad, W.P. (1978). Nature of adhesive material of *Colletotrichum graminicola* appressoria. Transactions of the British Mycological Society, *70*, 221-3.

Leach, C.M. (1971). A practical guide to the effects of visible and ultraviolet light on fungi. *In* Methods in Microbiology, 4, ed. C. Booth, pp. 609-64. London: Academic Press.

Leach, J.G. (1923). The parasitism of *Colletotrichum lindemuthianum*. University of Minnesota Agricultural Experimental Station, Technical Bulletin *14*.

Leben, C. (1974). Survival of plant pathogenic bacteria. Ohio Agricultural Research and Development Center, Special Circular *100*.

Lingappa, Y., Sussman, A.S. & Bernstein, I.A. (1963). Effect of light and media upon growth and melanin formation in *Aureobasidium pullulans* (= *Pullularia pullulans*). Mycopathologia et Mycologica Applicata, *20*, 109-28.

McDade, J.J. & Hall, L.B. (1964). Survival of gram-negative bacteria in the environment. I. Effect of relative humidity on surface-exposed organisms. American Journal of Hygiene, *80*, 192-204.

MacLeod, R.A. & Calcott, P.H. (1976). Cold shock and freezing damage to microbes. *In* The Survival of Vegetative Microbes, eds T.R.G. Gray & J.R. Postgate, pp. 81-109. Cambridge: Cambridge University Press.

Marks, G.C., Berbee, J.G. & Riker, A.J. (1965). Direct penetration of leaves of *Populus tremuloides* by *Colletotrichum gloeosporioides*. Phytopathology, *55*, 408-12.

Mazur, P. (1968). Survival of fungi after freezing and desiccation. *In* The Fungi. III, eds G.C. Ainsworth & A.S. Sussman, pp. 325-94. New York: Academic Press.

Menna, M.E. di (1960). Yeasts from soils under forest and under pasture. New Zealand Journal of Agricultural Research, *3*, 623-32.

Mishra, R.R. & Dickinson, C.H. (1981). Phylloplane and litter fungi of *Ilex aquifolium*. Transactions of the British Mycological Society, *77*, 329-37.

Morgan, H.R. & Beckwith, T.D. (1939). Mucoid dissociation in the colon-typhoid-*Salmonella* group. Journal of Infectious Disease, *65*, 113-24.

Morita, R.Y. (1976). Survival of bacteria in cold and moderate hydrostatic pressure environments with special reference to psychrophilic and barophilic bacteria. *In* The Survival of Vegetative Microbes, eds T.R.G. Gray & J.R. Postgate, pp. 279-98. Cambridge: Cambridge University Press.

Morris, E.R., Rees, D.A., Young, G., Walkinshaw, M.D. & Darke, A. (1977). Order-disorder transition for a bacterial polysaccharide in solution. A role for polysaccharide conformation in recognition between *Xanthomonas* pathogen and its plant host. Journal of Molecular Biology, *110*, 1-16.

Murray, G.M. & Maxwell, D.P. (1974). Ultrastructure of conidium germination of *Cochliobolus carbonum*. Canadian Journal of Botany, *52*, 2335-40.

Napoli, C.A., Dazzo, F.B. & Hubbel, D.H. (1975). Production of cellulose microfibrils by *Rhizobium*. Applied Microbiology *30*, 123-31.

Nurudeen, T.A. & Ahearn, D.G. (1979). Regulation of melanin production by *Cryptococcus neoformans*. Journal of Clinical Microbiology, *10*, 724-9.

Parbery, D.G. & Emmett, R.W. (1977). Hypotheses regarding appressoria, spores, survival and phylogeny in parasitic fungi. Revue de Mycologie, *41*, 429-47.

Park, D. (1982). Phylloplane fungi: tolerance of hyphal tips to drying. Transactions of the British Mycological Society, *79*, 174-8.

Phaff, H.J. (1971). Structure and biosynthesis of the yeast cell envelope. *In* The Yeasts. II, eds A.H. Rose & J.S. Harrison, pp. 135-210. London: Academic Press.

Potter, J.M., Tiffany, L.H. & Martinson, C.A. (1980). Substrate effect on *Helminthosporium maydis* race T conidium and germ tube morphology. Phytopathology, *70*, 715-9.

Pugh, G.J.F. & Buckley, N.G. (1971). The leaf surface as a substrate for colonization by fungi. *In* Ecology of Leaf Surface Micro-organisms, eds T.F. Preece & C.H. Dickinson, pp. 431-45. London: Academic Press.

Rees, G. & Jones, E.B.G. (1984). Observations on the attachment of spores of marine fungi. Botanica Marina, *27*, 145-60.

Rose, A.H. (1968). Physiology of micro-organisms at low temperatures. Journal of Applied Bacteriology, *31*, 1-11.

Rovira, A.D. & Campbell, R. (1974). Scanning electron microscopy of microorganisms on the roots of wheat. Microbial Ecology, *1*, 15-23.

Ruinen, J. (1961). The phyllosphere. I. An ecologically neglected milieu. Plant and Soil, *15*, 81-109.

Schnathorst, W.C. (1965). Environmental relationships in the powdery mildews. Annual Review of Phytopathology, *3*, 343-66.

Simpson, K.L., Chichester, C.O. & Phaff, H.J. (1971). Carotenoid pigments of yeasts. *In* The Yeasts. II, eds A.H. Rose & J.S. Harrison, pp. 493-515. London: Academic Press.

Sing, V.O. & Bartnicki-Garcia, S. (1975). Adhesion of *Phytophthora palmivora* zoospores: electron microscopy of cell attachment and cyst wall fibril formation. Journal of Cell Science, *18*, 123-32.

Sistrom, W.R., Griffiths, M. & Stanier, R.Y. (1956). The biology of a photosynthetic bacterium which lacks colored carotenoids. Journal of Cellular and Comparative Physiology, *48*, 473-515.

Slodki, M.E. (1966). The structure of extracellular phosphorylated galactans from *Sporobolomyces* yeasts. Journal of Biological Chemistry, *241*, 2700-6.

Spencer, J.F.T. & Gorin, P.A.J. (1966). Extracellular polysaccharides of yeast. Abhandlungen der Deutschen Akademie der Wissenschaften zu Berlin, Klasse für Medizin, 1966, 105-10.

Staples, R.C. (1985). The development of infection structures by the rusts and other fungi. Microbiological Sciences, *2*, 193-8.

Strange, R.E. & Cox, C.S. (1976). Survival of dried and airborne bacteria. *In* The Survival of Vegetative Microbes, eds T.R.G. Gray & J.R. Postgate, pp. 111-54. Cambridge: Cambridge University Press.

Sussman, A.S. (1968). Longevity and survivability of fungi. *In* The Fungi. III, eds G.C. Ainsworth & A.S. Sussman, pp. 447-86. New York: Academic Press.

Sussman, A.S., Lingappa, Y. & Bernstein, I.A. (1963). Effect of light and media upon growth and melanin formation in *Cladosporium mansoni*. Mycopathologia et Mycologia Applicata, *20*, 307-14.

Sutherland, I.W. (1982). Biosynthesis of microbial exopolysaccharides. Advances in Microbial Physiology, *23*, 80-150.

Sutter, H.-P., Jones, E.B.G. & Wälchli, O. (1984). Occurrence of crystalline hyphal sheaths in *Poria placenta* (Fr.) Cke. Journal of the Institute of Wood Science, *10*, 19-23.

Troy, F.A., II (1979). The chemistry and biosynthesis of selected bacterial capsular components. Annual Review of Microbiology, *33*, 519-60.

Valadon, L.R.G. (1976). Carotenoids as additional taxonomic characters in fungi: a review. Transactions of the British Mycological Society, *67*, 1-15.

Ward, J.B. & Berkeley, R.C.W. (1980). The microbial cell surface and adhesion. *In* Microbial Adhesion to Surfaces, eds R.C.W. Berkeley, J.M. Lynch, J. Melling, P.R. Rutter & B. Vincent, pp. 47-66. Chichester: Ellis Horwood.

Webb, S.J. (1960). Factors affecting the viability of air-borne bacteria. Canadian Journal of Microbiology, *6*, 89-105.

Wessels, J.G.H. & Sietsma, J.H. (1981). Fungal cell walls: a survey. *In* Plant Carbohydrates. II. Extracellular Carbohydrates, eds W. Tanner & F.A. Loewus, pp. 352-94. Berlin: Springer-Verlag.

Wheeler, H. (1975). Plant Pathogenesis. Berlin: Springer-Verlag.

Wheeler, H. & Gantz, D. (1979). Extracellular sheaths on hyphae of two species of *Helminthosporium*. Mycologia, *71*, 1127-35.

Wheeler, M.A. & Stipanovic, R.D. (1979). Melanin biosynthesis in *Thielaviopsis basicola*. Experimental Mycology, *3*, 340-50.

Woloshuk, C.P., Wolkow, P.M. & Sisler, H.D. (1981). The effect of three fungicides, specific for the control of rice blast disease, on the growth and melanin biosynthesis by *Pyricularia oryzae* Cav. Pesticide Science, *12*, 86-90.

# SOOTY MOULDS AND BLACK MILDEWS IN EXTRA-TROPICAL RAINFORESTS

I.H. Parbery and J.F. Brown

*Department of Botany, University of New England, Armidale, N.S.W. 2351, Australia*

## *INTRODUCTION*

Between the tropics there are three major areas of tropical rainforest. The largest is centred in the Amazon Basin and extends into Central America and the Caribbean islands to the north and south into coastal Brazil. The African rainforests radiate from the Congo Basin and especially into the West African region. The third area is more dispersed and includes the complex Indo-Malayan rainforests of South-East Asia with outliers in India, Australia and the Pacific Islands. Due to the efforts of various mycological collectors in these regions and their own (or others) taxonomic endeavours, we now have a fragmentary knowledge of some of the phylloplane fungi of these ecosystems. Outside the tropical rainforests there is a most diverse assemblage of vegetation types which are also referred to as 'rainforests'. In the context of this paper we shall mainly discuss southern hemisphere rainforests which are broadly classified as sub-tropical, warm temperate and cool temperate forms. Each of these is characterized by unique structural properties in addition to their floristic differences. There has been an obvious concentration in past studies of these rainforests on identifying the flora and establishing phyto-geographical relationships. During the last 50 years there has been increasing emphasis on ecological studies. However, the phylloplane microfloras of most of these communities have been virtually ignored, even though they are some of the most complex known.

The traditional concept of rainforest is mostly of a closed community dominated by various trees having broad, mesomorphic leaves. Within these communities there are characteristic life forms including lianas, epiphytes, stranglers, stem buttresses, etc. in addition to complex animal populations. Southern hemisphere extra-tropical rainforests appear to be distributed in response to the dominance of rainfall and temperature regimes. However, the dominant site factors of a community are really the result of complex interactions between climate, topography and soil factors. Quite often human activities such as clearing for agriculture, forestry and road construction drastically disturb the functioning of these systems. As a consequence, many of the extra-tropical rainforests are now scattered communities or 'refugia' and can be viewed as 'ecological islands' within other vegetation types (Webb & Tracey 1981). Alternatively, these rainforests represent moisture and nutrient-conserving niches which are relatively fire-proof. Between many of the communities there are effective barriers preventing dispersal of propagules. Present floristic differences give some indication of the degree of speciation in the past. The influence these processes have had on the associated phylloplane microfloras is uncertain.

Two groups of foliicolous ascomycete fungi with conspicuous, dark mycelia and pleomorphic reproductive structures are the sooty moulds and black mildews. The former group are superficial saprophytes utilizing either the secretions of sap-sucking insects or those from the host plant. Much of the earlier literature gave the impression that sooty moulds were similar to '*Fumago vagans*' (a *nomen confusum* for a mixture of hyphomycetes) or species of Capnodiaceae. Such concepts are entirely inappropriate for the sooty moulds of rainforest communities. Current knowledge of sooty moulds on rainforest plants largely applies to certain species of dematiaceous hyphomycetes and coelomycetes. For many of these taxa, teleomorph-anamorph connections have been established (Table 1) in which the teleomorph is usually in the Dothideales (Ascomycotina). In contrast, black mildews are a large, relatively homogeneous group of species in a single family, the Meliolaceae. All species are obligate parasites and are usually restricted to leaf surfaces. However, they differ from many of the more cosmopolitan sooty mould taxa in having a very narrow host range.

### *SOOTY MOULD FUNGI*

Sooty mould fungi probably occur in some form in most of the vegetation types of the world. They are particularly prevalent around the margins of rainforest communities and constitute a most conspicuous aspect of the 'mycological landscape' (Hughes 1976). In fact, available literature suggests that the sooty mould fungi probably reach their greatest development in association with rainforest and other vegetation in the countries within and bordering the Pacific Ocean. In these regions, many crops are also affected by sooty moulds, particularly citrus, coffee and mangoes (Wellman 1972). However, even though these fungi are both conspicuous and economically important, their ecological relationships are poorly understood. On the other hand, taxonomic studies of sooty mould fungi have been quite extensive (Arnaud 1910 to 1918; Fraser 1933 to 1936; Yamamoto 1954 to 1959; Batista & Ciferri 1962 to 1963; Hughes 1965 to 1983; Reynolds 1970 to 1979). Moreover, the extensive occurrence of pleomorphism in these taxa coupled with independent lines of taxonomic thought have created considerable confusion in nomenclature of species. We propose that the families and genera outlined in Table 1 contain most of the sooty mould species found in extra-tropical rainforests. The composition of the families has been largely based on proposals of Von Arx & Müller (1975), Hughes (1976), Eriksson (1981) and Hawksworth *et al.* (1983). It is also our intention to refrain from introducing further argument into an already confused taxonomy and rather attempt to highlight some areas of pertinent ecological investigation.

Most sooty mould fungi on rainforest plants tend to grow in species complexes. The structure of these phylloplane communities is extremely variable although some commmon features may be discerned. The initial colonizers are usually fungi with inconspicuous, branching hyphae which

Table 1. The principal sooty mould fungi of extra-tropical rainforests with established teleomorph-anamorph connections. Numbers in parentheses represent estimated number of species (after Hawksworth *et al.* 1983).

| Teleomorph | Anamorph/Synanamorphs |
|---|---|
| Antennulariellaceae Woron. | |
| *Achaetobotrys* Bat. & Cif. (2) | *Antennariella* Bat. & Cif. |
| *Antennulariella* Woron. (1) | *Antennariella* Bat. & Cif., *Capnodendron* S. Hughes |
| *Coccodiella* Hara (2) | |
| Capnodiaceae (Sacc.) Höhn. | |
| *Acrogenotheca* Bat. & Cif. (2)[1] | |
| *Anopeltis* Bat. & Peres (1) | |
| *Brooksia* Hansf. (1)[1] | *Capnogoniella* Bat. & Cif., *Hiospira* R.T. Moore, *Overeemia* Arn. |
| *Callebaea* Bat. (1) | |
| *Calyptra* Theiss. & Syd. (2) | |
| *Capnodinula* Bat. & Cif. (1) | |
| *Capnodium* Mont. (ca. 15) | *Fumagospora* Arn., *Phaeoxyphiella* Bat. & Cif., *Scolecoxyphium* Cif. & Bat. |
| *Capnophaeum* Speg. (3) | |
| *Ceramoclasteropsis* Bat. & Caval. (1) | |
| *Echinothecium* Zopf (2) | |
| *Elmerinula* Syd. (1) | |
| *Hyalocapnias* Bat. & Cif. (2) | |
| *Hyaloscolecostroma* Bat. & Oliv. (1) | |
| *Limacinula* (Sacc.) Höhn. (6) | |
| *Phragmocapnias* Theiss. & Syd. (1+) | *Conidiocarpus* Woron. |
| *Pseudomorfea* Punith. (1) | *Chaetasbolisia* Speg. |
| *Scoriadopsis* Mend. (1) | |
| *Scorias* Fr. (3) | *Conidiocarpus* Woron., *Polychaeton* (Pers.) Lev., *Scolecoxyphium* Cif. & Bat. |
| *Strigopodia* Bat. (3) | *Antennatula* Fr., *Capnophialophora* S. Hughes |
| Chaetothyriaceae Hansf. ex Barr | |
| *Actinocymbe* Syd. (4) | |
| *Aithaloderma* Syd. (4) | *Asbolisia* Bat. & Cif., *Ciferrioxyphium* Bat. & Maia, *Leptoxyphium* Speg., *Microxyphium* Sacc. |
| *Batistospora* Bez. & Herr.[1] | |
| *Ceramothyrium* Bat. & Maia (19) | |
| *Chaetothyrium* Speg. (25) | *Merismella* H. Sydow |
| *Limacinia* Neger[1] | *Capnocybe* S. Hughes |
| *Microcallis* H. Syd. (10) | |
| *Naetrocymbe* Bat. & Cif. (7) | |

Table 1 (continued)

| Teleomorph | Anamorph/Synanamorphs |
|---|---|
| Chaetothyriaceae (continued) | |
| *Phaeosaccardinula* Henn. (15) | |
| *Treubiomyces* Höhn. (2) | |
| *Yatesula* H. Syd. & Syd. (4) | |
| Euantennariaceae Hughes & Corl. | |
| *Euantennaria* Speg. (6) | *Antennatula* Fr., *Hormisciomyces* Bat. & Nascim. |
| *Trichopeltheca* Bat., Costa & Cif. (2) | *Plokamidomyces* Bat. & Nascim., *Trichothallus* F. Stev. |
| ? (sensu Hughes 1976) | *Capnokyma* S. Hughes, *Hormisciomyces* Bat. & Nascim. |
| Metacapnodiaceae Hughes & Corl. | |
| *Metacapnodium* Speg. (6) | *Capnobotrys* S. Hughes, *Capnophialophora* S. Hughes, *Capnosporium* S. Hughes, *Capnocybe* S. Hughes |
| *Ophiocapnocoma* Bat. & Cif. (3) | *Capnocybe* S. Hughes, *Capnophialophora* S. Hughes, *Hormiokrypsis* Bat. & Nascim. |
| Seuratiaceae Vuill. | |
| *Seuratia* Pat. (4) | *Atichia* Flotow |
| Triposporiopsidaceae Hughes | |
| *Triposporiopsis* Yamam. (1) | *Tripospermum* Speg. |

[1] *incertae sedis*.

are often associated with yeast-like cells. Various algal propagules, fungal conidia, fern spores and pollen may also be conspicuous. On other plants, the early colonists may be *Triposporiopsis*-like or the more thalloid species of the Euantennariaceae. Quite often these species are overgrown by more vigorous species such as the anamorphs of various Metacapnodiaceae and Antennulariellaceae. However, the species composition of any sooty mould complex is extremely difficult to predict for any particular 'host' or habitat. The initial fungal colonists at a particular stage of leaf development seem to be critical in determining subsequent fungal populations. We suggest that the chemical environment on the leaf surface, especially after the deposition of insect secretions, is probably of critical importance and further selection depends on the resultant microclimate. Whatever the species composition of sooty moulds, it indicates elaborate adaptations by the component species to grow intimately with others, yet compete for similar resources. Microscopic observations of the colonies on leaves will usually indicate the presence of at least six to eight different species. To add to this confusion a number of species of dematiaceous hyphomycetes, e.g. *Janetia* spp., have been identified as common associates of sooty moulds. These fungi are not fungicolous and their association has been termed 'capnophilic' (Hughes 1983). To what degree other fungi take advantage of the unique microhabitat within a sooty mould colony is unknown. Further research on these phenomena may help to provide significant advances in our understandings of fungal ecology.

Both light and scanning electron microscopy show important differences between the hyphal morphologies of sooty mould fungi. Characters such as wall ornamentation and the size and shape of cells (Fig. 1) have been used by some mycologists to distinguish taxa and establish teleomorph-anamorph relationships (Hughes 1972, 1974, 1976). Another school of thought (Reynolds 1971 *et seq.*) contends that these characters are taxonomically unreliable and conclusive evidence of relationships can only be obtained in conjunction with cultural studies. Notwithstanding these opinions, there are still characteristic differences between the hyphae of most families although all species seem to have hyphae with hygroscopic properties. The common occurrence of 'echinulate ornamentation' is thought to be an adaptive feature which deteriorates over time to supply the adhesive matrix associated with sooty mould hyphae (Reynolds 1971 c). This factor may account for significant mycelial differences observed between wet colonies on living leaves and similar material preserved as exsiccata.

The mycelium of a sooty mould fungus is usually structured as a weft (or subiculum) of apparently disorganized hyphae which may be several mm deep. Within the colony may be found terminal conidiogenous cells on hyphal branches producing one or more types of conidia. These conidia can vary from relatively large, pigmented and septate conidia, of e.g. *Antennatula*, *Capnocybe*, *Hormiokrypsis* and *Capnodendron* spp., to small, hyaline phialospores, of *Hormisciomyces* and *Capnophialophora* spp. (Figs 2 and 3). In contrast, other species have highly organized mycelia which can give them a characteristic appearance. *Trichopeltheca* spp. consist of a flat hyphal plate bearing *Trichothallus* phragmoconidia (Fig. 4) and/or *Plokamidomyces* phialospores. In many respects these fungi resemble some of the foliicolous trentepohlian algae. *Capnokyma*, *Acrogenotheca* (Fig. 5) and *Brooksia* spp. have erect, organized hyphae bearing various reproductive structures. The non-pigmented chaetothyriaceous fungi tend to adhere closely to the leaf surface and form characteristic branching patterns. Some species produce a thin pellicle of hyphae between the stomata or form conspicuous setae on dense mycelial cushions, e.g. *Dennisiella* spp. Functional roles for these setae (which are often hooked or bifurcate) are yet to be determined. Species of Metacapnodiaceae have tapering, moniliform hyphae which are visually recognizable. The colonies are usually dense, lustrous and thick (especially on petioles and stems) and consist of relatively rigid hyphae with smooth-walled cells (Fig. 6). The forms of hyphal organization in most sooty mould taxa have been comprehensively reviewed by Hughes (1976). The additional family included in this review (Seuratiaceae - including

Figs 1 to 4. 1. *Triposporiopsis* sp. hyphae showing hyphal ornamentation, opposite branching pattern and phialides (bar: 10 μm). 2. *Hormisciomyces* phialidic anamorphs associated with *Euantennaria* sp. (bar: 10 μm). 3. *Capnophialophora* phialidic anamorphs of Metacapnodiaceae (bar: 10 μm). 4. *Trichopeltheca* sp. (in foreground) consisting of a hyphal plate bearing *Trichothallus* phragmoconidia; the background thallus is *Enthallopycnidium* (?) sp. (bar: 100 μm).

*Atichia* anamorphs) is generally not included in the sooty mould fungi. However, the dark, gelatinous, lobed thalli of these fungi (Fig. 7) are common components of phylloplane sooty mould complexes and appear to obtain their nutrients from 'host' leachates. In relation to structure and reproductive processes, *Seuratia* spp. are quite dissimilar to all other sooty mould fungi (Meeker 1975 a,b).

Another interesting characteristic of some species in the Antennulariellaceae and Capnodiaceae is the production of elaborate pycnidial conidiomata. The *Antennariella* anamorphs of *Antennulariella* and *Achaetobotrys* produce small, globose meristogenous pycnidia. In contrast, the anamorphs of some Capnodiaceae (Table 1) produce highly variable, elongated pycnidia resulting from symphogenous development (Fig. 8). The spores are either hyaline or brown, septate or minute unicells. Very little is known of conidiogenous processes in the pycnidia and of mechanisms of spore dispersal. The small-spored species, e.g. *Polychaeton* and *Scolecoxyphium*, seem to deposit their spores in a sticky droplet at the ostiole of the pycnidium. This process suggests that either arthropods or water droplets are the primary dispersal agents. Even the large spores of *Phaeoxyphiella* spp. tend to adhere around the pycnidial neck. Attempts to organize the great array of fungi having these characters within Saccardoan classification concepts were made by Batista & Ciferri (1963 b). Their Asbolisiaceae contained 28 'form genera' and although many of their proposals are now discarded, the data provided have been most useful for other investigations of this group. Hughes (1976) has proposed that the pycnidia of some of the Asbolisiaceae are really compressed synnematous structures. This observation tends to further highlight the need for greater study of these sooty mould fungi to elucidate both taxonomic and ecological characters. Many of the teleomorphs also need to be reassessed in terms of ascoma and ascus structure to confirm their taxonomic position.

It has been stated (Reynolds 1975) that sooty moulds exhibit two principal modes of growth. 'Deciduous' growth forms consist of a thin layer of fungi mainly on the upper surface of leaves in association with insect honeydew. The fungal pellicle disintegrates in the dry season, falls from the leaf and regenerates in the next season. 'Persistent' growth forms are more common on stems and growth occurs from existing mycelia. Reproductive structures are continually present and the component species seem to prefer more temperate climates. Our experience of rainforest sooty moulds partially supports such growth form distinctions. However, both types may be found in any locality depending on the range of environmental variables which occur over time. Eriksson (1981) has also suggested that the sooty mould fungi evolved from lichen thalli. The symbiotic relationship with algae could have become redundant when a richer carbohydrate source in the form of insect honeydew became available. This is an interesting hypothesis but is difficult to reconcile in view of the very large populations of foliicolous lichens sharing the same phylloplane habitats as sooty mould fungi.

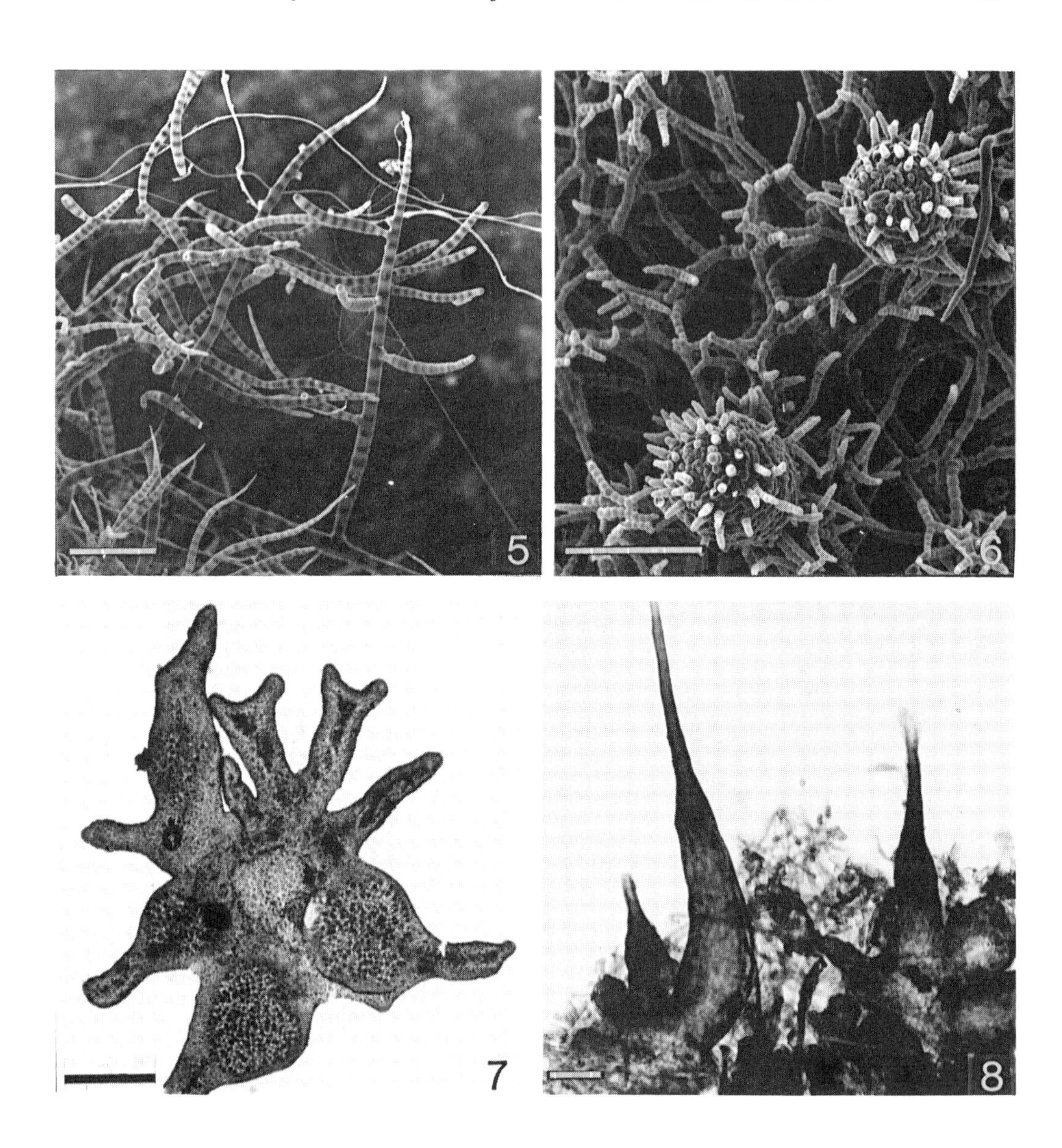


Figs 5 to 8. 5. *Acrogenotheca* sp. consisting of upright hyphae bearing conidial branches (bar: 100 μm). 6. *Metacapnodium* sp. showing moniliform hyphal organization and two ascomata with tapering hyphal appendages (bar: 100 μm). 7. *Atichia* anamorph of *Seuratia* sp.; darker regions in thallus are sporodochia containing conidia (bar: 100 μm). 8. Pycnidial conidiomata of *Polychaeton* (?) sp. (bar: 10 μm).

*BLACK MILDEW FUNGI (MELIOLACEAE)*

A conspicuous feature of the phylloplane mycoflora of various plants in tropical regions is the parasitic black mildew. Colonies of these fungi consist of very dark, superficial, hyphopodiate hyphae which tend to radiate from a central group of globose ascomata (Fig. 9). Setae are often present on the mycelium and give colonies a velvety appearance. Black mildew fungi are also common leaf parasites of many host plants in subtropical and temperate regions. Distribution records indicate occurrences in southern Africa, Chile, Canada, central Europe, Taiwan and Japan,with geographic limits in Scotland (58°N lat.) and southern Tasmania, Australia (42°S lat.). The majority of these records are from rainforest vegetation or other humid forest locations. Some of our recent observations suggest that black mildews are more widespread in wet heathland and swamp communities than host-pathogen checklists indicate. Arnaud (1946) reported the widespread occurrence of *Meliola* spp. on monocotyledonous hosts in some marshes of France. Because of various superficial similarities between black mildews, sooty moulds and other foliicolous ascomycete fungi, considerable confusion concerning taxonomic delimitation of these groups has occurred. The earliest valid descriptions of black mildew fungi were accommodated in the genus *Meliola* by Fries (1825). Later mycologists largely placed these and other species in the Perisporiales (Lindau 1897; Saccardo 1913; Theissen & Sydow 1917). Other than the contributions of Gaillard (1892) and Beeli (1920), the first monographic treatment of the 'true' black mildews was Stevens' "The Meliolineae" (1927, 1928). This taxon was derived from Arnaud (1918) and included seven genera: *Actinodothis* H. & P. Sydow, *Amazonia* Theiss., *Meliolina* H. & P. Sydow, *Irene* Theiss. & Syd., *Irenopsis* Stev., *Irenina* Stev. and *Meliola* Fr. emend. Bornet. Many taxonomists have placed the black mildews in the Meliolaceae and assigned the family to either i) Meliolales (Ainsworth *et al.* 1971; Müller & Von Arx 1973; Alexopoulos & Mims 1979), ii) Myriangiales (Hansford 1946) or iii) Erysiphales (Bessey 1950). However, the most significant contributions to the biology of these fungi were the monumental studies of Hansford (1961, 1963) which reduced the taxa and synonomy to: *Amazonia* Theiss. (syn. *Meliolaster* Doidge), *Appendiculella* Höhn. (syn. *Irene* Stev.), *Asteridiella* McAlp. (syn. *Irene* Theiss & Syd., *Irenina* Stev.), *Irenopsis* Stev. and *Meliola* (Fr.) Bornet (syn. *Amphitrichum* Nees ex Spreng., *Sphaeria* Fr., *Myxothecium* Kunze ex Fr., *Asteridium* Sacc.). Recent studies suggest that the Meliolaceae should be accommodated in the Dothideales (Eriksson 1981, 1982; Hawksworth *et al.* 1983). That is, with the ascolocular ascomycetes having $I^-$ (not staining with iodine) bitunicate (or secondarily pseudoprototunicate) asci in perithecioid ascomata or pseudothecia. Hawksworth *et al.* (1983) accepted 30 genera in the Meliolaceae. In this review, we wish to discuss only five genera (sensu Hansford 1961) as representative black mildews.

Black mildew colonies are mostly found on leaf laminae although some species commonly grow on petioles and young branches. Mature colonies are usually discrete and vary in size from less than 1 to 20 mm in diameter. Other species produce colonies which typically coalesce and may cover most of the leaf surface (Fig. 9). It is also common for these fungi to exhibit preferences for phylloplane colonization, species being described as either epiphyllous, hypophyllous or amphigenous on their hosts. The latter growth forms may also differ markedly according to the physical (and chemical?) environment of each surface. Close examination of colonies from both leaf surfaces is therefore necessary in any identification procedures. Although most colonies are clearly visible some of the larger diffuse forms are only seen by experienced observers. These differences are largely due to hyphal diameters, frequency of branching and the presence, size and density of setae. In contrast, thalloid *Meliola* spp. on conifers (Ellis 1974) produce extremely dense colonies in which structures are difficult to differentiate.

The influence of leaf age and position in the tree canopy on infection and subsequent development of black mildews is virtually unknown. Preliminary observations indicate that the younger leaves on many mature hosts rarely have visible colonies even though germinated spores may be present. However, adjacent seedlings of these host plants may have well developed colonies on any leaf. Data concerning phylloplane populations in rainforest canopies are almost non-existent. Observation of fallen branches and leaves occasionally confirms black mildew and other fungal colonization. In general, the sites of greatest black mildew infections in rainforests are in disturbed areas following natural tree or branch fall or human activities such as forestry and road construction.

Hansford (1961, 1963) has provided the most complete compilation of black mildews. Some 1814 species and varieties in five genera were described from *ca* 150 host plant families. Later authors have added 137 'new' species and varieties to this total (Commonwealth Mycological Institute, 1961-1985) including records from five additional host families. Table 2 outlines the distribution of black mildew species in the major host plant groups. Ninety-two % of all host plants currently known are dicotyledons and more than 72% of all described melioline taxa are

Table 2. Distribution of black mildew (melioline) species in the major plant host groups.

| Genus | Angiosperms | | Gymnosperms | Pteridophytes | Totals |
|---|---|---|---|---|---|
| | Dicots | Monocots | | | |
| *Amazonia* | 27 | 1 | 0 | 0 | 28 |
| *Appendiculella* | 49 | 0 | 5 | 0 | 54 |
| *Asteridiella* | 303 | 7 | 5 | 2 | 317 |
| *Irenopsis* | 126 | 1 | 0 | 4 | 131 |
| *Meliola* | 1296 | 107 | 5 | 1 | 1409 |
| Total | 1801 | 116 | 15 | 7 | 1939[1] |

[1] Plus *ca* 30 species on unknown hosts.

*Meliola* spp. *Asteridiella* spp. and *Irenopsis* spp. are less common (16.2 and 6.7%, respectively, of total species), while *Appendiculella* and *Amazonia* spp. (in total) account for less than 5%. Almost all records of black mildews on monocotyledonous hosts are *Meliola* spp., whereas other genera are relatively well represented in gymnosperm and pteridophyte hosts.

Cronquist (1981) recognized 383 families in the Magnoliophyta (cf. angiosperms). At least 150 of these families are known to provide host species for black mildew fungi. This represents about 40% of the total families and many of these host substantial numbers of Meliolaceae, viz. Rubiaceae (108 spp. or varieties), Euphorbiaceae (93), Fabaceae (84), Sapindaceae (82), Apocynaceae (70), Meliaceae (57), Rutaceae (52), Lauraceae (45), Myrtaceae (44) and Bignoniaceae (44). Elucidation of possible phylogenetic relationships in melioline taxa may also be approached at the host ordinal level (sensu Cronquist 1981). The most representative orders in terms of melioline species and component host families parasitized by these species are: Sapindales (259 species distributed in 10 host families), Fabales (158/3), Rubiales (108/1), Scrophulariales (107/6) and Gentianales (105/4). Many of the host families mentioned above are important components of the flora of extra-tropical rainforests. Moreover, examination of the data outlined above also highlights other important relationships:

a. Monocotyledons and the more 'primitive' dicotyledons (Magnoliales) are mostly parasitized by *Meliola* spp. (143).
b. *Appendiculella* spp. are absent from some large groups of host species, e.g. Rubiales, Scrophulariales, Fabales, Gentianales.
c. There is a relatively high incidence of *Irenopsis* spp. on hosts in the Malvales. Similarly, *Asteridiella* spp. are relatively common in the Solanales, Lamiales, Ericales and Myrtaceae.
d. Very small numbers of black mildew fungi have been reported from some large plant families in rainforests, e.g. Orchidaceae, Gesneriaceae and Araceae.

Initial observations of black mildews usually involve employing a hand or stereoscopic microscope. Colonies are then embedded in a transparent medium, e.g. collodion or cellulose acetate, for examination with a compound light microscope. Hyphae are 6-10 µm in diameter in most species although cell length varies between species. Mycelial branching is either opposite, alternate or mixed and when seen in conjuction with frequency and angle of branching provides a fairly characteristic growth habit. Perhaps the most distinguishing feature of melioline hyphae are the universal presence of short, 2-celled, lateral branches. These capitate hyphopodia (Fig. 10) are distributed in opposite, alternate or mixed arrangements at the distal ends of most hyphal cells. Important taxonomic characters also include the shape and dimensions on the 'foot' and 'head' cells which have also been termed 'stigmatopodium' and 'stigmatocyst' by Walker (1980) after Arnaud (1918). At some point beneath a head cell a fine hyphal filament or penetration peg *ca* 1-1.5 µm in diameter penetrates the cuticle and epidermal cell wall to produce a single saccate haustorium (Fig. 11). The fine structure of the host-parasite interface is currently unknown including possible adhesive

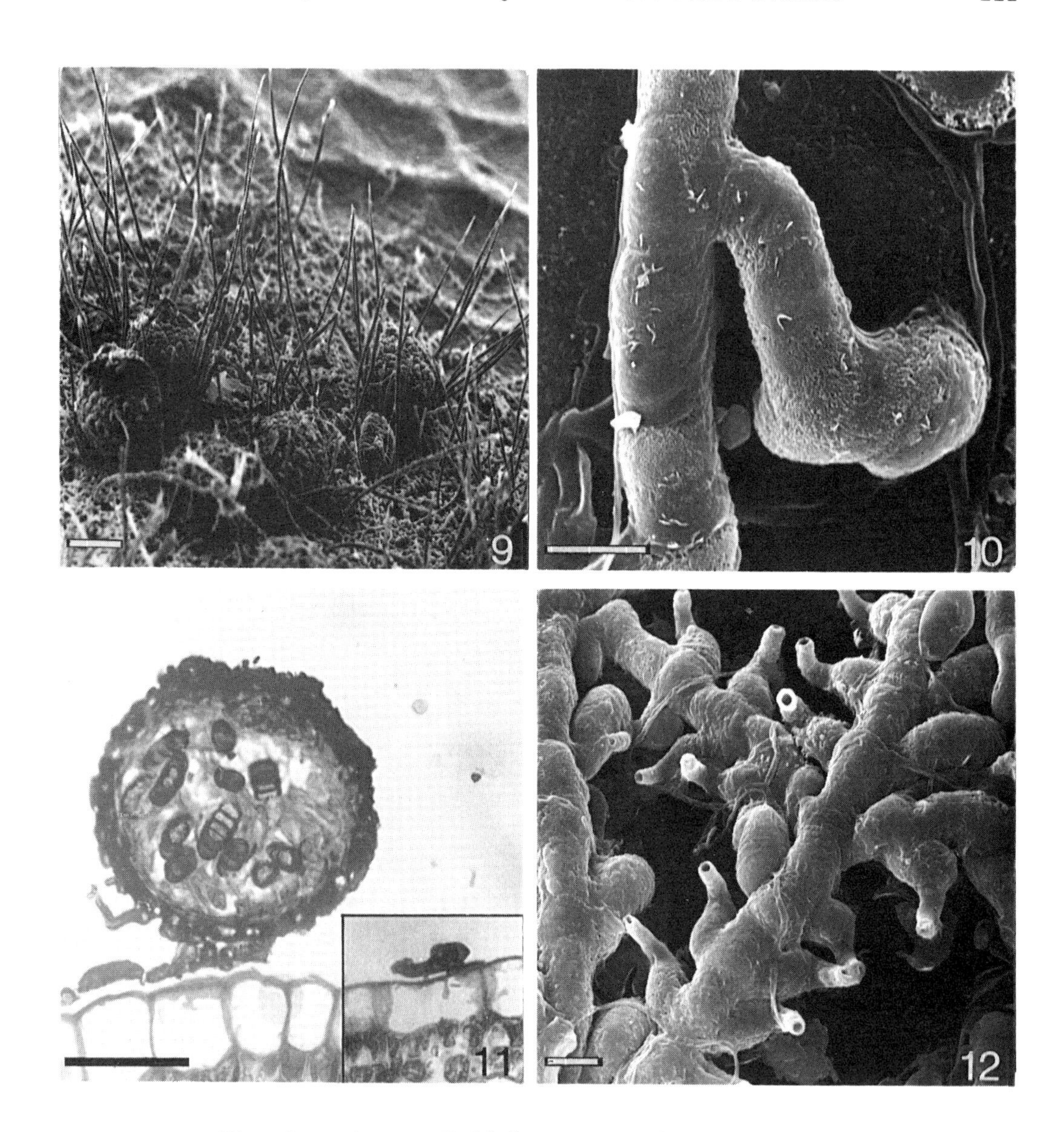


Figs 9 to 12. 9. *Meliola praetervisa* growing on adaxial surface of leaf of *Endiandra muelleri* (Lauraceae) (bar: 100 µm). 10. Capitate hyphopodium of *M. knowltoniae* on *Clematis glycinoides* (bar: 10 µm). 11. Transverse section of ascoma of *Irenopsis fieldiae* showing ascospores [inset: haustorium in leaf epidermis below hyphopodium] (bar: 100 µm). 12. Phialides ('mucronate hyphopodia') of *Asteridiella fraseriana* on *Cryptocarya meisneriana* (bar: 10 µm).

production as demonstrated in other hyphopodiate fungi (Onyile *et al.* 1982). Most black mildews also produce other short lateral branches, either on the same hyphae as capitate hyphopodia or on separate hyphae. These 'mucronate hyphopodia' usually consist of a single conoid to ampulliform cell with an open neck region directed away from the leaf surface (Fig. 12). Many workers have speculated on various functions for these structures. Hughes (1981) reported that they were actually phialides and produced phialoconidia. In addition, he proposed that these spores "probably function as spermatia" and have no propagative function. Our observations support the deletion of the term 'mucronate hyphopodium' because it neither attaches to the host nor has a known nutritional function. However, further research on the function(s) of phialides is warranted in relation to the amount of energy expended by many species in their formation.

More than 1400 *Meliola* spp. have been described and are differentiated from other species in the Meliolaceae by the presence of mycelial setae. These structures arise laterally from hyphal cells and immediately bend to assume a perpendicular position. Mycelial setae may vary in density within colonies, branching, apical morphology and length. The function of setae is largely unknown although a protective role against fungivorous arthropods could be suggested. Moreover, setose colonies tend to accumulate materials, e.g. dust, pollen, insect frass, etc. which absorb and retain water. Some developing *Meliola* spp. have septate, thin-walled setae with sunken apices (Fig. 13) which can give colonies a lustrous quality. The setae provide some evidence that they may be sites of conidial production similar to that described for some *Colletotrichum* spp. (Lenné *et al.* 1984). Small, finger-like branches on setal apices of *M. artabotrydis* were shown (Eriksson 1981) to produce conidium-like bodies. Further research is therefore necessary to determine whether anamorphs exist in a taxon which has been traditionally regarded as totally teleomorphic. It may also be pertinent to determine whether environmental factors are significant in the development of mycelial setae. The presence or absence of setae on mycelia and ascomata are the only taxonomic criteria for separation of the three largest melioline genera, *Meliola*, *Asteridiella* and *Irenopsis*. The perithecioid ascomata of black mildews are usually globose, superficial on a subiculum of non-hyphopodiate hyphae, non-ostiolate, black and mostly verrucose. *Appendiculella* spp. are characterized by having pallid, striated outgrowths of the ascomatal wall called larviform appendages (Fig. 14). *Irenopsis* and some *Meliola* spp. may have setae growing from the upper half of the wall (Fig. 15). *Amazonia* spp. differ from other genera in having a plate of radiating hyphae covering the ascomata. The 'perithecial' wall of all genera appears to consist of an inner, hyaline layer of small cells surrounded by a layer of large thick-walled cells (Fig. 16). In many species the outer cells tend to split in characteristic patterns and exhibit wall accretion patterns. Some studies of perithecial structure and development have been made (Ward 1883; Ryan 1926; Ragle 1930; Graff 1932) including aspects of ascosporogenesis (Thite 1974). However, it would seem that too many generalizations have been derived from these studies and a reassessment of a more representative group of species utilizing new techniques, is overdue.

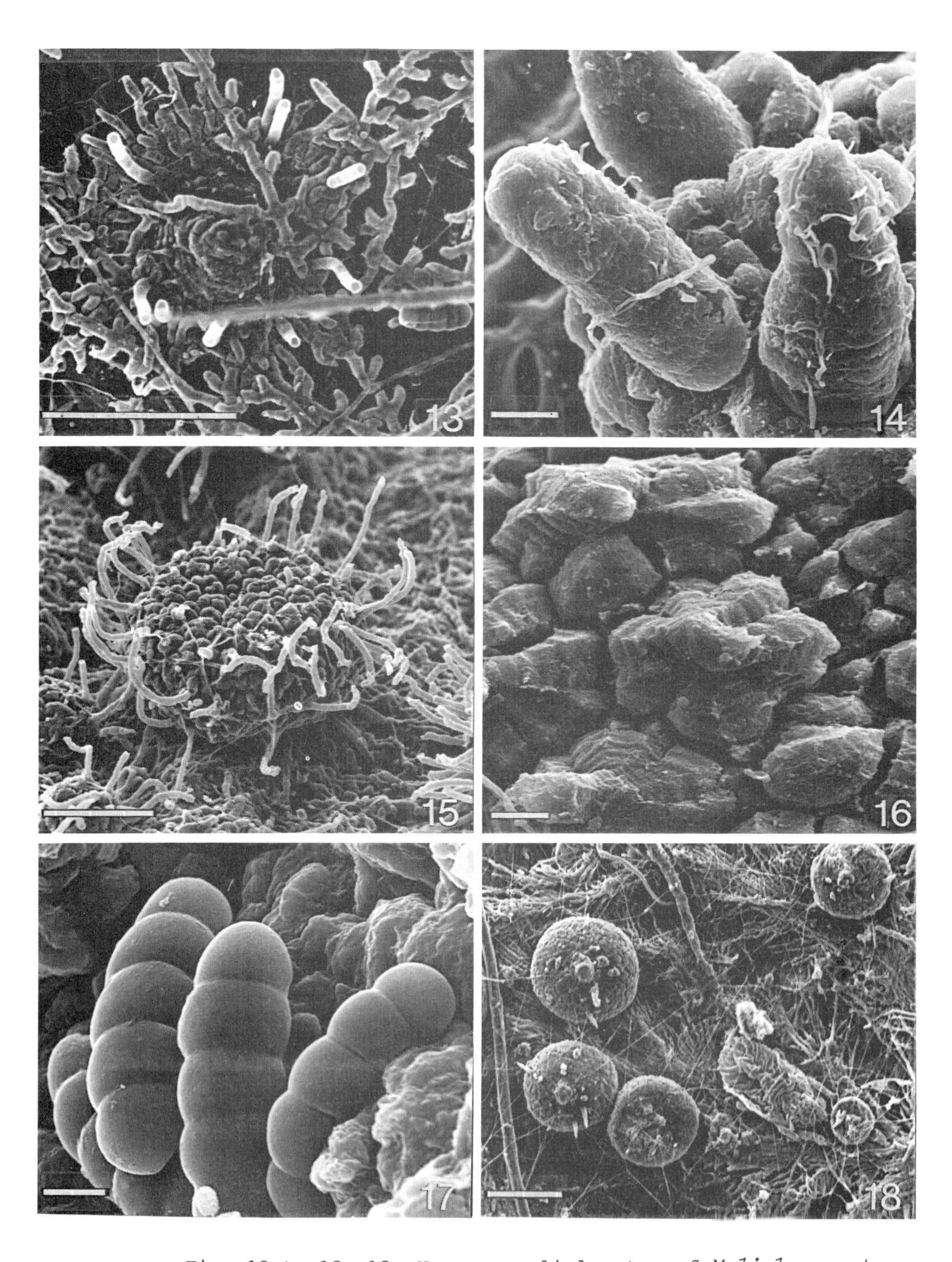


Figs 13 to 18. 13. Young mycelial setae of *Meliola praetervisa* around ascoma initial (bar: 100 µm). 14. Larviform appendages on 'perithecium' of *Appendiculella calostroma* (bar: 10 µm). 15. Setae on ascoma of *Irenopsis berggrenii*

Ascospore features are probably the least variable character of black mildews. Most species produce 4-septate ascospores (3-septate spores occur in <5% of the species) which are dark brown, smooth-walled and constricted at the septa (Fig. 17). The only record of spore ornamentation is from *Meliola codiae* (Huguenin 1969). Germinating ascospores are frequently observed on leaf surfaces of rainforest plants. The initial sequence of germ tube growth and hyphopodium development from the cells of the ascospore can give valuable clues regarding the identity of the fungus. Hansford (1961) claimed total inability to germinate ascospores either in the field or under laboratory conditions. Later workers (Goos 1974, 1978; Thite 1975) reported some success in germinating ascospores on agar media or by a hanging drop method. However, most of these ascospores produced germ tubes and no hyphopodia.

Hansford (1961) formalized a system of classifying black mildew fungi according to probable phylogenetic relationships of the host plants. Thus all species growing on hosts comprising a single family, e.g. Lauraceae, were treated as a relatively discrete group even though some fungi were morphologically identical with those parasitizing hosts in another family. In addition, host plant families were organized after Hutchinson (1926, 1934) although no attempts were made to compare the melioline taxa in closely related families. In the absence of cross-inoculation data from both controlled environments and forest habitats the species concept within these fungi is suspect. In fact, many of the species described in the literature "may be merely host-related phenotypes" (Pirozynski 1983). However, the obligate parasitism of black mildews coupled with their known antiquity (Dilcher 1965) and stable host-pathogen relationships suggests that Hansford's approach is still very useful. We suggest that future investigations could profit from i) adapting Hansford's (1961) scheme to more modern systems of host classification, e.g. Dahlgren (1980), Takhtajan (1980) or Cronquist (1981) and ii) determining relations between fungal taxa and various taxonomic categories of hosts. Both Stevens (1927, 1928) and Hansford (1961) also adopted a 'modified Beeli formula' to separate fungal species and/or varieties within a host family. This is an 8-digit numerical code used to 'split' taxa at the species level or below. In its accepted form it fails to adequately distinguish species and needs to be expanded for possible application in computerized numerical taxonomy.

The ecological relationships of black mildews are probably less understood than any other large group of parasitic fungi. They are known to be ectotrophic biotrophs with each species having a narrow range of host plants. However, our knowledge of colony longevity and spore dispersal and survival is extremely rudimentary. Undisturbed rainforest communi-

Figs 13 to 18 (continued)
(bar: 100 µm). 16. Outer 'perithecial' wall of *Meliola cyathodes* showing accretion patterns on surface cells (bar: 10 µm). 17. Ascospores of *Meliola polytricha* (bar: 10 µm). 18. Hyperparasites with pycnidia on *M. cyathodes* var. *trochocarpae* (bar: 100 µm).

ties provide a continual presence of susceptible hosts and water movements may account for some dispersal. In contrast, Pirozynski (1983) has suggested that dispersal "may be tied to a mycophagous factor" as ascospores are poorly adapted to aerial dispersal. We have already indicated that the infection processes of black mildews are poorly understood. Similarly, host resistance mechanisms are unknown even though many germinating ascospores are seen to be unable to penetrate the leaves of non-hosts. The only definite conclusions are that these fungi require wet and humid conditions and tend to grow best in shaded habitats provided suitable hosts are present (Toro 1952; Goos 1978).

Losses in production have been claimed for black mildew infections on a number of crop plants (Stevens & Dowell 1923; Wellman 1972). Rainforest plants, however, rarely exhibit any visible symptoms, even in severe infections. Sometimes the host tissues underlying colonies become chlorotic or some deformed epidermal cells are observed in leaf sections. Extensive colonization of the abaxial surface of *Orites excelsa* (Proteaceae) may produce some lumps on the adaxial surface and possibly induce early senescence. The only serious disease we have found is a petiole infection of *Morinda jasminoides* causing leaf drop.

Interactions between black mildews and other phylloplane populations on rainforest plants are poorly defined. Many other species of phylloplane fungi also inhabit rainforests - the majority being either Ascomycotina or Deuteromycotina. No reports of antagonism between any of these fungi and black mildews are known although some of the latter may be overgrown. In some communities one of the most characteristic features of black mildew colonies is their extensive parasitism by other fungi. These common associations were further exemplified by earlier mycologists who included hyperparasite structures such as conidiophores, conidia and conidiomata in descriptions of Meliolaceae. There is now an extensive literature on these largely conidial fungi (Stevens 1918; Hansford 1946; Deighton & Pirozynski 1972; Hawksworth 1981). Some biotrophic species appear to be confined to melioline hosts whilst others also attack other foliicolous ascomycetes (Hawksworth 1981). A number of hyperparasites apparently cause minimal damage to black mildew hosts; others either inhibit reproduction by preventing ascomatal development or destroy vegetative structures (Fig. 18). The widespread occurrence of algae in extra-tropical rainforests (especially discoid *Phycopeltis* spp. and filamentous Cyanophyta) seems to restrict some melioline species. Similar growth restrictions are apparent on leaves supporting very large numbers of foliicolous lichens. Leafy liverworts and other 'epiphyllae' tend to overgrow black mildews and eventually only fragments of hyphae and ascospores remain. Any specific relationships between these fungi and invertebrates (nematodes, mites, insects) are yet to be investigated.

## *REFERENCES*

Ainsworth, G.C., James, P.W. & Hawksworth, D.L. (1971). Ainsworth & Bisby's Dictionary of the Fungi, 6th ed. Kew: Commonwealth Mycological Institute.
Alexopoulos, C.J. & Mims, C.W. (1979). Introductory Mycology. New York: John Wiley

& Sons.

Arnaud, G. (1910). Contribution a l'étude des fumagines. 1re Partie (*Limacinia, Seuratia, Pleosphaeria*, etc). Annales de l'École Nationale d'Agriculture de Montpellier, Nouvelle Série, *9*, 239-77.

Arnaud, G. (1911). Contribution a l'étude des fumagines. 2me Partie. Systématique et organisation des espèces. Annales de l'École Nationale d'Agriculture de Montpellier, Nouvelle Série, *10*, 211-330.

Arnaud, G. (1912). Contribution a l'étude des fumagines. 3me Partie. Annales de l'École Nationale d'Agriculture de Montpellier, Nouvelle Série, *12*, 23-54.

Arnaud, G. (1918). Les Astérinées. Annales de l'École Nationale d'Agriculture de Montpellier, Nouvelle Série, *16*, 1-288.

Arnaud, G. (1946). Distribution géographique des champignons du genre *Meliola*. Comptes Rendus Hebdomadaires des Séances de l'Académie des Sciences, Paris, *223*, 1019-21.

Arx, J.A. von & Müller, E. (1975). A re-evaluation of the bitunicate Ascomycetes with keys to families and genera. Studies in Mycology, *9*, 1-159.

Batista, A.C. & Ciferri, R. (1962). The Chaetothyriales. Sydowia Annales Mycologici, Ser. II, Beihefte *3*, 1-129.

Batista, A.C. & Ciferri, R. (1963 a). Capnodiales. Saccardoa, *2*, 1-298.

Batista, A.C. & Ciferri, R. (1963 b). The sooty molds of the family Asbolisiaceae. Quaderno Laboratorio Crittogamico Istituto Botanico della Università Pavia, *31*, 1-229.

Beeli, M. (1920). Note sur le genre *Meliola*. Bulletin du Jardin Botanique de l'État, Bruxelles, *7*, 89-160.

Bessey, E.A. (1950). Morphology and Taxonomy of Fungi. Philadelphia: Blakiston.

Commonwealth Mycological Institute (1961-1985). Index of Fungi. Slough: Commonwealth Agricultural Bureaux.

Cronquist, A.J. (1981). An Integrated System of Classification of Flowering Plants. New York: Columbia University Press.

Dahlgren, R. (1980). A revised system of classification of the angiosperms. Botanical Journal of the Linnean Society, *80*, 91-124.

Deighton, F.C. & Pirozynski, K.A. (1972). Microfungi. V. More hyperparasitic hyphomycetes. Mycological Papers, *128*, 1-110.

Dilcher, D.L. (1965). Epiphyllous fungi from Eocene deposits in western Tennessee, U.S.A. Palaeontographica, *B 116*, 1-54.

Ellis, J.P. (1974). Some thalloid Meliolas. Transactions of the British Mycological Society, *63*, 93-8.

Eriksson, O. (1981). The families of bitunicate ascomycetes. Opera Botanica, *60*, 1-220.

Eriksson, O. (1982). Outline of the Ascomycetes - 1982. Mycotaxon, *15*, 203-48.

Fraser, L. (1933). An investigation of the sooty moulds of New South Wales. I. Historical and introductory account. Proceedings of the Linnean Society of New South Wales, *58*, 375-95.

Fraser, L. (1935 a). An investigation of the sooty moulds of New South Wales. III. The life histories and systematic positions of *Aithaloderma* and *Capnodium*, together with descriptions of new species. Proceedings of the Linnean Society of New South Wales, *60*, 97-118.

Fraser, L. (1935 b). An investigation of the sooty moulds of New South Wales. IV. The species of the Eucapnodieae. Proceedings of the Linnean Society of New South Wales, *60*, 159-78.

Fraser, L. (1935 c). An investigation of the sooty moulds of New South Wales. V. The species of the Chaetothyrieae. Proceedings of the Linnean Society of New South Wales, *60*, 280-90.

Fraser, L. (1936). Notes on the occurrence of the Trichopeltaceae and Atichiaceae in New South Wales, and on their mode of nutrition, with a description of a new species of *Atichia*. Proceedings of the Linnean Society of New South Wales, *61*, 277-84.

Fries, E.M. (1825). Systema Orbis Vegetabilis, Lund.

Gaillard, A. (1892). Le Genre *Meliola*. Paris: Klincksieck.
Goos, R.D. (1974). A scanning electron microscope and *in vitro* study of *Meliola palmicola*. Proceedings of the Iowa Academy of Science, *81*, 23-7.
Goos, R.D. (1978). Field and laboratory studies of meliolaceous fungi in Hawaii. Mycologia, *70*, 995-1006.
Graff, P.W. (1932). The morphological and cytological development of *Meliola circinans*. Bulletin of the Torrey Botanical Club, *59*, 241-65.
Hansford, C.G. (1946). The foliicolous Ascomycetes, their parasites and associated fungi. Mycological Papers, *15*, 1-240.
Hansford, C.G. (1961). The Meliolineae. Sydowia Annales Mycologici, Ser. II, Beihefte *2*, 1-806.
Hansford, C.G. (1963). Iconographia meliolinearum. Sydowia Annales Mycologici, Ser. II, Beihefte *5*.
Hawksworth, D.L. (1981). A survey of the fungicolous conidial fungi. *In* Biology of Conidial Fungi. Vol. I, eds G.T. Cole & B. Kendrick, pp. 171-244. New York: Academic Press.
Hawksworth, D.L., Sutton, B.C. & Ainsworth, G.C. (1983). Ainsworth & Bisby's Dictionary of the Fungi, 7th ed. Kew: Commonwealth Mycological Institute.
Hughes, S.J. (1965). New Zealand fungi. 5. *Trichothallus* and *Plokamidomyces* states of *Trichopeltheca*. New Zealand Journal of Botany, *3*, 320-32.
Hughes, S.J. (1966). New Zealand fungi. 7. *Capnocybe* and *Capnophialophora*, new form genera of sooty moulds. New Zealand Journal of Botany, *4*, 333-53.
Hughes, S.J. (1967 a). New Zealand fungi. 9. *Ophiocapnocoma* with *Hormiokrypsis* and *Capnophialophora* states. New Zealand Journal of Botany, *5*, 117-33.
Hughes, S.J. (1967 b). New Zealand fungi. 10. *Acrogenotheca elegans*. New Zealand Journal of Botany, *5*, 504-18.
Hughes, S.J. (1970). New Zealand fungi. 14. *Antennaria*, *Antennularia*, *Antennatula*, *Hyphosoma*, *Hormisciella* and *Capnobotrys* gen. nov. New Zealand Journal of Botany, *8*, 153-209.
Hughes, S.J. (1972). New Zealand fungi. 17. Pleomorphism in Euantennariaceae and Metacapnodiaceae, two new families of sooty moulds. New Zealand Journal of Botany, *10*, 225-42.
Hughes, S.J. (1974). New Zealand fungi. 22. *Euantennaria* with *Antennatula* and *Hormisciomyces* states. New Zealand Journal of Botany, *12*, 299-356.
Hughes, S.J. (1975). New Zealand fungi. 24. *Capnokyma corticola* gen. nov., sp. nov., a hyphomycetous sooty mould. New Zealand Journal of Botany, *13*, 637-44.
Hughes, S.J. (1976). Sooty moulds. Mycologia, *68*, 693-820.
Hughes, S.J. (1981 a). New Zealand fungi. 31. *Capnobotrys*, an anamorph of Metacapnodiaceae. New Zealand Journal of Botany, *19*, 193-226.
Hughes, S.J. (1981 b). Mucronate hyphopodia of Meliolaceae are phialides. Canadian Journal of Botany, *59*, 1514-7.
Hughes, S.J. (1983). New Zealand fungi. 32. *Janetia capnophila* sp. nov. and some allies. New Zealand Journal of Botany, *21*, 177-82.
Huguenin, B. (1969). Micromycètes du Pacifique Sud. VII. Méliolinées de Nouvelle-Calédonie. Revue de Mycologie, *34*, 23-61.
Hutchinson, J. (1926). The Families of Flowering Plants. Vol. 1. London: MacMillan.
Hutchinson, J. (1934). The Families of Flowering Plants. Vol. 2. London: MacMillan.
Lenné, J.M., Sonoda, R.M. & Parbery, D.G. (1984). Production of conidia by setae of *Colletotrichum* species. Mycologia, *76*, 359-62.
Lindau, G. (1897). Perisporiales. *In* Die Natürlichen Pflanzenfamilien. Vol. 1.1, eds A. Engler & K. Prantl, p. 325. Leipzig: Engelmann.
Meeker, J.A. (1975 a). Revision of Seuratiaceae. I. Morphology of *Seuratia*. Canadian Journal of Botany, *53*, 2462-82.
Meeker, J.A. (1975 b). Revision of Seuratiaceae. II. Taxonomy and nomenclature of *Seuratia*. Canadian Journal of Botany, *53*, 2483-96.
Müller, E. & Arx, J.A. von (1973). Pyrenomycetes: Meliolales, Coronophorales, Sphaeriales. *In* The Fungi. Vol. IVA, eds G.C. Ainsworth, F.K. Sparrow & A.S. Sussman, pp. 87-132. New York: Academic Press.

Onyile, A.B., Edwards, H.H. & Gessner, R.V. (1982). Adhesive material of the hyphopodia of *Buergenerula spartinae*. Mycologia, *74*, 777-84.

Pirozynski, K.A. (1983). Pacific mycogeography: an appraisal. Australian Journal of Botany (Supplementary Series), *10*, 137-59.

Ragle, M.R. (1930). The structure of the perithecium in the Meliolineae. Mycologia, *22*, 312-5.

Reynolds, D.R. (1970). Notes on capnodiaceous fungi. I. *Capnodiopsis*. Bulletin of the Torrey Botanical Club, *97*, 253-5.

Reynolds, D.R. (1971 a). Notes on capnodiaceous fungi. II. *Leptocapnodium*. Bulletin of the Torrey Botanical Club, *98*, 151-4.

Reynolds, D.R. (1971 b). The sooty mold ascomycete genus, *Limacinula*. Mycologia, *63*, 1173-209.

Reynolds, D.R. (1971 c). On the use of hyphal morphology in the taxonomy of sooty mold Ascomycetes. Taxon, *20*, 759-68.

Reynolds, D.R. (1975). Observations on growth forms of sooty mold fungi. Nova Hedwigia, *26*, 179-93.

Reynolds, D.R. (1978). Foliicolous ascomycetes: 1. The capnodiaceous genus *Scorias* reproduction. Contributions in Science, Natural History Museum of Los Angeles County, *288*, 1-16.

Reynolds, D.R. (1978). Foliicolous ascomycetes: 2. *Capnodium salicinum* Montagne emend. Mycotaxon, *7*, 501-7.

Reynolds, D.R. (1979). Foliicolous ascomycetes: 3. The stalked capnodiaceous species. Mycotaxon, *8*, 417-45.

Ryan, R.W. (1926). The development of the perithecium in the Microthyriaceae and a comparison with *Meliola*. Mycologia, *18*, 100-10.

Saccardo, P.A. (1913). Sylloge Fungorum Omnium hucusque Cognitorum, *22*, 19. Pavia.

Stevens, F.L. (1918). Some meliolicolous parasites and commensals from Porto Rico. Botanical Gazette, *65*, 227-49.

Stevens, F.L. (1927). The Meliolineae. I. Annales Mycologici, *25*, 405-69.

Stevens, F.L. (1928). The Meliolineae. II. Annales Mycologici, *26*, 165-384.

Stevens, F.L. & Dowell, R.I. (1923). A *Meliola* disease of cacao. Phytopathology, *13*, 247-50.

Takhtajan, A.L. (1980). Outline of the classification of flowering plants (Magnoliophyta). Botanical Review, *46*, 225-359.

Theissen, J. & Sydow, H. (1917). Synoptische Tafeln. Annales Mycologici, *15*, 389-491.

Thite, A.N. (1974). Ascosporogenesis in *Meliola jasminicola*. Transactions of the British Mycological Society, *63*, 189-93.

Thite, A.N. (1975). Ascospore germination in *Meliola jasminicola*. Indian Phytopathology, *28*, 94-6.

Toro, R.A. (1952). A study of the tropical American black mildews. University of Puerto Rico Journal of Agriculture, *36*, 24-87.

Walker, J. (1980). *Gaeumannomyces*, *Linocarpon*, *Ophiobolus* and several other genera of scolecospored ascomycetes and *Phialophora* conidial states, with a note on hyphopodia. Mycotaxon, *11*, 1-129.

Ward, H.M. (1883). On the morphology and development of the perithecium of *Meliola*, a genus of tropical epiphyllous fungi. Philosophical Transactions of the Royal Society, London, *174*, 583-97.

Webb, L.T. & Tracey, J.G. (1981). Australian rainforests: patterns and change. *In* Ecological Biogeography of Australia. Vol. 1, ed. A. Keast, pp. 605-94. The Hague: Junk.

Wellman, F.L. (1972). Tropical American Plant Disease. Metuchen: The Scarecrow Press.

Yamamoto, W. (1954). Taxonomic studies on the Capnodiaceae. II. On the species of the Eucapnodieae. Annals of the Phytopathological Society of Japan, *19*, 1-5.

Yamamoto, W (1955). Spore types in the imperfect stage of some genera of the Capnodiaceae. Annals of the Phytopathological Society of Japan, *20*, 83-8.

Yamamoto, W. (1956). Taxonomic studies on the Capnodiaceae. III. On the species of the Chaetothyrieae. Annals of the Phytopathological Society of Japan, *21*, 167-70.

Yamamoto, W. (1959). Revision of the species of the Capnodiaceae. Science Reports, Hyogo University of Agriculture, *4*, 17-22.

# LEAF YEASTS AS INDICATORS OF AIR POLLUTION

P. Dowding

*Department of Botany, Trinity College, Dublin 2, Ireland*

## *INTRODUCTION*

Leaves of higher plants have evolved as gas-exchange organs and in their relatively exposed position function as effective particulate traps as well. Organisms living on leaf surfaces are therefore in a very exposed position with regard to both gaseous and particulate pollutants. Several authors (reviewed by Huttunen & Soikkeli 1983) have observed that normally rough and/or folded cuticular surfaces become much smoother in areas of moderate to high levels of pollution by acid or alkaline gases or by rain below the normal range of pH (5 to 6). This smoothing will, among other things, reduce what little shelter there is on the leaf surface for phylloplane micro-organisms.

Phylloplane organisms spend much of the time in a dry and presumably inactive state. Many possess adaptive mechanisms to survive dry periods without damage (Park 1982) as do lichens. It has been found that lichens are much more resistant to the effects of acid gases, in particular sulphur dioxide, when they are dry than when they are wet. There is no direct evidence about the relative sensitivities of phylloplane organisms in the dry and wet states, but it is reasonable to assume that in the cases of water-soluble gases such as sulphur dioxide and ammonia their effects would be greater in the hydrated state. Particulate pollutants, which include the crystals of water-soluble compounds such as ammonium sulphate (Martin & Barber 1978) and sodium chloride can have the following range of effects on neighbouring microbial cells depending on their chemical and physical nature and on the amount of acid or alkaline gases present in solution around the particles:

1. Unreactive particles, such as silica, may offer some shelter to micro-organisms.
2. Surface-active but chemically neutral particles such as soot can catalyze chemical reactions and can act as adsorbents for organic gases.
3. Biological particles (spores and pollens) often leak nutrients on rewetting and after death and can therefore provide substrates for growth.
4. Crystals of water-soluble salts either derived from direct impaction or from the evaporation of solution on the leaf surface can provide mineral nutrition (i.e. ammonium salts, phosphate dust) and may have an osmotic effect. The latter phenomenon is especially true of sodium chloride on leaves near the sea-coast. Some of these compounds may have a buffering effect (e.g. ammonium salts).
5. Industrially derived biological particles such as starch grains will

provide nutrients.

6. Ash and lime particles are alkaline hydroxides of various metals and can neutralize acid solutions on leaves. They are especially prevalent in towns. The metals involved (iron, calcium and magnesium principally) are not very toxic to micro-organisms.
7. Clay particles are mostly derived from rain splash and are most common on leaves near the ground. They possess considerable buffering and waterholding capacity and can also provide a limited amount of mineral nutrients.

The situation with regard to the activity of pollutants in the phyllosphere is therefore rather complex even if one only considers the raw materials. If the possible changes in state (wetting/drying cycles and temperature cycles) are taken into consideration as well, the complexity of the possible chemical and biological reactions increases.

This review will first present the results of synoptic surveys done in three cities and around two isolated industrial sites. Such experimental evidence as there is to support the hypotheses derived from the synoptic surveys will then be considered before the theory of the mechanisms involved is discussed. Lastly the development of the method away from synoptic surveys toward estimation of actual sulphur dioxide levels will be outlined.

## *SYNOPTIC SURVEYS OF LEAF YEAST POPULATIONS*

### *Methods*

Dowding & Carvill (1980) were the first to report a numerical reduction of *Sporobolomyces* spp. which was associated with urban conditions (Fig. 1) in Dublin. They selected *Fraxinus excelsior* as an ubiquitous host with large leaves and a consistent leaf age on short shoots and on the first two pairs of leaves on long shoots of different plants. The sample collection for this work from between 40 and 70 sites in and around Dublin was done on two consecutive mornings to minimize variations in counts caused by changes in weather conditions but yet allow rapid processing of samples for an overnight spore-fall phase of 16 h. At this point the method was limited to synoptic surveys that could be accomplished by 8 to 10 h sampling and could only take place in the summer (June to October inclusive). The inability to use the method in the winter when higher pollution emissions from urban rather than industrial sources could be expected was a great limitation.

The spore-fall method for isolating ballistosporous phylloplane yeasts established by Last (1955) and improved by Dickinson (1971) is a robust technique. A schedule suitable for use by schools is described by Richardson *et al.* (1985). The method is relatively insensitive to environmental variations during incubation and not greatly affected by contamination by other organisms. Discs or uniform lengths of undamaged mature leaves are cut on a clean surface. Between seven and ten of these pieces are then stuck to the inside of a petri dish lid by means of petroleum jelly or cellulose paste (water-based and non-fungicidal type). It is important to be consistent about which surface is left exposed as the density of sporulating yeast colonies on the upper and lower surfaces can be quite different. The lid is then inverted over the base of the dish which contains 10 to 15 ml of cold sterile 0.5 to 2% malt extract agar. The dishes are then incubated in the dark with their lids uppermost for 16 to 24 h. During this time the near-saturated humidity in the closed dish encourages maximal sporulation by the yeast colonies on the leaf surface facing the agar. The ballistospores are shot off and fall to the agar surface. Consistent timing of this first phase of the isolation procedure is rather more critical if periods of less than 24 h are used (Pennycook & Newhook 1974; Johnston 1983). After the first phase the dishes are inverted to prevent further inoculation of the agar by spores, and are incubated until colonies arising from the deposited spores are visible.

The variations in counts caused by differences in leaf age and physiological status, and by weather events prior to sampling are generally greater than those caused by microsite variations on the leaf or by the isolation method itself. Bashi & Fokkema (1976) have demonstrated that under favourable controlled conditions not all colonies of *Sporobolomyces* spp. developed sterigmata and spores, so it is likely that the count obtained by the spore-fall method is only a relative measure of the abundance of cells of *Sporobolomyces* spp. on a leaf surface.

There is another source of variations in counts which is the host species. Generally, counts are higher on unhairy deciduous leaf surfaces than on hairy or evergreen leaves, all other circumstances being equal. The surface of the thick smooth cuticle of many evergreen leaves is a much harsher environment in terms of variations in water status and in temperature than is the surface of the thinner undulating cuticle of many deciduous leaves (Gates & Papian 1971). Subsequent work with common evergreens such as *Hedera helix* confirmed that the highest counts of *Sporobolomyces* spp. on evergreen leaves were so low as to reduce urban/rural differences to non-significant levels (Barnes 1979). Work is currently in progress on the use of grass leaves such as those of *Lolium perenne*, but there are sampling difficulties associated with the use of grasses. They are tedious to identify in their winter state and more importantly all grass leaves grow from their bases and are difficult to age with any confidence in the field.

### *Synoptic surveys of cities*

DUBLIN: Several synoptic surveys of populations of *Sporobolomyces* spp. on leaves of *Fraxinus excelsior* were carried out by P.H. Carvill and the results of one of these were reported by Dowding & Carvill (1980). They demonstrated that there was a significant diminution in colony density from rural to suburban sites and from suburban to urban sites. Fig. 1 shows that the distribution in the suburban zone was not uniform; sites with low counts were located in satellite town centres, while sites with high counts were found along the sea-coast and near recreational areas. The results reported by Dowding & Carvill (1980) were repeated in subsequent surveys within the expected range of sampling error. The distribution of counts followed a lognormal pattern and all statistical calculations were done on transformed data ($\log_{10}$ (count+1)). Both N.J. Fokkema and B. Schippers (this volume) and S.S. Hirano and C.D. Upper (this volume) have reported lognormal patterns for other groups of phylloplane organisms.

CORK: A synoptic survey was carried out by secondary-school children in October 1982 and the results were published by Ni Lambha *et al.* (1983). The distribution of counts of pink mirror yeasts followed a more compli-

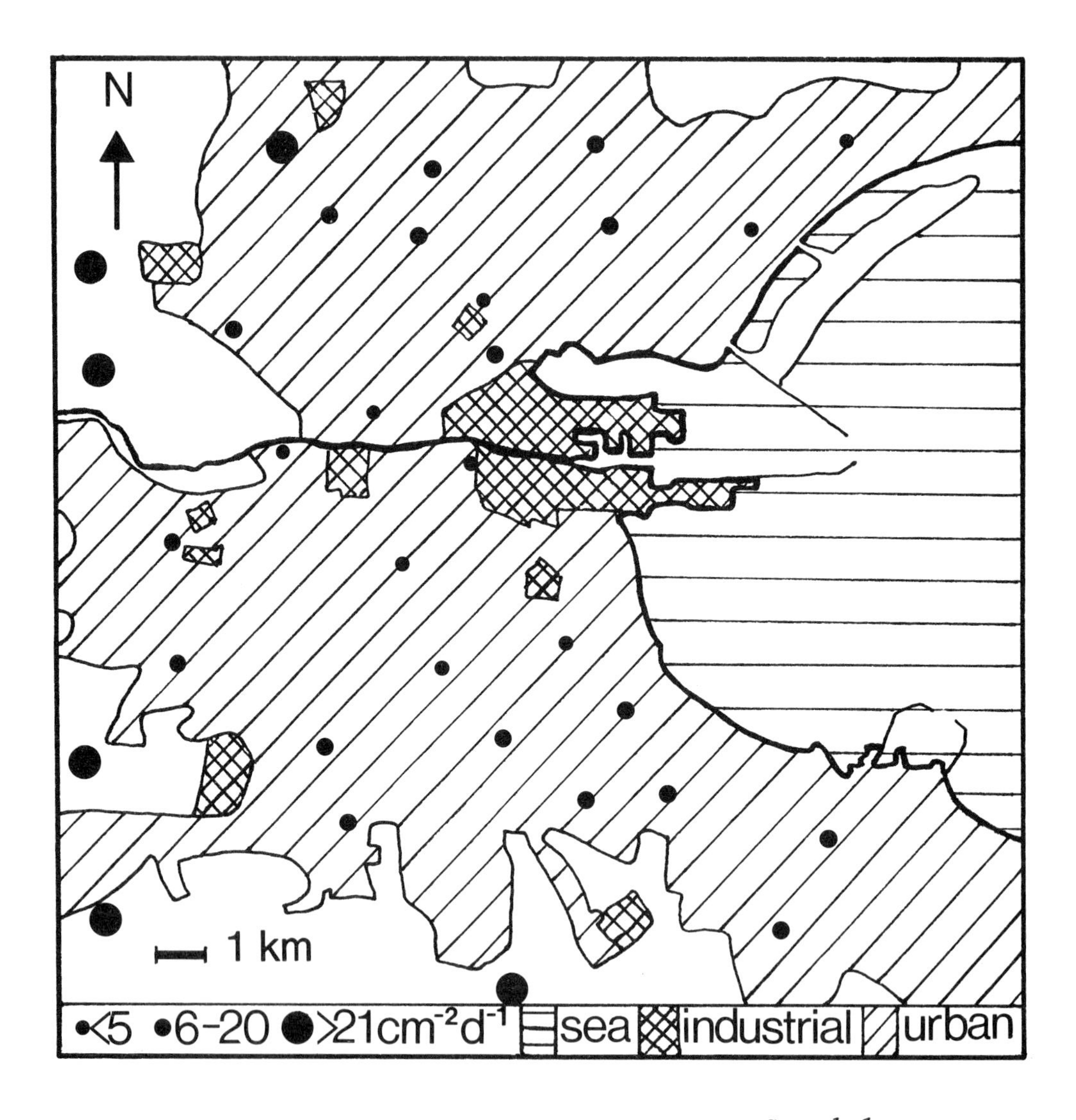


Fig. 1. Average numbers of colonies of *Sporobolomyces* spp. $cm^{-2}$ $day^{-1}$ recorded from discs from ash leaves in Dublin from 1976 to 1978. Redrawn after Dowding & Carvill (1980). Three leaves each yielding ten 0.8 $cm^2$ discs were used at each site on each sampling occasion.

cated pattern than that of the lichens but this was readily explicable by reference to the weather of the previous week and to the principal sites of emissions that week. Areas of low counts were found downwind of the large industrial/power generation complex by the docks and in three heavily urbanized valleys where inversions were frequent.

SHANNON: The results of a synoptic survey carried out by school students on 14 October 1983 were published by Ni Lambha *et al.* (1984). Much of the area is rural/estuarine but there are significant concentrations of heavy industry and a number of towns, the largest of which is Limerick. The lowest counts were found in Limerick associated with the centre and with the refuse disposal centre to the NW of the town. Moderate reductions in counts were found at Shannon industrial estate and in the smaller towns such as Ennis. A similar survey was carried out on 25 and 26 September 1984 over 1450 $km^2$ from Cork to Waterford (Ni Lambha *et al.* 1985) and another one for the counties of Wexford and Wicklow in 1985. The logistics of using school students for surveys of this type are discussed by Richardson *et al.* (1985).

The conclusion from these and other partial studies of urban areas is that there is some inhibitory factor of the urban environment on sporulation by *S. roseus* as estimated by the spore-fall method. This factor could be one or several of the acid gases, toxic particulates or even the lower relative humidity (r.h.) associated with urban microclimates. It is unlikely that a lower inoculum density is the cause as, except in very large polluted regions (over 100 km across), very large numbers of spores are blown into cities each night from rural areas outside the polluted area.

### *Synoptic surveys of industrial sites*

MALLOW: Two synoptic surveys were carried out by P.H. Carvill in 1977 on the flat ground within 3 km of the sugar-beet-processing factory 4 km SW of Mallow, Co. Cork. One took place 2 d before the factory commenced its seasonal operation and the second was carried out 2 d after it had started processing the sugar-beets. The principal emissions from the plant were ascertained to be water vapour and the combustion products of heavy fuel oil containing 3.5% sulphur. Both *Sporobolomyces* spp. and *Tilletiopsis* spp. were enumerated and two measures of abundance were employed: a total colony count, as before, and a simple count of the number of leaf discs out of a maximum of 24 producing any colonies at all. It is obvious that the latter procedure would be much quicker and less tedious than the former in rural areas where the density of sporulating colonies is often so high as to make counting of discrete colonies difficult unless it is done at a very early stage when the colonies are very small. In Fig. 2 isolines surround sites with significantly smaller numbers than the average.

The results for *Tilletiopsis* spp. (Fig. 2 A,B) are clear even though a general reduction of 59% in colony count and of 25% in disc score occurred in the time interval between samplings. There was a much greater reduction in both the total colony count and in the number of leaf discs producing colonies in the area 1 km to the N and E of the factory than in the equivalent area to the S and W of the factory and in the more distant sites. *Tilletiopsis* spp. were not enumerated by the school-children and were only found at the most outlying sites around Dublin in September and October.

As the presence or absence of certain lichen species was already accepted as an integrating (and long-term) measure of sulphur dioxide concentrations, a concurrent lichen survey was carried out. The results of this survey (Fig. 2 C) showed a more even distribution of affected sites immediately around the factory. The results for *Sporobolomyces* spp. (Fig. 3 A-D) showed a small non-significant increase in the disc score overall (8%) with a clear but statistically non-significant pattern of change between samplings (Fig. 3 A). The changes in colony counts over most of the survey area were significant (Fig. 3 D). There were large decreases close to and NE and at one site less than 1 km S of the factory. Five sites at intermediate distances showed a significant increase in counts with the most unaffected sites showing a slight decrease.

The greater sensitivity and smaller numbers of *Tilletiopsis* spp. make them attractive candidates for surveys of this type, but their greater sensitivity to urban influences and their more restricted season compared to *Sporobolomyces* spp. rather restrict their use. The results of the Mallow survey revealed that the reduction in sporulation density occurs very soon after exposure to combustion products. Lichen communities take a long time, possibly 10 years, to respond, so are unable as in this case to show that the pollution is seasonal.

AVONMOUTH: Bewley & Campbell (1980) reported the results of two synoptic surveys of the phylloplane microflora of hawthorn (*Crataegus monogyna* and *C. oxyacantha*) at 20 sites to the NNE of a heavy metal smelter at Avonmouth, U.K. *S. roseus* was isolated by the spore-fall method from the upper surface of five leaves from each site. Counts of *S. roseus* were related to distance from the smelter and fell to zero within 2 km of the plant. However, more than 70% of the variation in counts was statistically unrelated to variation in heavy metal concentrations on leaves from the same sites, and Bewley & Campbell (1980) speculated that this could be due to a non-linear relationship between metal concentration and *S. roseus* counts or to the interference by other pollutants such as sulphur dioxide. Finally, the authors concluded that within-site variation in counts of *S. roseus* was so great as to render it useless as an indicator of specific heavy metal pollution, though they thought that it might be of use in studies of more general environmental pollution. It was not clear how the authors expressed the counts of *S. roseus* as there was no indication that they measured the area of the leaves from which *S. roseus* was isolated. Some of the variation in counts that they reported could have been due to variation in leaf size and shape. The number of replicate leaves or leaflets used by Dowding & Carvill (1980) and by Ni Lambha *et al.* (1983, 1984, 1985) reported above were much greater than the five whole leaves used by Bewley & Campbell (1980). Up to 10 leaflets from each of three leaves and one standard 1 cm diam (0.8 $cm^2$) disc was cut from each leaflet.

*MECHANISMS OF REDUCTION OF YEAST POPULATIONS BY AIR POLLUTANTS*

*Experimental findings*

Bashi & Fokkema (1977) reported that under controlled conditions populations of *S. roseus* inoculated onto wheat leaves increased for 3 to 11 days at 95% r.h. Vegetative growth rates were observed to increase with the interpolation of an 8 h period of dew in each 24 h. At

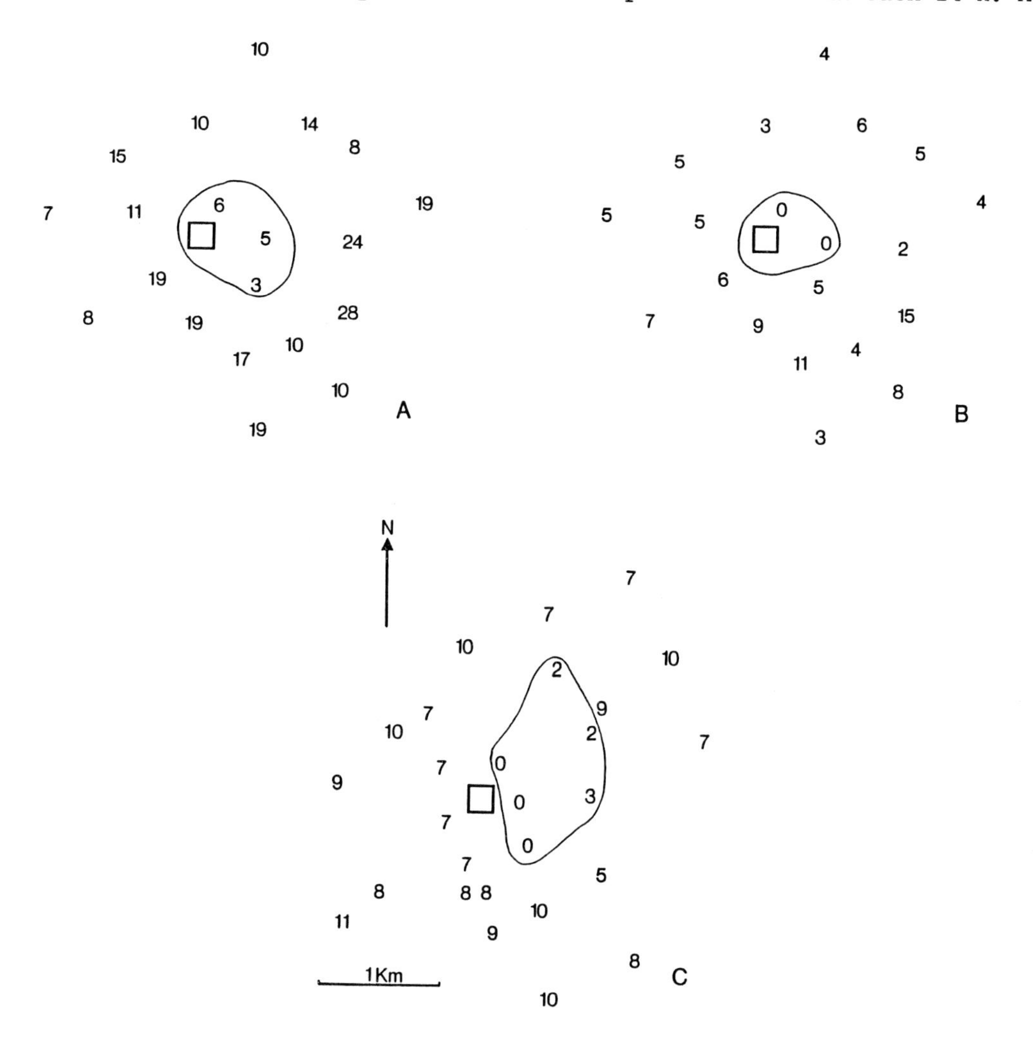


Fig. 2. Occurrence of *Tilletiopsis* spp. and of lichen species on ash trees around a sugar-beet-processing factory (open square) near Mallow, Co. Cork. A, B. The mean numbers of colonies of *Tilletiopsis* spp. $cm^{-2}$ leaf $d^{-1}$ recorded from ash leaves before (A) and after (B) the start of processing. C. The mean number of lichen species per trunk. Scale, symbols and orientation in A, B and C are the same. The isolines enclose counts that are significantly lower than the mean count for the area.

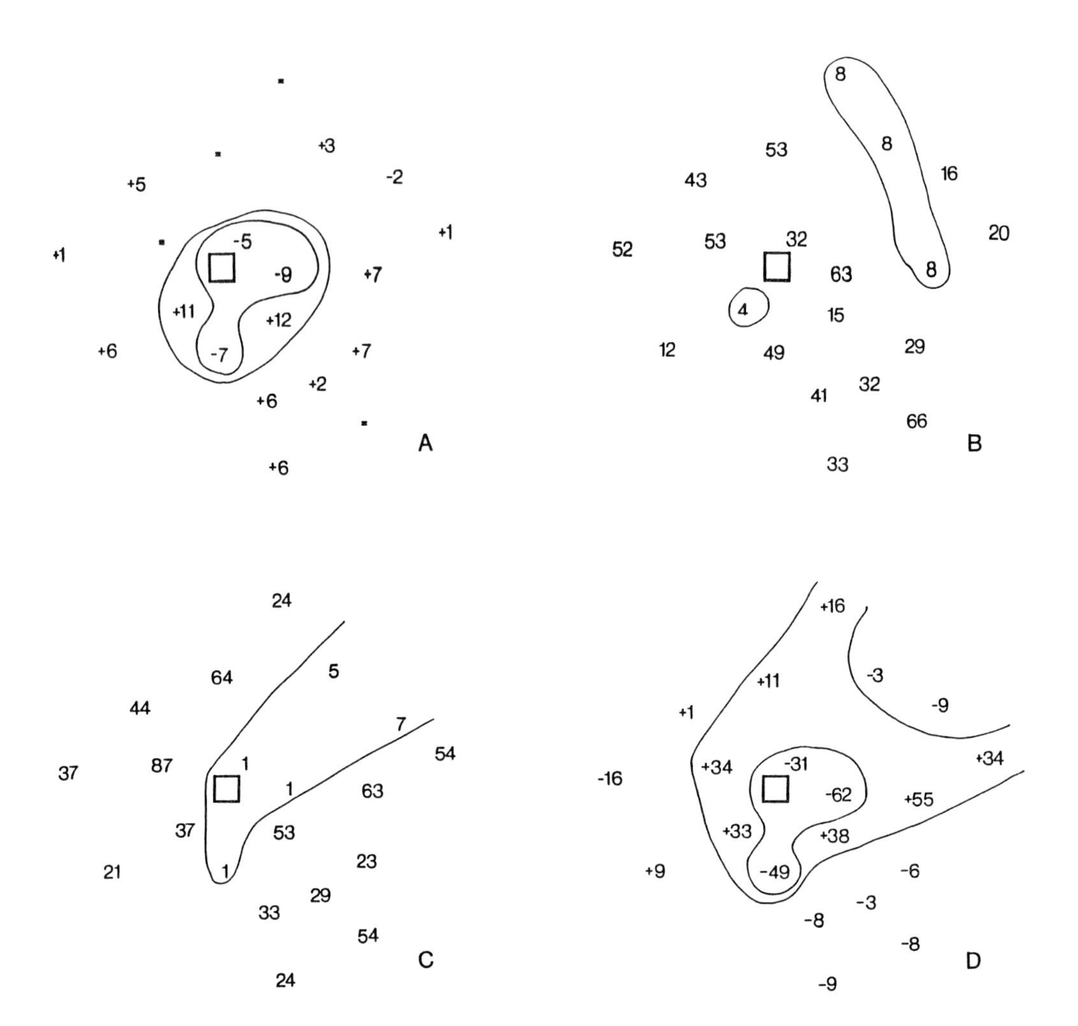


Fig. 3. Occurrence of *Sporobolomyces* spp. around a sugar-beet-processing factory near Mallow, Co. Cork. Scale, symbols and orientation as in Fig. 2. The isolines in B and C enclose counts that are significantly lower than the mean for the area. The inner isoline in A and D encloses significant diminutions in counts and the outer isoline encloses significant increase in counts between the two sampling periods. A. The differences between numbers of discs (out of 24) producing colonies of *Sporobolomyces* spp. $cm^{-2}$ leaf $d^{-1}$ from ash leaves sampled 2 d before and 2 d after the start of processing. B, C. The mean numbers of colonies of *Sporobolomyces* spp. $cm^{-2}$ leaf $d^{-1}$ arising from discs of ash leaves sampled 2 d before (B) and 2 d after (C) the start of processing. D. The differences in mean numbers of colonies of *Sporobolomyces* spp. $cm^{-2}$ leaf $d^{-1}$ arising from discs of ash leaves sampled 2 d before and 2 d after the start of processing.

lower humidities, without dew, the populations either did not change significantly (75% r.h.) or decreased by 10% $d^{-1}$ (65% r.h.). Mishra & Dickinson (1984) concluded that on holly leaves there was little evidence for active vegetative growth at 95% r.h. and below. Flannigan & Campbell (1977) observed a maximum count of 150,000 *S. roseus* cells $cm^{-2}$ of wheat flag leaves after a dry spell in August; however, dew may have occurred. Importantly, it does not seem that *S. roseus* requires free water to grow on the leaf surface, though, as Burrage (1971) and Bashi & Fokkema (1977) pointed out, the conditions at the leaf surface are likely to be more humid than in the air surrounding the sensors responsible for giving the r.h. measurements. That supposition will only be true so long as the leaf temperature is the same as or lower than that of the air surrounding it (Gates & Papian 1971). In both experiments reported the leaf surface temperatures would have tended to be higher than the air surrounding them as they were under artificial illumination and were shielded from cooler surfaces to which they could radiate. In any case Garland & Branson (1977) estimated that in temperate latitudes in summer, leaf surfaces are wet for about 30% of the time. It is possible that leaf surfaces in towns are drier than those in rural areas, but any difference is within the range of humidities tolerated by *Sporobolomyces* spp., and moreover the results from the three rural synoptic surveys show that something other than wetness or humidity is exerting a very strong effect on populations of *Sporobolomyces* spp. Dew certainly occurs in central Dublin but conventional meteorological measurements do not include dew duration or allow its calculation. Fluctuations in counts caused by natural phenomena have to be taken into account before pollutants can be identified as causative agents in observed diminutions of counts.

P.H. Carvill (personal communication) conducted a field experiment in which he burnt 1.35 kg elemental sulphur over 12 h (1400 to 0200 BST) towards the west edge of a field surrounded by a tall hedge with many ash trees. He sampled leaves from 15 trees around the field at 0800, before and after the sulphur was burnt, and was able to demonstrate a significant fall in counts of *S. roseus* on the leaves from three trees NE of (downwind) and three trees very close to and upwind of the source only 6 h after the end of combustion (Fig. 4). He was unable to measure the peak or average concentrations of sulphur dioxide at the time of exposure but concluded that the burning rate was not excessive for a point source of sulphur dioxide, being only equivalent to a rate of 38.5 kg (48 l) of heavy heating oil in 12 h (3.5% sulphur being permitted for all industrial installations in Ireland at that time).

Curran (1984) repeated the experiment with 250 g sulphur and attempted simultaneous measurement of the absorption of sulphur dioxide by exposed hydrogen peroxide solutions along the line of *F. excelsior* saplings from which the leaf samples were taken. On another occasion a nearby hawthorn hedge was exposed to smoke from burning plant material for *ca* 2 h. Curran (1984) observed a small but significant decrease in counts of *Sporobolomyces* spp. 1 and 4 days after exposure (Table 1) to the smoke of burning sulphur but the exposed solution had picked up very little sulphur dioxide, possibly because the open dishes were placed on the ground for the exposure period. There was no decrease observed after the subjectively much more intense exposure to 'natural' smoke.

The evidence to date from field experiments is suggestive that exposure to sulphur dioxide rapidly reduces the sporulating ability of populations of *Sporobolomyces* spp. on deciduous tree leaves but does not exclude the possibility of other air pollutants having similar and possibly synergistic effects.

*Possible modes of action of sulphur dioxide*

The oxidation product of sulphur dioxide, sulphuric acid, may be the most important synergist of sulphur dioxide in polluted situations. The acidity of the solution of sulphur dioxide determines the proportions of the ions of sulphurous acid (Puckett *et al.* 1973).

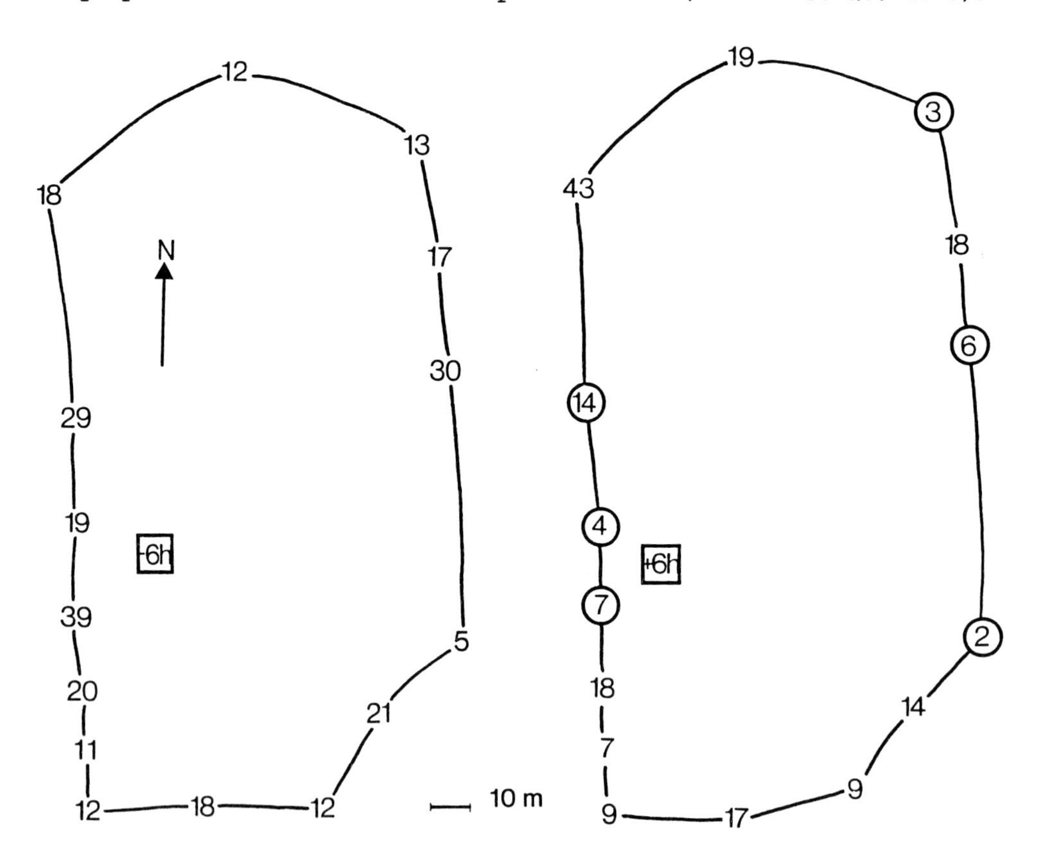


Fig. 4. The mean numbers of colonies of *Sporobolomyces* spp. $cm^{-2}$ leaf $d^{-1}$ arising from discs of ash leaves sampled 6 h before and 6 h after a 12-h period in which 1.35 kg elemental sulphur was completely combusted in the open at the spot marked with an open square. Numbers which are circled show a statistically significant (P<0.001) decrease from the numbers recorded at the same site earlier.

Although the most toxic compound of sulphur dioxide with water is believed to be undissociated sulphurous acid ($H_2SO_3$) (Saunders 1966), it comprises less than 1% of the dissolved sulphur dioxide above pH 4.0, and forms a slightly increasing proportion below pH 4.0. Garsed (1983) stated that the resistance to solution of sulphur dioxide into water was minimal above pH 4. As more sulphur dioxide is taken up and particularly if its oxidation rate in the solution is accelerated by an initially high pH (Baddeley *et al.* 1971), catalytic cations (Dlugi *et al.* 1982) or soot particles and/or higher than normal ozone concentrations (Martin & Barber 1985), the resistance to sulphur dioxide entering the solution will increase. Theoretically it will become infinite when the equilibrium constant between gaseous and dissolved undissociated sulphur dioxide has been attained. This is unlikely in nature as water films are never pure water. Studies that have been done on the chemistry of water on surfaces and in droplets have revealed a complicated chemistry in which neutralization of sulphuric acid by bases is as important in towns (Waller 1963) as in less polluted areas. Brimblecombe (1978) observed that dew was able to dissolve more sulphur dioxide than did distilled water because of dissolved cations in water films on leaf surfaces. Leaf surfaces and particles on them are reactive and the films and droplets are evanescent, so that the concentration of any soluble gas depends as much on the relative humidity of the air as on the concentration of the gas in the air surrounding the droplet or surface water film. Bewley & Campbell (1978) observed that micro-organisms were more abundant on large amorphous particles on leaves polluted with heavy metals than on the leaf surface itself. Their results showed that most of the heavy metal contamination consisted of particles less than 1 μm in diameter, and concluded that the large particles must have been providing additional nutrients. It is much more likely, however, that the large particles were basic and were neutralizing the acid and/or precipitating the metal ions in solution because of the low pH associated with emissions from the Avonmouth smelter. Commins (1963) found appreciable amounts of insoluble bases in London air. Martin & Barber (1978) reported that

Table 1. Mean counts (numbers $cm^{-2}$ $d^{-1}$) of *Sporobolomyces* spp. from discs of ash leaves before and after exposure to smoke from burning sulphur for 1 hour. After Curran (1984).

| Treatment | Time after exposure | | | |
|---|---|---|---|---|
| | 0 h | 1 h | 4 d | 1 d |
| Control | 2.8 *[1] | 5.2 * | 2.4 | 3.2 * |
| Exposed | 3.2 * | 2.7 * | 0 | 1.0 |

[1] Values with an asterisk are significantly greater than zero at $P<0.05$.

calcium sulphate forms a substantial part of deposited sulphate in the U.K. They speculated that this might have been derived directly from soil, but it is more likely that it derived from the reaction in the air of sulphurous and sulphuric acid in droplets with particles of lime (mixtures of calcium carbonate and calcium hydroxide) which became airborne as a result of both agricultural operations and building construction. Clay dust would also be neutralizing, but the cations would be a mixture of aluminium, iron, potassium and ammonium principally.

In rural areas, ammonia gas derived from biodegradation of high nitrogen wastes such as slurry from intensive cattle and pig farms would also neutralize some of the sulphate. Martin & Barber (1985) found that most of the ammonium in rural U.K. was in the small ($<2.5$ µm) particle fraction but that the amount of ammonium ions present in air was not sufficient to neutralize all the sulphate ions present.

Leaf surfaces themselves can affect the chemical properties of the water on them. Rowlatt *et al.* (1978) found that the pH of acidic rain was increased by two units after passage over senescent tree leaves in autumn. In polluted situations increased exudation of buffering capacity onto the senescing leaf surface may well permit populations of pollution-sensitive phylloplane organisms, such as *Sporobolomyces* spp., to increase for a short time. Huttunen & Soikkeli (1983) using scanning electron microscopy found that in polluted atmospheres minute roughnesses on the surface of leaf cuticles of conifers, *Rosa rugosa* and *Vaccinium* spp. were lost and that small epidermal folds in the surface of the leaves of *Aesculus hippocastanum* had disappeared. It is possible that such changes in microhabitat make the urban leaf a less suitable place for phylloplane organisms, quite apart from the sensitivity of the phylloplane community to any particular pollutant.

The absorption of sulphur dioxide by leaf surfaces is greatly affected by the wetness of the leaf surface as well as by the opening of the stomata. Fowler & Unsworth (1979) reported that if leaves were covered in dew cuticular resistance to uptake of sulphur dioxide was reduced to one-tenth of that of dry leaves. Fowler (1981) found that the canopy resistance of a senescent wheat crop to sulphur dioxide fell to zero when the leaves were wet. He estimated that in a wheat crop absorption of sulphur dioxide by the various surfaces was as follows: by wet cuticle 5 mm $s^{-1}$, through open stomata 5 mm $s^{-1}$, through dry live surface 2 mm $s^{-1}$ and into dry dead leaves 3 mm $s^{-1}$. Taylor & Tingey (1983) estimated that the rate of sulphur dioxide uptake by the surface of mesophyll cells was 3 µM $m^{-2}$ $h^{-1}$. If the figures of Taylor & Tingey (1983) are combined with those of Fowler (1981) and an external/internal leaf surface ratio of 1/16 is accepted, then an uptake rate of 0.05 mM $SO_2$ $m^{-2}$ $h^{-1}$ into the surface water film appears to be possible. If the water film evaporates the dissolved sulphur dioxide in it in whatever form will be concentrated. As so many possible reactions might occur to the dissolved sulphur dioxide it is probably not profitable to speculate on the concentrations of sulphurous acid that might be attained in the absence of impurities and of oxidation. However, in order-of-magnitude terms and in the absence of further reactions, a surface water film 20 μm thick on average will come to equilibrium with a sulphur dioxide concentration in air of 100 μg $m^{-3}$ and a solution concentration of *ca* 35 mg sulphurous acid $l^{-1}$ (0.5 mM) in 1 hour. Saunders (1966) found by a combination of laboratory and field experimentation that these were threshold concentrations for toxicity to the infection of rose leaves by the spores of *Diplocarpon rosae*. He speculated that yeasts (as a class, rather than phyllosphere yeasts in particular) were more sensitive than *D. rosae* on the basis of work by Ingram (1948) who had found that fruit-spoilage yeasts were inhibited by sulphur dioxide concentrations of 10 mg $l^{-1}$ in pure water (equivalent to 30 μg $m^{-3}$ in air). By contrast, *Botrytis cinerea* (Couey & Uota 1961) and *Alternaria* sp. (Couey 1965) conidia were inhibited from germination when exposed to sulphur dioxide in air at concentrations above 2800 μg $m^{-3}$, nearly two orders of magnitude greater than the concentration toxic for yeasts. Both *B. cinerea* and *Alternaria* sp. were more sensitive to sulphur dioxide at 100% r.h. than at lower humidities.

Indirect evidence therefore allows one to conclude that populations of *Sporobolomyces* spp. might well be inhibited from growth at environmentally realistic levels of sulphur dioxide in air. Various experiments have been attempted in the author's laboratory with populations of *Sporobolomyces* spp. in pure liquid culture. P.H. Carvill (personal communication) found that there was a stimulation of cell division in liquid culture and on the surface of semi-solid nutrient medium at low concentrations of sulphur dioxide (less than 10 mg $l^{-1}$ in liquid) and progressive inhibition at higher concentrations. Similar results were found by Curran (1984) (Table 2), but the sensitivity of the cells of *Sporobolomyces* spp. was also a function of the concentration of malt extract in the medium, which suggested that there could have been some reaction between the malt and the sulphur dioxide. Curran's experiments were also carried out at pH 6, which was the optimum for vegetative growth, but which is not a good pH for the formation of undissociated sulphurous acid.

### *EVALUATION OF THE SPORE-FALL METHOD*

The use of the spore-fall method for synoptic surveys by secondary school students is growing in popularity. It has been used for a synoptic survey of Aberdeen and is undergoing trials in 12 other European cities in 1985 (Richardson *et al.* 1985). The short response time of sporulation of *Sporobolomyces* spp. to sulphur dioxide pollution frequently gives a different pattern to that given by mapping the presence of differentially sensitive lichen species which are slow to succumb to and even slower to recover from chronic air pollution. Provided that good instructions are given and that the sampling is synchronized over the whole area surveyed the results have been consistent. Biology teachers have welcomed the introduction of a practical microbiological technique that neither demands a complicated and expensive culture medium nor is prone to contamination. There are no problems for identification at the level that is required for a synoptic pollution survey.

The use of the method as an estimator of past concentration of sulphur dioxide in the air at a particular locality is made difficult by several factors:

1. The seasonality of colony density and of sporulation potential, and the sensitivity of natural populations to relatively short-term (several days to several weeks) changes in the duration of leaf wetness and in temperature. This is particularly true of rainfall frequency and amount. Even if host and leaf age are standardized the background unpolluted count level will be set by the stage in the season and by the wetness of the preceding days. It is possible that the use of a biological simulation model would allow fairly good estimates of background count levels once a body of count and weather data had been built up for an area. The values of rate parameters used in such a model could not be transferred without possible large error for use in other areas. It would not be worth the effort if the following problems cannot be resolved at the same time.

2. The response of *Sporobolomyces* spp. to different concentrations of sulphur dioxide is dose-dependent (i.e. there is a time function as well as a concentration function) and is non-linear at least in the vegetative phase on agar and in liquid culture and possibly in the sporulating phase. The augmentation of cell duplication and possibly of sporulation at low concentrations of sulphur dioxide is parallelled by Saunders' (1966) observations on *D. rosae* and makes the estimation of an average concentration of sulphur dioxide in the preceding time interval very difficult if the concentrations are subject to variation, as is the rule in the outside environment. Without further experiments it would be impossible to say if the augmentation effect occurs in the cell growth and pre-sporulation phases and at what level of sulphur dioxide it will occur. Direct observation of changes in spore-fall counts after the exposure of plants to controlled combinations of different sulphur dioxide concentrations, humidities, temperatures and leaf wetness durations would give an empirical answer to this problem.

Table 2. Maximum cell densities (numbers $ml^{-1}$) of *Sporobolomyces* spp. in 50-ml malt extract solutions with added sulphur dioxide at pH 6. Results are the mean of three replicates. After Curran (1984).

| Concentration of malt extract ($g\ l^{-1}$) | Sulphur dioxide concentration ($mg\ l^{-1}$) | | | | | |
|---|---|---|---|---|---|---|
| | 0 | 5 | 10 | 20 | 40 | 60 |
| 2 | 200 | 64 | 46 | 82 | 79 | 40 |
| 6 | 690 | 400 | 330 | 325 | 310 | 290 |
| 10 | 680 | 1280 | 450 | 560 | 350 | 240 |

3. The extent of synergism and of antagonism from other airborne materials and from the substratum itself has not been investigated experimentally. If *Sporobolomyces* spp. are to be used for anything more than an indicator of general pollution levels (Bewley & Campbell 1980), these relationships will have to be addressed, possibly at the same time as empirical trials.

## *REFERENCES*


Baddeley, M.S., Ferry, B.W. & Finegan, E.J. (1971). A new method of measuring lichen respiration: response of selected species to temperature, pH and sulphur dioxide. Lichenologist, *5*, 18-25.

Barnes, F. (1979. Laboratory exposures of *Sporobolomyces roseus* and *S. salmonicolor* to sulphur dioxide, and a study of winter populations of *Sporobolomyces* spp. on ivy leaves. B.A. (Mod.) thesis, University of Dublin.

Bashi, E. & Fokkema, N.J. (1976). Scanning electron microscopy of *Sporobolomyces roseus* on wheat leaves. Transactions of the British Mycological Society, *67*, 500-5.

Bashi, E. & Fokkema, N.J. (1977). Environmental factors limiting growth of *Sporobolomyces roseus*, an antagonist of *Cochliobolus sativus*, on wheat leaves. Transactions of the British Mycological Society, *68*, 17-25.

Bewley, R.J.F. & Campbell, R. (1978). Scanning electron microscopy of oak leaves contaminated with heavy metals. Transactions of the British Mycological Society, *71*, 508-11.

Bewley, R.J.F. & Campbell, R. (1980). Influence of zinc, lead and cadmium on the microflora of hawthorn leaves. Microbial Ecology, *6*, 227-40.

Brimblecombe, R. (1978). 'Dew' as a sink for sulphur dioxide. Tellus, *30*, 151-8.

Burrage, S.W. (1971). The micro-climate at the leaf surface. *In* Ecology of Leaf Surface Micro-organisms, eds T.F. Preece & C.H. Dickinson, pp. 91-101. London: Academic Press.

Commins, B.T. (1963). Determination of particulate acid in town air. Analyst, *88*, 364-7.

Couey, H.M. (1965). Inhibition of germination of *Alternaria* spores by sulfur dioxide under various moisture conditions. Phytopathology, *55*, 525-7.

Couey, H.M. & Uota, M. (1961). Effect of concentration, exposure time, temperature, and relative humidity on the toxicity of sulfur dioxide to the spores of *Botrytis cinerea*. Phytopathology, *51*, 815-9.

Curran, E.B. (1984). The effects of sulphur dioxide on the reproduction and growth of *Sporobolomyces roseus*. B.A. (Mod.) thesis, University of Dublin.

Dickinson, C.H. (1971). Cultural studies of leaf saprophytes. *In* Ecology of Leaf Surface Micro-organisms, eds T.F. Preece & C.H. Dickinson, pp. 129-37. London: Academic Press.

Dlugi, R., Jordan, S. & Lindemann, E. (1982). Influence of particle properties on heterogeneous $SO_2$ reactions. *In* Physico-chemical Behaviour of Atmospheric Pollutants, eds B. Versino & H. Ott, pp. 308-18. Dordrecht: D. Reidel.

Dowding, P. & Carvill, P.H. (1980). A reduction of counts of *Sporobolomyces roseus* Kluyver on ash (*Fraxinus excelsior* L.) leaves in Dublin city. Irish Journal of Environmental Science, *1*, 65-8.

Flannigan, B. & Campbell, I. (1977). Pre-harvest mould and yeast floras on the flag-leaf, bracts and caryopsis of wheat. Transactions of the British Mycological Society, *69*, 485-94.

Fowler, D. (1981). Turbulent transfer of sulphur dioxide to cereals; a case study. *In* Plants and their Atmospheric Environment, eds J. Grace, E.B. Ford & P.G. Jarvis, pp. 139-46. Oxford: Blackwell.

Fowler, D. & Unsworth, M.H. (1979). Turbulent transfer of sulphur dioxide to a wheat crop. Quarterly Journal of the Royal Meteorological Society,

*105*, 767-84.
Garland, J.A. & Branson, J.R. (1977). The deposition of sulphur dioxide to pine forest assessed by radioactive tracer method. Tellus, *29*, 445-54.
Garsed, R. (1983). Uptake and distribution of pollutants in the plant and residence time of the active species. *In* Gaseous Air Pollutants and Plant Metabolism, eds M.J. Koziol & F.R. Whatley, pp. 83-9. London: Butterworths.
Gates, D.M. & Papian, L.E. (1971). Atlas of Energy Budgets of Plant Leaves. London: Academic Press.
Huttunen, M. & Soikkeli, M. (1983). The effects of various gaseous pollutants on plant cell ultrastructure. *In* Gaseous Air Pollutants and Plant Metabolism, eds M.J. Koziol & F.R. Whatley, pp. 117-27. London: Butterworths.
Ingram, M. (1948). The germicidal effects of free and combined sulphur dioxide. Journal of the Society of Chemical Industry, *67*, 18-21.
Johnston, B. (1983). An investigation into some of the external factors affecting phylloplane populations of *Sporobolomyces roseus*, a pigmented leaf yeast. B.A. (Mod.) thesis, University of Dublin.
Last, F.T. (1955). Seasonal incidence of *Sporobolomyces* on cereal leaves. Transactions of the British Mycological Society, *38*, 221-39.
Martin, A. & Barber, F.R. (1978). Some observations of acidity and sulphur in rainwater from rural sites in Central England and Wales. Atmospheric Environment, *12*, 1481-7.
Martin, A. & Barber, F.R. (1985). Particulate sulphate and ozone in rural air: preliminary results from three sites in central England. Atmospheric Environment, *19*, 1091-102.
Mishra, R.R. & Dickinson, C.H. (1984). Experimental studies of phylloplane and litter fungi on *Ilex aquifolium*. Transactions of the British Mycological Society, *82*, 595-604.
Ni Lambha, E., Richardson, D.H.S., Dowding, P. & Kiang, S. (1985). An Air Quality Survey of County Waterford Carried out by Schoolchildren. Dublin: An Foras Forbartha.
Ni Lambha, E., Richardson, D.H.S., Dowding, P. & O'Sullivan, A. (1984). An Air Quality Survey of Limerick and of the Shannon Region Carried out by Schoolchildren. Dublin: An Foras Forbartha.
Ni Lambha, E., Richardson, D.H.S., Dowding, P. & Wells, J.M. (1983). An Air Quality Survey of Cork City and of Great Island Carried out by Schoolchildren. Dublin: An Foras Forbartha.
Park, D. (1982). Phylloplane fungi: tolerance of hyphal tips to drying. Transactions of the British Mycological Society, *79*, 174-8.
Pennycook, S.R. & Newhook, F.J. (1974). Diel periodicity and circadian rhythm of ballistospore discharge in the Sporobolomycetaceae. Transactions of the British Mycological Society, *63*, 237-48.
Puckett, K.J., Nieboer, E., Flora, W.P. & Richardson, D.H.S. (1973). Sulphur dioxide: its effect on $^{14}C$ fixation in lichens and suggested mechanisms of toxicity. New Phytologist, *72*, 141-54.
Richardson, D.H.S., Dowding, P. & Ni Lambha, E. (1985). Monitoring air quality with leaf yeasts. Journal of Biological Education, *19*, 289-303.
Rowlatt, S., Crawford, D.B. & Unsworth, M.H. (1978). Sulphur cycle in wheat and other farm crops. *In* Sulphur in Forages, ed. J.C. Brogan, pp. 1-14. Dublin: An Foras Taluntais.
Saunders, P.J.W. (1966). The toxicity of sulphur dioxide to *Diplocarpon rosae* Wolf causing blackspot of roses. Annals of Applied Biology, *58*, 103-14.
Taylor, G.E., Jr. & Tingey, D.T. (1983). Sulfur dioxide flux into leaves of *Geranium carolinianum* L. Evidence for a nonstomatal or residual resistance. Plant Physiology, *72*, 237-44.
Waller, R.E. (1963). Acid droplets in town air. International Journal of Air and Water Pollution, *7*, 773-8.

# PHYLLOSPHERE VERSUS RHIZOSPHERE AS ENVIRONMENTS FOR SAPROPHYTIC COLONIZATION

N.J. Fokkema and B. Schippers

*Willie Commelin Scholten Phytopathological Laboratory (Department of Plant Pathology: State University Utrecht; University of Amsterdam; Free University, Amsterdam), Javalaan 20, 3742 CP Baarn, The Netherlands*

## *INTRODUCTION*

Surfaces of plants are attractive for micro-organisms as well as for microbiologists. Microbial ecologists are mostly studying the ecology of a particular component of the microflora on a particular part of the plant, with emphasis on bacteria or fungi but seldom on both. The environments of the surfaces of various plant parts are defined by the suffixes '-sphere' or '-plane' which results in terms such as rhizosphere (Hiltner 1904), rhizoplane (Clark 1949), phyllosphere (Last 1955; Ruinen 1956), spermosphere (Verona 1958), phylloplane (Last & Deighton (1965) wrongly attributed this term to Kerling (1958)), gemmisphere (Leben 1971), cauliplane (Sivak & Person 1973) and palynosphere (Diem 1973). Considering the environmental differences it may be advantageous to distinguish between rhizosphere and phyllosphere only, as the environments of subterranean and aerial plant surfaces, respectively. At first sight the differences between these two environments seem enormous and scientists are generally devoted to studying either rhizosphere or phyllosphere phenomena but not both.

The enormous research input in rhizosphere microbiology makes it worth considering to what extent these environments are really different as far as microbial colonization of these habitats is concerned. Phyllosphere microbiologists have adopted sampling techniques and concepts from rhizosphere studies, but although the number of phyllosphere microbiologists is relatively small, techniques and thoughts can be exchanged in both directions. Because of the great breadth of subject matter encompassed by the topic, we have selected the following aspects of phyllosphere and rhizosphere colonization for detailed comparison. First, we will compare microbial colonization of the phyllosphere and rhizosphere in relation to either substrate. Second, the spatial variability of micro-organisms in the rhizosphere and phyllosphere, especially the differences between yeasts and bacteria, will be discussed. As an example, the discussion of microbial activities occurring both in phyllosphere and rhizosphere will focus on siderophore production and $N_2$ fixation by free-living bacteria.

## *PLANT SURFACE COLONIZATION: TEMPORAL EFFECTS*

### *Phyllosphere*

In the phyllosphere, bacteria and yeasts develop exponentially on newly unfolded leaves till a steady state is reached; sometimes the population density declines at the end of the season (e.g. Hislop & Cox 1969; Fokkema 1971; Flannigan & Campbell 1977; Pennycook & Newhook 1981; Rabbinge *et al.* 1984). Buds may have a considerable population of micro-organisms, which partly maintains itself on the developing leaves (Leben 1971; Warren 1976; Andrews & Kenerley 1980). Techniques used for estimating population densities are usually based on culturing washing fluids, and densities are preferably expressed as numbers of colony-forming units (CFU) per unit area (Parbery *et al.* 1981). Graphs such as Fig. 1, showing the seasonal development of total bacteria, *Pseudomonas* spp. and yeasts on wheat flag leaves, are well known. In that particular season, there was little difference between the numbers of yeasts and bacteria. Such a similarity in colonization patterns has also been noticed in the phyllosphere of e.g. apple (Andrews & Kenerley 1978; Pennycook & Newhook 1981), but it is not necessarily a strict rule. Filamentous fungi, such as *Cladosporium* spp., seldom amount to more than 10% of the yeast (including *Aureobasidium pullulans*) counts (e.g. Rabbinge *et al.* 1984). Although it can be argued that the plating technique is less suitable for filamentous fungi than for unicellular yeasts, microscopical observations of leaves of *Acer platanoides* (Breeze & Dix 1981) revealed that the total biomass of yeasts in July (summer) was often 50 times greater than that of hyphal fungi. On leaves, *Pseudomonas* spp. comprise rarely more than 5 to 10% of the total numbers of bacteria (Fig. 1; Andrews & Kenerley 1978).

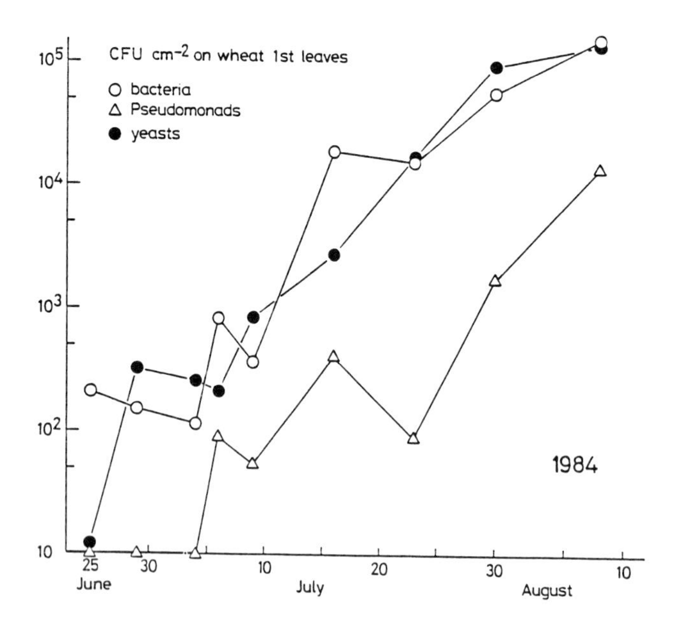


Fig. 1. Seasonal development of total bacteria, *Pseudomonas* spp. and yeasts on field-grown wheat flag leaves. Mean numbers of CFU $cm^{-2}$ leaf were estimated from culturing leaf washings (Fokkema *et al.* 1979) of eight individual leaves (G.J. van Schijndel, B. Schippers and N.J. Fokkema, unpublished results).

Leaf surface colonization observed as a variable of time, involves the following factors: trapping time or duration of exposure, 'incubation' time of the microflora, changing substrates as the leaf ages, exposure to incidental deposits of pollen, honeydew and dust, and changing weather conditions. The loss of nutrients (Tukey 1970; Frossard 1981) from the leaf to the phyllosphere seems quantitatively limited (Blakeman 1973; Fokkema & Lorbeer 1974). Only about 0.7% of the total amount of $^{14}C$-labelled photosynthates in leaf tissue was found in leaf leachates (Frossard *et al.* 1983). Death of leaves is, in contrast to that of roots, generally a quick process, it involves all tissues, and the nutrients released benefit mainly to the litter microflora. Compared to roots, however, leaves as a substrate show little dramatic changes until the start of the senescence.

### *Rhizosphere*

The root surface is a rapidly changing substrate, particularly during the first 10 days after the emergence of the root from the seed (Van Vuurde 1978; Fig. 2). In the rhizosphere there is a tremendous loss of organic compounds (Foster & Bowen 1982) in the form of exudates, secretions, plant mucilages and lysates from dying epidermal and cortex cells. Haller & Stolp (1985) estimated that 25% of the organic matter of maize roots is lost to the environment; for field-grown wheat a seasonal loss of 1000 kg $ha^{-1}$ has been calculated (Martin & Puckridge 1982).

Van Vuurde & Schippers (1980) investigated the microbial colonization of daily formed root parts (day segments) of individual wheat roots grown in soil. Microbial development (Fig. 3) is subject to an increasing colonization period as well as to changing nutrient supply (Fig. 2). For each individual root the same, though not always synchronous, trends in colonization can be recognized. Day segment 1 to 2, comprising the root tip, is almost uncolonized; on day segment 3 to 5 there is a sharp increase in bacterial density which could be correlated with the emergence of lateral roots and increasing lysis of epidermal and cortex cells.

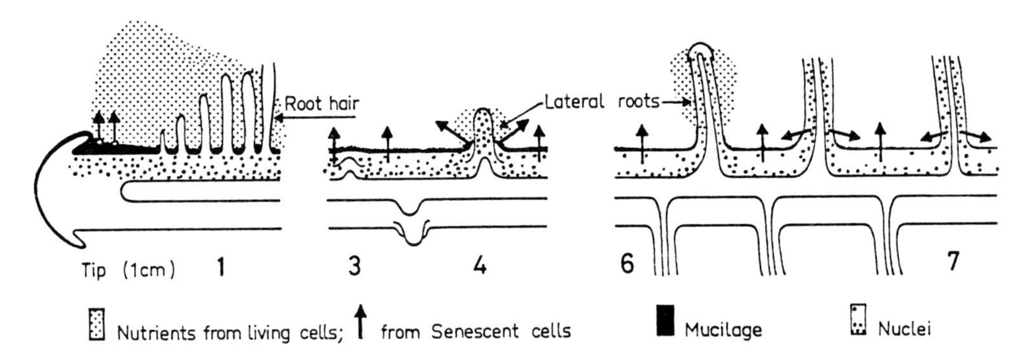


Fig. 2. Diagram of sources of substrate for microbial growth in the rhizosphere of a seminal wheat root. Figures indicate age of root parts in days after emergence from the seed (from Van Vuurde 1978).

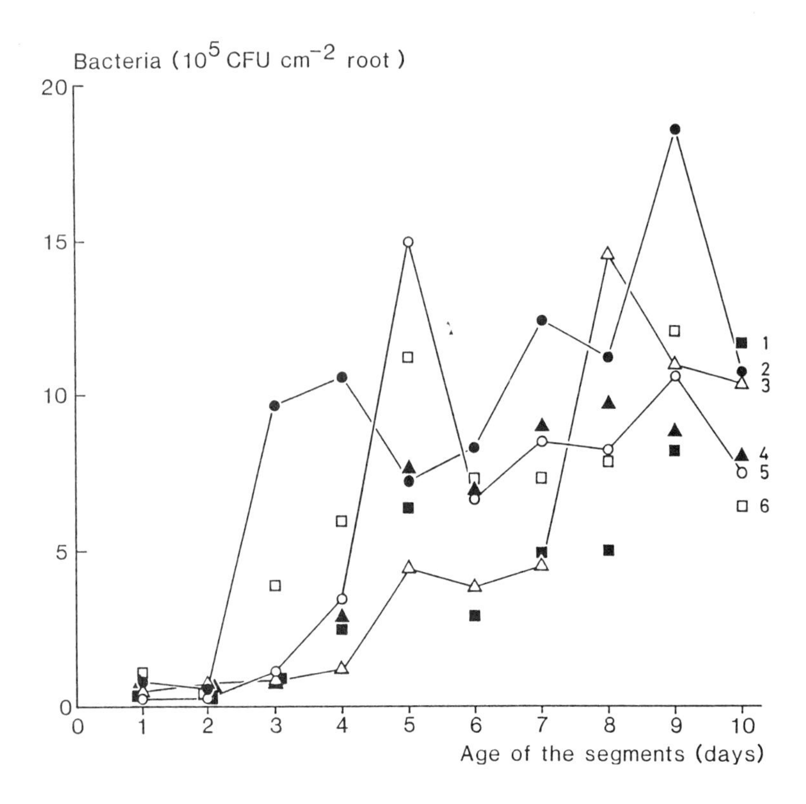


Fig. 3. Bacterial colonization of day segments from root tip to root base of six seminal roots. Data from dilution plating (adapted from Van Vuurde 1978).

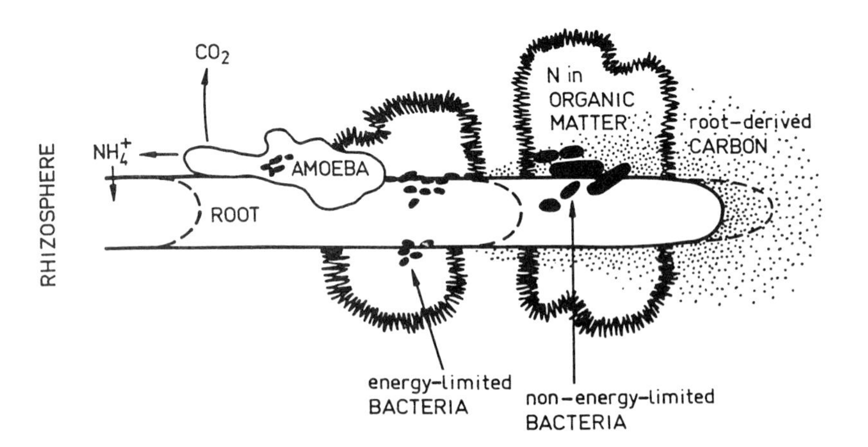


Fig. 4. Diagram of the role of bacteria and predating amoebae in the carbon and nitrogen flow in the rhizosphere (courtesy of Dr M. Clarholm, adapted from Clarholm 1983).

This lysis is delayed around segment 6 which is reflected in a retarded bacterial growth, while on older day segments the numbers of bacteria increase again till on segment 9 and 10 a steady state has been reached. By that time most of the epidermal and outer cortex cells have disappeared and nutrient release will diminish. Fungal colonization is still poor on the older segments but actinomycetes occur regularly (Van Vuurde & Schippers 1980).

For comparison of the bacterial densities found on wheat leaves (Fig. 1) with those on wheat seminal roots (Fig. 3), we converted the root colonization figures which are originally expressed per 10 mm root section (Van Vuurde & Schippers 1980) to numbers $cm^{-2}$ root using the dimensions given in their paper. This resulted in an average maximal density of $10^6$ bacteria $cm^{-2}$, a value occasionally also found on leaves (Hirano *et al.* 1982, assuming that 100 $cm^2$ corresponds to 1 g; Ercolani 1985). One may argue that the wheat roots were only 10 days old and that higher bacterial densities might have occurred on older roots. Bacterial densities on 12-week-old roots of *Lolium perenne* (Table 1), however, amounted to an average density of $6.4 \times 10^5$ bacteria $cm^{-2}$, which is not exceptional in the phyllosphere. Rovira *et al.* (1974) also mention the occurrence of yeasts in the rhizosphere at a density of $1.6 \times 10^4$ cells $cm^{-2}$, a value also normally found on leaves. Older wheat roots seemed to have a much greater abundance of fungi than roots of seedlings, which were almost free from fungal hyphae (Van Vuurde & Schippers 1980). Direct microscopic determination of the bacterial density revealed percentages cover varying from 2 to 8 (Rovira *et al.* 1974; Van Vuurde & Schippers 1980). Calculations of actual numbers from percentage cover resulted in densities about 10 times higher than those obtained by plating techniques (Table 1). For proper comparison of bacterial densities in the rhizosphere with those in the phyllosphere, figures obtained from the same technique, i.e. dilution plating, need to be used.

Considering the dramatic cell death, a phenomenon also described by Deacon & Henry (1980), and the abundant nutrient exudation in young roots, the bacterial density on the roots is not extraordinarily high compared to densities encountered in the phyllosphere. This suggests that epidermal cell death may not provide such an abundant organic nutrient source for bacteria as would be expected, or that other factors such as availability of water (Cook & Papendick 1972) or of $Fe^{3+}$ ions (Schippers *et al.* 1986 c) are limiting. Another possibility, which is often overlooked, might be predation of bacteria by amoebae and subsequent remineralization of root exudates (Coleman *et al.* 1977; Clarholm 1983). Fig. 4 illustrates the following events in the rhizosphere (Clarholm 1983). Amoebae are attracted to the root by increasing numbers of bacteria on the root tip. As a result of the grazing activity of amoebae, a substantial part of bacterial nitrogen will be released as ammonium, which can be taken up by the root again (Coleman *et al.* 1977; Clarholm 1983).

*Conclusions*

The following conclusions can be drawn:

1. Bacteria colonize roots rapidly; within two weeks the maximal density can be reached. Older roots do not carry a much higher density of bacteria.
2. Bacteria colonize leaves relatively slowly, but ultimately, under favourable conditions, densities similar to those on roots can be reached.
3. Yeasts are the main fungal colonizers on leaves. On roots, however, although yeasts are present, hyphal fungi seem to comprise most of the fungal biomass.

Differences in the start of the microbial colonization of leaves and roots probably reflect the manner in which micro-organisms are trapped, the environment in which the plant organ is growing and the available substrate. Evidently more bacteria will be encountered in the soil by a growing root than in the air by a growing leaf. Furthermore, roots will initially release a lot of nutrients, whereas on leaves nutrient release is more gradual, while, in addition, exogenous nutrients may accumulate. Bacteria and fungi may compete for the same nutrients but differences in water requirements may eventually determine the population densities of both groups (Bashi & Fokkema 1977).

Table 1. Root surface microflora of 12-week-old *Lolium perenne* plants (adapted from Rovira *et al.* (1974)).

| Parameters of population density | Total bacteria | Gram-neg. bacteria | Filament. fungi | Yeasts |
|---|---|---|---|---|
| *Plate counts*: | | | | |
| Number (x $10^3$) $mg^{-1}$ root | 107 | 48 | 0.35 | 2.6 |
| Number (x $10^3$) $cm^{-2}$ root[1] | 642 | 288 | 2.1 | 15.6 |
| *Direct observations*: | | | | |
| Covered area (%) | 7.7 | - | *ca* 3[2] | 0.1[3] |
| Number (x $10^3$) $mm^{-3}$ root | 1090[4] | - | - | - |
| Length of hyphae (mm $mm^{-2}$) | - | - | 12.1 | - |

[1] Calculated from sample dimensions presented in Rovira *et al.* (1974), and assuming that 1 mg root equals 1 $mm^3$ and the shape of the root is a smooth cylinder.

[2] Estimation from hyphal density (Rovira *et al.* 1974).

[3] Estimation from numbers $cm^{-2}$, assuming a one-cell-deep layer of yeasts each covering 6 $\mu m^2$.

[4] Estimation from % cover, assuming a one-cell-deep layer of bacteria each covering 1 $\mu m^2$ (Rovira *et al.* 1974).

*PLANT SURFACE COLONIZATION: SPATIAL VARIABILITY*

*Phyllosphere versus rhizosphere*

Colonization of rhizosphere and phyllosphere can also be compared by considering the quantitative variability of micro-organisms on individual leaves or roots within a large sample of comparable items. Hirano *et al.* (1982) demonstrated that phyllosphere bacteria of various field-grown crops are generally lognormally distributed. This means that population densities on individual leaves expressed as e.g. the number of CFU $g^{-1}$ leaf are not normally distributed but the $\log_{10}$-transformed values are. Graphic tests (e.g. Sokal & Rohlf 1981) to demonstrate normal distribution are based on plotting individual values against cumulative probability values or against ranked normal deviates (rankits). In both instances, when normally distributed, points should lie on a straight line. A more formal statistical procedure for testing normality has been described by Shapiro & Wilk (1965). Such graphs also nicely illustrate the variation of individual observations (Hirano *et al.* 1982;

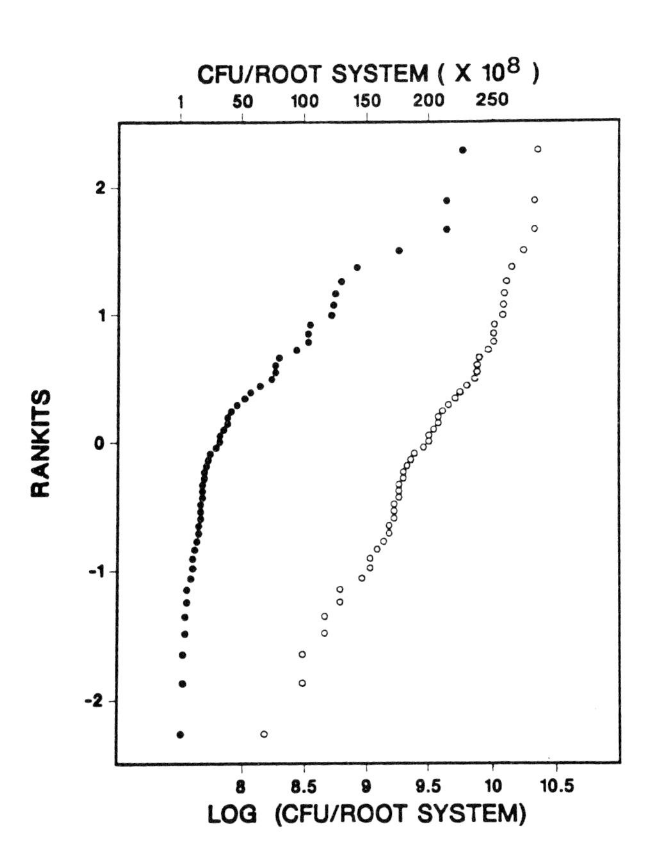


Fig. 5. Rankit diagrams of total aerobic bacteria in the rhizosphere, determined by dilution plating from a set of 53 barley root systems. Symbols: ●, CFU root $system^{-1}$; o, $\log_{10}$ (CFU root $system^{-1}$). Reproduced with permission from Loper *et al.* (1984).

S.S. Hirano and C.D. Upper, this volume, Fig. 1). Interestingly, Loper *et al.* (1984) described the lognormal distribution of bacterial populations in the rhizosphere in a paper with almost the same set-up as that by Hirano *et al.* (1982), which greatly facilitates comparison. The distribution of total bacteria on roots of 53 barley seedlings is visualized in a rankit diagram (Fig. 5). It is obvious that the bacteria are lognormally distributed because only $\log_{10}$-transformed values lie on a straight line. Hirano *et al.* (1982) analyzed the distribution of phyllosphere bacteria from 53 sets of data and Loper *et al.* (1984) used 43 sets of data. They concluded that bacterial populations associated with plant surfaces are approximated by a lognormal distribution. The variation of the total bacterial populations in the phyllosphere and rhizosphere was less than that of a given component of the population such as ice nucleation-active bacteria, applied bacterial antagonists (Hirano *et al.* 1982) or applied plant growth-promoting strains of *Pseudomonas* spp. (Loper *et al.* 1984).

The underlying factors causing this enormous variation between individual samples are not fully understood, but may be related to physical factors influencing bacterial development. In the phyllosphere, the effect of free water on multiplication of bacteria is so well known, that one would imagine that on leaves differences in duration of leaf wetness (rain and dew) may be responsible for the lognormality. This may be an oversimplification since, surprisingly, epiphytic bacteria of submerged leaves of freshwater plants are neither normally nor lognormally distributed (Baker & Orr 1986). It seems unlikely that nutrients in the phyllosphere are lognormally distributed over individual leaves since leaves collected were always similar (same sampling time, equivalent age and position) (Hirano *et al.* 1982). Availability of nutrients to bacteria, however, is certainly mediated by the presence of water films on the leaves. Water on leaves increases the leaching of nutrients from the underlying tissue and redistributes nutrients in the phyllosphere, with the possibility that the nutrients are ultimately washed off the leaves. The striking resemblance of the variability of bacteria on individual roots and leaves does not necessarily imply that similar factors are governing the variability. However, if water is the main common factor determining the lognormality, then availability of water in both environments apparently does not differ as much as is commonly believed.

### *Yeasts versus bacteria*

Rankit diagrams are also a useful tool for comparison of the suitability or fitness of different groups of resident or introduced micro-organisms for colonization of plant surfaces. We compared the distribution of phyllosphere yeasts and *Cladosporium* spp. with that of bacteria on wheat flag leaves and some interesting preliminary conclusions can be drawn. Leaf washings of 24 field-grown wheat leaves were plated and rankit diagrams were made of untransformed and $\log_{10}$-transformed numbers of total bacteria (Fig. 6 A), total yeasts (Fig. 6 B) and *Cladosporium* spp. (Fig. 6 C). A striking difference in the spatial variability of the bacteria to that of the yeasts was observed. Bacteria follow a lognormal distribution since only the transformed values form a

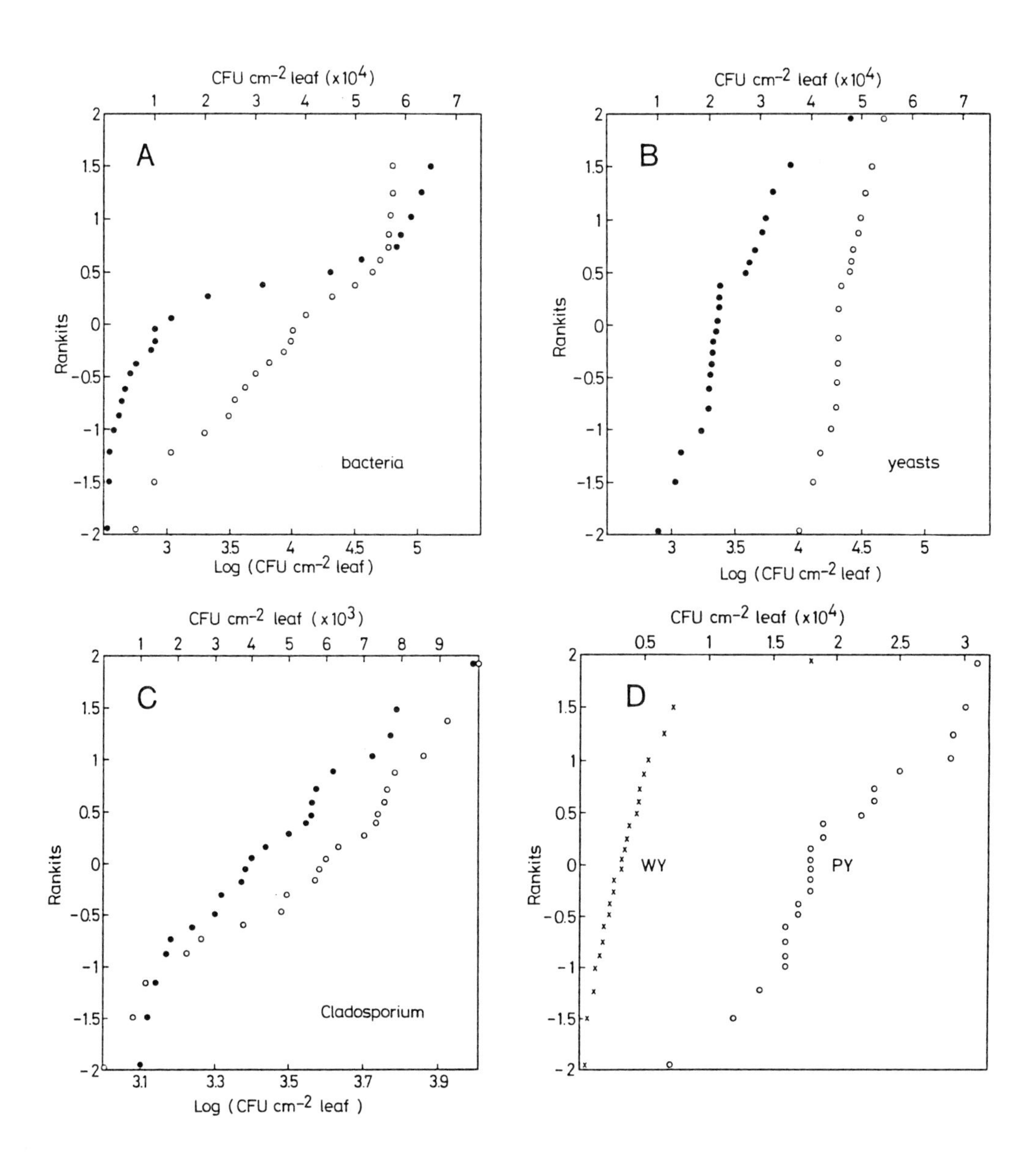


Fig. 6. Rankit diagrams of four groups of phyllosphere micro-organisms determined by dilution plating from one set of 24 wheat flag leaves. A: Total bacteria; B: Total yeasts; C: *Cladosporium* spp.; D: White yeasts (WY) and pink yeasts (PY). Symbols in A, B and C: ●, CFU $cm^{-2}$ leaf; o, $log_{10}$(CFU $cm^{-2}$ leaf) (N.J. Fokkema and J.W. Meijer, unpublished results).

straight line. $Log_{10}$ transformation does not have much effect on the shape of the rankit diagram of yeasts and of *Cladosporium* spp. which already approximates a straight line before transformation. Therefore, yeasts and *Cladosporium* spp. seem to be normally distributed. Yeasts comprise two distinct populations: white yeasts (mainly *Cryptococcus* spp.) and pink yeasts (mainly *Sporobolomyces* spp.), each apparently normally distributed (Fig. 6 D). The same trends were found on other sets of data from another wheat field. Bacteria were always lognormally distributed, but yeasts on the same leaves seem to be mostly normally distributed. We applied the Shapiro-Wilk test for normality (Shapiro & Wilk 1965) to these data and also to data sets from deciduous trees (Table 2). If the calculated statistic W approaches 1 and the corresponding P values are larger than 0.01, the distribution cannot be considered as non-normal. The test largely confirmed the indications obtained from graphic tests and the striking difference between distribution of yeasts and that of bacteria was occasionally also found on plants other than wheat.

Table 2. The Shapiro-Wilk test (statistic W) for normality performed on untransformed and $log_{10}$-transformed numbers of total bacteria and yeasts (CFU $cm^{-2}$) of different leaf samples.

| Data set | Transformation | Bacteria | | | Yeasts | | |
|---|---|---|---|---|---|---|---|
| | | CFU[1] | W | P[2] | CFU | W | P |
| Wheat 1 (n=24) | untransformed | 11 | 0.809 | a | 22 | 0.899[3] | b |
| | $log_{10}$-transformed | | 0.921 | c | | 0.939 | d |
| Wheat 2 (n=23) | untransformed | 152 | 0.881 | a | 23 | 0.912 | b |
| | $log_{10}$-transformed | | 0.939 | d | | 0.981 | c |
| Apple 1 (n=24) | untransformed | 137 | 0.724 | a | 42 | 0.914 | b |
| | $log_{10}$-transformed | | 0.947 | d | | 0.953 | d |
| Apple 2 (n=24) | untransformed | 146 | 0.813 | a | 33 | 0.815 | a |
| | $log_{10}$-transformed | | 0.962 | d | | 0.935 | d |
| Elm (n=24) | untransformed | 11 | 0.575 | a | 4 | 0.867 | a |
| | $log_{10}$-transformed | | 0.939 | d | | 0.964 | d |

[1] Geometric mean ($10^3$ $cm^{-2}$ leaf).
[2] P values: a, ≤ 0.01; b, 0.02-0.05; c, 0.05-0.10; d, 0.10-0.50.
[3] If without one outlying value, then W: 0.939; P: d.

## *CONSEQUENCES OF MICROBIAL COLONIZATION*

Having discussed how rhizosphere and phyllosphere are colonized by saprophytic micro-organisms, we arrive at the discussion of the consequences of microbial colonization for the plant which are either directly or indirectly beneficial, neutral or detrimental. For some groups of micro-organisms this has been well established. There is no doubt about the beneficial role of mycorrhizal fungi and rhizobia for the plant. On the other hand, the detrimental role of micro-organisms is well demonstrated by plant diseases caused by well defined microbial pathogens. Indirectly beneficial are those saprophytes who antagonize plant pathogens (Baker & Cook 1974; Blakeman & Fokkema 1982; Cook & Baker 1983; Windels & Lindow 1985). A dual role of some saprophytes, however, cannot be completely ruled out (Fokkema 1981; Schippers *et al.* 1986 a; V. Smedegaard-Petersen and K. Tolstrup, this volume). Although the properties of some groups of micro-organisms are generally acknowledged, micro-organisms with a neutral or perhaps a 'not yet determined' status form an enormous source for future research.

It is only in the last decade that ice nucleation activity of bacteria has been discovered (see S.E. Lindow, this volume), which will have great implications for the control of frost damage to plants. Further, the beneficial role of the endophytic fungi to the host population has only recently been put forward (see reviews by O. Petrini, K. Clay, G.C. Carroll, and J.D. Miller, this volume). Some properties of micro-organisms of possible interest, such as production of plant hormones (Libbert & Manteuffel 1970; Buckley & Pugh 1971; Diem 1971; Greene 1980) and the fixation of atmospheric nitrogen by free-living bacteria, have been known for a long time but the importance for the plant, as far as phyllosphere-related interactions are concerned, has not conclusively been determined. Recent literature about $N_2$-fixing free-living bacteria in the phyllosphere and rhizosphere will be reviewed. A relatively newly discovered attribute of rhizosphere bacteria is their plant growth-promoting activity, which seems to be the result of antagonism towards harmful micro-organisms not being ordinary plant pathogens (Suslow 1982; Schippers *et al.* 1985, 1986 c). This antagonism is correlated with the microbial production of siderophores, which are iron-chelating compounds produced by e.g. *Pseudomonas* spp. The role of siderophores in the phyllosphere was first discussed by Swinburne (1981). In the last five years, research about the importance of siderophore-producing bacteria in the rhizosphere has increased explosively. Since *Pseudomonas* spp. may equally well colonize both phyllosphere and rhizosphere, the current knowledge about siderophores in these environments will be discussed.

### *Nitrogen fixation by free-living bacteria*

*Phyllosphere.* Pioneer work by Ruinen (1965, 1970, 1974) on the occurrence of free-living $N_2$-fixing bacteria in the phyllosphere and their possible role as providers of organic nitrogen to the plant has triggered a lot of research with variable results. Reported experiments deal either with the role of naturally occurring $N_2$-fixing micro-organisms or with the effects of inoculation of selected $N_2$-fixing bacteria on agricultural crops and environmental conditions influencing nitrogen fixation. Nitrogenase activity is usually determined by measuring reduction of acetylene to ethylene; the reliability of the method has been proven by $^{15}N$ isotope incorporation (Bentley & Carpenter 1984). With the acetylene method nitrogenase activity of the organisms is determined but the fate of the fixed nitrogen remains unanswered. This nitrogen may be recycled in the phyllosphere micro-organisms, taken up by the leaves or washed from the leaves, and may eventually benefit the plant through uptake by the roots. Exposing leaves or whole plants to $^{15}N_2$ and tracing the label in the plant tissue gives direct evidence of the profit for the plant.

Under natural conditions it is impossible to discriminate between fixation by bluegreen algae and $N_2$-fixing bacteria. In the tropics such an epiphyll complex can provide in situ 10 to 25% of the N content of leaves of *Welfia georgii* (Palmae) as was determined with $^{15}N_2$ (Bentley & Carpenter 1984). Wet leaves fixed about 30 times more $N_2$ than dry leaves, and rehydration rapidly restored the $N_2$ fixation in which mainly bluegreen algae were involved (Bentley & Carpenter 1980). On the other hand, nitrogen fixation by epiphylls on coffee leaves was insignificant, ranging from 0.7 to 1.4 g $N_2$ ha$^{-1}$ year$^{-1}$. The coffee plantation, however, was annually fertilized with nitrogen which is known to interfere with $N_2$ fixation (Roskoski 1980). Micro-organisms in the phyllosphere of tropically grown guatemala grass (*Tripsacum laxum*) fixed 260 to 400 g $N_2$ ha$^{-1}$ year$^{-1}$ ($^{15}N_2$ method). The fixation mainly occurred in the leaf sheaths by *Klebsiella* spp. (Bessems 1973). A lower oxygen pressure and a high C/N ratio (>100) of the water in the leaf sheath favours $N_2$ fixation (Bessems 1973). Murty (1983, 1984) demonstrated $N_2$ fixation (acetylene reduction) on leaves of tropically grown sugarcane, sorghum, ragi (*Eleusine coracana*), bamboo, mulberry and cotton, mainly by naturally occurring *Beijerinckia* spp. It was estimated that for cotton this amounted to 1.6 to 3.2 kg N ha$^{-1}$ during the entire growth period.

In temperate climates the amounts of nitrogen fixed are generally smaller than in the tropics. In the leaf sheath of maize, an environment similar to that in guatemala grass, hardly any $N_2$ fixation could be detected (Bessems 1973). Jones (1976) reported a rather high $N_2$-fixing ability of unidentified micro-organisms colonizing Douglas fir needles, but this could not be confirmed in a later study (Jones 1982).

The attractive idea to use $N_2$-fixing phyllosphere bacteria as biofertilizers has been thoroughly investigated under tropical conditions in India. Pati & Chandra (1981) sprayed wheat at fortnightly intervals with suspensions of *Beijerinckia indica* or *Azotobacter chroococcum* and obtained significant yield increases with both bacteria of about 70% over the unfertilized control treatment, which was near to that obtained by normal N fertilization. Heat-killed bacteria and the medium had no effect. Sen Gupta & Sen (1982) compared the effect of spraying various isolates of *Klebsiella pneumoniae*, two *Azotobacter* spp. and *Derxia gummosa* on unfertilized wheat. Only isolates of *K. pneumoniae* increased yield from a very low control level of 10 to 200 g m$^{-2}$, a level almost similar to that obtained with urea (105 kg N ha$^{-1}$). When rice (Sen Gupta *et al.* 1982) was used as a test crop similar results were obtained, which could be correlated with simultaneously measured $N_2$ fixation by *Klebsiella* spp. The treatments resulted in six to seven-fold yield increases and urea treatment resulted in a seven to eight-fold increase. Treatments with *Klebsiella* spp. had an additional effect if applied together with half the dose of fertilizer but not with the full dose. It was demonstrated that fertilization reduced the nitrogenase activity (Sen Gupta *et al.* 1982). Similar successful treatments are reported for cauliflower, cabbage, tomato, pumpkin, eggplant, potato, onion, radish, spinach and *Trichosantes dioica* (Nandi *et al.* 1983). As culture filtrates of the strains of *Klebsiella* spp. did not contain auxins, gibberellins or cytokinins (Nandi *et al.* 1983), the yield increases should be attributed to $N_2$ fixation. Considering the number of micro-organisms likely to be involved, Pati & Chandra (1981), however, pointed out that the amount of fixed $N_2$ is unlikely to directly account for the observed yield increases. Furthermore, it is questionable whether treatments of aerial plant parts with suspensions of $N_2$ fixers is limited to an effect in the phyllosphere only, because unintentionally, some of the bacteria may have reached the rhizosphere as well.

Still, $N_2$ fixation in the phyllosphere is relatively unimportant when compared with the situation in the rhizosphere, which is clearly illustrated by the fact that $N_2$ fixation in the phyllosphere has not been discussed at all at the 5th International Symposium on Nitrogen Fixation (Veeger & Newton 1984).

*Rhizosphere.* We will only mention briefly the free-living $N_2$-fixing micro-organisms because of their similarity in behaviour with their counterparts in the phyllosphere. A wealth of information can be obtained from the Canadian Journal of Microbiology, volume 29 (8) (1983) and from the reports edited by Skinner & Uomala (1986), which comprise the proceedings of the second and third International Symposium on Nitrogen Fixation with Nonlegumes, respectively. Vose (1983) gave a general review on the most important plant-microbe associations. The main genera involved in $N_2$ fixation in the rhizosphere are: *Achromobacter, Azospirillum, Bacillus, Beijerinckia, Derxia, Klebsiella* and *Pseudomonas*. From various reports it was suggested that mixed populations of $N_2$-fixing bacteria and non-fixers had higher nitrogenase activity than pure cultures of $N_2$ fixers (Vose 1983). It is interesting to note that also fluorescent *Pseudomonas* spp. can fix nitrogen (e.g. Barraquio *et al.* 1983; Haahtela *et al.* 1983). Okon (1984) reviewed recent achievements in increasing cereal yields as a result of inoculation with *Azospirillum* spp. Five to 30% increases of cereal yield over untreated controls are reported and the greatest effect occurs if normal fertilization is reduced, an interaction also observed for $N_2$ fixation in the phyllosphere. However, Reynders & Vlassak (1982) obtained 9 to 15% yield increase in wheat irrespective of nitrogen gifts.

### *Siderophore production*

*Phyllosphere.* In the phyllosphere siderophores may originate from the plant or from colonizing micro-organisms. Swinburne (1981) reviewed the stimulating effect of aqueous diffusates from banana and apple fruits, their iron-chelating compounds, and artificial iron chelators on spore germination and appressoria formation of *Colletotrichum musae* and *Diaporthe perniciosa*. Iron present in the conidia seems to inhibit germination, because conidia harvested from $Fe^{3+}$-containing media showed poor germination in contrast to those from $Fe^{3+}$-deficient media (Harper *et al.* 1980). The reported stimulation of germination by chelating agents was explained by their complexation (inactivation) of inhibitory iron at a site within the conidia (Graham & Harper 1983). In addition to siderophores derived from fruits, *Pseudomonas* sp. isolate UV3 and its siderophore also stimulated spore germination (McCracken & Swinburne 1979) and infection of bananas (Brown & Swinburne 1981). Chelating agents also stimulated the formation of a pigmented type of appressoria of *C. musae* which leads to successful infection (Swinburne & Brown 1983). Iron-chelating agents also stimulated spore germination and appressorium formation of *Botrytis cinerea* and subsequent infection of broad bean leaves (Brown & Swinburne 1982).

These observations may be of great plant-pathological importance considering the omnipresence of leaf surface pseudomonads which are potentially capable of producing siderophores. Two main questions have to be answered:
1. Are siderophores actually produced by bacteria in the phyllosphere?
2. How general is the sensitivity of natural inoculum to iron chelators?
The striking results discussed above apply to conidia which germinate poorly in water, a response which may have been the result of unnatural culture conditions. Moreover, *B. cinerea* strains may differ in their spore germinability. Vedie & Le Normand (1984) performed similar experiments with well germinating conidia of *B. cinerea* and *B. fabae* on broad bean and found no stimulation of infection in the presence of a chelating agent. A $Fe^{3+}$-chelating agent also had no effect on germination of well germinating conidia of *Septoria nodorum* and *Pyrenophora teres*, nor on the subsequent appressoria formation and infection of wheat and barley, respectively (Brown & Sharma 1985). In contrast, $Fe^{2+}$-chelating compounds acted like fungicides (Brown & Sharma 1985).

In general, there are more reports about antagonistic interactions between pseudomonads and fungal pathogens in the phyllosphere (Blakeman & Fokkema 1982) than about synergistic interactions. Nevertheless, the latter occur (Clark & Lorbeer 1977) but may be underestimated because in biocontrol studies these interactions tend to be neglected. A synergistic interaction, however, is not necessarily caused by the provision of siderophores, because bacteria may also produce phytotoxins (Harrison 1983). It remains questionable whether the phyllosphere as an environment is $Fe^{3+}$-deficient to a degree which triggers siderophore production by the micro-organisms. Indirect studies by Lindow *et al.* (1984) could not demonstrate siderophore production by *Pseudomonas syringae* pv. *syringae* on bean leaves. Although Swinburne (1981) speculated that the iron concentration in the phyllosphere may be low enough to allow bacterial siderophore production, it is surprising that experiments demonstrating that *Pseudomonas* sp. UV3 can stimulate infection by *C. musae* and *B. cinerea* are lacking. Moreover, the stimulation of appressoria formation of *Colletotrichum acutatum* by *Pseudomonas* sp. UV3 could also be explained by nutrient competition (Blakeman & Parbery 1977). On beetroot leaves similar bacteria reduced germination of *B. cinerea*, *Phoma betae* and *Cladosporium herbarum* spores (Blakeman & Brodie 1977); this suggests that siderophores, as a stimulating agent, were not involved in these experiments. An additional complicating factor is that treatment of leaves with chelating agents may suppress fluorescent pseudomonads and favour other bacteria and this may also have consequences for the development of fungal pathogens (Vedie & Le Normand 1984; Brown 1986).

*Rhizosphere*. Stimulation of plant growth and increase of yield by treatment of seeds or tubers with selected siderophore-producing pseudomonads are likely to be based on the suppression of plant growth-inhibiting micro-organisms which are not necessarily plant pathogens (Kloepper *et al*. 1980; Schroth & Hancock 1982; Suslow 1982; Geels & Schippers 1983 b; Schippers *et al*. 1985, 1986 a). With increasing cropping frequency, the effect of crop-specific harmful micro-organisms may increase over seasons and this effect can partly be overcome by siderophore-producing pseudomonads (Schippers *et al*. 1985). In iron-deficient environments, fluorescent pseudomonads produce siderophores of the pyoverdine type (Neilands 1981, 1984) which show a high affinity to $Fe^{3+}$ ions. The production of siderophores enables the bacteria to compete successfully with other, e.g. harmful, bacteria for $Fe^{3+}$ ions, because the $Fe^{3+}$-siderophore complex can primarily be taken up by the strain which produced the siderophore. Evidence for the involvement of siderophores in plant growth promotion comes from experiments in which dissolved $Fe^{3+}$ ions, added to the environment, nullified the plant growth-promoting effect of pseudomonads (Kloepper *et al*. 1980). Furthermore, mutants of growth-promoting pseudomonads defective in siderophore production had also lost their plant growth-promoting effect, whereas their ability to colonize the rhizosphere was not affected (Kloepper & Schroth 1981; Schippers *et al*. 1986 b). Considering our paragraph on nitrogen fixation it may be worth to check growth-promoting rhizobacteria for $N_2$-fixing abilities.

Apart from the suppression of non-parasitic harmful micro-organisms, seed bacterization with isolates of *Pseudomonas* spp. also reduced infection of wheat by *Gaeumannomyces graminis* (Weller & Cook 1983; B. Schippers, unpublished results), of cotton by *Pythium* and *Rhizoctonia* spp. (Howell & Stipanovic 1979, 1980), of potato by *Erwinia carotovora* (Kloepper 1983) and of flax, radish and cucumber by *Fusarium oxysporum* (Scher & Baker 1980, 1982). In addition to siderophores the production of antibiotics may also be involved in these interactions.

The plant growth-promoting rhizobacteria reported so far are all fluorescent pseudomonads which may colonize the phyllosphere as well. Whether there is substrate specificity needs to be investigated. Currently nothing is known about a possible plant growth-stimulating effect in the phyllosphere by the inhibition of harmful non-pathogenic leaf micro-organisms, although bacterial antagonism against ice nucleation-active bacteria shows a certain similarity (S.E. Lindow, this volume). Siderophores may also be involved in antagonistic interactions between leaf surface bacteria and bacterial and fungal pathogens as a part of the nutrient competition between these organisms (Blakeman & Fokkema 1982). Very intriguing is the presence of pseudomonads in the phyllosphere in view of possible synergistic effects with plant pathogens as discussed before.

The availability of siderophore-negative mutants obtained by Tn*5* transposon mutagenesis (Marugg *et al.* 1985) and the ability to mark isolates genetically (see also J.H. Andrews, this volume) open excellent possibilities for studying the involvement of siderophores in microbial interactions. The crucial question about the in situ production of siderophores in rhizosphere and phyllosphere has been answered for the rhizosphere. P.A.H.M. Bakker and B. Schippers (unpublished results) demonstrated that colonization of the rhizosphere by a siderophore-negative mutant of *Pseudomonas putida* isolate WCS358 was significantly increased if simultaneously its wild-type or the wild-type of *P. fluorescens* isolate WCS374 was added whose siderophores can be utilized by isolate WCS358. Isolate WCS374, however, cannot make use of the siderophore of isolate WCS358, and therefore simultaneous introduction in the rhizosphere of the wild-type of isolate WCS358 did not stimulate colonization of the siderophore-negative mutant of isolate WCS374. These observations can only be explained by the in situ production of siderophores. A similar approach for demonstrating in situ production of siderophores is currently followed at our laboratory for demonstrating siderophore production in the phyllosphere.

## *FUTURE PROSPECTS AND CONCLUSIONS*

Knowledge of the ecology of micro-organsims in the phyllosphere and rhizosphere is a prerequisite for the practical use in agriculture of beneficial microbial colonizers of plant surfaces. Although our generalizations about plant surface colonization are based on limited observations, they hopefully provide a framework for further exploration. To our surprise, the quantitative differences in population densities of bacteria in rhizosphere and phyllosphere are smaller than we expected. Great daily fluctuations, however, may occur in the phyllosphere (S.S. Hirano and C.D. Upper, this volume) and seem to be characteristic for this environment. This does not necessarily imply that bacteria are less suitable antagonists in the phyllosphere than in the rhizosphere. As long as the target micro-organisms are equally affected by environmental conditions and the operating mechanisms of antagonism are not seriously affected by daily fluctuations of bacterial numbers, antagonism may be significant for practical application.

The great spatial variability of bacteria in rhizosphere and phyllosphere can be a major disavantage for reliable practical use of the various beneficial properties of plant surface bacteria. A conclusive answer about the underlying factors is urgently needed. Loper *et al.* (1985) observed a positive relationship between osmotolerance of selected bacterial strains applied to seed potatoes and their success in root colonization. This indicates that water availability may be a crucial limiting factor. Selection for osmotolerance of the bacteria to be applied may therefore be rewarding.

Tracing applied micro-organisms is impossible when they cannot be recognized by their colony phenomenology after plating. The use of biological markers (J.H. Andrews, this volume), however, has opened new and promising means of monitoring colonization of introduced micro-organisms. This enables us to investigate the reasons why introduction of micro-organisms often failed (Spurr & Knudsen 1985). Investigations on population dynamics of marked *Pseudomonas* spp. applied to cereal seeds or seed potatoes demonstrated successful establishment on the entire root system (Geels & Schippers 1983 a; Weller 1984; Loper *et al.* 1985). However, a decline in population density of the applied bacteria on the root system during the growing season was always observed. The fluorescent antibody technique (Schank *et al.* 1979; Bohlool & Schmidt 1980) allows the exact localization of the applied micro-organisms in situ, and may reveal specific sites for colonization.

Manipulation of naturally occurring micro-organisms in the phyllosphere by spraying nutrients (Fokkema *et al.* 1979) does not have much effect without simultaneously added micro-organisms. Fertilization of the soil does not seem to increase the nutrient status in the phyllosphere (Bessems 1973; Frossard & Fokkema 1982). A more detailed study of the nutrient requirements of the micro-organisms to be manipulated (Morris & Rouse 1985) and of the fate of the added nutrients in the phyllosphere is needed to develop successful applications. In the rhizosphere of rye grass N and P fertilization enhanced the ratio between fungi and bacteria (Turner *et al.* 1985). Their results emphasize that bacterial ecology may be better understood if bacteria and fungi are studied together, and this may equally well apply to phyllosphere studies.

Finally, one should be aware that in commercially grown crops beneficial micro-organisms already present on the plant surface are regularly exposed to agrochemicals which may reduce their population densities (Hislop 1976; Andrews 1981; Smiley 1981; Becker 1982). In fact, the best evidence that biological control of plant pathogens operates in nature originates from occasional observations that application of certain agrochemicals resulted in more disease (Bollen 1979; Blakeman & Fokkema 1982). Often alternative chemicals can be used without affecting the beneficial micro-organisms. Avoiding unnecessary loss of naturally occurring beneficial organisms (Fokkema 1983) is one of the simplest means of taking advantage of them.

*ACKNOWLEDGEMENTS*

We gratefully acknowledge Ms G.J. van Schijndel, Ms C.F. Tebra, J.W. Meijer and Ms W.A.J. Jansen for experimental assistance and help in statistical analyses. We thank Ms M.A. Williamson for correcting the English text.

*REFERENCES*

Andrews, J.H. (1981). Effects of pesticides on non-target micro-organisms on leaves. *In* Microbial Ecology of the Phylloplane, ed. J.P. Blakeman, pp. 283-304. London: Academic Press.

Andrews, J.H. & Kenerley, C.M. (1978). The effects of a pesticide program on non-target epiphytic microbial populations of apple leaves. Canadian Journal of Microbiology, *24*, 1058-72.

Andrews, J.H. & Kenerley, C.M. (1980). Microbial populations associated with buds and young leaves of apple. Canadian Journal of Botany, *58*, 847-55.

Baker, K.F. & Cook, R.J. (1974). Biological Control of Plant Pathogens. San Francisco: Freeman & Co.

Baker, J.H. & Orr, D.R. (1986). Distribution of epiphytic bacteria on freshwater plants. Journal of Ecology, *74*, 155-65.

Barraquio, W.L., Ladha, J.K. & Watanabe, I. (1983). Isolation and identification of $N_2$-fixing *Pseudomonas* associated with wetland rice. Canadian Journal of Microbiology, *29*, 867-73.

Bashi, E. & Fokkema, N.J. (1977). Environmental factors limiting growth of *Sporobolomyces roseus*, an antagonist of *Cochliobolus sativus*, on wheat leaves. Transactions of the British Mycological Society, *68*, 17-25.

Becker, J.O. (1982). Zum Einfluss von Herbiziden und Fungiziden auf die Mikroflora der Phylloplane der Weizenhalmbasis unter Berücksichtigung potentieller Antagonisten von *Pseudocercosporella herpotrichoides*. Dissertation, Georg-August-Universität, Göttingen.

Bentley, B.L. & Carpenter, E.J. (1980). Effects of desiccation and rehydration on nitrogen fixation by epiphylls in a tropical rainforest. Microbial Ecology, *6*, 109-13.

Bentley, B.L. & Carpenter, E.J. (1984). Direct transfer of newly-fixed nitrogen from free-living epiphyllous microorganisms to their host plant. Oecologia, *63*, 52-6.

Bessems, E.P.M. (1973). Nitrogen fixation in the phyllosphere of Gramineae. Agricultural Research Reports *786*. Wageningen: Pudoc.

Blakeman, J.P. (1973). The chemical environment of leaf surfaces with special reference to spore germination of pathogenic fungi. Pesticide Science, *4*, 575-88.

Blakeman, J.P. & Brodie, I.D.S. (1977). Competition for nutrients between epiphytic micro-organisms and germination of spores of plant pathogens on beet-root leaves. Physiological Plant Pathology, *10*, 29-42.

Blakeman, J.P. & Fokkema, N.J. (1982). Potential for biological control of plant diseases on the phyllophane. Annual Review of Phytopathology, *20*, 167-92.

Blakeman, J.P. & Parbery, D.G. (1977). Stimulation of appressorium formation in *Colletotrichum acutatum* by phylloplane bacteria. Physiological Plant Pathology, *11*, 313-25.

Bohlool, B.B. & Schmidt, E.L. (1980). The immunofluorescence approach in microbial ecology. *In* Advances of Microbial Ecology. Vol. 4, ed. M. Alexander, pp. 203-41. New York: Plenum Press.

Bollen, G.J. (1979). Side-effects of pesticides on microbial interactions. *In* Soil-Borne Plant Pathogens, eds B. Schippers & W. Gams, pp. 451-81. London: Academic Press.

Breeze, E.M. & Dix, N.J. (1981). Seasonal analysis of the fungal community on *Acer platanoides* leaves. Transactions of the British Mycological Society, *77*, 321-8.

Brown, A.E. (1986). Ferrous complexes and chelating compounds in suppression of fungal diseases of cereals. *In* Iron, Siderophores and Plant Pathogens, ed. T.R. Swinburne. New York: Plenum Press (in press).

Brown, A.E. & Sharma, H.S.S. (1985). A possible role for ferrous complexes in fungal disease suppression: glume blotch and net blotch of cereals. Phytopathologische Zeitschrift, *113*, 178-88.

Brown, A.E. & Swinburne, T.R. (1981). Influence of iron and iron chelators on formation of progressive lesions by *Colletotrichum musae* on banana fruits. Transactions of the British Mycological Society, *77*, 119-24.

Brown, A.E. & Swinburne, T.R. (1982). Iron-chelating agents and lesion development by *Botrytis cinerea* on leaves of *Vicia faba*. Physiological Plant Pathology, *21*, 13-21.

Buckley, N.G. & Pugh, G.J.F. (1971). Auxin production by phylloplane fungi. Nature (London), *231*, 332.

Clarholm, M. (1983). Dynamics of soil bacteria in relation to plants, protozoa and inorganic nitrogen. Dissertation, Uppsala, Swedish University of Agricultural Sciences, Department of Microbiology Report 17.

Clark, C.A. & Lorbeer, J.W. (1977). The role of phyllosphere bacteria in pathogenesis by *Botrytis squamosa* and *B. cinerea* on onion leaves. Phytopathology, *67*, 96-100.

Clark, F.E. (1949). Soil microorganisms and plant roots. Advances in Agronomy, *1*, 241-88.

Coleman, D.C., Cole, C.V., Anderson, R.V., Blaha, M., Campion, M.K., Clarholm, M., Elliott, E.T., Hunt, H.W., Shaefer, B. & Sinclair, J. (1977). An analysis of rhizosphere-saprophage interactions in terrestrial ecosystems. Ecological Bulletins (Stockholm), *25*, 299-309.

Cook, R.J. & Baker, K.F. (1983). The Nature and Practice of Biological Control of Plant Pathogens. St. Paul: Am. Phytopath. Soc.

Cook, R.J. & Papendick, R.I. (1972). Influence of water potential of soils and plants on root disease. Annual Review of Phytopathology, *10*, 349-74.

Deacon, J.W. & Henry, C.M. (1980). Age of wheat and barley roots and infection by *Gaeumannomyces graminis* var *tritici*. Soil Biology & Biochemistry, *12*, 113-8.

Diem, H.G. (1971). Production de l'acide indolyl-3-acétique par certaines levures épiphylles. Comptes Rendus Hebdomadaires des Séances de l'Académie des Sciences, Paris, Série D, *272*, 941-3.

Diem, H.G. (1973). Phylloplan et phyllosphère. Canadian Journal of Botany, *51*, 1079-80.

Ercolani, G.L. (1985). Factor analysis of fluctuation in populations of *Pseudomonas syringae* pv. *savastanoi* on the phylloplane of the olive. Microbial Ecology, *11*, 41-9.

Flannigan, B. & Campbell, I. (1977). Pre-harvest mould and yeast floras on the flag leaf, bracts and caryopsis of wheat. Transactions of the British Mycological Society, *69*, 485-94.

Fokkema, N.J. (1971). The effect of pollen in the phyllosphere of rye on colonization by saprophytic fungi and on infection by *Helminthosporium sativum* and other leaf pathogens. Netherlands Journal of Plant Pathology, *77* (suppl. 1), 1-60.

Fokkema, N.J. (1981). Fungal leaf saprophytes, beneficial or detrimental? *In* Microbial Ecology of the Phylloplane, ed. J.P. Blakeman, pp. 433-54. London: Academic Press.

Fokkema, N.J. (1983). Naturally-occurring biological control in the phyllosphere. Les Colloques de l'INRA, *18*, 71-9.

Fokkema, N.J., Houter, J.G. den, Kosterman, Y.J.C. & Nelis, A.L. (1979). Manipulation of yeasts on field-grown wheat leaves and their antagonistic effect on *Cochliobolus sativus* and *Septoria nodorum*. Transactions of the British Mycological Society, *72*, 19-29.

Fokkema, N.J. & Lorbeer, J.W. (1974). Interactions between *Alternaria porri* and the saprophytic mycoflora of onion leaves. Phytopathology, *64*, 1128-33.

Foster, R.C. & Bowen, G.D. (1982). Plant surfaces and bacterial growth: the rhizosphere and rhizoplane. *In* Phytopathogenic Prokaryotes. Vol. I, eds M.S. Mount & G.H. Lacy, pp. 159-85. London: Academic Press.

Frossard, R. (1981). Effect of guttation fluids on growth of micro-organisms on leaves. *In* Microbial Ecology of the Phylloplane, ed. J.P. Blakeman, pp. 213-26. London: Academic Press.

Frossard, R. & Fokkema, N.J. (1982). Effect of nitrogen fertilization on the growth of *Sporobolomyces roseus* on wheat leaves. Canadian Journal of Botany, *60*, 2741-4.

Frossard, R., Fokkema, N.J. & Tietema, T. (1983). Influence of *Sporobolomyces roseus* and *Cladosporium cladosporioides* on leaching of $^{14}C$-labelled assimilates from wheat leaves. Transactions of the British Mycological Society, *80*, 289-96.

Geels, F.P. & Schippers, B. (1983 a). Selection of antagonistic fluorescent *Pseudomonas* spp. and their root colonization and persistence following treatment of seed potatoes. Phytopathologische Zeitschrift, *108*, 193-206.

Geels, F.P. & Schippers, B. (1983 b). Reduction of yield depressions in high frequency potato cropping soil after seed tuber treatments with antagonistic fluorescent *Pseudomonas* spp. Phytopathologische Zeitschrift, *108*, 207-14.

Graham, A.H. & Harper, D.B. (1983). Distribution and transport of iron in conidia of *Colletotrichum musae* in relation to the mode of action of germination stimulants. Journal of General Microbiology, *129*, 1025-34.

Greene, E.M. (1980). Cytokinin production by microorganisms. Botanical Review, *46*, 25-74.

Haahtela, K., Helander, I., Nurmiaho-Lassila, E.-L. & Sundman, V. (1983). Morphological and physiological characteristics and lipopolysaccharide composition of $N_2$-fixing ($C_2H_2$-reducing) root-associated *Pseudomonas* sp. Canadian Journal of Microbiology, *29*, 874-80.

Haller, Th. & Stolp, H. (1985). Quantitative estimation of root exudation of maize plants. Plant and Soil, *86*, 207-16.

Harper, D.B., Swinburne, T.R., Moore, S.K., Brown, A.E. & Graham, H. (1980). A role for iron in germination of conidia of *Colletotrichum musae*. Journal of General Microbiology, *121*, 169-74.

Harrison, J.G. (1983). Growth of lesions caused by *Botrytis fabae* on field bean leaves in relation to foliar bacteria, non-enzymic phytotoxins, pectic enzymes and osmotica. Annals of Botany, *52*, 823-38.

Hiltner, L. (1904). Über neuere Erfahrungen und Probleme auf dem Gebiet der Bodenbacteriologie und unter besonderer Berücksichtigung der Gründingung und Brache. Arbeiten der Deutschen Landwirtschaftsgesellschaft, *98*, 59-78.

Hirano, S.S., Nordheim, E.V., Arny, D.C. & Upper, C.D. (1982). Lognormal distribution of epiphytic bacterial populations on leaf surfaces. Applied and Environmental Microbiology, *4*, 695-700.

Hislop, E.C. (1976). Some effects of fungicides and other agrochemicals on the microbiology of the aerial surfaces of plants. *In* Microbiology of Aerial Plant Surfaces, eds C.H. Dickinson & T.F. Preece, pp. 41-74. London: Academic Press.

Hislop, E.C. & Cox, T.W. (1969). Effects of captan on the non-parasitic microflora of apple leaves. Transactions of the British Mycological Society, *52*, 223-35.

Howell, C.R. & Stipanovic, R.D. (1979). Control of *Rhizoctonia solani* on cotton seedlings with *Pseudomonas fluorescens* and with an antibiotic produced by the bacterium. Phytopathology, *69*, 480-2.

Howell, C.R. & Stipanovic, R.D. (1980). Suppression of *Pythium ultimum*-induced damping-off of cotton seedlings by *Pseudomonas fluorescens* and its antibiotic, pyoluteorin. Phytopathology, *70*, 712-5.

Jones, K. (1976). Nitrogen fixing bacteria in the canopy of conifers in a temperate forest. *In* Microbiology of Aerial Plant Surfaces, eds C.H. Dickinson & T.F. Preece, pp. 451-63. London: Academic Press.

Jones, K. (1982). Nitrogen fixation in the canopy of temperate forest trees: a reexamination. Annals of Botany, *50*, 329-34.

Kerling, L.C.P. (1958). De microflora op het blad van *Beta vulgaris* L. (with English summary). Tijdschrift over Planteziekten, *64*, 402-10.

Kloepper, J.W. (1983). Effect of seed piece inoculation with plant growth promoting rhizobacteria on populations of *Erwinia carotovora* on potato roots and in daughter tubers. Phytopathology, *73*, 217-9.

Kloepper, J.W., Leong, J., Teintze, M. & Schroth, M.N. (1980). Enhanced plant growth by siderophores produced by plant growth-promoting rhizobacteria. Nature (London), *286*, 885-6.

Kloepper, J.W. & Schroth, M.N. (1981). Relationship of in vitro antibiosis of plant growth-promoting rhizobacteria to plant growth and the displacement of root microflora. Phytopathology, *71*, 1020-4.

Last, F.T. (1955). Seasonal incidence of *Sporobolomyces* on cereal leaves. Transactions of the British Mycological Society, *38*, 221-39.

Last, F.T. & Deighton, F.C. (1965). The non-parasitic microflora on the surfaces of living leaves. Transactions of the British Mycological Society, *48*, 83-99.

Leben, C. (1971). The bud in relation to the epiphytic microflora. *In* Ecology of Leaf Surface Micro-organisms, eds T.F. Preece & C.H. Dickinson, pp. 117-27. London: Academic Press.

Libbert, E.& Manteuffel, R. (1970). Interactions between plants and epiphytic bacteria regarding their auxin metabolism. VII. Influence of epiphytic bacteria on amount of diffusable auxin from corn coleoptiles. Physiologia Plantarum, *23*, 93-8.

Lindow, S.E., Loper, J.E. & Schroth, M.N. (1984). Lack of evidence for *in situ* fluorescent pigment production by *P. syringae* on leaf surfaces. Phytopathology, *74*, 825-6 (Abstr.).

Loper, J.E., Haack, C. & Schroth, M.N. (1985). Population dynamics of soil pseudomonads in the rhizosphere of potato (*Solanum tuberosum* L.). Applied and Environmental Microbiology, *49*, 416-22.

Loper, J.E., Suslow, T.V. & Schroth, M.N. (1984). Lognormal distribution of bacterial populations in the rhizosphere. Phytopathology, *74*, 1454-60.

McCracken, A.R. & Swinburne, T.R. (1979). Siderophores produced by saprophytic bacteria as stimulants of germination of conidia of *Colletotrichum musae*. Physiological Plant Pathology, *15*, 331-40.

Martin, J.K. & Puckridge, D.W. (1982). Carbon flow through the rhizosphere of wheat crops in South Australia. *In* The Cycling of Carbon, Nitrogen, Sulfur and Phosphorus in Terrestrial and Aquatic Ecosytems, eds I.E. Galbally & J.R. Freney, pp. 77-82. Canberra: Australian Academy of Science.

Marugg, J.D., Spanje, M. van, Hoekstra, W.P.M., Schippers, B. & Weisbeek, P.J. (1985). Isolation and analysis of genes involved in siderophore biosynthesis in the plant growth-stimulating *Pseudomonas putida* strain WCS358. Journal of Bacteriology, *164*, 563-70.

Morris, C.E. & Rouse, D.I. (1985). Role of nutrients in regulating epiphytic bacterial populations. *In* Biological Control on the Phylloplane, eds C.E. Windels & S.E. Lindow, pp. 63-82. St. Paul: Am. Phytopath. Soc.

Murty, M.G. (1983). Nitrogen fixation (acetylene reduction) in the phyllosphere of some economically important plants. Plant and Soil, *73*, 151-3.

Murty, M.G. (1984). Phyllosphere of cotton a habitat for diazotrophic microorganisms. Applied and Environmental Microbiology, *48*, 713-8.

Nandi, A.S., Sen Gupta, B. & Sen, S.P. (1983). Utility of phyllosphere $N_2$-fixing microorganisms in the improvement of growth of some vegetables. Journal of Horticultural Science, *58*, 547-54.

Neilands, J.B. (1981). Microbial iron compounds. Annual Review of Biochemistry, *50*, 715-31.

Neilands, J.B. (1984). Siderophores of bacteria and fungi. Microbiological Sciences, *1*, 9-14.

Okon, Y. (1984). Response of cereal and forage grasses to inoculation with $N_2$-fixing bacteria. *In* Advances in Nitrogen Fixation Research, eds C. Veeger & W.E. Newton, pp. 303-9. The Hague/Wageningen: Martinus Nijhoff & Dr W. Junk Publishers/Pudoc.

Parbery, I.H., Brown, J.F. & Bofinger, V.J. (1981). Statistical methods in the

analysis of phylloplane populations. *In* Microbial Ecology of the Phylloplane, ed. J.P. Blakeman, pp. 47-65. London: Academic Press.
Pati, B.R. & Chandra, A.K. (1981). Effect of spraying nitrogen-fixing phyllospheric bacterial isolates on wheat plants. Plant and Soil, *61*, 419-27.
Pennycook, S.R. & Newhook, F.J. (1981). Seasonal changes in the apple phylloplane microflora. New Zealand Journal of Botany, *19*, 273-83.
Rabbinge, R., Brouwer, A., Fokkema, N.J., Sinke, J. & Stomph, T.J. (1984). Effects of the saprophytic leaf mycoflora on growth and productivity of winter wheat. Netherlands Journal of Plant Pathology, *90*, 181-97.
Reynders, L. & Vlassak, K. (1982). Use of *Azospirillum brasilense* as biofertilizer in intensive wheat cropping. Plant and Soil, *66*, 217-23.
Roskoski, J.P. (1980). $N_2$ fixation ($C_2H_2$ reduction) by epiphylls on coffee, *Coffea arabica*. Microbial Ecology, *6*, 349-55.
Rovira, A.D., Newman, E.I., Bowen, H.J. & Campbell, R. (1974). Quantitative assessment of the rhizoplane microflora by direct microscopy. Soil Biology & Biochemistry, *6*, 211-6.
Ruinen, J. (1956). Occurrence of *Beijerinckia* species in the phyllosphere. Nature (London), *177*, 220-1.
Ruinen, J. (1965). The phyllosphere. III. Nitrogen fixation in the phyllosphere. Plant and Soil, *22*, 375-94.
Ruinen, J. (1970). The phyllosphere. V. The grass sheath, a habitat for nitrogen-fixing micro-organisms. Plant and Soil, *33*, 661-71.
Ruinen, J. (1974). Nitrogen fixation in the phyllosphere. *In* The Biology of Nitrogen Fixation, ed. A. Quispel, pp. 121-69. Amsterdam: North-Holland Publishing Co.
Schank, S.C., Smith, R.L., Weiser, G.C., Zuberer, D.A., Bouton, J.H., Quesenberry, K.H., Tyler, M.E., Milam, J.R. & Littell, R.C. (1979). Fluorescent antibody technique to identify *Azospirillum brasilense* associated with roots of grasses. Soil Biology & Biochemistry, *11*, 287-95.
Scher, F.M. & Baker, R. (1980). Mechanism of biological control in a *Fusarium*-suppressive soil. Phytopathology, *70*, 412-7.
Scher, F.M. & Baker, R. (1982). Effect of *Pseudomonas putida* and a synthetic iron chelator on induction of soil suppressiveness to *Fusarium* wilt pathogens. Phytopathology, *72*, 1567-73.
Schippers, B., Geels, F.P., Hoekstra, O., Lamers, J.G., Maenhout, C.A.A.A. & Scholte, K. (1985). Yield depressions in narrow rotations caused by unknown microbial factors and their suppression by selected pseudomonads. *In* Ecology and Management of Soilborne Plant Pathogens, eds C.A. Parker, A.D. Rovira, K.J. Moore, P.T.W. Wong & J.F. Kollmorgen, pp. 127-30. St. Paul: Am. Phytopath. Soc.
Schippers, B., Bakker, A.W., Bakker, P.A.H.M., Weisbeek, P.J. & Lugtenberg, B. (1986 a). Plant growth-inhibiting and stimulating rhizosphere micro-organisms. *In* Microbial Communities in Soil, eds V. Jensen, A. Kjøller & L.H. Sørensen. Amsterdam: Elsevier Publishers (in press).
Schippers, B., Geels, F.P., Bakker, P.A.H.M., Bakker, A.W., Weisbeek, P.J. & Lugtenberg, B. (1986 b). Methods of studying plant growth-stimulating pseudomonads: problems and progress. *In* Iron, Siderophores and Plant Pathogens, ed. T.R. Swinburne. New York: Plenum Press (in press).
Schippers, B., Lugtenberg, B. & Weisbeek, P.J. (1986 c). Plant growth control by fluorescent pseudomonads. *In* Non-conventional Approaches to Disease Control, ed. I. Chet. New York: John Wiley & Sons (in press).
Schroth, M.N. & Hancock, J.G. (1982). Disease-suppressive soil and root-colonizing bacteria. Science, *216*, 1376-81.
Sen Gupta, B., Nandi, A.S. & Sen, S.P. (1982). Utility of phyllosphere $N_2$-fixing micro-organisms in the improvement of crop growth. I. Rice. Plant and Soil, *68*, 55-67.
Sen Gupta, B. & Sen, S.P. (1982). Utility of phyllosphere $N_2$-fixing micro-organisms in the improvement of crop growth. II. Wheat. Plant and Soil, *68*, 69-74.

Shapiro, S.S. & Wilk, M.B. (1965). An analysis of variance test for normality (complete samples). Biometrika, *52*, 591-611.

Sivak, B. & Person, C.O. (1973). The bacterial and fungal flora of the bark, wood, and pith of alder, black cottonwood, maple, and willow. Canadian Journal of Botany, *51*, 1985-8.

Skinner, F.A. & Uomala, P., eds (1986). Nitrogen Fixation with Non-Legumes. Plant and Soil, *90*, 1-460.

Smiley, R.W. (1981). Nontarget effects of pesticides on turfgrasses. Plant Disease, *65*, 17-23.

Sokal, R.R. & Rohlf, F.J. (1981). Biometry. The Principles and Practice of Statistics in Biological Research, 2nd edition. San Francisco: W.F. Freeman & Co.

Spurr, H.W., Jr. & Knudsen, G.R. (1985). Biological control of leaf diseases with bacteria. *In* Biological Control on the Phylloplane, eds C.E. Windels & S.E. Lindow, pp. 45-62. St. Paul: Am. Phytopath. Soc.

Suslow, T.V. (1982). Role of root-colonizing bacteria in plant growth. *In* Phytopathogenic Prokaryotes. Vol 1, eds M.S. Mount & G.H. Lacy, pp. 187-223. New York: Academic Press.

Swinburne, T.R. (1981). Iron and iron chelating agents as factors in germination, infection and aggression of fungal pathogens. *In* Microbial Ecology of the Phylloplane, ed. J.P. Blakeman, pp. 227-43. London: Academic Press.

Swinburne, T.R. & Brown, A.E. (1983). Appressoria development and quiescent infections of banana fruit by *Colletotrichum musae*. Transactions of the British Mycological Society, *80*, 176-8.

Tukey, H.B., Jr. (1970). The leaching of substances from plants. Annual Review of Plant Physiology, *21*, 305-24.

Turner, S.M., Newman, E.I. & Campbell, R. (1985). Microbial population of ryegrass root surfaces: influence of nitrogen and phosphorus supply. Soil Biology & Biochemistry, *17*, 711-5.

Vedie, R. & Le Normand, M. (1984). Modulation de l'expression du pouvoir pathogène de *Botrytis fabae* Sard. et de *Botrytis cinerea* Pers. par des bactéries du phylloplan de *Vicia faba* L. Agronomie, *4*, 721-8.

Veeger, C. & Newton, W.E., eds (1984). Advances in Nitrogen Fixation Research. The Hague/Wageningen: Martinus Nijhoff & Dr W. Junk Publishers/Pudoc.

Verona, O. (1958). La spermosphère. Annales de l'Institut Pasteur, *95*, 795-8.

Vose, P.B. (1983). Developments in nonlegume $N_2$-fixing systems. Canadian Journal of Microbiology, *29*, 837-50.

Vuurde, J.W.L. van (1978). Ecology of the root microflora of wheat. Dissertation, State University Utrecht.

Vuurde, J.W.L. van & Schippers, B. (1980). Bacterial colonization of seminal wheat roots. Soil Biology & Biochemistry, *12*, 559-65.

Warren, R.C. (1976). The occurrence of microbes among buds of deciduous trees. European Journal of Forest Pathology, *6*, 38-45.

Weller, D.M. (1984). Distribution of a take-all suppressive strain of *Pseudomonas fluorescens* on seminal roots of winter wheat. Applied and Environmental Microbiology, *48*, 897-9.

Weller, D.M. & Cook, R.J. (1983). Suppression of take-all of wheat by seed treatments with fluorescent pseudomonads. Phytopathology, *73*, 463-9.

Windels, C.E. & Lindow, S.E., eds (1985). Biological Control on the Phylloplane. St. Paul: Am. Phytopath. Soc.

# YIELD-REDUCING EFFECT OF SAPROPHYTIC LEAF FUNGI IN BARLEY CROPS

V. Smedegaard-Petersen[1] and K. Tolstrup[2]

[1]*Department of Plant Pathology, Royal Veterinary and Agricultural University, Thorvaldsensvej 40, DK-1871 Copenhagen V, Denmark*
[2]*Danish Potato Breeding Foundation, DK-7184 Vandel, Denmark*

## *INTRODUCTION*

In nature the aerial parts of growing plants are exposed to a large number of different micro-organisms including bacteria, yeasts and filamentous fungi. Most of these organisms are saprophytes living on dead organic matter. Only a relatively small number are pathogens, causing visible disease symptoms. In temperate climates the saprophytic mycoflora of the cereal phyllosphere is dominated by a few filamentous fungal species including *Cladosporium* spp., *Aureobasidium pullulans*, *Alternaria* spp. and among the yeasts *Sporobolomyces* spp. (pink yeasts) and *Cryptococcus* spp. (white yeasts) (Fokkema 1971; Flannigan & Campbell 1977; Tolstrup & Smedegaard-Petersen 1984).

It is often assumed that saprophytic micro-organisms may be indirectly beneficial to plants by acting as antagonists to plant pathogens. In the phyllosphere yeasts and bacteria can reduce conidial germination and prepenetration mycelial growth of a number of necrotrophic pathogens, and consequently reduce infection (Fokkema 1981; Blakeman & Fokkema 1982).

Antagonistic interactions between saprophytic micro-organisms and plant-pathogenic fungi in soils, as well as on leaf surfaces include the production of inhibitory substances and competition for nutrients (Lockwood & Filonow 1981). Phyllosphere nutrients may originate from pollen grains and aphid honeydew or leachates from spores of the pathogen (Fokkema 1984).

Although the concept of antagonism seems to be in favour of plants, there is little experimental evidence that saprophytes promote plant growth (Rabbinge *et al.* 1984). On the contrary, evidence has been put forward that *Sporobolomyces* and *Cryptococcus* spp. may be harmful to plants by degrading the cuticle (Ruinen 1966; MacNamara & Dickinson 1981) and filamentous saprophytic fungi may promote senescence of mature cereal leaves (Dickinson 1981). Very recent investigations have suggested that saprophytic, filamentous leaf fungi can cause significant yield losses in barley crops without causing visible disease symptoms (Tolstrup 1984; Tolstrup & Smedegaard-Petersen 1984). The yield-limiting effect of these saprophytes seems to result from active energy-requiring defense reactions which drain the host plants of stored energy, leading to reduced plant growth and yield, in much the same way as avirulent races of barley powdery mildew can reduce the yield of barley crops without causing visible symptoms (Smedegaard-Petersen & Stølen 1981; Smedegaard-Petersen & Tolstrup 1985).

The present review does not intend to cover the mycological and floristic aspects of the phyllosphere mycoflora but will concentrate on recent research on the yield-limiting effect of saprophytic leaf fungi on barley crops and the underlying physiological and histochemical reactions leading to this effect.

## *THE OCCURRENCE OF SAPROPHYTIC LEAF FUNGI IN BARLEY CROPS*

Densely growing cereal crops provide a suitable microclimate for the growth and multiplication of saprophytic micro-organisms. When the lower leaves die, they rapidly become colonized by a large variety of fungi which upon sporulation deposit numerous spores on the upper green leaves (Warnock 1973). Floristic studies dealing specifically with the leaf mycoflora of barley crops suggest that, in the temperate climate, *Sporobolomyces* spp. (pink yeasts) and *Cladosporium* spp. are dominant in the fungal population and a variety of other fungi are present in varying frequencies (Dickinson 1973; Diem 1974).

The mycoflora of the green aerial parts of spring barley crops was examined in Danish field trials during 1981 to 1983 (Tolstrup 1984; Tolstrup & Smedegaard-Petersen 1984). The microbiological methods employed included leaf incubation in moist chambers, leaf printing, spore fall and spore trapping with a Schwarzbach spore sampler. Direct microbiological observations were made on leaf impression films. The composition of the mycoflora on green leaves in 1982 appears in Table 1. *Cladosporium* spp. and *Sporobolomyces roseus* (pink yeasts) were the dominant inhabitants occurring with very high frequencies. Other saprophytic species including *Alternaria* spp., *Aureobasidium pullulans* and *Epicoccum nigrum* were typically present in low frequencies early in the growth season but increased in number from the beginning of July (Fig. 1). A total of 34 different fungal genera and species were identified (Table 1).

## *PHYSIOLOGICAL DEFENSE-LIKE RESPONSES OF BARLEY LEAVES TO INOCULATION WITH SAPROPHYTES*

Spores of *Cladosporium macrocarpum*, *C. herbarum*, *C. cladosporioides* and *Alternaria alternata*, inoculated onto green barley leaves, readily germinate and attempt to penetrate the outer cell wall. Germ tubes of spores of *Cladosporium* spp. produce appressoria-like hyphal swellings but there is no obvious association between these structures and the attempted penetration. In contrast, the attempted penetration of *A. alternata* typically occurs from appressoria (Tolstrup 1984).

By the application of staining techniques and fluorescence microscopy, green barley leaves were found to react to attempted penetration by pronounced histochemical and physiological changes, including the formation of cell wall appositions, 'papillae', between the cell wall and the plasmalemma and disc-shaped areas, 'haloes', surrounding the sites of attempted penetration (Tolstrup 1984; Tolstrup & Smedegaard-Petersen

1984). Thus, all four fungal species mentioned induced papillae and haloes within 3 days after inoculation (Figs 2 and 3). *Cladosporium* spp., especially *C. cladosporioides*, preferentially induced the production of papillae and haloes in the lateral epidermal cell walls, while *A. alternata* usually induced the formation of papillae and haloes in the

Table 1. The composition of the mycoflora on green barley leaves in untreated field plots in 1982 (Tolstrup & Smedegaard-Petersen 1984).

| Frequency | Fungal name |
|---|---|
| Very common | *Cladosporium cladosporioides* |
| | *C. herbarum* |
| | *C. macrocarpum* |
| | *Sporobolomyces roseus* (pink yeasts) |
| Common | *Alternaria alternata* |
| | *A. tenuissima* |
| | *Ascochyta* sp. |
| | *Aureobasidium pullulans* |
| | *Bipolaris sorokiniana* (syn.: *Helminthosporium sativum*) |
| | *Botrytis cinerea* |
| | *Cephalosporium* sp. |
| | *Cryptococcus albidus* var. *albidus* |
| | *C. laurentii* var. *laurentii* |
| | *Cryptococcus* spp. (and other white yeasts) |
| | *Drechslera teres* |
| | *Epicoccum nigrum* |
| | *Erysiphe graminis* |
| | *Fusarium nivale* |
| | *Fusarium* spp. |
| | *Mycosphaerella* spp. |
| | *Septoria nodorum* |
| | *Stemphylium botryosum* |
| | *Ulocladium atrum* |
| | Mycelia sterilia |
| Uncommon | *Itersonilia* sp. |
| | *Gonatobotrys* sp. |
| | *Periconia byssoides* |
| | *Pleospora* sp. |
| | *Phyllosticta* sp. |
| | *Puccinia hordei* |
| | *Rhynchosporium secalis* |
| | *Septoria passerinii* |
| | *Tilletiopsis minor* |
| | *Torula herbarum* |
| | other unidentified species |

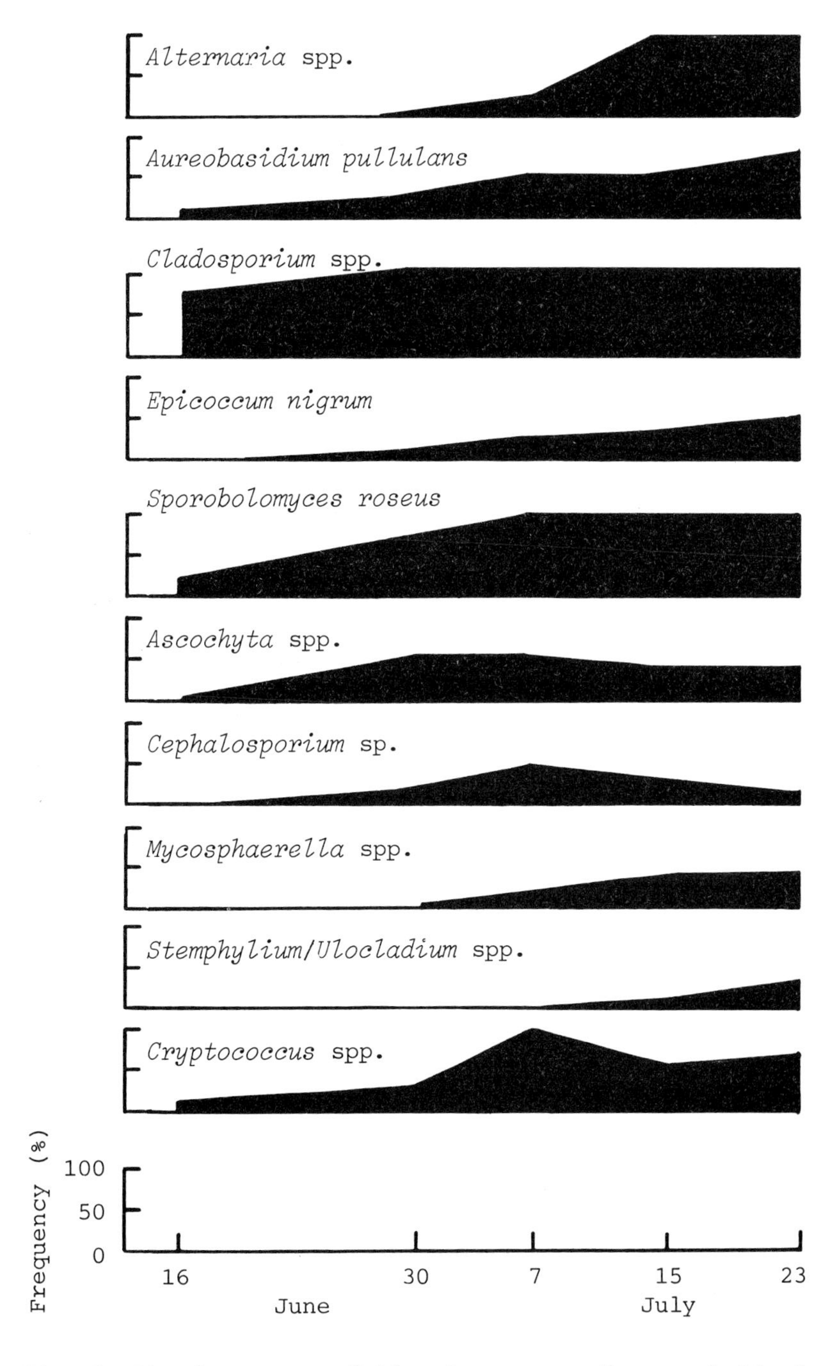


Fig. 1. The frequency of 10 main groups of saprophytic leaf fungi on green penultimate leaves of field-grown barley in 1982. Frequency is expressed as percentage of leaf discs (7 mm) colonized; n = 40 (Tolstrup 1984).

outer horizontal cell walls. After inoculation with *C. macrocarpum*, the first papillae and haloes were apparent in the lateral epidermal cell walls after only 8 hours and in the outer epidermal cell walls after 19 hours. In case of *C. herbarum* and *A. alternata* attempted penetration was detected after 24 and 15 hours, respectively.

For each of the four fungal species *C. macrocarpum*, *C. herbarum*, *C. cladosporioides* and *A. alternata*, at least 500 sites of attempted penetration were microscopically and histochemically examined. Only in one single case with *A. alternata* was the infection hypha capable of penetrating through the papilla and a short distance into the mesophyll before the growth ceased. Since these fungi are unable to cause successful infection and to establish a nutritional relationship with green metabolically active barley leaves, they must be considered as true saprophytes and not as weak parasites as indicated by Dickinson (1981).

There is good evidence to suggest that the histochemical and physiological changes in barley leaves inoculated with saprophytes are part of a successful defense system preventing their entry into the host. The inability of *Cladosporium* spp. to penetrate through the papillae is consistent with previous studies suggesting a close relationship between the failure of fungal infection and papillae incidence (Aist *et al.* 1979 a; Ride & Pearce 1979; Sherwood & Vance 1980), papillae fluorescence (Mayama & Shishiyama 1978; Kita *et al.* 1980), papillae size (Skou 1982), and increases in the rate of respiration (Smedegaard-Petersen 1980, 1982). Although studies on papillae involvement in resistance are not conclusive (Bushnell & Bergquist 1975; Aist 1976; Aist *et al.* 1979 b), convincing evidence for a role of papillae as a resistance factor against non-pathogenic fungi, appears from experiments in which treatment of leaf tissue with cycloheximide, an inhibitor of protein synthesis, blocked papillae formation and permitted penetration of originally non-pathogenic fungi into leaves of gramineous species (Sherwood & Vance 1980). Characteristically, the formation of papillae and haloes around the sites of attempted penetrations coincides with a marked increase in the rate of plant respiration, the appearance of autofluorescence in the surrounding host cells (Smedegaard-Petersen 1982) and the occurrence of hypersensitively reacting epidermal cells.

It appears from the previous discussion that green, metabolically active barley plants are not unaffected by abundantly occurring leaf saprophytes such as *C. macrocarpum*, *C. herbarum*, *C. cladosporioides* and *A. alternata*. Although the plants do not react to saprophyte inoculation with visible symptoms, it will become apparent from the following that the highly elevated physiological activity, probably defense reactions, occur at the expense of stored host energy and therefore ultimately may lead to reduction in the yield (Tolstrup 1984; Smedegaard-Petersen & Tolstrup 1985).

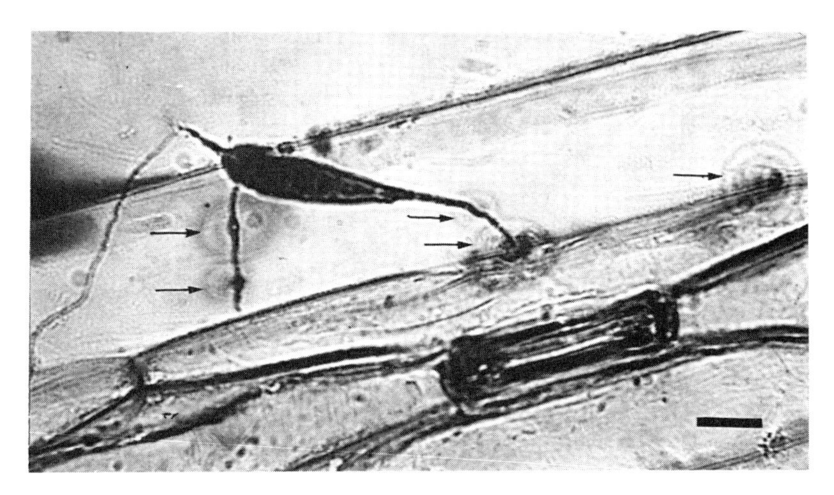

Fig. 2. *Alternaria alternata* spore attempting penetration into a green barley leaf thus eliciting the formation of papillae and haloes around the site of the unsuccessful penetration. Bar represents 20 µm.

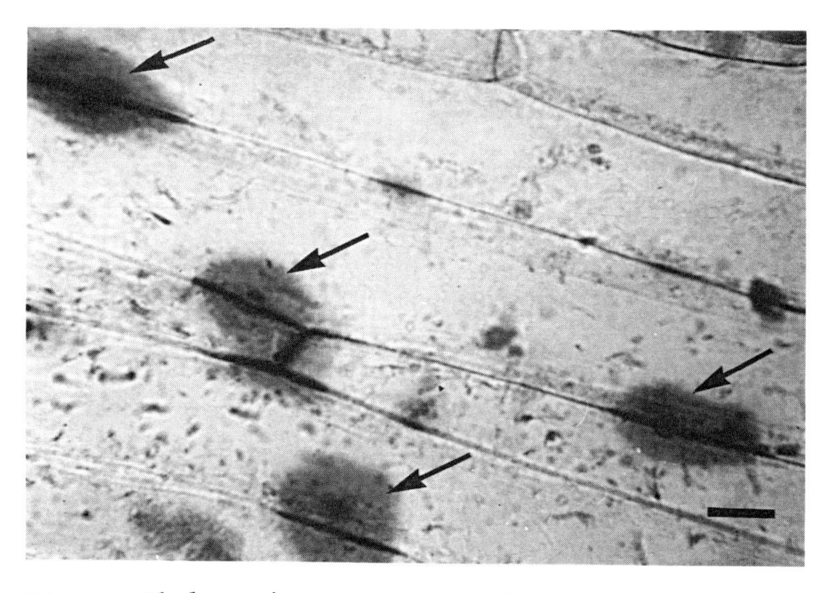

Fig. 3. *Cladosporium macrocarpum* attempting penetration into a green barley leaf thereby eliciting the formation of papillae and haloes. Bar represents 20 µm.

### *CONTROL OF LEAF SAPROPHYTES WITH FUNGICIDES INCREASES GRAIN YIELD*

To examine the effect of leaf saprophytes on growth and yield of barley crops, field trials were conducted in which attempts were made to keep the leaves 'clean' by repeated applications of the fungicides captafol, triadimefon, propiconazole and prochloraz (Tolstrup 1984; Tolstrup & Smedegaard-Petersen 1984). Since the prerequisite for such experiments is that they are carried out in nearly disease-free plots, the trials were located in 1982 in a field of oats and in 1983 in a field of pea to avoid inoculum from neighbouring barley crops. As a consequence of isolation, only little disease occurred in the plots in both years. The activity of the leaf mycoflora was monitored and plants from different treatments were analyzed for senescence and chlorophyll content regularly during the growth season.

In 1982 treatments with captafol, propiconazole and prochloraz increased the grain yield significantly by 10 to 16% (Table 2). The strongest effect was obtained with repeated treatments with captafol which reduced *Cladosporium* spp. to almost nil (Fig. 4) and increased grain yield significantly by 16% (Table 2). The kernel weight showed the same pattern of differences although not significantly. Triadimefon had less effect on *Cladosporium* spp. and no significant effect on the yield. In trials the reduction in *Cladosporium* spp. and the increase in yield correlated well with a delay in senescence and a high chlorophyll content (Fig. 5). An exception was triadimefon that on some occasions did

Table 2. The effect of fungicide treatment on grain yield in field trials with little or no disease in the plots (Tolstrup & Smedegaard-Petersen 1984).

| Treatment | 1982 | | | 1983 | | |
|---|---|---|---|---|---|---|
| | Grain yield | | Kernel weight (%) | Grain yield | | Kernel weight (%) |
| | $10^2$ kg $ha^{-1}$ | % | | $10^2$ kg $ha^{-1}$ | % | |
| a. untreated (control) | 44.5 | 100 | 100 | 38.6 | 100 | 100 |
| b. captafol 3 x early | - | - | - | 40.0 | 104 | 103 |
| c. captafol 3 x late | - | - | - | 39.6 | 103 | 104 |
| d. captafol 5 x | 51.8*[1] | 116* | 105 | 40.4* | 105* | 103 |
| e. triamedifon 2 x | 47.3 | 106 | 102 | 38.4 | 99 | 100 |
| f. propiconazole 2 x | 49.0* | 110* | 103 | 40.9* | 106* | 103 |
| g. prochloraz 2 x | 48.9* | 110* | 105 | 38.9 | 101 | 103 |

[1] *: Significantly different from the control ($P \leq 0.05$).

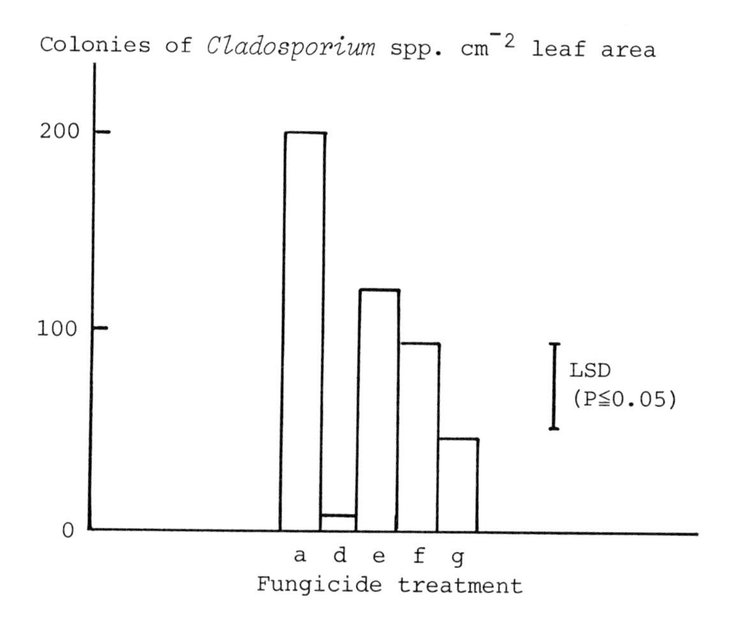


Fig. 4. The effect of four fungicides on the occurrence of *Cladosporium* spp. in barley in field trials in 1982, determined on 15 July by the leaf print method. a: control, d: captafol 5x, e: triadimefon 2x, f: propiconazole 2x, g: prochloraz 2x (Tolstrup 1984).

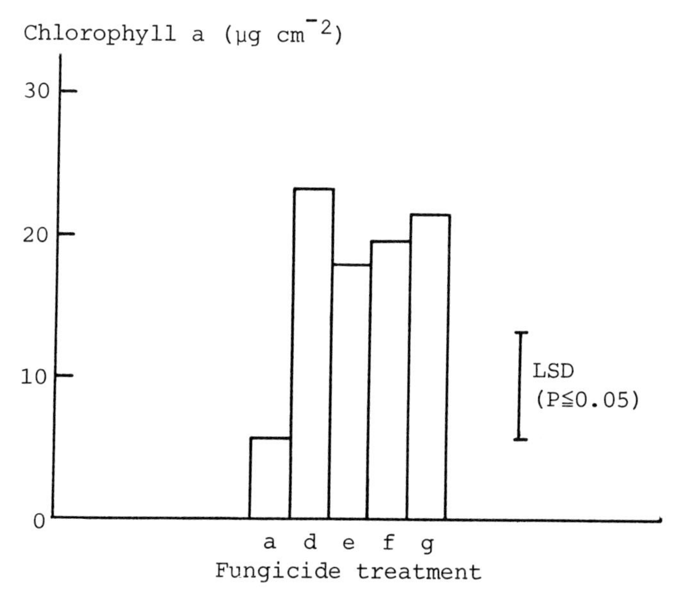


Fig. 5. The effect of fungicide treatments on the chlorophyll content of barley leaves in field trials, 21 July, 1982. a: control, d: captafol 5x, e: triadimefon 2x, f: propiconazole 2x, g: prochloraz 2x (Tolstrup & Smedegaard-Petersen 1984).

delay senescence without having a significant effect on *Cladosporium* spp. and yield. A similar delaying effect on senescence has been reported for benomyl, a benzimidazole compound with cytokinin-like properties which are believed to account for its antisenescence behaviour (Thomas 1974). In 1983, the trials showed the same pattern as in 1982 but with lower yield increases, probably due to the late sowing date and the extremely dry weather, unfavourable for fungal development.

A complete evaluation of the results strongly suggests that the increases in yield in the fungicide-treated plots resulted from the control of filamentous saprophytic surface fungi, especially *Cladosporium* spp., on the green leaves, although it cannot be excluded that some of the fungicides may have had a direct physiological effect on the plants. Reduction or elimination of the saprophytes may save the plant for energy-consuming defense reactions and thus make more host energy available for grain production.

*REPEATED INOCULATION OF BARLEY PLANTS WITH CLADOSPORIUM SPP. DURING THE GROWTH PERIOD REDUCES GRAIN YIELD*

The previous section describes field trials designed to reveal the effect of saprophytic surface fungi on the yield of barley by controlling these organisms with fungicides in nearly disease-free field plots. In subsequent yield experiments, we attempted to obtain additional and more direct proof for such an effect in growth chamber experiments by inoculating barley plants (cv. Rupal) repeatedly during the growth season with *Cladosporium* spp. (Tolstrup 1984; Tolstrup & Smedegaard-Petersen 1984). Each of the 12 chambers employed had a size of 1 $m^2$ and contained each 180 plants in eight pots. The temperature was regulated to *ca* 15°C during the day and 10°C at night, and the light was adjusted to a photoperiod of 18 h of light (570 W $m^{-2}$) and 6 h of darkness. The plants in 6 of the 12 chambers were inoculated six times

Table 3. The effect of saprophytic *Cladosporium* spp. on the yield of barley plants in growth chamber yield experiments. Plants were inoculated six times during the growth period with a mixture of *C. macrocarpum* and *C. herbarum* (Tolstrup 1984).

| Treatment | Grain yield | | Kernels per spike (%) | Kernel weight (%) | Straw yield (%) | Plant height (cm) | Tillers per plant (nrs) |
|---|---|---|---|---|---|---|---|
| | g/pot | % | | | | | |
| Control | 16.9 | 100 | 100 | 100 | 100 | 79.2 | 6.5 |
| Inoculated | 15.4**[1] | 91** | 93** | 98*[1] | 97 | 77.6** | 6.4 |

[1] **: Significantly different from the control (P≦0.01); *: significantly different from the control (P≦0.05).

during the growth period with a mixture of spores of *C. macrocarpum* and *C. herbarum* isolated from a barley field. The inoculation dates were chosen so that every newly expanded leaf was inoculated at least once. Regularly during the growth period, leaves were sampled for analyses of chlorophyll a.

Although the inoculated plants did not show any visible disease symptoms during the growth period, the grain yield was reduced by 9%, the kernel weight by 2%, the number of kernels per spike by 7% and the straw length by 1.6 cm (Table 3). In addition, there was an insignificant reduction in the number of tillers per plant. Also, the chlorophyll content of successive leaves of inoculated plants was reduced, most markedly in the flag leaves.

## *CONCLUSIONS*

Floristic investigations on the qualitative and quantitative composition of the saprophytic leaf mycoflora in Danish barley crops demonstrated that *Cladosporium herbarum*, *C. cladosporioides*, *C. macrocarpum* and *Alternaria alternata* are dominant among the filamentous fungi. Under favourable climatic conditions especially *Cladosporium* spp. are present on the plant surfaces in large densities from mid June onwards. All the species were strongly inhibited in their growth by the fungicides captafol, propiconazole and prochloraz, while triadimefon had a lesser effect.

None of the leaf saprophytes mentioned are able to infect and cause disease of green, metabolically active barley plants. However, upon attempted penetration they provoke a highly increased biosynthetic activity in the plants, probably defense reactions, which seem to occur on the expense of stored host energy and under favourable environmental conditions for fungal development led to significant reductions in grain yield. Under such conditions the control of saprophytic leaf fungi with fungicides resulted in significant increases in grain yield even though little or no disease was present in the crop. In yield experiments in growth chambers inoculation with *Cladosporium* spp. six times during the growth period caused significant decreases in growth and grain yields of barley plants without causing visible disease symptoms.

## *ACKNOWLEDGEMENT*

Much of the research described was supported by grants from the Danish Agricultural and Veterinary Research Council.

## *REFERENCES*

Aist, J.R. (1976). Papillae and related wound plugs of plant cells. Annual Review of Phytopathology, *14*, 145-63.

Aist, J.R., Kunoh, H. & Israel, H.W. (1979 a). Challenge appressoria of *Erysiphe graminis* fail to breach preformed papillae of a compatible barley cultivar. Phytopathology, *69*, 1245-50.

Aist, J.R., Waterman, M.A. & Israel, H.W. (1979 b). Papillae and penetration: some problems, procedures and perspectives. *In* Recognition and Specificity in Plant Host-Parasite Interactions, eds J.M. Daly & I. Uritani, pp. 85-95. Baltimore: University Park Press.

Blakeman, J.P. & Fokkema, N.J. (1982). Potential for biological control of plant diseases on the phylloplane. Annual Review of Phytopathology, *20*, 167-92.

Bushnell, W.R. & Bergquist, S.E. (1975). Aggregation of host cytoplasm and the formation of papillae and haustoria in powdery mildew of barley. Phytopathology, *65*, 310-8.

Dickinson, C.H. (1973). Effects of ethirimol and zineb on phylloplane microflora of barley. Transactions of the British Mycological Society, *60*, 423-31.

Dickinson, C.H. (1981). Biology of *Alternaria alternata*, *Cladosporium cladosporioides* and *C. herbarum* in respect of their activity on green plants. *In* Microbial Ecology of the Phylloplane, ed. J.P. Blakeman, pp. 169-84. London: Academic Press.

Diem, H.G. (1974). Micro-organisms of the leaf surface: estimation of the mycoflora of the barley phyllosphere. Journal of General Microbiology, *80*, 77-83.

Flannigan, B. & Campbell, I. (1977). Pre-harvest mould and yeast floras on the flag leaf, bracts and caryopsis of wheat. Transactions of the British Mycological Society, *69*, 485-94.

Fokkema, N.J. (1971). The effect of pollen in the phyllosphere of rye on colonization by saprophytic fungi and on infection by *Helminthosporium sativum* and other leaf pathogens. Netherlands Journal of Plant Pathology, *77* (suppl. 1), 1-60.

Fokkema, N.J. (1981). Fungal leaf saprophytes, beneficial or detrimental? *In* Microbial Ecology of the Phylloplane, ed. J.P. Blakeman, pp. 433-54. London: Academic Press.

Fokkema, N.J. (1984). Competition for endogenous and exogenous nutrients between *Sporobolomyces roseus* and *Cochliobolus sativus*. Canadian Journal of Botany, *62*, 2463-8.

Kita, N., Toyoda, H., Yano, T. & Shishiyama, J. (1980). Correlation of fluorescent appearance in papilla with unsuccessful penetration attempts in susceptible barley inoculated with *Erysiphe graminis* f. sp. *hordei*. Annals of the Phytopathological Society of Japan, *46*, 594-7.

Lockwood, J.L. & Filonow, A.B. (1981). Responses of fungi to nutrient limiting conditions and to inhibitory substances in natural habitats. Advances in Microbial Ecology, *5*, 1-61.

MacNamara, O.C. & Dickinson, C.H. (1981). Microbial degradation of plant cuticle. *In* Microbial Ecology of the Phylloplane, ed. J.P. Blakeman, pp. 455-73. London: Academic Press.

Mayama, S. & Shishiyama, J. (1978). Localized accumulation of fluorescent and u.v.-absorbing compounds at penetration sites in barley leaves infected with *Erysiphe graminis hordei*. Physiological Plant Pathology, *13*, 347-54.

Rabbinge, R., Brouwer, A., Fokkema, N.J., Sinke, J. & Stomph, T.J. (1984). Effects of the saprophytic leaf mycoflora on growth and productivity of winter wheat. Netherlands Journal of Plant Pathology, *90*, 181-97.

Ride, J.P. & Pearce, R.B. (1979). Lignification and papilla formation at sites of attempted penetration of wheat leaves by nonpathogenic fungi. Physiological Plant Pathology, *15*, 79-92.

Ruinen, J. (1966). The phyllosphere IV. Cuticle decomposition by microorganisms in the phyllosphere. Annales de l'Institut Pasteur, *111* (suppl.), 342-6.

Sherwood, R.T. & Vance, C.P. (1980). Resistance to fungal penetration in Gramineae. Phytopathology, *70*, 273-9.

Skou, J.P. (1982). Callose formation responsible for the powdery mildew resistance in barley with genes in the ml-o locus. Phytopathologische Zeitschrift, *104*, 90-5.

Smedegaard-Petersen, V. (1980). Increased demand for respiratory energy of barley leaves reacting hypersensitively against *Erysiphe graminis*, *Pyrenophora teres* and *Pyrenophora graminea*. Phytopathologische Zeitschrift, *99*, 54-62.

Smedegaard-Petersen, V. (1982). The effect of defence reactions on the energy balance and yield of resistant plants. *In* Active Defense Mechanisms in Plants, ed. R.K.S. Wood, pp. 299-315. New York: Plenum Press.

Smedegaard-Petersen, V. & Støen, O. (1981). Effect of energy-requiring defense reactions on yield and grain quality in a powdery mildew-resistant barley cultivar. Phytopathology, *71*, 396-9.

Smedegaard-Petersen, V. & Tolstrup, K. (1985). The limiting effect of disease resistance on yield. Annual Review of Phytopathology, *23*, 475-90.

Thomas, T.H. (1974). Investigations into the cytokinin-like properties of benzimidazole-derived fungicides. Annals of Applied Biology, *76*, 237-41.

Tolstrup, K. (1984). Saprophytic fungi in the phyllosphere of barley and their effects on the plants growth and grain yield. Dissertation, Royal Vet. & Agric. Univ., Copenhagen.

Tolstrup, K. & Smedegaard-Petersen, V. (1984). Saprophytic leaf fungi on barley and their effect on leaf senescence and grain yield. Växtskyddsnotiser, *48*, 66-75.

Warnock, D.W. (1973). Origin and development of fungal mycelium in grains of barley before harvest. Transactions of the British Mycological Society, *61*, 49-56.

# SECTION III

# ENDOPHYTIC LEAF FUNGI

# TAXONOMY OF ENDOPHYTIC FUNGI OF AERIAL PLANT TISSUES

O. Petrini

*Bodenacherstrasse 86, CH-8121 Benglen, Switzerland*

## *INTRODUCTION*

### *General considerations*

Fungi live in mutualistic, antagonistic or neutral symbiosis with a wide range of both autotrophic and heterotrophic organisms. The nature of these associations is diverse and exhibits varying degrees of intimacy and nutritional interdependence. Topographically, however, a fungus can either live on or within its host. De Bary (1866) defined the fungi living on the surface of their host as epiphytes, while those living inside the plant tissue he termed endophytes. The delimitation of these two states is not always very sharp.

The taxonomy and distribution of endophytic antagonistic symbionts (the so-called endophytic pathogens, e.g. the rusts) have been extensively studied in plant pathology. However, it was nearly forty years after Neill's (1939) first report on an unidentified endophyte of ryegrass (*Lolium perenne* L.) and his (Neill 1941) tentative identification of the fungus as *Endoconidium temulentum* Prillieux & Delacroix (possibly the anamorph of a *Gloeotinia* sp.) before extensive research work started on the ecology of endophytic fungi living as mutualistic and neutral symbionts (e.g. Bernstein & Carroll 1977; Carroll & Carroll 1978; Petrini & Müller 1979; Luginbühl & Müller 1980). This paper aims to discuss some methodological problems related to the isolation, sporulation and identification of endophytes as well as to give some information on the taxonomic position of the fungi known to live endophytically within healthy-looking plant tissues.

### *Isolation of endophytes*

In order to eliminate spores or mycelial fragments attached to the surface of the plant material investigated, an effective surface sterilization technique has been developed (Table 1). It can be assumed, however, that a small number of spores or hyphal fragments present on the epidermis of the host could survive the sterilization procedures, although spore survival rate experiments have shown the efficacy of the techniques used (Petrini 1984; O. Petrini, unpublished results). On the other hand, epiphytes can penetrate into the host tissues, as was demonstrated by O'Donnell & Dickinson (1980). The presence in the endophyte populations of fungi known to be common epiphytes (e.g. *Alternaria alternata* (Fr.) Keissler, *Cladosporium cladosporioides* (Fr.) De Vries) or contaminants (e.g. *Aspergillus* spp., *Penicillium* spp.) can be partly explained by the reasons mentioned above, although their endophytic character cannot be excluded a priori.

*Induction of sporulation in endophytic fungi*

Most endophytes will sporulate after a few weeks at 15 to 21°C, either in darkness or in daylight on 1% malt extract agar. Only a small proportion of isolates will remain sterile; the number of sterile strains varies but is roughly constant for a given host.

Numerous methods can be used to induce sporulation in sterile cultures.

Table 1. Isolation of endophytes: sterilization times and concentration of the Na-hypochlorite solutions used for the surface sterilization. The following sequential steps are involved: A, first dip in 96% ethanol for 1 minute (lichens and mosses for ½ minute); B, sterilization in Na-hypochlorite; C, second dip in 96% ethanol for ½ minute. Experimental values for some plant species so far investigated.

| Plant species | Sterilization[1] in Na-hypochlorite[2] | | References |
|---|---|---|---|
| | Minutes | Dilution | |
| Lichens | 1 | 1:5 | O. Petrini (unpublished) |
| Mosses | 1 | 1:5 | O. Petrini (unpublished) |
| Ferns | 3 | 1:5 | Dreyfuss & Petrini (1984) |
| Conifers a) needles | 5 | 1:2 | Carroll & Carroll (1978) |
| b) twigs | 7 | 1:2 | Petrini & Müller (1979) |
| Monocotyledons: | | | |
| *Triticum aestivum* L. (leaves and culms) | 3 | 1:5 | Riesen & Sieber (1985) |
| Dicotyledons: | | | |
| *Arctostaphylos uva-ursi* | 3 | 1:5 | Widler & Müller (1984) |
| *Brassica napus* L. | 3 | 1:5 | H. Ruckstuhl & O. Petrini (unpublished) |
| *Erica carnea* a) leaves | 3 | 1:5 | Oberholzer (1982) |
| b) stems | 5 | 1:5 | Oberholzer (1982) |
| *Rhododendron* spp. | 3 | 1:5 | O. Petrini (1985) |
| *Vaccinium* spp. | 3 | 1:5 | O. Petrini (1985) |

[1] If not otherwise stated, times and concentrations refer to the sterilization of leaves.

[2] Commercially available Chloros containing 15% available chlorine.

Exposure at 8°C under fluorescent or near-ultraviolet light with a 12 h dark-light cycle or incubation at 4°C in the darkness can induce the sporulation in sterile cultures. The inoculation of sterile strains on natural media (corn meal agar, potato/carrot agar, vegetable V-8 agar, etc.), on chemically defined media (Czapek Dox agar, peptone dextrose agar, starch agar, etc.) or on pieces of the host plant sterilized for 20 minutes at 120°C can yield good results. The most important factor in the induction of sporulation, however, is time: isolates of endophytes may fruit only two or three months after inoculation, and some require up to 12 to 14 months. Endophytes often cease to sporulate after two to three transfers. An induction of sporulation can then be achieved by inoculating the recalcitrant strain on pieces of the host plant sterilized as described above and transferring it subsequently onto 0.1% malt extract agar.

### *Problems related to the identification of endophytes*

The identification of endophytic fungi has proved to be extremely difficult, mostly because of the lack of information on the cultural characters of species already described, but also because very little is known about the peculiar microhabitat represented by plant tissues. Ascomycetes, Basidiomycetes, Deuteromycetes, and very few Oomycetes have so far been isolated as endophytes (Table 2). The rare occurrence of Basidiomycetes can be possibly attributed at least in part to the use of unspecific culture media, not suitable to slow-growing fungi or to organisms requiring defined growth factors.

Fungi forming their teleomorph in culture can usually be identified easily to genus and to species level; new species of Ascomycetes and Basidiomycetes can therefore be detected and described without too much trouble. Within the Deuteromycetes, however, the situation is complicated by the lack of reliable descriptions of known hyphomycetous or coelomycetous species in culture, as well as by the poor knowledge of the pleomorphic states which can occur in a single species during its life cycle. Some strains of coelomycetes or hyphomycetes, for instance, show a broader variation in shape and size of their conidiomata, conidiophores and conidia in culture than they usually do on their hosts; for example, *Septoria tritici* Rob., the anamorph of *Phaeosphaeria tritici* (Garov) Hedjaroude, shows a yeast-like growth during the first weeks in culture and forms conidiomata only after a few months. *Epichloë typhina* (Pers.) Tul. has been studied in detail by Sampson (1933), Neill (1941), Morgan-Jones & Gams (1982) and Latch *et al.* (1984). The anamorph of this species has been described extensively by Morgan-Jones & Gams (1982) and by Bacon *et al.* (1977); when *E. typhina* is grown in pure culture, the *Sphacelia* anamorph known from the host is not formed and conidiophores and conidia do not aggregate into stromata. On the other hand, although it is comparatively easy to distinguish between an acervulus and a sporodochium found on the host, it is almost impossible to do so when dealing with cultures, thus complicating considerably the decision-making process. Particularly within the coelomycetes it is difficult to come to a definitive identification, because a large number of species and genera which can be easily distinguished on the host are virtually indistinguishable in culture.

Table 2. Synopsis of published and unpublished results on the proportion of Ascomycetes, Basidiomycetes, Deuteromycetes and members of other fungal classes in % of the total number of species isolated as endophytes from the leaves and stems of phanerogams. Due to the often small sample size used in the investigations, these figures have only an indicative character.

| Host plant | Ascomycetes | Basidiomycetes | Coelomycetes | Hyphomycetes | Others | References |
|---|---|---|---|---|---|---|
| *Abies alba* | 10 | 0 | 38 | 47 | 5 | I. Brunner (unpublished) |
| *Arctostaphylos uva-ursi* | 22 | 1 | 40 | 36 | 1 | Widler & Müller (1984) |
| *Brassica napus* | 18 | 3 | 15 | 60 | 4 | H. Ruckstuhl & O. Petrini (unpublished) |
| *Calluna vulgaris* | 11 | 0 | 43 | 45 | 1 | Petrini (1985) |
| *Erica carnea*[1] | 11 | 0 | 34 | 49 | 6 | Oberholzer (1982) |
| *Juniperus communis* | 22 | 2 | 30 | 44 | 2 | Petrini & Müller (1979) |
| *Loiseleuria procumbens* | 22 | 0 | 45 | 31 | 2 | Petrini (1985) |
| *Pinus cembra* | 28 | 3 | 39 | 28 | 2 | O. Petrini *et al.* (unpublished) |
| *Triticum aestivum* | 36 | 1 | 11 | 47 | 5 | Riesen & Sieber (1985) |
| *Ulex europaeus* | 27 | 0 | 23 | 50 | 0 | Fisher *et al.* (1986) |
| *U. gallii* | 28 | 0 | 24 | 48 | 0 | Fisher *et al.* (1986) |
| *Vaccinium myrtillus* | 18 | 3 | 33 | 44 | 2 | Petrini (1985) |
| *V. vitis-idaea* | 17 | 0 | 52 | 30 | 1 | Petrini (1985) |

[1] Some root isolates are also considered.

*THE DISTRIBUTION OF ENDOPHYTES WITHIN THE PLANT KINGDOM*

*General considerations*

More than 200 species of phanerogams and cryptogams from tropical, temperate and alpine regions have been investigated for the presence of endophytes. Endophytic fungi have been isolated from all species of phanerogams studied, and some reports indicate that endophytes are also present in mosses, liverworts and ferns (Boullard 1951, 1957, 1979; Schuster 1966; Dreyfuss & Petrini 1984). On the basis of these considerations it can be postulated that all living plants probably host endophytic fungi. Three plant families, the Coniferae, Ericaceae and Gramineae have been intensively sampled; therefore, I shall

Table 3. Distribution frequencies of the most commonly observed endophyte taxa in the leaves of some conifers. The distribution frequency is computed as the proportion of all collecting sites sampled in which a given taxon was found on a given host.

| Host plant | Fungal taxon | Distribution frequency (%) |
|---|---|---|
| *Abies alba*[1 2] | *Cryptocline abietina* | 100 |
| *A. amabilis*[3] | *Phyllosticta* sp. 1 | 56 |
| *A. grandis*[3] | *Phyllosticta* sp. 1 | 72 |
| | *Cryptocline abietina* | 33 |
| *A. lasiocarpa*[3] | *C. abietina* | 80 |
| *A. magnifica*[3] | *Phyllosticta* sp. 1 | 100 |
| | *C. abietina* | 33 |
| *A. procera*[3] | *Phyllosticta* sp. 1 | 58 |
| *Calocedrus decurrens*[4] | *Linodochium* sp. | 100 |
| *Chamaecyparis lawsoniana*[4] | *Chloroscypha seaveri* f. *lawsoniana* | 60 |
| | anamorph of *Xylaria* | 80 |
| *Juniperus communis*[5] | *Anthostomella formosa* | 80 |
| | anamorph of *Holmiella sabina* | 80 |
| | *Kabatina juniperi* | 90 |
| *J. occidentalis*[4] | anamorph of *Retinocyclus abietis* | 100 |
| *Pinus attenuata*[3] | *Naemacyclus* sp. | 50 |
| *P. cembra*[2] | *Brunchorstia pinea* | 100 |
| | *Sirodothis* sp. | 100 |
| *P. lambertiana*[3] | *Naemacyclus minor* | 25 |
| *Pseudotsuga menziesii*[3] | *Rhabdocline parkeri* | 83 |
| *Sequoia sempervirens*[3] | *Chloroscypha chloromela* | 100 |
| *Thuja plicata*[4] | *C. seaveri* | 100 |

[1] I. Brunner (unpublished).
[2] F. Canavesi (unpublished).
[3] Carroll & Carroll (1978).
[4] Petrini & Carroll (1981).
[5] Petrini & Müller (1979).

limit the following considerations to these three plant families.

*The endophytes of conifers*

The endophytes of conifer needles have been extensively studied by Carroll *et al.* (1977), Carroll & Carroll (1978), Petrini & Müller (1979), Petrini & Carroll (1981) and in a number of unpublished investigations by I. Brunner, F. Cuny, F. Canavesi, G. Schnell and O. Petrini. A synopsis of the fungal species most commonly isolated is presented in Table 3. Only those species which accounted for at least 10% of the total overall infection in a given host have been included. For extensively sampled conifers, such as *Juniperus communis* L. (Petrini & Müller 1979) or *Pseudotsuga menziesii* (Mirb.) Franco (Carroll & Carroll 1978) up to 110 (mean value: 60) fungal species for a given host could be isolated; the majority (80 to 90%) were seen infrequently or only

Table 4. Distribution frequencies of the most commonly observed endophyte taxa in some Ericaceae. The computation of the distribution frequency is defined in Table 3.

| Host plant | Fungal taxon | Distribution frequency (%) |
|---|---|---|
| *Arctostaphylos uva-ursi*[1 2 3] | *Coccomyces arctostaphyli* | 72 |
| | *Cryptocline arctostaphyli* | 90 |
| | anamorph of *Godronia callunigera* | 5 |
| | *Phyllosticta pyrolae* | 36 |
| *Calluna vulgaris*[2 4] | anamorph of *Godronia callunigera* | 75 |
| *Erica carnea*[2 5] | *Cryptocline arctostaphyli* | 5 |
| | *Phyllosticta pyrolae* | 88 |
| *Gaultheria shallon*[3] | *Pezicula* sp. | 80 |
| | *Phyllosticta pyrolae* | 100 |
| | *Ramularia* sp. | 80 |
| *Rhododendron ferrugineum*[2] | *Cryptocline arctostaphyli* | 14 |
| | *Cryptocline* sp. | 100 |
| *Vaccinium myrtillus*[2 4] | *Cryptocline arctostaphyli* | 15 |
| | anamorph of *Godronia callunigera* | 5 |
| | *Pezicula myrtillina*[6] | 100 |
| | *Phyllosticta pyrolae* | 5 |
| *V. vitis-idaea*[2 4] | *Coleophoma empetri* | 10 |
| | *Cryptocline arctostaphyli* | 28 |
| | *Phomopsis* sp. | 28 |
| | *Phyllosticta pyrolae* | 28 |

[1] Widler & Müller (1984). [2] Petrini (1985).
[3] Petrini *et al.* (1982). [4] Petrini (1984).
[5] Oberholzer (1982).
[6] Isolated only from the twigs of *Vaccinium myrtillus*.

once.

Although fungal endophytes are widespread, many of them either sporulate rarely on their hosts or form inconspicuous fruiting bodies. As a result, many common endophytes belong to taxa which are seldom collected in the field. Many fungi isolated are anamorphs of known conifer-inhabiting Ascomycetes (e.g. *Brunchorstia pinea* (Karst.) Höhn., the *Corniculariella* anam. of *Holmiella sabina* Petrini *et al.*, the anam. of *Retinocyclus abietis* (Croun) Groves & Wells, *Sirodothis* spp.), while others are Ascomycetes previously reported from coniferous hosts (*Chloroscypha* spp., *Cryptocline abietina* Petrak, *Naemacyclus* spp.).

*The endophytes of Ericaceae*

Ten species of Ericaceae have been sampled (Oberholzer 1982; Petrini *et al.* 1982; Petrini 1984; Widler & Müller 1984; Petrini 1985) and all of them host a large number of endophyte species. For instance, up to 197 fungal taxa have been isolated from *Arctostaphylos uva-ursi* (L.) Sprengel collected in two sites during a study period of two years (Widler & Müller 1984). Most of the species isolated have not been recorded hitherto from ericaceous hosts, but they have been isolated from a number of other plants. Some other taxa are already known from Ericaceae and show a degree of specificity even at the species level (e.g. *Pezicula myrtillina* (Karst.) Karst., *Phomopsis* sp.). However, most species recorded in Table 4 are common to two or more hosts (e.g. *Cryptocline arctostaphyli* Petrini, *Phyllosticta pyrolae* Ell. & Ev.) within the family. Petrini (1985) has demonstrated that a correlation exists between host plant and fungal species. For instance, *Coccomyces arctostaphyli* (Rehm) B. Erikss. and *Cryptocline arctostaphyli* appear to prefer *Arctostaphylos uva-ursi*; *Calluna vulgaris* (L.) Hull is mostly colonized by *Godronia callunigera* Karst., whilst *Anthostomella tomicum* (Lév.) Sacc. and *Phyllosticta pyrolae* are most frequently isolated from *Erica carnea*.

*Grass endophytes*

The incidence of grass endophytes and their taxonomy have been the object of extensive studies, mainly because of the impact the fungi involved in the symbiosis have on the ecology of grass populations. Table 5 presents a synopsis of the fungal endophytes most commonly isolated from Gramineae. Diehl (1950) has investigated both taxonomically and ecologically the Balansiae (Clavicipitaceae), a group of fungi that parasitize grasses and, to a lesser extent, sedges and rushes. A high degree of host specificity has been shown for *Balansia strangulans* (Mont.) Diehl, which is found in a given site almost invariably on only one host, although other grasses known as hosts elsewhere may be growing in the immediate vicinity of the infected plant. Other members of the Balansiae are host-specific at least at the tribe level, and will infect only closely related host plants (Diehl 1950).

Riesen & Sieber (1985) have demonstrated that known pathogens of wheat are already present in healthy plants over the whole vegetation period.

*Idriella bolleyi* (Sprague) von Arx, a root pathogen, can be isolated routinely from wheat roots; *Epicoccum purpurascens* Ehrh. ex Schlecht is a common inhabitant of wheat leaves, stems and seeds. *Didymella exitialis* (Morini) Müller and *Phaeosphaeria nodorum* (Müller) Hedjaroude, two common pathogens of wheat leaves and culms, have been isolated abundantly from apparently healthy-looking wheat plants.

### *Host specificity*

Endophytic fungi can be roughly divided into two groups. Some species are almost ubiquitous and can be isolated from a number of hosts belonging to different plant families and growing under different ecological and geographical conditions (e.g. *Geniculosporium* spp., *Cladosporium* spp., *Nodulisporium* spp., *Pleospora herbarum* Rabh., coprophilous and xylariaceous fungi). Other species show a fair degree of host specificity, which follows essentially the same patterns discussed by Hijwegen (1979) and Savile (1979) for some obligate antagonistic symbionts (particularly Uredinales).

*Coccomyces arctostaphyli* and *Cryptocline arctostaphyli* are usually isolated from *Arctostaphylos uva-ursi*; *Phyllosticta pyrolae* is the most

Table 5. Distribution frequencies of the most commonly observed endophyte taxa in some grasses. The computation of the distribution frequency is defined in Table 3.

| Host plant | Fungal taxon | Distribution frequency (%) |
|---|---|---|
| *Danthonia spicata*[1] | *Atkinsoniella hypoxylon* | n.g.[7] |
| *Festuca arundinacea*[2 3] | *Epichloë typhina* | n.g. |
| | *Acremonium coenophialum* | n.g. |
| *Festuca rubra*[3 4] | *Epichloë typhina* | n.g. |
| *Lolium perenne*[3] | *Acremonium loliae* | n.g. |
| | *Gliocladium*-like | n.g. |
| *Oryza* sp. (seeds)[5] | *Rhynchosporium oryzae* | n.g. |
| *Triticum aestivum*[6] | *Didymella exitialis* | 100 |
| | *Epicoccum purpurascens* | 100 |
| | *Fusarium culmorum* | 100 |
| | *Gibberella zeae* | 100 |
| | *Idriella bolleyi* | 100 |
| | *Monographella nivalis* | 100 |
| | *Phaeosphaeria nodorum* | 100 |

[1] Clay & Jones (1984). [2] Neill (1941).
[3] Latch *et al.* (1984). [4] Sampson (1933).
[5] Thomas (1984). [6] Riesen & Sieber (1985).
[7] n.g.: not given in the original publication.

important endophyte of leaves of *Erica* spp. (Oberholzer 1982; Widler & Müller 1984; Petrini 1985). A strong host specificity of endophytes at the species level has been so far demonstrated only for *Chloroscypha chloromela* (Phill. & Hark.) Seaver (on *Sequoia sempervirens* (D. Don) Endl.; Petrini 1982) and *Pezicula myrtillina* (on *Vaccinium myrtillus* L.; Petrini 1984). Further investigations, however, could show that these fungi are not as specific as they appear. Current observations (Petrini 1985; H. Ruckstuhl and O. Petrini, unpublished results) tend to support the hypothesis that endophytes are specific only at the family level. If different species within the same plant family are present at the same site, colonization of all species by the same endophyte will occur. On the other hand, family-specific endophytic fungi will seldom be isolated from representatives of taxonomically unrelated plant families: for instance, endophytes specific to Pinaceae will almost never be found in Ericaceae or Gramineae. A given endophyte, however, can be isolated from hosts belonging to closely related families: *Coleophoma empetri* (Rostr.) Petrak and *Phyllosticta pyrolae*, for instance, have been recorded from both Ericaceae and Pyrolaceae, two very closely related families in the subclass Ericales (Petrini 1985). On the whole, the degree of host specificity observed among endophytes can hardly permit endophytic distribution to be used as a measure of taxonomic affinity among various members of the same plant family, although in some cases the comparison of endophyte populations can yield useful taxonomic informations (Carroll & Carroll 1978).

The formation of host-specific strains in fungi common to different hosts could be a possible explanation for the variability observed in some fungi which show different cultural characteristics and growth forms depending upon their origin (Petrini *et al.* 1982). No significant morphological differences can usually be found in mature sporulating cultures and therefore it is impossible to assign the isolates to different species. Carroll & Petrini (1983) reported that strains of the same fungus isolated from different parts of the same host differ in their ability to utilize different substrates. Genetic variability within the same species could account for these observations; however, the formation of host-specific strains in endophytes is a hypothesis which requires further investigations.

### *Organ specificity of endophytic fungi*

Carroll & Caroll (1978) detected a certain degree of specificity with respect to the location of endophyte species within the needles (petiole or blade) of conifers in the Pacific Northwest. Analysis of the endophyte distribution in needles and twigs of *Juniperus communis* by Petrini & Müller (1979) did not show any drastic differences in species composition either wihtin a needle or between needles and twigs. Oberholzer (1982) did not succeed in demonstrating any diversity in species composition between the leaf and twig endophytes of *Erica carnea*.

J.K. Stone (personal communication) has shown that *Rhabdocline parkeri* Stone *et al.* and *Phyllosticta* sp. co-exist in needles of Douglas fir (*Pseudotsuga menziesii*). *R. parkeri* is confined to epidermal and hypodermal cells, while *Phyllosticta* sp. occurs intercellularly in the mesophyll.

Riesen & Sieber (1985) have most frequently isolated *Cladosporium oxysporum* Berk. & Curt., *Didymella exitialis*, *Hypoxylon serpens* (Pers.: Fr.) Kickx, and *Rhizoctonia solani* Kühn. from wheat leaves; *Fusarium oxysporum* Schlecht, *Gibberella zeae* (Schw.) Peck and, to a lesser extent, *Phaeosphaeria nodorum* are typical colonizers of the wheat culms. Thus, a certain degree of organ or even tissue specificity, as suggested by Carroll & Petrini (1983), may exist.

### *Geographical distribution of host-specific endophytes*

The geographical distribution of host-specific endophytic fungi appears to be related to the distribution of their hosts. One of the most commonly isolated endophytes of *Arctostaphylos uva-ursi* in Switzerland, *Cryptocline arctostaphyli* (Widler & Müller 1984; Petrini 1985), has often been isolated from the same host in Western Oregon (Petrini *et al.* 1982, as *Aureobasidium ribis*). Carroll & Carroll (1978) isolated from Douglas fir in Oregon a number of endophyte genera previously reported by Caroll *et al.* (1977) from the same host in Europe.

## *TAXONOMY OF SOME UBIQUITOUS ENDOPHYTES*

### *Plant tissues as ecological niches*

A large number of genera and species of fungi belonging to different families and classes is able to live endophytically inside the plant tissue. Some of them are characteristic components of the epiphytic fungal population and are probably only secondary invaders of the internal plant tissues. On the other hand, some fungal species seem primarily to fit the endophyte niche. Species of *Chloroscypha* Seaver, *Cryptocline* Petrak, *Cryptosporiopsis* Bubák & Kabát, *Lophodermium* Chevallier, *Phomopsis* Sacc. and *Phyllosticta* Pers., have adapted to endophytic life and are routinely isolated as endophytes from a number of hosts (Luginbühl & Müller 1980; Petrini & Carroll 1981; Petrini 1982; Petrini 1985). The same pattern of colonization can be postulated for other fungal genera such as *Pleospora* (Crivelli 1983) and *Phaeosphaeria* (Leuchtmann 1984). How many genera of fungi have adapted to live primarily endophytically is unclear and will require further study.

### *Xylariaceae as endophytes*

The occurrence of xylariaceous fungi among the endophytes has been discussed in detail by Petrini & Petrini (1985). Some species of Xylariaceae which were reported to fruit only on a given host (e.g. *Hypoxylon fragiforme* (Pers.:Fr.) Kickx on *Fagus silvatica* L.) have been isolated from plants belonging to different families; thus, while most Xylariaceae can grow endophytically in many hosts, the physiological conditions required for the formation of the teleomorph are fulfilled only on certain host plants. Endophytic Xylariaceae can be divided roughly into two groups. Representatives of the genus *Anthostomella*, for instance, and a few species from *Hypoxylon* and *Rosellinia* seem to be fairly host-specific. Several *Xylaria* spp. and *Hypoxylon* spp., on the other hand, are widely distributed within the plant kingdom; for instance, members of the *H. serpens* complex are apparently quite unspecific.

The limited number of investigations on endophytic fungi does not allow any conclusions to be drawn about the geographical distribution of xylariaceous endophytes. However, some patterns have so far been detected. *Xylaria* spp. have frequently been isolated from tropical Araceae, Bromeliaceae, Orchidaceae and Pteridophyta (Petrini & Dreyfuss 1981; Dreyfuss & Petrini 1984); their rather rare records from hosts in the temperate regions indicate the tropical distribution of species belonging to this genus, a feature already known for the teleomorphs (e.g. Dennis 1956, 1957).

### *Coprophilous fungi*

Coprophilous fungi belonging to the genera *Ascobolus*, *Coprinus*, *Delitschia*, *Gelasinospora*, *Lasiobolus*, *Podospora*, *Preussia*, *Sordaria* and *Sporormiella* have often been isolated from living plant tissues (Petrini 1985). Their occurrence as endophytes is obviously difficult to explain; however, spore survival rate experiments (Petrini 1984) have demonstrated the endophytic character of these fungi. Very likely all coprophilous fungi are able to live endophytically during at least part of their life cycle: the isolation of *Quasiconcha reticulata* Barr & Blackwell and its anamorph from the roots of *Pinus halepensis* Mill. and *Thuja occidentalis* L. (Blackwell & Gilbertson 1985) seems to confirm this hypothesis. The low overall frequencies of infections so far reported indicate on the other hand that the occurrence of coprophilous fungi has to be considered casual and very likely of no great significance for the host plants.

## *ACKNOWLEDGEMENTS*

The author would like to thank Prof. G.C. Carroll (Eugene, Oregon, USA), Dr S.M. Francis (CMI, Kew, U.K.) and Dr L.E. Petrini (Zürich) for helpful comments and for critically reading the manuscript. The technical help by Miss C. Steinbrück during the preparation of the manuscript is gratefully acknowledged.

## *REFERENCES*

Bacon, C.W., Porter, J.K., Robbins, J.D. & Luttrell, E.S. (1977). *Epichloë typhina* from toxic tall fescue grasses. Applied and Environmental Microbiology, *34*, 576-81.

Bernstein, M.E. & Carroll, G.C. (1977). Internal fungi in old-growth Douglas fir foliage. Canadian Journal of Botany, *55*, 644-53.

Blackwell, M. & Gilbertson, R.L. (1985). *Quasiconcha reticulata* and its anamorph from conifer roots. Mycologia, *77*, 50-4.

Boullard, B. (1951). Champignons endophytes de quelques fougères indigènes et observations relatives à *Ophioglossum vulgatum* L. Botaniste, *35*, 257-80.

Boullard, B. (1957). La mycotrophie chez les Ptéridophytes. Sa fréquence, ses charactères, sa signification. Botaniste, *41*, 1-185.

Boullard, B. (1979). Considérations sur la symbiose fongique chez les Ptéridophytes. Syllogeus, *19*, 22-3.

Carroll, F.E., Müller, E. & Sutton, B.C. (1977). Preliminary studies on the incidence of needle endophytes in some European conifers. Sydowia, *29*, 87-103.

Carroll, G.C. & Carroll, F.E. (1978). Studies on the incidence of coniferous needle endophytes in the Pacific Northwest. Canadian Journal of Botany, *56*, 3034-43.

Carroll, G.C. & Petrini, O. (1983). Patterns of substrate utilization by some fungal endophytes from coniferous foliage. Mycologia, *75*, 53-63.

Clay, K. & Jones, J.P. (1984). Transmission of *Atkinsoniella hypoxylon* (Clavicipitaceae) by cleistogamous seeds of *Danthonia spicata* (Gramineae). Canadian Journal of Botany, *62*, 2893-5.

Crivelli, P.G. (1983). Ueber die heterogene Ascomycetengattung *Pleospora* Rabh.; Vorschlag für eine Aufteilung. Dissertation ETH No. 7318, Zürich: ADAG Druck.

De Bary, A. (1866). Morphologie und Physiologie der Pilze, Flechten und Myxomyceten. Leipzig: Engelmann.

Dennis, R.W.G. (1956). Some Xylarias of tropical America. Kew Bulletin, 401-44.

Dennis, R.W.G. (1957). Further notes on tropical American Xylariaceae. Kew Bulletin, 297-332.

Diehl, W.W. (1950). *Balansia* and the Balansiae in America. Agriculture Monograph No. 4. Washington, D.C.: U.S. Department of Agriculture.

Dreyfuss, M. & Petrini, O. (1984). Further investigations on the occurrence and distribution of endophytic fungi in tropical plants. Botanica Helvetica, *94*, 33-40.

Fisher, P.J., Anson, A.E. & Petrini, O. (1986). Fungal endophytes in *Ulex europaeus* and *Ulex gallii*. Transactions of the British Mycological Society, *86*, 153-6.

Hijwegen, T. (1979). Fungi as plant taxonomists. Symbolae Botanicae Upsalienses, *22*, 146-65.

Latch, G.C.M., Christensen, M.J. & Samuels, G.J. (1984). Five endophytes of *Lolium* and *Festuca* in New Zealand. Mycotaxon, *20*, 535-50.

Leuchtmann, A. (1984). Ueber *Phaeosphaeria* Miyake und andere bitunicate Ascomyceten mit mehrfach querseptierten Ascosporen. Sydowia, *37*, 75-194.

Luginbühl, M. & Müller, E. (1980). Endophytische Pilze in den oberirdischen Organen von vier gemeinsam an gleichen Standorten wachsenden Pflanzen (*Buxus*, *Hedera*, *Ilex*, *Ruscus*). Sydowia, *33*, 185-209.

Morgan-Jones, G. & Gams, W. (1982). Notes on hyphomycetes. XLI. An endophyte of *Festuca arundinacea* and the anamorph of *Epichloë typhina*, new taxa in one of two new sections of *Acremonium*. Mycotaxon, *15*, 311-8.

Neill, J.C. (1939). Blind-seed disease of rye-grass. New Zealand Journal of Science and Technology, *20*, 281-301.

Neill, J.C. (1941). The endophytes of *Lolium* and *Festuca*. New Zealand Journal of Science and Technology, *23*, 185-93.

Oberholzer, B. (1982). Untersuchungen über endophytische Pilze von *Erica carnea* L. Dissertation ETH No. 7198. Zürich: ADAG Druck.

O'Donnell, J. & Dickinson, C.H. (1980). Pathogenicity of *Alternaria* and *Cladosporium* isolates on *Phaseolus*. Transactions of the British Mycological Society, *74*, 335-42.

Petrini, L.E. & Petrini, O. (1985). Xylariaceous fungi as endophytes. Sydowia, *38* (in press).

Petrini, O. (1982). Notes on some species of *Chloroscypha* endophytic in Cupressaceae of Europe and North America. Sydowia, *35*, 206-22.

Petrini, O. (1984). Endophytic fungi in British Ericaceae: a preliminary study. Transactions of the British Mycological Society, *83*, 510-2.

Petrini, O. (1985). Wirtsspezifität endophytischer Pilze bei einheimischen Erica-

ceae. Botanica Helvetica, *95*, 213-8.
Petrini, O. & Carroll, G.C. (1981). Endophytic fungi in the foliage of some *Cupressaceae* in Oregon. Canadian Journal of Botany, *59*, 629-36.
Petrini, O. & Dreyfuss, M. (1981). Endophytische Pilze von epiphytischen Araceae, Bromeliaceae und Orchidaceae. Sydowia, *34*, 135-48.
Petrini, O. & Müller, E. (1979). Pilzliche Endophyten am Beispiel von *Juniperus communis* L. Sydowia, *32*, 224-51.
Petrini, O., Stone, J.K. & Carroll, F.E. (1982). Endophytic fungi in evergreen shrubs in Western Oregon: a preliminary study. Canadian Journal of Botany, *60*, 789-96.
Riesen, T. & Sieber, T. (1985). Endophytic fungi in winter wheat (*Triticum aestivum* L.). Endophytische Pilze von Winterweizen (*Triticum aestivum* L.). Dissertationen ETH. Zürich: ADAG Druck.
Sampson, K. (1933). The systemic infection of grasses by *Epichloë typhina* (Pers.) Tul. Transactions of the British Mycological Society, *18*, 30-47.
Savile, D.B.O. (1979). Fungi as aids to plant taxonomy: methodology and principles. Symbolae Botanicae Upsalienses, *22*, 135-45.
Schuster, R.M. (1966). The Hepaticae and Anthocerotae of North America. Vol. I. New York & London: Columbia University Press.
Thomas, M.D. (1984). Dry-season survival of *Rhynchosporium oryzae* in rice leaves and stored seeds. Mycologia, *76*, 1111-3.
Widler, B. & Müller, E. (1984). Untersuchungen über endophytische Pilze von *Arctostaphylos uva-ursi* (L.) Sprengel (Ericaceae). Botanica Helvetica, *94*, 307-37.

# GRASS ENDOPHYTES

K. Clay

*Department of Botany, Louisiana State University, Baton Rouge, LA 70803, USA*

## *INTRODUCTION*

Flowering plants often exist interdependently with microorganisms. Many grasses form symbiotic associations with clavicipitaceous fungi that exist perennially within their hosts. Long of interest only to mycologists, studies of fungal endophytes of grasses have recently taken on broader significance following the discovery of their toxic effects on animals.

Research on fungal endophytes of grasses dates back nearly 200 years when Persoon (1798) described the species *Sphaeria typhina*, later referred as *Epichloë typhina* (Pers.) Tul. (Tulasne & Tulasne 1861). De Bary (1863) demonstrated that the external fructifications of *E. typhina* arose from internal hyphae found abundantly within the tissues of the host grass. Subsequently, a number of new genera and species were described, culminating with the monograph of the tribe Balansiae (excepting *Epichloë*) by Diehl (1950). Many studies have focused on *E. typhina* as a model system for understanding endophyte reproduction, physiology and host teratology (Sampson 1933; Ingold 1947; Western & Cavett 1959; Kirby 1961; Thrower & Lewis 1973). The enigmatic endophytes of *Lolium* and *Festuca* spp., which do not have a perfect state and whose hosts are asymptomatic, have been subject to numerous studies focusing on their host relations and their possible taxonomic relationship to *Epichloë* (Sampson 1935, 1937, 1939; Neill 1940, 1941; Lloyd 1959). Most recent studies have focused on the toxic effects of endophytes on mammalian and insect herbivores (Bacon *et al.* 1975, 1977; Hoveland *et al.* 1980; Funk *et al.* 1983; Clay *et al.* 1985 a). The purpose of this paper is to review the knowledge of endophytes in the tribe Balansiae (Ascomycetes, Clavicipitaceae) and their imperfect relatives with an emphasis on their agricultural and ecological significance.

## *ENDOPHYTE TAXONOMY*

The tribe Balansiae is placed in the family Clavicipitaceae, subfamily Clavicipitoideae, based on the structure of the ascostromata and asci (Diehl 1950; Rogerson 1970). This tribe is distinct from the tribe Clavicipiteae (and the genus *Claviceps*) based on differences in conidia and sclerotia. The Balansiae, as originally circumscribed by Diehl (1950), consist of the genera *Atkinsonella*, *Balansia*, *Balansiopsis* and *Epichloë*, with the genus *Myriogenospora* later added by Luttrell & Bacon (1977). The mycelium of *M. atramentosa* (Berk. & Curt.). Diehl is superficial rather than endophytic, as in the rest of the tribe (Luttrell & Bacon 1977). The five genera are differentiated from each other primarily on the basis of conidial fructifications; vegetative hyphae are similar (Diehl 1950; Luttrell & Bacon 1977). The genus *Balansia* is the largest with 15 to 20 species while the other genera are each represented by only one or a few species (Diehl 1950; Bacon *et al.* 1986).

The anamorphic or imperfect states of the Balansiae have been assigned to two or three genera. The name *Ephelis* has been assigned to the conidial state of *Balansia* as well as one of the conidial states of *Atkinsonella*. Ephelidial conidia were observed in *Myriogenospora* (Rykard & Luttrell 1982), supporting its placement in the Balansiae. The name *Sphacelia* has been assigned to the conidial state of *Epichloë* and one of the conidial states of *Atkinsonella* (Diehl 1950; Rykard *et al.* 1984). The endophytes of tall fescue (*Festuca arundinacea*) and perennial ryegrass (*Lolium perenne*) have been considered as *E. typhina* (Sampson 1935, 1937; Neill 1941; Bacon *et al.* 1977). More recently, Morgan-Jones & Gams (1982) have rejected the name *Sphacelia* for the imperfect state of *E. typhina* and have assigned it to the genus *Acremonium*. Based on conidial differences they distinguished the anamorphic (imperfect) state of *E. typhina* from the imperfect endophytes of tall fescue and perennial ryegrass, which have been named *A. coenophialum* Morgan-Jones & Gams and *A. loliae* Latch, Christensen & Samuels, respectively (Morgan-Jones & Gams 1982; Latch *et al.* 1984). An immunological study (Johnson *et al.* 1985 b) found no differences in reactivity among the endophytes of tall fescue and perennial ryegrass, and *E. typhina* when tested against an antiserum prepared from the tall fescue endophyte. The imperfect endophytes of tall fescue and perennial ryegrass may be biotypes of *E. typhina*. A fungal endophyte is known from annual ryegrass (*Lolium temulentum*) (Freeman 1904, 1906; McClennan 1920; Sampson 1935) which resembles the endophyte of perennial ryegrass in certain respects but differs significantly in many ways, precluding its placement with the endophytes considered here. The taxonomy has been clarified by the growth of many endophytes in pure culture (Mantle 1967; Ullasa 1969; Rykard & Luttrell 1982; Clark *et al.* 1983; Bacon 1985; White & Cole 1985 b).

## *ENDOPHYTE REPRODUCTION*

Endophyte reproduction is related to the effect of the endophyte on its host's reproductive system. Two main modes of reproduction are recognized. The endophytes of tall fescue and perennial ryegrass, which typify one type of reproduction, are completely internal and do not produce external fruiting bodies (see Fig. 1). Reproduction occurs by vegetative growth of hyphae into the developing ovules of the host so that seeds produced by an infected maternal plant also contain the endophyte (Sampson 1933, 1935, 1937; Neill 1940, 1941; Bacon *et al.* 1977; Funk *et al.* 1983; Siegel *et al.* 1984 a). Flowering and seed set of infected host plants are the same as in uninfected plants. Endophyte viability in seeds decreases more rapidly than the viability of the seeds themselves, representing a potential control method (Neill 1940; Lloyd 1959; Harvey *et al.* 1982; Latch & Christensen 1982; Williams *et al.* 1984 a,b; Siegel *et al.* 1984 b, 1985). Control of endophytes in adult plants requires treatment with fungicides (Harvey *et al.* 1982; Latch & Christensen 1982; Williams *et al.* 1984 b).

A second type of endophyte reproduction occurs where fruiting bodies develop on inflorescences or leaves of their hosts. Host sterility results from the growth of stromatic tissue around the developing inflorescence or from the inhibition of floral initiation. *E. typhina*, *A. hypoxylon*, *B. obtecta* and *B. cyperi* are endophytes that typically cause mechanical abortion of host inflorescences (see Figs 2, 3 and 4). Inflorescences are grossly deformed, consisting primarily of stromatic tissue of the endophyte, and do not produce viable seed (Edgerton 1919; Sampson 1933; Diehl 1950; Clay 1984). The common name 'choke' has been applied to these symptoms. The endophytes produce conidia and ascospores from the stromatic tissue, in effect substituting spore production for seed production by the host. Other endophytes, as exemplified by *B. epichloë* and *B. henningsiana*, inhibit floral development (Diehl 1950; Bacon *et al.* 1975). Fruiting bodies bearing conidia and ascospores are procuced on the leaves of their hosts (see Figs 5, 6 and 7). It appears that the endophytes disrupt the normal hormonal balance of the plant thereby inhibiting flowering. Phytohormone analogs have been detected in pure cultures of *Balansia* spp. (Porter *et al.* 1977, 1985). Nutritional stresses of their hosts may also inhibit floral development (Bacon *et al.* 1983).

The infective role of conidia and ascospores is not clear (Rykard *et al.* 1985). Diehl (1950) inoculated stigmas of uninfected plants of the grass *Cenchrus echinatus* with conidia of *B. obtecta*, germinated the seed, and reared the seedlings. Although most plants were normal, a few were infected by *B. obtecta*, indicating that the grass stigma may be a potential route for infection. He also detected hyphae in the seeds of healthy plants growing in the vicinity of infected plants for a number of species, further supporting the stigma as a route of infection. Western & Cavett (1959) suggested that ascospores and conidia of *E. typhina* might germinate on cut stubble in mown or grazed fields. They had some success inoculating plants of *Dactylis glomerata* by this method but this work has not been repeated with any degree of success (Rykard *et al.* 1985). The work of Kohlmeyer & Kohlmeyer (1974) suggests that insects might be a vector for spore dispersal. For those endophytes producing fruiting bodies and causing host sterility, spores must occasionally infect new plants but the frequency and the mechanism of new infections await further data.

The effects on host reproduction and the external manifestation of infection often varies among hosts of the same endophytes. Sampson (1933) showed that plants of *Festuca rubra* infected by *E. typhina* were of three types. Most plants produced completely choked inflorescences but on other plants a few normal inflorescences were produced along with a majority of choked inflorescences. However, seeds from the normal inflorescences were found to contain the endophyte. The third type of plants were apparently uninfected but were found to contain abundant mycelium (termed latent infection by Sampson 1933). The occasional production of 'healthy' inflorescences by infected plants has been observed in many

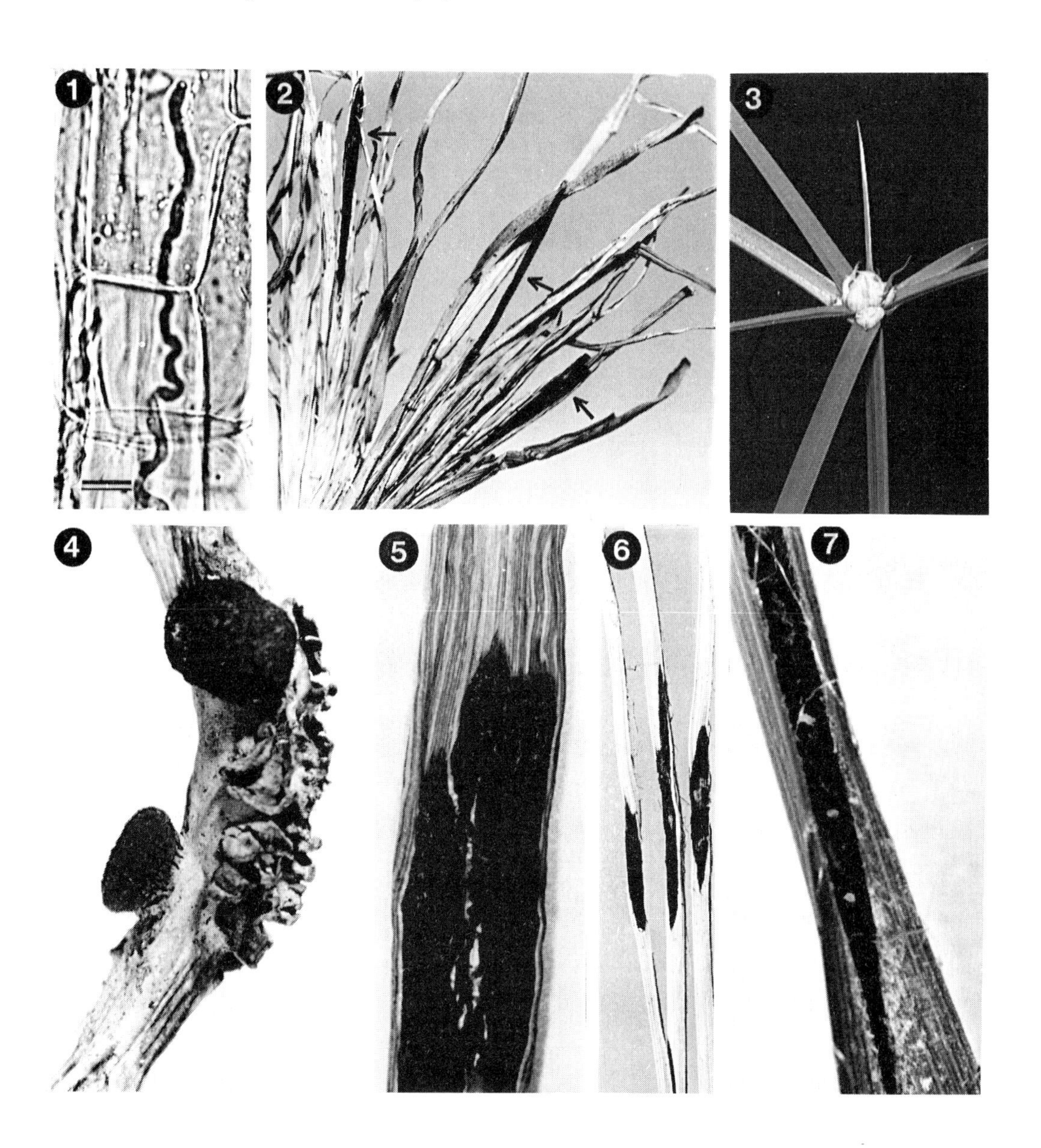


Fig. 1. *Acremonium coenophialum* hypha in leaf sheath of tall fescue (bar is 20 µm). Fig. 2. *Cenchrus echinatus* infected by *Balansia obtecta* (arrows point to sclerotia), x 0.5. Fig. 3. Sedge *Cyperus virens* infected by *B. cyperi*, x 0.5. Fig. 4. Sclerotium of *Atkinsonella hypoxylon* on aborted inflorescence of *Danthonia spicata*, x 8.5. Fig. 5. Stroma of *B. epichloë* on adaxial leaf surface of *Sporobolus poiretii*, x 4. Fig. 6. Stroma of *B. henningsiana* on abaxial leaf surface of *Andropogon scoparius*, x 0.5. Fig. 7. Stroma of *Myriogenospora atramentosa* on adaxial leaf surface of *Andropogon scoparius*, x 4.

species (Diehl 1950; K. Clay, unpublished results). If seeds from these inflorescences were infected then seed transmission of endophytes may occur even when hosts are normally sterilized by infection. Clay & Jones (1984) reported that *A. hypoxylon* infecting *Danthonia spicata* causes abortion of most flowers but is seed-borne in other flowers. Uninfected individuals produce wind-pollinated flowers at the apex of the flowering culm and self-pollinated flowers in the lower leaf axils of the same culm; infected plants bear only self-pollinated flowers (Clay 1984). This situation represents a transitional form between the seed-borne imperfect endophytes and the perfect endophytes that cause host sterility. The imperfect endophytes of perennial ryegrass and tall fescue may represent an extreme form found regularly in endophyte species that usually produce external fruiting bodies.

## *HOST RANGE*

With one or two exceptions, all known hosts of Balansiae and their imperfect relatives are grasses (Gramineae) or sedges (Cyperaceae). Kilpatrick *et al.* (1961) reported *E. typhina* occurring on the rush *Juncus effusus* and Diehl (1950) relates a questionable report of *E. bertonii* Speg. on the composite *Mikania scandens*.

Compilations of hosts indicate that many grasses are infected by endophytes. U. S. Department of Agriculture (1960) listed 34 genera of grasses infected by Balansiae endophytes while Diehl (1950) listed 43 grass and sedge genera as hosts to Balansiae endophytes in the Americas, excluding *Epichloë*. There are about 15 host genera each for *B. epichloë* (Weese) Diehl, *B. henningsiana* (Moell.) Diehl, and *B. claviceps* Speg. *Balansia aristidae* (Atk.) Diehl and *B. strangulans* (Mont.) Diehl have many host species restricted primarily to the grass genera *Aristida* and *Panicum*, respectively. Two species, *B. cyperi* Edg. and *B. cyperaceum* (Berk. & Curt.) Diehl, are known only from a few species of sedges (Edgerton 1919; Diehl 1950). The species *B. oryzae* Sydow is reported from rice in China, India and the United States (U. S. Department of Agriculture 1960; Ou 1972). *Atkinsonella hypoxylon* (Pk.) Diehl, the only species in the genus, is known only from *Danthonia compressa*, *D. spicata* and *Stipa leucotricha*, and the three species of *Balansiopsis* have a total of only five or six host genera (Diehl 1950). The one species of *Myriogenospora* is quite widespread, occurring on a large number of warm season grasses (Diehl 1950; Luttrell & Bacon 1977). *E. typhina* is probably the most widespread endophyte. Sprague (1950) reported 18 host grass genera in the United States while 26 host grasses are known in Britain (Sampson & Western 1954). Kohlmeyer & Kohlmeyer (1974) listed 26 host genera for *E. typhina* from specimens collected primarily in North America and Europe. The species *E. cinerea* is reported from two grasses in India by Mhaskar & Rao (1976) but otherwise little information is available on other *Epichloë* species. Imperfect endophytes have been found in 11 species of *Festuca* and 3 species of *Lolium* (White & Cole 1985 a). Given the difficulties in detection, endophytic fungi with no external symptoms may be more common than realized in other grasses.

The ability of an endophyte to colonize additional host species appears to be restricted as evidenced by the limited host range of many endophytes. For example, most hosts of *E. typhina* and *Atkinsonella* spp. are cool-season grasses whereas the hosts of *Balansia* spp., *Balansiopsis* spp. and *Myriogenospora* spp. primarily are warm-season grasses. The species of *Balansia* that attack grasses are distinct from the species that attack sedges. Many endophytes are restricted to just a single host genus and may be found on interspecific hybrids (Sampson 1940; Bradshaw 1959; Theis *et al.* 1961). However, there is some evidence that host shifts can occur. For example, smut grass (*Sporobolus poiretii*) has been introduced to North America where it is frequently attacked by *B. epichloë*, which is not known from the grasses native range. In Louisiana many populations of purple nutsedge (*Cyperus rotundus*), an introduced weed, are parasitized by *B. cyperi* (K. Clay, unpublished results). Purple nutsedge has not been reported to be a host for Balansiae endophytes elsewhere. It is interesting to note that the infected populations fall within the geographic range of *B. cyperi* on its native host *C. virens* in the southeastern United States.

## *FREQUENCY OF INFECTION*

An accurate estimate of the frequency of infection depends on knowledge of the proportion of host populations with infected plants and the proportion of infected plants within the population. Given the complete lack of symptoms of some endophytes and the ephemeral nature of symptoms of other endophytes, it is not suprising that there are relatively few published data on the frequency of infection. The available data suggest that *E. typhina* often occurs at high frequency in many European grasses. For example, 25% of the red fescue seed (*Festuca rubra*) imported into the U.S.A. from Hungary was found to be infected by *E. typhina* (Wernham 1942). Most fields of *Dactylis glomerata* in Britain raised for seed contained plants infected by *E. typhina* with the proportion of infected plants increasing with age of the field (Large 1952). Most populations of *Agrostis tenuis* and *A. stolonifera* in Britain were free from infection but up to 50% of the plants in some populations were infected by *E. typhina* (Bradshaw 1959). Other species of Balansiae can be found in high frequencies. Diehl (1950) reported many cases of infection levels approaching 100% in populations of various host species of *Balansia* spp. and *Atkinsonella* spp. in North America. Clay (1984) reported that the incidence of infection of *Danthonia spicata* by *A. hypoxylon* was 15% in a North Carolina population. The same endophyte was found infecting over 80% of the plants of *Stipa leucotricha* occurring under a woodland canopy but less than 10% of the plants in the neighbouring open prairie in Central Texas (K. Clay, unpublished results). *B. cyperi* occurs at high levels in populations of *Cyperus virens* in Louisiana (Edgerton 1919; Clay *et al.* 1985 b). There are anecdotal reports of various *Balansia* spp. commonly attacking *Aristida glauca* in Texas (Diehl 1930), *Cenchrus echinatus* in Florida (Weber 1924; Diehl 1950), *Panicum* spp. in Trinidad (Baker 1934), and rice in India and China (Ou 1972). *M. atramentosa* often occurs in high frequency on species of *Paspalum* and *Andropogon* in the southeastern United States (Atkinson 1905; Luttrell & Bacon 1977).

The imperfect endophytes of tall fescue and perennial ryegrass occur in nearly 100% of their potential hosts in certain areas. Neill (1940) found 100% of the samples of certified perennial ryegrass from New Zealand to be infected by *A. loliae* while a later survey (Lloyd 1959) also revealed very high levels of infection in New Zealand pastures. More recently Latch & Christensen (1982) found that a number of cultivars of perennial ryegrass in New Zealand had high levels of endophyte infection with other cultivars endophyte-free. Funk *et al.* (1983) found that 27 of 48 perennial ryegrass cultivars in the United States were highly infected. Most tall fescue pastures in the southeastern United States consisted of plants infected by *A. coenophialum*. A survey of 200 fields throughout Kentucky revealed that 97% contained infected plants and that the average level of infection within fields was 64% (Siegel *et al.* 1984 a). A survey of tall fescue seed crops from Missouri (Rycyk & Sharpe 1984) revealed that at least some seed from each of 193 different fields was infected and that 91% of the fields had greater than 90% infection. Long (1985) reported that 30% of tall fescue plants sampled in Tennessee were infected. Surveys of herbarium specimens have also shown high levels of infection in tall fescue and in other fescue species (White & Cole 1985 a).

## *HOST ECOLOGY*

Host grasses typically exist in mixed populations of infected and uninfected plants where they interact with each other and other members of the community. It is therefore of interest to consider whether endophyte-infected and uninfected grasses are ecologically equivalent. A number of studies have shown that endophyte-infected and endophyte-free grasses may occur in different habitats. *E. typhina* is most common in old, grazed pastures (Bradshaw 1959; Sampson & Western 1954). Diehl (1950) suggested that infection by *Balansia* species is also greater in older host populations. *Stipa leucotricha* was highly infected by *A. hypoxylon* when it occurred in mattes (tree-islands in otherwise open prairie) but rarely was infected in open prairie in Texas. Similarly, the grass *Ctenium aromaticum* was highly infected by *B. epichloë* in Louisiana woodland habitats but was rarely infected in more open habitats (K. Clay, unpublished results). Further work is required to determine whether different environments are more or less conducive to new infections or whether the success of infected plants varies with environment.

Hosts are often sterilized by endophyte infection. Baudoin (1975) has discussed the concept of parasitic castration in animals whereby a parasite destroys the reproductive organs of the host. Castrated hosts (and their parasites) often exhibit greater survival and growth due to reallocation of energy away from reproduction. Endophyte-induced sterility may result in more vigorous vegetative growth by infected grasses. Bradshaw (1959), referring to the infection of *Agrostis* spp. by *E. typhina*, reported that infected plants had a greater density of vegetative tillers compared to uninfected plants. He suggested that the lack of seed production and vigorous vegetative spread of infected plants may be advantageous in situations where seeds are not important sources of reproduction. Similar observations were made on *D. spicata* infected by *A. hypoxylon* (Clay 1984). In his review of the Balansiae, Diehl (1950) reported that many endophytes cause brooming of their hosts, an extreme increase in tiller density.

Endophyte-infected perennial ryegrass and tall fescue, which flower and set seed normally, may also exhibit greater vegetative vigour. Read *et al.* (1983) reported that forage production of endophyte-infected tall fescue pastures was greater than in uninfected pastures. Mortimer & Di Menna (1983) and Gaynor & Hunt (1983) both reported a similar situation in perennial ryegrass pastures. Endophyte-infected perennial ryegrass plants produced significantly more dry matter than uninfected plants when reared in a controlled environment (Latch *et al.* 1985 b). However, Siegel *et al.* (1984 a) and Neill (1941) observed no differences in growth and forage production of infected and uninfected fescue and ryegrass, respectively.

Differences in growth and reproduction between infected and uninfected grasses may lead to long-term differences in survival and persistence. In demographic experiments with infected or uninfected plants planted into natural plant communities, tillers of *D. spicata* infected by *A. hypoxylon* had significantly greater survival and growth than uninfected tillers in a grassland community (Clay 1984). Moreover, tillers of *D. spicata* infected by *A. hypoxylon* were more competitive versus the grass *Anthoxanthum odoratum* than uninfected tillers. Similar ongoing experiments with tall fescue, smut grass and the sedge *C. virens* show similar trends. Enhanced growth and competitive ability of endophyte-infected grasses would help explain the widespread occurrence of endophytes in nature.

## *EFFECTS ON HERBIVORES*

Researchers have long considered the possibility that endophytes might be toxic to herbivores in light of their taxonomic similarity to *Claviceps* but early attempts to demonstrate toxicity of infected grasses were unsuccessful (Neill 1941; Diehl 1950; Cunningham 1958; Lloyd 1959). It has long been known that cattle grazing tall fescue in the southeastern United States are subject to a malady whose symptoms include lameness, loss of extremities (rear feet, tail), rough hair coat, increased body temperature, weight loss or reduced rate of gain, reduced milk production, spontaneous abortion, and gangrene of the extremities (Yates 1962; Bacon *et al.* 1975; Hoveland *et al.* 1983; Wallner *et al.* 1983; Bond *et al.* 1984). Symptoms are most evident in the hot summer months, perhaps when cattle are under the greatest physiological stress (Jackson *et al.* 1984). Other animals may also be affected (Garrett *et al.* 1980; Daniels *et al.* 1983). Bacon *et al.* (1975) demonstrated that a number of Balansiae-infected grasses were found in association with tall fescue but were not being grazed when symptoms were observed. They found later that pastures associated with symptoms of fescue toxicity were highly infected with an endophyte then identified as *E. typhina* (Bacon *et al.* 1977). In contrast, low levels of infection were found in pastures with no history of toxicity. Subsequently, a number of studies have shown that fescue toxicosis occurs only in the presence of the fungal endophyte (Hoveland *et al.* 1980; Schmidt *et al.* 1983; Siegel *et al.* 1984 a). Carlson (1983) has estimated minimal losses of 200 million dollars per year in the southeastern U.S.A. due to the fescue endophyte.

The association of the ryegrass endophyte with 'ryegrass staggers' followed a similar course. Ryegrass staggers are observed most commonly in sheep in New Zealand; symptoms include twitching, difficulty in movement, and, at worst, inability to stand (Mortimer 1983). Other mammals (cattle, horses, deer) may be affected, mostly in the hotter, drier summer months (Aasen *et al.* 1969; Fletcher & Harvey 1981; Mackintosh *et al.* 1982). Early studies by Neill (1940, 1941), Cunningham (1958) and Lloyd (1959) detected no toxic effect of endophyte-infected ryegrass on herbivores. More recently, Fletcher & Harvey (1981) found that the level of infection was positively correlated with the intensity of ryegrass staggers in controlled experiments. Related studies have supported their work (Fletcher 1983; Mortimer & Di Menna 1983), indicating that infected perennial ryegrass may be toxic to mammals. In New Zealand, Everest (1983) has estimated economic losses of approximately five dollars per year per sheep due to ryegrass staggers. Considering the whole country, losses must total in the tens of millions of dollars.

Poisonings of cattle and sheep grazing *Balansia*-infected grasses in India were reported by Nobindro (1934). Symptoms of ergot poisoning have been frequently observed in cattle grazing dallisgrass (*Paspalum dilatatum*) and bahiagrass (*P. notatum*) which are often infected by *Myriogenospora atramentosa* (Luttrell & Bacon 1977; Clay *et al.* 1985 a; Bacon *et al.* 1986). Many of the other host grasses of fungal endophytes are weedy and widespread (Bacon *et al.* 1975, 1986). They may be of some forage value and contribute substantially to animal diets, but they rarely occur in pure stands so toxic effects may be minor and difficult to assign to a particular grass.

Recent studies have shown that endophyte-infected grasses are resistant to many insect herbivores, confounding the economic impact of endophytes. As in the case of mammalian toxicoses, most work has focused on the imperfect endophytes of perennial ryegrass and tall fescue. Prestidge *et al.* (1982), Mortimer & Di Menna (1983) and Gaynor & Hunt (1983) found that endophyte-infected perennial ryegrass plots sustained less damage and supported fewer number of Argentine stem weevils (*Listronotus bonariensis*) than non-infected plots. Argentine stem weevils fed less and deposited fewer eggs on endophyte-infected perennial ryegrass compared with uninfected grass in controlled choice experiments (Barker *et al.* 1984). Funk *et al.* (1983) found that plots of endophyte-infected perennial ryegrass had less damage by sod webworms (*Crambus* spp.) and fewer eggs in the soil compared to uninfected plots. In laboratory choice experiments, the fall armyworm (*Spodoptera frugiperda*) significantly preferred uninfected plants of perennial ryegrass over infected plants (Hardy *et al.* 1985). Siegel *et al.* (1985), Johnson *et al.* (1985 a), and Latch *et al.* (1985 a) have demonstrated a similar relationship between the endophyte of tall fescue and insect damage. Laboratory experiments with several aphid species and the large milkweed bug (*Oncopeltus fasciatus*) demonstrated that the survival of the insects was significantly reduced when confined to infected plants.

Antiherbivore effects are also found in the perfect Balansiae endophytes. Larvae of the fall armyworm, a generalist insect herbivore, had reduced survival, lower weight gains, and increased developmental time when fed leaves from grasses in the genera *Paspalum*, *Stipa* and *Cenchrus* which were infected by endophytes in the genera *Myriogenospora*, *Atkinsonella* and *Balansia*, respectively, compared to uninfected grasses (Clay *et al.* 1985 a), paralleling the results from fall armyworm reared on infected tall fescue and perennial ryegrass. The anti-insect properties of endophytes are not limited to grasses. In a study with sedges (*Cyperus* spp.) the survival, growth and development of fall armyworm larvae were significantly retarded when they were fed leaves from plants infected by *B. cyperi* compared to those fed leaves from uninfected plants (Clay *et al.* 1985 b).

## *ALKALOID PRODUCTION*

The similarity between the symptoms of ergot poisoning and the symptoms of fescue and ryegrass toxicity naturally led researchers to suspect that an alkaloid-producing clavicipitaceous fungus might be involved (Bacon *et al.* 1977; Aasen *et al.* 1969; Yates 1962, 1971). The alkaloid perloline was considered as a possible cause of mammalian toxicoses (Aasen *et al.* 1969; Bacon *et al.* 1977; Bush *et al.* 1982), but further studies did not substantiate the toxic role of this alkaloid. Ergot alkaloids have been found in tall fescue (Maag & Tobiska 1956) but the researchers were not aware of endophytes. Porter *et al.* (1979 a) demonstrated that the fescue endophyte produced a number of ergot alkaloids in vitro but whether they were involved in fescue toxicity was not known. Two pyrrolizidine alkaloids, N-formyl loline and N-acetyl loline, have been isolated and identified from infected tall fescue (Robbins *et al.* 1972; Bush *et al.* 1982; Jones *et al.* 1983) and, unlike perloline, the alkaloids were not found in uninfected fescue. Feeding trials with cattle showed that a diet high in these alkaloids caused symptoms of fescue toxicity whereas a diet low in these alkaloids did not. In perennial ryegrass attention has been focused on toxins called lolitrems (Gallagher *et al.* 1981, 1982, 1984; Siegel *et al.* 1985). Structural elucidation of these compounds confirms that they also are alkaloids (Gallagher *et al.* 1981, 1984) and resemble other known tremorgenic mycotoxins. Feeding trials with sheep have shown that symptoms of ryegrass staggers could be induced by a diet high in lolitrems.

*Balansia* species are known to produce a variety of ergot alkaloids in vivo and in vitro, as is the related *E. typhina* (Bacon *et al.* 1977, 1979, 1981; Porter *et al.* 1977, 1979 a,b, 1981; Bacon 1985). The specific alkaloids produced varies among endophyte species, among hosts of one species, and among isolates from single host-endophyte combinations (Porter *et al.* 1977, 1981; Bacon *et al.* 1979, 1981). Ergot alkaloids have been shown to be toxic to chick embryos (Bacon *et al.* 1975, 1977; Porter *et al.* 1977). The four other genera in the Balansiae have not been critically examined but it is likely that they also produce alkaloids. There are reports of domestic mammals showing symptoms resembling ergot poisoning when grazing on grasses infected by species of *Balansia*, *Epichloë* and *Myriogenospora* (Nobindro 1934; Walker 1970; Bacon *et al.* 1975, 1979; Luttrell & Bacon 1977).

The correspondence between mammalian toxicoses and insect resistance suggests that a similar mechanism is responsible for both effects. Johnson *et al.* (1985 a) found that infected tall fescue was subject to reduced feeding by insects compared to uninfected grass. Serial extracts indicated that the feeding deterrent was found primarily in the extract that had the highest concentration of N-formyl and N-acetyl loline alkaloids. In contrast, Prestidge *et al.* (1985) demonstrated that the feeding deterrents to Argentine stem weevil were produced by pure cultures of the endophyte of perennial ryegrass and that these deterrents were not lolitrems. In a study unrelated to endophytes, Corcuera (1984) showed that naturally occurring indole alkaloids had antifeedant and toxic effects on aphids. He suggested that this class of alkaloid may play a role in protecting plants from herbivory.

While there is a strong positive correlation between endophytes, alkaloids, and mammalian toxicoses and insect resistance, the functional relationships are not known with certainty. In particular, it is not clear whether alkaloids are the basis of mammalian toxicoses and insect resistance or whether they are correlated with some other factor actually responsible for the toxic effects. Moreover, it is not clear whether alkaloids are produced by the endophytes or whether they are produced by the hosts independently or incombination with the endophytes. However, the data on alkaloid production by endophytes in vitro and the induction of toxic symptoms by purified chemical compounds suggest that alkaloids are produced by the endophytes and that they are indeed responsible for the observed effects on animals.

## *CONCLUSIONS*

The fungi considered here have in common a number of features including their endophytic habitat, their association with grasses, and their potential for producing alkaloids toxic to mammalian and insect herbivores. Compared to other fungal pathogens, detrimental effects of endophytes on their hosts are minimal. Endophytes may actually enhance the survival and growth of their hosts. The agricultural impact of endophytes is both positive and negative. The economic benefits of increased resistance to insects of endophyte-infected grasses may or may not outweigh the economic losses resulting from the toxic effects on domestic animals. Attempts to eliminate fungal endophytes from pasture grasses may be unsuccesful if ecological interactions between grasses and their environment favour endophyte-infected plants. Conversely, breeding for endophyte-infected grasses is relatively easy, especially for seed-borne endophytes. A number of highly infected commercial cultivars of perennial ryegrass are now available and more are being developed. Future attempts to control or promote endophyte-infected grasses in agronomic situations will, in large part, depend on the results of ongoing reserach. Of particular importance will be the elucidation of the biochemical and physiological bases for mammalian toxicoses and insect resistance associated with infected grasses. If the effects on mammals and insects are caused by different factors it may be possible to develop infected grasses that are resistant to insects but are nontoxic to mammals.

More basic information is required before we fully understand the biology and evolutionary interactions between endophytes and grasses. Virtually all of our current knowledge is based on studies on a few endophytes (the endophytes of tall fescue and perennial ryegrass, viz. *E. typhina* and *Balansia* spp.) in just a few geographical areas (primarily the United States, Europe and New Zealand). There is a need for systematic searches for grasses with endophytes, especially imperfect endophytes, and for poorly known tropical grasses that might have endophytes. It would be surprising if endophytes were not more common than we currently know. More information is required on the ecological status of endophytes. Are they parasites or mutualists? Many studies suggest that infected grasses are more successful than uninfected grasses. If generally true, the grass-endophyte association may represent an ongoing case of evolution and co-evolution towards mutual dependency. Long-term field studies will be an important part of future endophyte research.

Over 50% of the references cited here have been published since 1980, many in 1985. The number of published research papers has increased dramatically following the recent discoveries that endophyte-infected grasses are often toxic to mammalian herbivores and resistant to insect herbivores. There are no signs that the volume of research will abate in the future. Besides their obvious and important economic significance, endophytes of grasses represent a model system for understanding the role of symbiotic micro-organisms in the biology of higher plants.

## *REFERENCES*

Aasen, A.J., Culvenor, C.C.J., Finnie, E.P., Kellock, A.W. & Smith L.W. (1969). Alkaloids as a possible cause of ryegrass staggers in grazing livestock. Australian Journal of Agricultural Research, *20*, 71-86.

Atkinson, G.F. (1905). The genera *Balansia* and *Dothichloe* in the United States with a consideration of their economic importance. Journal of Mycology, *11*, 248-67.

Bacon C.W. (1985). A chemically defined medium for the growth and synthesis of ergot alkaloids by species of *Balansia*. Mycologia, *77*, 418-23.

Bacon, C.W., Porter, J.K. & Robbins, J.D. (1975). Toxicity and occurrence of *Balansia* on grasses from toxic fescue pastures. Applied Microbiology, *29*, 553-6.

Bacon, C.W., Porter, J.K. & Robbins, J.D. (1979). Laboratory production of ergot alkaloids by species of *Balansia*. Journal of General Microbiology, *113*, 119-26.

Bacon, C.W., Porter, J.K. & Robbins, J.D. (1981). Ergot alkaloid biosynthesis by isolates of *Balansia epichloë* and *B. henningsiana*. Canadian Journal of Botany, *59*, 2534-8.

Bacon, C.W., Porter J.K. & Robbins, J.D. (1986). Ergot toxicity from endophyte infected weed grasses: a review. Agronomy Journal, *78*, 106-16.

Bacon, C.W., Porter, J.K., Robbins, J.D. & Luttrell, E.S. (1977). *Epichloë typhina* from toxic tall fescue grasses. Applied and Environmental Microbiology, *34*, 576-81.

Bacon, C.W., Robbins, J.D. & Porter, J.K. (1983). Toxins from fungal endophytes of pasture weed grass species as additional sources of grass toxicities. *In* Proceedings Forage and Turfgrass Endophyte Workshop, pp. 35-46. Corvallis: Oregon State University Extension Service.

Baker, R.E.D. (1934). *Balansia trinitensis* (Cooke and Massee). An interesting Trinidad fungus. Tropical Agriculture, *11*, 293.

Barker, G.M., Pottinger, R.P., Addison, P.J. & Prestidge, R.A. (1984). Effect of *Lolium* endophyte fungus infections on behaviour of adult Argentine stem weevil. New Zealand Journal of Agricultural Research, *27*, 271-7.

de Bary, A. (1863). Ueber die Entwickelung der *Sphaeria typhina* Pers. und Bail's "Mycologische Studien". Flora, *46*, 401-9.

Baudoin, M. (1975). Host castration as a parasitic strategy. Evolution, *29*, 335-52.

Bond, J., Powell, J.B. & Weinland, B.T. (1984). Behaviour of steers grazing several varieties of tall fescue during summer conditions. Agronomy Journal, *76*, 707-9.

Bradshaw. A.D. (1959). Population differentiation in *Agrostis tenuis* Sibth. II. The incidence and significance of infection by *Epichloe typhina*. New Phytologist, *58*, 310-5.

Bush, L.P., Cornelius, P.L., Buckner, R.C., Varney, D.R., Chapman, R.A., Burrus, P.B., II, Kennedy, C.W., Jones, T.A. & Saunders, M.J. (1982). Association of N-acetyl loline and N-formyl loline with *Epichloe typhina* in tall fescue. Crop Science, *22*, 941-3.

Carlson, G.E. (1983). The tall fescue problem - past and present. *In* Proceedings Forage and Turfgrass Endophyte Workshop, pp. 3-6. Corvallis: Oregon State University Extension Service.

Clark, E.M., White, J.F., Jr. & Patterson, R.M. (1983). Improved histochemical techniques for the detection of *Acremonium coenophialum* in tall fescue and methods of in vitro culture of the fungus. Journal of Microbiological Methods, *1*, 149-55.

Clay, K. (1984). The effect of the fungus *Atkinsonella hypoxylon* (Clavicipitaceae) on the reproductive system and demography of the grass *Danthonia spicata*. New Phytologist, *98*, 165-75.

Clay, K., Hardy, T.N. & Hammond, A.M., Jr. (1985 a). Fungal endophytes of grasses and their effects on an insect herbivore. Oecologia, *66*, 1-6.

Clay, K., Hardy, T.N. & Hammond, A.M., Jr. (1985 b). Fungal endophytes of *Cyperus* and their effects on an insect herbivore. American Journal of Botany, *72*, 1284-9.

Clay, K. & Jones, J.P. (1984). Transmission of *Atkinsonella hypoxylon* (Clavicipitaceae) by cleistogamous seed of *Danthonia spicata* (Gramineae). Canadian Journal of Botany, *62*, 2893-5.

Corcuera, L.J. (1984). Effects of indole alkaloids from Gramineae on aphids. Phytochemistry, *23*, 539-41.

Cunningham, I.J. (1958). Non-toxicity to animals of ryegrass endophyte and other endophytic fungi of New Zealand grasses. New Zealand Journal of Agricultural Research, *1*, 489-97.

Daniels, L.B., Nelson, T.S., Piper, E.L., Beasley, J.N., Stidham, D. & Brown, C.J. (1983). Physiological and reproductive performances of cattle and horses grazing fescue. *In* Proceedings Tall Fescue Toxicosis Workshop, pp. 19-24. Athens, GA: Cooperative Extension Service, University of Georgia.

Diehl, W.W. (1930). *Ephelis*-like conidia and floret sterility in *Aristida*. Phytopathology, *20*, 673-5.

Diehl, W.W. (1950). *Balansia* and the Balansiae in America. Agriculture Monograph 4. Washington, DC: United States Department of Agriculture.

Edgerton, G.W. (1919). A new *Balansia* on *Cyperus*. Mycologia, *11*, 259-61.

Everest, P.G. (1983). Ryegrass staggers: an overview of the North Canterbury situation and possible costs to the farmer. Proceedings of the New Zealand Grassland Association, *44*, 228-9.

Fletcher, L.R. (1983). Effects of presence of *Lolium* endophyte on growth rates of weaned lambs, growing on to hoggets, on various ryegrasses. Proceedings of the New Zealand Grassland Association, *44*, 237-9.

Fletcher, L.R. & Harvey, I.C. (1981). An association of a *Lolium* endophyte with ryegrass staggers. New Zealand Veterinary Journal, *29*, 185-6.

Freeman, E.M. (1904). The seed-fungus of *Lolium temulentum* L., the darnel. Philosophical Transactions of the Royal Society, London, B*214*, 1-28.

Freeman, E.M. (1906). The affinities of the fungus of *Lolium temulentum* L. Annals of Mycology, *4*, 32-4.

Funk, C.R., Halisky, P.M., Johnson, M.C., Siegel, M.R., Stewart, A.V., Ahmad, S., Hurley, R.H. & Harvey, I.C. (1983). An endophytic fungus and resistance to sod webworms: association in *Lolium perenne* L. Bio/Technology, *1*, 189-91.

Gallagher, R.T., Campbell, A.G., Hawkes, A.D., Holland, P.T., McGaveston, D.A., Pansier, E.A. & Harvey, I.C. (1982). Ryegrass staggers: the presence of lolitrem neurotoxins in perennial ryegrass seed. New Zealand Veterinary Journal, *30*, 183-4.

Gallagher, R.T., Hawkes, A.D., Steyn, P.S. & Vleggaar, R. (1984). Tremorgenic neurotoxins from perennial ryegrass causing ryegrass staggers disorder of livestock: structure elucidation of lolitrem B. Journal of the Chemical Society, Chemical Communications, 1984, 614-6.

Gallagher, R.T., White, E.P. & Mortimer, P.H. (1981). Ryegrass staggers: isolation of potent neurotoxins lolitrem A and lolitrem B from staggers-producing pastures. New Zealand Veterinary Journal, *29*, 189-90.

Garrett, L.W., Heiman, E.D., Pfander, W.H. & Wilson, L.L. (1980). Reproductive problems of pregnant mares grazing fescue pastures. Journal of Animal Science, *51*, 237 (Abstr.).

Gaynor, D.L. & Hunt, W.F. (1983). The relationship between nitrogen supply, endophytic fungus, and Argentine stem weevil resistance in ryegrasses. Proceedings of the New Zealand Grassland Association, *44*, 257-63.

Hardy, T.N., Clay, K. & Hammond, A.M., Jr. (1985). Fall armyworm (Lepidoptera: Noctuidae): a laboratory bioassay and larval preference study for the fungal endophyte of perennial ryegrass. Journal of Economic Entomology, *78*, 571-5.

Harvey, I.C., Fletcher, L.R. & Emms, L.N. (1982). Effects of several fungicides on the *Lolium* endophyte in ryegrass plants, seeds, and in culture. New Zealand Journal of Agricultural Research, *25*, 601-6.

Hoveland, C.S., Haaland, R.L., King, C.C., Jr., Anthony, W.B., Clark, E.M., McGuire, J.A., Smith, L.A., Grimes, H.W. & Holliman, J.L. (1980). Association of *Epichloë typhina* fungus and steer performance on tall fescue pasture. Agronomy Journal, *72*, 1064-5.

Hoveland, C.S., Schmidt, S.P., King, C.C., Jr., Odum, J.W., Clark, E.M., McGuire, J.A., Smith, L.A., Grimes, H.W. & Holliman, J.L. (1983). Steer performance and association of *Acremonium coenophialum* fungal endophyte on tall fescue pasture. Agronomy Journal, *75*, 821-4.

Ingold, C.T. (1947). The water-relations of spore discharge in *Epichloe*. Transactions of the British Mycological Society , *31*, 277-80.

Jackson, J.A., Hemken, R.W., Boling, J.A., Harmon, R.J., Buckner, R.C. & Bush, L.P. (1984). Loline alkaloids in tall fescue hay and seed and their relationship to summer fescue toxicosis in cattle. Dairy Science, *67*, 104-9.

Johnson, M.C., Dahlman, D.L., Siegel, M.R., Bush, L.P., Latch, G.C.M., Potter, D.A. & Varney, D.R. (1985 a). Insect feeding deterrents in endophyte-infected tall fescue. Applied and Environmental Microbiology, *49*, 568-71.

Johnson, M.C., Siegel, M.R. & Schmidt, B.A. (1985 b). Serological reactivities of endophytic fungi from tall fescue and perennial ryegrass and of

*Epichloe typhina*. Plant Disease, *69*, 200-2.
Jones, T.A., Buckner, R.C., Burrus, P.B., II & Bush, L.P. (1983). Accumulation of pyrrolizidine alkaloids in benomyl-treated tall fescue parents and their untreated progenies. Crop Science, *23*, 1135-40.
Kilpatrick, R.A., Rich, A.E. & Conklin, J.G. (1961). *Juncus effusus*, a new host for *Epichloë typhina*. Plant Disease Reporter, *45*, 899.
Kirby, E.J.M. (1961). Host-parasite relations in the choke disease of grasses. Transactions of the British Mycological Society, *44*, 493-503.
Kohlmeyer, J. & Kohlmeyer, E. (1974). Distribution of *Epichloë typhina* (Ascomycetes) and its parasitic fly. Mycologia, *66*, 77-86.
Large, E.C. (1952). Surveys for choke (*Epichloe typhina*) in cocksfoot seed crops, 1951. Plant Pathology, *1*, 23-8.
Latch, G.C.M. & Christensen, M.J. (1982). Ryegrass endophyte, incidence, and control. New Zealand Journal of Agricultural Research, *25*, 443-8.
Latch, G.C.M., Christensen, M.J. & Gaynor, D.L. (1985 a). Aphid detection of endophyte infection in tall fescue. New Zealand Journal of Agricultural Research, *28*, 129-32.
Latch, G.C.M., Christensen, M.J. & Samuels, G.J. (1984). Five endophytes of *Lolium* and *Festuca* in New Zealand. Mycotaxon, *20*, 535-50.
Latch, G.C.M., Hunt, W.F. & Musgrave, D.R. (1985 b). Endophytic fungi affect growth of perennial ryegrass. New Zealand Journal of Agricultural Research, *28*, 165-8.
Lloyd, A.B. (1959). The endophytic fungus of perennial ryegrass. New Zealand Journal of Agricultural Research, *2*, 1187-94.
Long, E.A. (1985). Occurrence of *Acremonium coenophialum* in tall fescue in Tennessee. Plant Disease, *69*, 467-8.
Luttrell, E.S. & Bacon, C.W. (1977). Classification of *Myriogenospora* in the Clavicipitaceae. Canadian Journal of Botany, *55*, 2090-7.
Maag, D.D. & Tobiska, J.W. (1956). Fescue lameness in cattle. II. Ergot alkaloids in tall fescue grass. American Journal of Veterinary Research, *17*, 201-4.
McClennan, E. (1920). The endophytic fungus of *Lolium*, Part I. Proceedings of the Royal Society of Victoria, *11*, 252-301.
Mackintosh, C.G., Orr, M.B., Gallagher, R.T. & Harvey, I.C. (1982). Ryegrass staggers in Canadian Wapiti deer. New Zealand Veterinary Journal, *30*, 106-7.
Mantle, P.G. (1967). Growth of *Atkinsonella hypoxylon* in pure culture. Transactions of the British Mycological Society, *50*, 497-506.
Mhaskar, D.N. & Rao, V.G. (1976). Development of the ascocarp in *Epichloë cinerea* (Clavicipitaceae). Mycologia, *68*, 994-1001.
Morgan-Jones, G. & Gams, W. (1982). Notes on hyphomycetes. XLI. An endophyte of *Festuca arundinacea* and the anamorph of *Epichloe typhina*, new taxa in one of the two new sections of *Acremonium*. Mycotaxon, *15*, 311-8.
Mortimer, P.H. (1983). Ryegrass staggers: clinical, pathological and aetiological aspects. Proceedings of the New Zealand Grassland Association, *44*, 230-3.
Mortimer, P.H. & di Menna, M.E. (1983). Ryegrass staggers: further substantiation of a *Lolium* endophyte aetiology and the discovery of weevil resistance of ryegrass pastures infected with *Lolium* endophyte. Proceedings of the New Zealand Grassland Association, *44*, 240-3.
Neill, J.C. (1940). The endophyte of ryegrass (*Lolium perenne* L.). New Zealand Journal of Science and Technology, *21*, 280-91.
Neill, J.C. (1941). The endophytes of *Lolium* and *Festuca*. New Zealand Journal of Science and Technology, *23*, 185-93.
Nobindro, U. (1934). Grass poisoning among cattle and goats in Assam. Indian Veterinary Journal, *10*, 235-6.
Ou, S.H. (1972). Rice Diseases. London: Eastern Press Ltd.
Persoon, C.H. (1798). *Sphaeria typhina*. *In* Icones et Descriptiones Fungorum Minus Cognitorum, p. 21. Lipsiae.

Porter, J.K., Bacon, C.W., Cutler, H.G., Arrendale, R.F. & Robbins, J.D. (1985). *In vitro* auxin production by *Balansia epichloë*. Phytochemistry, *24*, 1429-31.

Porter, J.K., Bacon, C.W. & Robbins, J.D. (1979 a). Ergosine, ergosinine, and chanoclavine I from *Epichloe typhina*. Journal of Agricultural and Food Chemistry, *27*, 595-8.

Porter, J.K., Bacon, C.W. & Robbins, J.D. (1979 b). Lysergic acid derivatives from *Balansia epichloe* and *B. claviceps* (Clavicipitaceae). Journal of Natural Products, *42*, 309-14.

Porter, J.K., Bacon, C.W., Robbins, J.D. & Betowski, D. (1981). Ergot alkaloid identification in Clavicipitaceae systemic fungi of pasture grasses. Journal of Agricultural and Food Chemistry, *29*, 653-7.

Porter, J.K., Bacon C.W., Robbins, J.D., Himmelsbach, D.S. & Hignam, H.C. (1977). Indole alkaloids from *Balansia epichloe* (Weese). Journal of Agricultural and Food Chemistry, *25*, 88-93.

Prestidge, R.A., Lauren, D.R., Van der Zijpp, S.G. & Menna, M.E. di (1985). Isolation of feeding deterrents to Argentine stem weevil in cultures of endophytes of perennial ryegrass and tall fescue. New Zealand Journal of Agricultural Research, *28*, 87-92.

Prestidge, R.A., Pottinger, R.P. & Barker, G.M. (1982). An association of *Lolium* endophyte with ryegrass resistance to Argentine stem weevil. Proceedings of the 35th New Zealand Weed and Pest Control Conference, pp. 199-222.

Read, J.C., Davis, C., Giroir, E. & Camp, B.J. (1983). The effect of fungal endophyte *Epichloe typhina* on animal performance. Agronomy Abstracts, p. 113.

Robbins, J.D., Sweeny, J.G., Wilkinson, S.R. & Burdick, D. (1972). Volatile alkaloids of Kentucky 31 tall fescue seed (*Festuca arundinacea* Schreb.). Journal of Agricultural and Food Chemistry, *29*, 1040-3.

Rogerson, C.T. (1970). The Hypocrealean fungi (Ascomycetes, Hypocreales). Mycologia, *62*, 865-910.

Rycyk, F., Jr. & Sharpe, D. (1984). Infection of tall fescue seed in Missouri by the endophyte *Epichloë typhina*. Plant Disease, *68*, 1099.

Rykard, D.M., Bacon, C.W. & Luttrell, E.S. (1985). Host relations of *Myriogenospora atramentosa* and *Balansia epichloë* (Clavicipitaceae). Phytopathology, *75*, 950-6.

Rykard, D.M. & Luttrell, E.S. (1982). Development of the conidial state of *Myriogenospora atramentosa*. Mycologia, *74*, 648-54.

Rykard, D.M., Luttrell, E.S. & Bacon, C.W. (1984). Conidiogenesis and conidiomata in the Clavicipitoideae. Mycologia, *76*, 1095-103.

Sampson, K. (1933). The systemic infection of grasses by *Epichloe typhina* (Pers.) Tul. Transactions of the British Mycological Society, *18*, 30-47.

Sampson, K. (1935). The presence and absence of an endophytic fungus in *Lolium temulentum* and *L. perenne*. Transactions of the British Mycological Society, *19*, 337-43.

Sampson, K. (1937). Further observations on the systemic infection of *Lolium*. Transactions of the British Mycological Society, *21*, 84-97.

Sampson, K. (1939). Additional notes on the systemic infection of *Lolium*. Transactions of the British Mycological Society, *23*, 316-9.

Sampson, K. & Western, J.H. (1954). Diseases of British Grasses and Herbage Legumes, 2nd edition. London: Cambridge University Press.

Schmidt, S.P., Hoveland, C.S., Clark, E.M., Davis, N.D., Smith, L.A., Grimes, H.W. & Holliman, J.L. (1982). Association of an endophytic fungus with fescue toxicity in steers fed Kentucky 31 tall fescue seed or hay. Journal of Animal Science, *55*, 1259-63.

Siegel, M.R., Johnson, M.C., Varney, D.R., Nesmith, W.C., Buckner, R.C., Bush, L.P., Burrus, P.B., II, Jones, T.A. & Boling, J.A. (1984 a). A fungal endophyte in tall fescue: incidence and dissemination. Phytopathology, *74*, 932-7.

Siegel, M.R., Latch, G.C.M. & Johnson, M.C. (1985). *Acremonium* fungal endophytes of tall fescue and perennial ryegrass: significance and control. Plant Disease, *69*, 179-83.

Siegel, M.R., Varney, D.R., Johnson, M.C., Nesmith, W.C., Buckner, R.C., Bush, L.P., Burrus, P.B., II & Hardison, J.R. (1984 b). A fungal endophyte of tall fescue: evaluation of control methods. Phytopathology, *74*, 937-41.

Sprague, R. (1950). Diseases of Cereals and Grasses (Fungi, except Smuts and Rusts). New York: Ronald Press Inc.

Theis, T., Sotomayor-Rios, A. & Calpouzos, L. (1961). Inflorescence blight on a napier grass x cattail millet hybrid caused by *Ephelis trinitensis* (*Balansia claviceps*). Plant Disease Reporter, *45*, 72-3.

Thrower, L.B. & Lewis, D.H. (1973). Uptake of sugars by *Epichloë typhina* (Pers. ex Fr.) Tul. in culture and from its host, *Agrostis stolonifera*. New Phytologist, *72*, 501-8.

Tulasne, L.R. & Tulasne, C. (1861). Selecta Fungorum Carpologia. Vol. 1, p. 24, Paris.

Ullasa, B.A. (1969). *Balansia claviceps* in artificial culture. Mycologia, *61*, 572-9.

U. S. Department of Agriculture (1960). Index of Plant Diseases in the United States. Agriculture Handbook 165. Washington, DC: U. S. Government Printing Office.

Walker, J. (1970). Systemic fungal parasite of *Phalaris tuberosa* in Australia. Search, *1*, 81-3.

Wallner, B.M., Booth, N.H., Robbins, J.D., Bacon, C.W., Porter, J.K., Kiser, T.E., Wilson, R. & Johnson, B. (1983). Effect of an endophytic fungus isolated from toxic pasture grass on serum prolactin concentrations in the lactating cow. American Journal of Veterinary Research, *44*, 1317-22.

Weber, G.F. (1924). *Ephelis mexicana* Fr., *Balansia hypoxylon* (Pk.) Atk. on sandbur (*Cenchrus echinatus* L.). Phytopathology, *14*, 66 (Abstr.).

Wernham, C.C. (1942). *Epichloë typhina* on imported fescue seed. Phytopathology, *32*, 1093.

Western, J.H. & Cavett, J.J. (1959). The choke disease of cocksfoot (*Dactylis glomerata*) caused by *Epichloe typhina* (Fr.) Tul. Transactions of the British Mycological Society, *42*, 298-307.

White, J.F., Jr. & Cole, G.T. (1985 a). Endophyte-host associations in forage grasses. I. Distribution of fungal endophytes in some species of *Lolium* and *Festuca*. Mycologia, *77*, 323-7.

White, J.F., Jr. & Cole, G.T. (1985 b). Endophyte-host associations in forage grasses. II. Taxonomic observations on the endophyte of *Festuca arundinacea*. Mycologia, *77*, 483-6.

Williams, M.J., Backman, P.A., Clark, E.M. & White, J.F., Jr. (1984 a). Seed treatments for control of the tall fescue endophyte *Acremonium coenophialum*. Plant Disease, *68*, 49-52.

Williams, M.J., Backman, P.A., Crawford, M.A., Schmidt, S.P. & King, C.C., Jr. (1984 b). Chemical control of the tall fescue endophyte and its relationship to cattle performance. New Zealand Journal of Experimental Agriculture, *12*, 165-71.

Yates, S.G. (1962). Toxicity of tall fescue forage: a review. Economic Botany, *16*, 295-303.

Yates, S.G. (1971). Toxin-producing fungi from fescue pasture. *In* Microbial Toxins. Vol. 7, eds S. Kadis, A. Ciegler & S.J. Ajl, pp. 191-206. New York: Academic Press.

# THE BIOLOGY OF ENDOPHYTISM IN PLANTS WITH PARTICULAR REFERENCE TO WOODY PERENNIALS

G.C. Carroll

*Department of Biology, University of Oregon, Eugene, OR 97403, USA*

## *INTRODUCTION*

The term 'endophyte' has been extant in the literature for well over 100 years. As originally defined by De Bary (1866), the term denoted any fungus whose hyphae invaded tissues or cells of living autotrophic organisms. O. Petrini (this volume) has noted that such a broadly defined category includes everything from virulent foliar pathogens to mycorrhizal root symbionts and serves little purpose except to contrast internal fungi with fungal epiphytes, which subsist entirely on the surfaces of plants. In fact years of usage, if not outright declaration, have restricted the term in important ways. Fungi which cause visible symptoms of plant disease are now usually referred to as pathogens, although virtually all such fungi penetrate the host tissue and exist endophytically. Fungi which occur both within and outside the plant, such as mycorrhizal fungi, are usually also excluded from the endophyte category. By default, then, endophytes comprise fungi which cause inapparent asymptomatic infections entirely within the tissues of plants. In many cases mutualism has been suspected in endophytic symbiosis, and in a few it has been demonstrated. Such known or suspected mutualistic symbioses will be the major focus of the discussion below.

## *DETECTION OF ENDOPHYTES*

Since endophytes cause asymptomatic infections, their presence within host tissues cannot be determined simply by casual inspection. Censuses of endophytic fungi have typically relied on vigorous surface sterilization of samples of host tissue followed by incubation of the sterile samples on agar medium, and scoring and isolation of the fungi which grow out (O. Petrini, this volume). This approach offers several advantages. A large number of samples can be processed and scored for fungi easily, and consequently, the investigation of a variety of ecological problems becomes possible. These have included the following: 1) the relationship between frequency of endophytic infection and physical parameters of the site where the host grows (Carroll & Carroll 1978; Petrini & Carroll 1981; Petrini *et al.* 1982); 2) the relationship between infection frequency and leaf age (Bernstein & Carroll 1977; Petrini & Carroll 1981); 3) seasonal increases in infection frequency (Bernstein & Carroll 1977); 4) the interaction between host genotype and endophyte infection frequency (Todd 1984); 5) field studies on the interaction between endophytes, fungal pathogens and insect pests (Todd 1984). Because large numbers of samples as well as samples with large physical dimensions can easily be surface sterilized, rare endophytes and those occurring in cytologically inaccessible tissues (e.g. heavily lignified cells) can be detected. In addition, a majority of endophytes ultimately sporulate in culture and can thus be identified. Studies based entirely on culture work are subject to several defects, however. Firstly, no matter how rigorous the surface sterilization procedure, occasional fungal propagules or hyphae on the sample surface may escape death and subsequently grow in the culture medium. This problem becomes particularly acute when hairy stems or leaves are involved. Further, certain fungi have been reported to form subcuticular appressoria or latent hyphae; although external to the plant tissue, these also give positive readings when surface-sterilized samples are scored for endophytic infection (McOnie 1967; Verhoeff 1974; Muirhead & Deverall 1981; Weidemann & Boone 1984). Beyond this, work on surface-sterilized leaf or stem samples provides little evidence on the location of the endophyte within the host tissue or on the number and extent of individual infections. Work in my latoratory on the endophytes of Douglas fir (*Pseudotsuga menziesii*) needles (Todd 1984; Stone 1985) has shown that estimation of infection frequency from culture of needle segments may underestimate actual frequency of separate infections by a factor of 10 to 100. If samples are infected by two or more endophytes simultaneously, negative interactions in culture or differences in growth rates may lead to further errors.

Microscopic observation of cleared, stained samples can answer many of the questions left open by studies based on culture work alone. Many of the early studies on leaf and stem endophytes were based largely on cytological examination (Rayner 1915, 1929; Sampson 1933, 1935, 1937, 1939; Bose 1947; Boursnell 1950). A few recent studies have also demonstrated endophytes visually in plant stem and leaf tissue (Tokunaga & Ohira 1973; Bernstein & Carroll 1977; Nathaniels & Taylor 1983; Andrews *et al.* 1985; Stone 1985; D. Cabral and J. Stone, unpublished results). Detection of endophytes under the microscope can be a laborious and difficult task, particularly when they occur intracellularly in recalcitrant tissues such as leaf epidermis or wood. When the tissue is colonized by two or more endophytes simultaneously, visual detection of the separate species may be uncertain unless characteristic patterns of colonization have been shown for each endophyte species individually. Careful investigation of the biology of particular host-endophyte systems requires both culture work and cytological examination. The great majority of endophytes, however, is known only from culture work (Petrini 1984); for broad censuses of the endophyte flora, pure culture work will continue to be the method of choice.

Where endophytic fungi are of great economic significance, sophisticated biochemical techniques have been developed to quantify the degree of infection. Johnson *et al.* (1982) have adapted an enzyme-linked immunosorbent assay (ELISA) to the detection of *Epichloë typhina* in tall fescue (see also K. Mendgen, this volume). This test detected the fungus in concentrations as low as 100 ng ml$^{-1}$ and was 100% effective in distinguishing infected from uninfected samples (as determined under a microscope).

*ENDOPHYTES AS MUTUALISTIC SYMBIONTS*

Carroll & Carroll (1978) proposed that many of the endophytes recovered from coniferous needles might be mutualistic symbionts. They suggested decreased palatability for grazing insects and antagonism towards needle pathogens as possible benefits for the host trees. Several casual observations as well as a few more detailed studies have now verified that endophytes may confer resistance against herbivores and antagonism towards pathogens on host plants.

*Herbivore resistance*

Looking first at the casual observations, Cubit (1974) showed that thalli of the marine green alga, *Enteromorpha vexata*, growing along the coast of southern Oregon were heavily grazed by marine molluscs during the summer unless infected by the endophytic fungus *Turgidiosculum* sp. Where fungi were present only 0 to 20% of the thalli were grazed, and these thalli were able to survive the summer months. Diamandis (1981) has observed that first-year needles of *Pinus brutia* in Greece are heavily grazed by the pine processionary caterpillar unless infected with *Elytroderma torres-juanii*.

In three cases the host-endophyte-grazer system has been studied intensively. In all three situations the life cycle of the grazer and the infection cycle of the endophyte are known in detail. The precise location of the endophytes within host tissues has been established, and something is known about the incidence of endophytes in natural populations of host plants. In all three cases the chemical basis for protection from herbivores is being studied. The best documented case history to date is that for the grasses. The endophytes are allied to the Clavicipitaceae and form systemic infections which are transmitted from one generation to the next through the seed. Endophytism in grasses is reviewed at length by K. Clay in this volume and will not be dealt with further here.

The second case history to emerge from recent literature involves the interaction among elm bark beetles, the endophyte *Phomopsis oblonga*, and elm trees. Elms are attacked by bark beetles which carry with them spores of *Ceratocystis ulmi*, causative agent of the Dutch elm disease. Webber (1981) has shown that the inner bark of elm trees attacked by *Scolytus* spp. is often colonized by *P. oblonga* and that the presence of *P. oblonga* there is clearly associated with a disruption of beetle breeding and a general decline of the beetle populations. Experiments in which beetles were forced to breed in *P. oblonga*-infected logs resulted in a drastic curtailment in the number of emergent beetle progeny. When beetles were offered a choice between burrowing in *P. oblonga*-infected versus control logs, they clearly favoured the control logs. The chemical basis for beetle antagonism has recently been elucidated (Claydon *et al*. 1985). The widespread occurrence of *P. oblonga* as a natural inhabitant in the outer bark of healthy elms in northern and western England and in northern Wales may explain the slow spread of Dutch elm disease in these localities (Webber & Gibbs, 1984).

A third case history involving endophyte-mediated antagonism towards insects arises out of work in my laboratory over the past 12 years. Douglas fir needles are commonly infected by two endophytes. One of them, *Rhabdocline parkeri* (Sherwood-Pike *et al.* 1986) is truly ubiquitous in trees in moist habitats, having been found in 83% of all sites sampled in the Pacific Northwest and virtually 100% of the sites where precipitation as rainfall exceeds 100 cm $yr^{-1}$. *R. parkeri* produces an anamorphic state, *Meria parkeri*, which is very similar to *Meria laricis*, a needle pathogen on larch (Funk 1985). A second endophyte, *Phyllosticta* sp., is much less common than *R. parkeri*, although still frequently isolated. Carroll & Carroll (1978) found it in 45% of the localities where Douglas fir was sampled. The same species of *Phyllosticta* is the dominant endophyte on true firs (*Abies* spp.) in the Pacific Northwest. Conidia of *M. parkeri* penetrate epidermal cells directly, each conidium forming a single infection hypha which is limited to a single epidermal cell until the needles senesce and abscise (Stone 1985). The number of separate infections per needle increases as the needles age, with very few recorded from needles in their first year and as many as 800 separate infections seen in 5-yr needles just prior to abscission (Stone 1985). *Phyllosticta* sp. penetrates between the epidermal cells and produces sparse intercellular hyphae within the mesophyll of the needle. Infections may become intracellular as the needles age (Stone 1985).

Of the two endophytes in Douglas fir *R. parkeri* has been studied most intensively, and subsequent discussion will be confined to this species. The infection cycle and role of the fungus as an insect antagonist have only just been elucidated; the account that follows is based on a Ph.D. dissertation by Stone (1985) and unpublished work by me.

*Infection cycle of R. parkeri.* The *Meria* conidial state first appears on senescent, freshly fallen needles from mid-October through November. *M. parkeri* conidia are slimy and are dispersed from needles lodged in the canopy by rain drops which fall abundantly during the fall and winter months in the Pacific Northwest. During this period conidia from fallen needles initiate new endophytic infections in needles. Conidia also infect needle galls produced by Cecidomyiid flies and sporulate prolifically there within one month. Infected galls continue to produce conidia for a period of several months and in fact represent the major source of inoculum for new needle infections. Early in December minute ascocarps of the *Rhabdocline* state form on newly fallen needles beneath infected trees. Germination of ascospores either in culture or on natural substrates has not been observed; their role in the infection cycle is still unknown.

*Insect antagonism by R. parkeri.* Douglas fir needles may be attacked by any of four species of gall midges in the genus *Contarinia* (Condrashoff 1961 a,b). The adult flies emerge just at the time of bud burst in late spring. After mating female flies lay small clusters of minute red eggs among the very young needles as they emerge from the buds. The eggs hatch, producing small highly mobile larvae. Each larva chews a small hole in a young needle and immediately ensconses itself within the needle tissue. Those that survive early predation by parasitoids and antagonistic responses by the host, feed on needle tissue throughout the summer. By September a substantial gall has formed around each surviving larva.

Infection of the galls by *R. parkeri* occurs in October with the advent of fall rains. Evidence for antagonism between the fungus and insect comes from an evaluation of larval mortality in infected galls compared with uninfected galls. In the fall of 1983 paired samples (infected versus uninfected) of approximately 50 galls each were taken from each of eight trees at a Bureau of Land Management progeny test site in northern Lane County, Oregon. Unexplained larval mortality (that not due to parasitoids) was compared in each of the paired samples. The results are shown in Table 1. Clearly larval mortality is much higher in the galls which are infected with the fungus ($P<0.001$). The same experiment was repeated with a larger sample size and more elaborate experimental design in the fall of 1984. Fly galls on 40 trees were sprayed with a suspension of *R. parkeri* conidia four times at intervals of one week in late September and October. Subsequent examination of the galls in November and December revealed that the inoculated galls were much more heavily infected than uninoculated galls and that again larval mortality was higher in the infected galls. Because the fungus does not invade the bodies of living larvae, fungal toxins are thought to be responsible for larval death (see J.D. Miller, this volume).

Table 1. Comparison of gall midge larval mortality in endophyte-infected and endophyte-free galls. Number of galls in each sample shown in parentheses.

| Tree | Larval mortality (%) | |
|---|---|---|
| | Endophyte-infected | Endophyte-free |
| 1315 | 71 (55) | 10 (58) |
| 1323 | 69 (54) | 10 (50) |
| 1331 | 77 (52) | 19 (53) |
| 1448-1 | 68 (38) | 15 (48) |
| 1448-2 | 91 (51) | -[1] |
| 1452 | 61 (56) | 29 (52) |
| 1538 | 45 (56) | 11 (54) |
| 1547 | 57 (53) | 4 (51) |
| 1588 adj. | 69 (55) | 2 (50) |

[1] -: Too few galls found.

### *Antagonism towards fungal pathogens*

A large phytopathological literature exists on microbial infection of uncongenial hosts and protection against virulent pathogens which such infection may provide (e.g. Matta 1971; Kuč & Hammerschmidt 1978). Some of the situations described by Matta (1971), e.g. asymptomatic *Fusarium* sp. infections, certainly would qualify as examples of endophytism. The ability of such infections to 'immunize' host plants against pathogens suggests that mutualism is involved.

A recent, unequivocal example comes from work on *Lophodermium* spp. in pine needles. Minter *et al.* (1978) have shown that *L. pinastri*, *L. conigenum* and *L. seditiosum* are the major colonists on Scots pine (*Pinus sylvestris*) needles, each species characteristically colonizing and fruiting on needles in a distinct microhabitat. *L. pinastri* fruits only on senescent needles; *L. conigenum* fruits on prematurely killed healthy needles, and *L. seditiosum* fruits on young needles on seedlings in young plantations (Minter & Millar 1980 a). The three species are culturally distinct (Minter & Millar 1980 b) and can all be isolated from green apparently healthy needles (Millar & Richards 1975). *L. seditiosum* is pathogenic in young trees. The other two species fruit only after needles have fallen or died from another cause, and hence must be considered true endophytes. Minter (1981) reported that *L. seditiosum* fruits on needles attached to fallen branches only when *L. conigenum* is absent from the forest. When *L. conigenum* is present, the pathogenic species is excluded from this habitat. In populations of Scots pine of mixed age, then, we might expect that the presence of *L. conigenum* in needles of older trees would prevent infection of needles in seedling and young trees by *L. seditiosum*. Here the benefits of endophytism would be experienced at the population level rather than at the level of individual trees.

Preliminary results on the possible antagonism of the grass endophyte, *Acremonium coenophialum* towards a few non-specific soil fungi as well as grass pathogens have recently been reported (White & Cole 1985). Culture filtrates of *A. coenophialum* inhibited in vitro growth of the recognized grass pathogens *Nigrospora sphaerica*, *Periconia sorghina* and *Rhizoctonia cerealis*. Here the cause of in vitro inhibition is clearly direct antibiosis. In many of the cases cited by Matta (1971) and Kuč & Hammerschmidt (1978), however, the protective effect of the endophyte may involve stimulation of general plant defenses against fungal attack, defenses which include the production of enzymes against fungal cell walls and the synthesis of phytoalexins.

## *OTHER PROBABLE MUTUALISTIC ASSOCIATIONS*

The examples of mutualism described above represent a small fraction of the endophytes which have been reported from culture studies. How widespread are mutualistic associations among the total endophyte flora in plants? Proof of mutualism based on antagonism to grazers and/ or pathogens requires knowledge of the important pests prevalent on a given host plant and rigorous field and laboratory experiments. Consequently, answers to the question of generality may not come quickly.

However, I predict that mutualistic endophytes will prove as widespread as mycorrhizal associations.

I propose the following criteria to identify likely candidates as mutualists:
1. The fungus causes no apparent symptoms of disease in the host plant.
2. The endophyte is transmitted through the seed. If transmission does not occur through the seed, transmission should occur reliably from one adult plant to another (lateral transmission).
3. The fungus is widely dispersed through the host tissues. If the infection units are small, they should be numerous. If the fungus is confined to leaves a high proportion of leaves should be infected.
4. The endophyte should be ubiquitous on a given host.
5. The fungus produces peculiar secondary metabolites of likely antibiotic or toxic nature.
6. The fungus shows an infection pattern similar to that of known endophytic mutualists.

Given these criteria, what are likely candidates for endophytic mutualism? Certainly the seed-borne endophytes of *Calluna vulgaris* (Rayner 1915), *Casuarina equisetifolia* (Bose 1947), *Helianthemum chamaecistus* (Boursnell 1950), *Vaccinium* spp. (Rayner 1929), and possibly *Glycine max* (Kulik 1984) should be reinvestigated. *Acremonium curvulum* endophytic in *Myriophyllum spicatum* (Andrews *et al.* 1985) should be checked for production of antibiotic or toxic substances. The species described by Roll-Hansen & Roll-Hansen (1979 a,b,c) and Huse (1981) from healthy wood of *Picea abies*, particularly *Ascocoryne* spp., *Neobulgaria premnophila* and *Nectria fuckeliana*, may well show antagonism towards wood-rotting fungi or insect pests. The isolation of a unique terphenylquinone, ascocorynin, from *Ascocoryne sarcoides* makes a mutualistic role for this fungus likely (Quack *et al.* 1982). Endophytes which show a pattern of infection similar to that of *Rhabdocline parkeri* in Douglas fir may well play a similar role. Compare, for instance, the pattern of intracellular infections by *R. parkeri* in the epidermis of Douglas fir needles (Fig. 1) with similar patterns produced by *Chloroscypha chloromela* in *Sequoia sempervirens* (Fig. 2) and by *Stagonospora* sp. in the leaf epidermal cells of *Juncus* spp. (Fig. 3).

Finally, fungi in the same genera or families as known mutualists should be scrutinized. Endophytic members of the Hypodermataceae and Hemiphacidiaceae as well as other species of *Phomopsis* and *Acremonium* beg to be investigated. Genera which are especially well represented among the endophytes, e.g. *Phyllosticta*, *Cryptocline* and *Cryptosporiopsis*, should also be singled out for attention.

## *STRATEGIES OF ENDOPHYTIC MUTUALISM*

When all the examples of endophytic associations discussed above are considered, two fundamentally different strategies emerge. The first of these, here termed constitutive mutualism, is typified by the grasses and their endophytes. The endophyte is carried through the seed, is always present in the lines that carry it, and develops a systemic

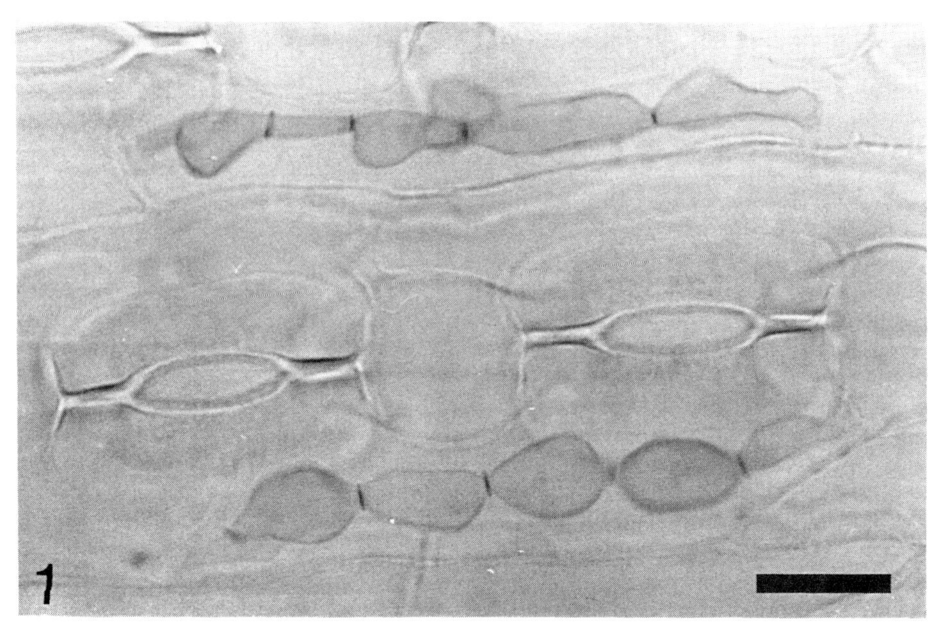


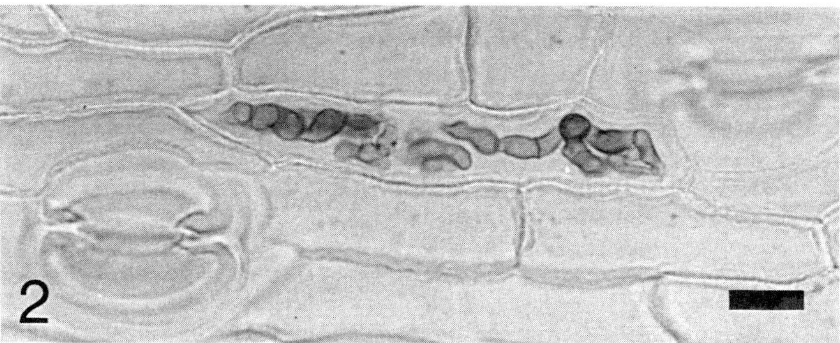


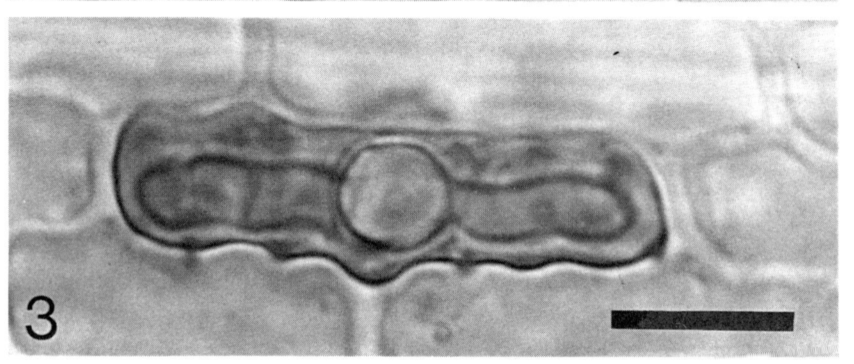


Figs 1 to 3. Endophytic fungi within leaf epidermal cells. 1. *Rhabdocline parkeri* in *Pseudotsuga menziesii*. 2. *Chloroscypha chloromela* in *Sequoia sempervirens*. 3. *Stagonospora* sp. in *Juncus effusus* var. *pacificus*. Scale bars represent 20 µm. (Figs 1 and 2 courtesy of J.K. Stone; Fig. 3 courtesy of D. Cabral).

infection with a substantial biomass of internal mycelium. The fungus produces potent toxins which show impressive deterrent effect on herbivorous insects (Funk *et al.* 1983; Clay *et al.* 1985; K. Clay, this volume). The plant harbouring the endophyte derives direct benefit from the presence of the fungus because of lessened grazing by herbivores. The costs to the host may be correspondingly high. In some such associations the fungus sterilizes its host (K. Clay, this volume). The amount of mycelium within the plant suggests that metabolic costs may also be high. One should expect that any seed-borne endophyte will conform to this pattern.

A second strategy, *inducible mutualism*, involves a much looser association between endophyte and host plant. In these situations, typified by *Phomopsis oblonga* in elm and *Rhabdocline parkeri* in Douglas fir, the endophyte is spread from senescent or dead host tissue to the healthy plant. Distribution of the endophyte may be decidedly spotty and can be affected greatly by the age of the host and its geographical location (Carroll & Carroll 1978; Webber & Gibbs 1984). The fungi occur in tissues or cells which are senescent or metabolically inactive such as bark or epidermal cells. Where living leaves are colonized, individual infections are small and sparsely distributed (Stone 1985). These infections serve not as a direct deterrent to herbivores, but rather as an inoculum source when tissues are killed or stressed by herbivore feeding. Subsequent infection of the injured tissue results in deterrence or death of the herbivore and a resulting general decrease in herbivore populations but only after damage to the host plant has already occurred. In this situation the benefits to the individual host are diffuse, and the metabolic costs are probably correspondingly small. The effect on the host population, however, may be very important.

In fact the real peculiarity of these situations bears further examination. Natural selection is thought to act primarily on individuals. Consequently, mutualistic symbioses usually operate at the level of individual organisms: the benefits which are exchanged increase the fitness of the associated individuals and no others. With the mutualism described above, however, the plant which is host to endophytes may not be the only or even the primary beneficiary of the association. Thus, with the elm-*P. oblonga* example, no individual tree is protected from beetle attack by harbouring the endophyte. By the time *P. oblonga* becomes active the tree is likely to be infected with *Ceratocystis ulmi* and moribund. The benefits of the association accrue to neighbouring trees in the form of depleted beetle populations. In Douglas fir, *Contarinia* spp. flies are a negligible problem for old-growth trees and become numerous enough to cause serious damage only in stands of seedlings or young trees. Nevertheless, needles on old trees are heavily infected with the endophyte, and such infections presumably serve as an inoculum source for adjacent stands of young trees in recently burned areas or clearcuts. Again the benefits to the Douglas fir appear to accrue to populations, not to individuals.

Perhaps the clearest example occurs with the *Lophodermium* species on Scots pine (Minter 1981). Here, the virulent pathogen on needles of young trees and seedlings, *L. seditiosum*, is excluded from needles still attached to dead branches of older trees by the endophyte *L. conigenum*. Where *L. conigenum* is absent, the increased fungal inoculum from the pathogen can cause a significant increase in needle blight in adjacent seedlings or young trees. Here the pathogen does not cause significant damage to adult trees, yet the trees harbour an endophyte which suppresses the pathogen.

The evolution of such associations may be understood in terms of kin selection, a form of genic selection whereby organisms which share a high proportion of the same genes may show altruistic behaviour. Such behaviour can be favourably selected if the beneficiaries of the altruistic behaviour share high proportions of their genomes with the self-sacrificial altruist (Futuyma 1979; Volpe 1985). Wind dispersal of seed and pollen in trees is relatively inefficient (Spurr & Barnes 1980). For species which rely on wind for pollination and seed dispersal adjacent trees may show a high degree of genetic similarity, and mutualisms dependent on kin selection could be expected to evolve and to work effectively. Whether kin selection actually occurs in the situations described above can only be determined after painstaking dissections of the population dynamics of host, endophyte and pest in small stands of trees.

### *WHY ENDOPHYTES?*

In the few situations which have been analyzed in detail endophytic mutualisms appear to be based on acquired chemical defenses (K. Clay, this volume; J.D. Miller, this volume). Higher plants are known to produce multifarious potent chemical defenses in their own right (e.g. Bell 1981; Bailey & Mansfield 1982). Why should some plants form mutualistic associations to accomplish the same end? One answer suggests that biochemical pathways for synthesizing toxigenic secondary metabolites are particularly likely to arise in the fungi and are a natural consequence of a filamentous growth form (Moss 1984).

A further explanation for such situations emerges from a consideration of the generation times of endophyte host plants compared with those of the insect pests or disease organisms which attack them (Price 1980). A Douglas fir tree may live 500 to 800 years in tracts which escape frequent fire. Trees such as Scots pine or elm are more short-lived, but may easily live 100 years. During a given generation a tree's genotype is fixed; the tree will not be able to alter the particular mix of allelochemicals and other defenses allotted to it at the time of fertilization prior to seed development. On the other hand, most insects go through one generation a year, and some insects and many disease organisms go through multiple generations in a year. As a result a pest can specialize in overcoming the defenses of a particular tree. Edmunds & Alstad (1978) have shown such extreme specialization on a single host individual for a scale insect on pine. Indeed, the pressures for specialization were thought to be so great that extinction was predicted for the insect deme when its host tree died. I propose that association with a short-cycle endophyte represents a response of long-lived plants to the challenge presented by short-cycle pests. Some evidence for this hypothesis comes from work in my laboratory (D. Cabral and G.C. Carroll, unpublished results) on the endophytes of annual and perennial *Juncus* spp. The hypothesis predicts that the perennial species should have one or two prevalent endophytes, presumed to be mutualists active against short-cycle insects, while the annual species should have few or no endophytes because such associations would be neither advantageous nor necessary. The annual recombination events during flowering should provide ample opportunity for rearrangements of chemical defenses against insects. Four species of *Juncus* were examined. The three perennial species, *J. effusus* var. *pacificus*, *J. patens* and *J. bolanderi*, each had a single dominant endophyte which could be isolated from segments of culms with high frequency (70 to 100%) as early as July following the period of new culm production. Although the details of the infection cycle have yet to be worked out, microscopy showed infection of single epidermal cells (Fig. 3). *J. bufonius*, an annual species, showed a much lower infection frequency (15 to 25%) by all endophytes, and no single endophyte proved dominant; about a dozen species were found, but no species occurred with an infection frequency greater than 5%.

Inducible mutualism may be particularly appropriate for challenges arising from the rapid evolution of pests because each host may harbour multiple genetic strains of the endophyte, each with a particular repertoire of mycotoxins. Indeed, in some cases separate endophyte species may have been accommodated as defenses against separate pests. In Douglas fir *Rhabdocline parkeri* and *Phyllosticta* sp. are both common in needles from older trees. Although preliminary evidence suggests that *Phyllosticta* sp. produces toxic secondary metabolites active against insects (J.D. Miller, this volume), the fungus infects galls of *Contarinia* spp. only rarely. If *Phyllosticta* sp. is indeed active against insects, some other insect herbivore is presumably the target pest. The predominance of *Phyllosticta* sp. as an endophyte in the true firs, a group not subject to attack by *Contarinia* spp., further supports this conclusion.

## *EVOLUTION OF ENDOPHYTES*

Evolutionary theory predicts that mutualism should evolve whenever an organism faces an unavoidable antagonistic challenge from a parasite (Thompson 1982). All available evidence suggests that endophytes have evolved from plant-pathogenic fungi. Many endophytes on coniferous hosts show the limited substrate utilization capabilities typical of biotrophic pathogens (Carroll & Petrini 1983; Stone 1985). In the field the distinction between endophyte and pathogen may become fuzzy. Some pathogens pass through a phase of extensive asymptomatic growth in the host before disease symptoms appear (Nathaniels & Taylor 1983; Kulik 1984). In other cases endophytes may cause pathogenic symptoms only when the host is stressed (Merrill *et al.* 1979; Millar 1980; Andrews *et al.* 1985). Pathogens on an economically important crop may exist endophytically on undomesticated plants growing among crop plants as weeds (Hepperly *et al.* 1985). The above situations may represent stages in the evolution of truly mutualistic associations. Alternatively, some may be artifacts of modern agricultural practice in which genetically uniform monocultures are grown over large tracts, and pests are controlled chemically. Under these conditions cultivars which tolerate endophytic fungi, at some cost in productivity, would be selected against. Plants would derive no benefit from endophytes in the face of external chemical pest control, and pathogenic strains of former endophytes might be expected to appear. The best evidence for the origin of endophytes arises from a consideration of their taxonomic affinities with known pathogens. Frequently endophytes are sister species to virulent pathogens on the same or closely related hosts. *Acremonium coenophialum*, a major grass endophyte, is probably closely related to *Epichloë typhina*. In pine the endophytic species *Lophodermium pinastri* and *L. conigenum* belong to the same genus as the pathogenic *L. seditiosum* (Minter *et al.* 1978). In Douglas fir *Rhabdocline parkeri* is closely related to two other virulent pathogens, *R. wierii* and *R. pseudotsugae* (Parker & Reid 1969). Indeed, all three species can coexist in the same tree.

The question of co-evolution among members of ecological communities and between individual species, even those existing in a mutualistic relationship, is at present hotly debated among evolutionists. To what extent are host plant, endophyte and pest truly co-evolved and to what extent have they evolved merely in parallel? A comparison of endophyte and host taxonomy certainly suggests that parallel evolution has taken place. For example, the fungal genus *Chloroscypha* (Helotiales) appears to be restricted to an endophytic habitat within the needles of various members of the Cupressaceae (Carroll & Carroll 1978; Petrini & Carroll 1981). *Meria parkeri* (= anamorph of *Rhabdocline parkeri*) is very closely related to *Meria laricis*, a pathogen on larch (Funk 1985). Douglas fir and larch are thought to be among the most closely related genera in the Pinaceae, despite profound differences in pattern of needle retention, i.e. deciduous versus evergreen (Christiansen 1972; Prager *et al.* 1976). In cases such as these it is tempting to suppose that speciation in the host plants has been mirrored by speciation in the endophytes. In many other cases, however, endophytes show broad host preferences (Carroll & Carroll 1978; Petrini *et al.* 1982; Petrini 1984), and these situations are not good candidates even for parallel evolution.

True co-evolution involves stepwise reciprocal adaptations on the part of each member of an association to the other members. The current consensus among evolutionists says that few examples of co-evolution can be found among mutalistic symbionts (Futuyma & Slatkin 1983; Nitecki 1983). Certainly co-evolution has occurred among plants and pathogenic fungi; Barrett (1983) has presented a recent synopsis of the literature which demonstrates this fact. If endophytes have evolved from pathogens, one might expect true co-evolution between host plants and endophytes as well. Detailed studies on the genetics of host and endophyte populations have not yet been carried out however; at this point, co-evolution has yet to be proved for any endophytic association.

In concluding this section one can ask whether endophytic mutualisms have evolved repeatedly or just once. If the endophytic habit evolved only once, all endophytic lineages must be very ancient, indeed must have evolved at the same time that land plants and Ascomycetes appeared. Most endophytes are Ascomycetes; however, very distantly related groups are represented among the endophyte flora, including Loculoascomycetes, Pyrenomycetes and Discomycetes. Any theory which ascribes a single origin to the endophytic habit must explain the presence in the endophyte flora of taxa as distantly related as *Rhabdocline*, *Chloroscypha*, *Diaporthe*, *Hypoxylon*, *Leptosphaeria* and members of the Clavicipitaceae. Luttrell (1974) argues persuasively that plant pathogens have evolved repeatedly, at different times and on different hosts. He further suggests that the present frequency of pathogens represents a historical accumulation over the course of evolution. If this view is correct, and if indeed endophytes are derived from pathogens, then the endophytic habit has evolved repeatedly.

## *THE FUTURE OF ENDOPHYTE RESEARCH*

Microbial symbioses with higher plants have proved of great economic significance because of their role in the nutrition of certain crop plants. The importance of microbial root symbioses was first recognized for the nitrogen-fixing bacteria and more recently for mycorrhizal fungi. In 25 years the study of mycorrhizae has developed from an arcane activity carried out in a few scattered laboratories, largely in the United States, Great Britain and Sweden to a vital, truly international research area engaging the attentions of several hundreds of biologists. Endophytes represent a major category of newly appreciated microbial symbiosis. Their economic importance has already been demonstrated in the grasses (K. Clay, this volume). One can confidently predict that, when the generality of mutualistic symbiosis between endophytes and higher plants is more broadly proven and recognized, the endophyte research bandwagon will become very crowded.

What kinds of research may one expect? Proof of the generality of acquired chemical defense as a basis for endophytic symbiosis will certainly take a high priority in any research program. Five to ten years from now dozens of examples of both constitutive and inducible mutualism, now typified by endophytes in grasses and Douglas fir, respectively, will probably have been reported, and the chemical basis of pest deterrence will be known.

Many endophytic mutualisms are comparatively loose. Unlike most other fungal symbioses, both fungus and host can be grown and manipulated separately with ease. Because of this endophytic mutualisms may well provide excellent model systems for answering broader questions about the evolution of mutalistic symbiosis. Schemske (1983) has listed several questions which demand attention if our understanding of the evolution of mutualism is to advance. Briefly, these are:

1. Is there genetic variation in traits that influence interactions among mutalists?
2. What proportion of the total variation in fitness among individuals in a given mutualistic taxon is attributable to mutualistic interaction?
3. How are positive effects of a mutualism associated with particular fitness components?

4. What ecological characteristics influence the evolution of mutualistic structure and specificity?
5. How important is co-evolution in the evolution of mutualistic interaction? How might the study of endophytes bear on these questions? Considering them in order, I suggest the following possibilities:

1. Genetic variation among mutualists. Among endophytic mutualisms traits which influence interactions among the mutualists would include at least the following for the fungus: a) infectivity of the spores; b) specificity of the spores for a particular host genome; c) degree of damage inflicted on the host; d) degree of deterrence towards pests; e) quantity and number of toxigenic secondary metabolites produced. Relevant host traits might include innate resistance to pests and susceptibility to infection by particular strains of the endophyte. Endophytic mutualisms involving economic plants should prove particularly rewarding experimental systems because numerous strains of genetically uniform host plants may be available. Where the host plants are trees, progeny test sites associated with timber improvement projects can provide test plots in a natural situation where at least something is known about the genetic background of the host. Variation in the relevant traits can be detected by testing multiple strains of fungi against known host genotypes and scoring for the characteristics of interest.

2. Mutualism and fitness. Endophytic symbioses should lend themselves well to the kind of cost-benefit analyses carried out so frequently with the mycorrhizae. In situations where the host plant can be propagated clonally and the endophyte is spread by lateral transmission, the growth and fitness of the same host clone with and without the endophyte can be compared under a number of different environmental regimes. Where lateral transmission of the endophyte does not occur, endophyte-free clones can be produced by treatment of infected clones with systemic fungicides.

3. Positive effects of mutualism and individual fitness components. In the case of endophytes in plants, the influence of the fungus on the partitioning of host resources between growth and reproduction will prove of great interest (see K. Clay, this volume).

4. Ecological characteristics and mutualist specificity and structure. This type of question probably cannot be answered by a series of experiments on a single host plant-endophyte combination. Rather, a whole body of data from a variety of situations must be considered. I propose the following hypotheses to account for several aspects of endophyte distribution and life cycle:
a. Long-lived perennials will contain endophytes while annuals do not.
b. Inducible endophytic mutualisms which involve kin selection will be common among trees with inefficient seed dispersal or clonal growth (e.g. aspen). Conversely, constitutive mutualism should be the rule for endophytes in trees in the tropics where efficient seed dispersal and pollination is mediated by animals.
c. Lateral transmission of endophyte spores will be common where host plants occur in moist warm climates which favour dispersal of fungal spores. Seed-borne transmission should occur in plants from dry or cold

habitats. The sparse evidence available to date supports these hypotheses. Future studies should look for ways to test them definitively.

5. Endophytic mutualism and co-evolution. The question of co-evolution has been dealt with above. Situations where closely related species of endophytes occur on closely related hosts and act against closely related pests should be examined. Co-evolution would seem especially probable where pathogenic species also occur in an endophyte species-complex. In that light, a worldwide census of the species of *Pseudotsuga* for endophytes and pathogens in the genus *Rhabdocline* would be a profitable line of investigation. In particular, the pathogenic species of *Rhabdocline* should be examined for the production of metabolites similar to those produced by the endophytic species.

The discriminating microbial ecologist will perceive other basic issues in ecological and evolutionary theory for which endophytic mutualism may have more than an imitative, 'me-too', role to play. However, the research suggested above should suffice for a long time.

## *REFERENCES*


Andrews, J.H., Hecht, E.P. & Bashirian, S. (1985). Association between the fungus *Acremonium curvulum* and Eurasian water milfoil, *Myriophyllum spicatum*. Canadian Journal of Botany, *60*, 1216-21.

Bailey, J.A. & Mansfield, J.W., eds (1982). Phytoalexins. Glasgow: Blackie.

Barrett, J.A. (1983). Plant-fungus symbioses. *In* Coevolution, eds D.J. Futuyma & M. Slatkin, pp. 137-60. Sunderland: Sinauer Associates.

de Bary, A. (1866). Morphologie und Physiologie der Pilze, Flechten, und Myxomyceten. Vol. II. Hofmeister's Handbook of Physiological Botany. Leipzig.

Bell, A.A. (1981). Biochemical mechanisms of disease resistance. Annual Review of Plant Physiology, *32*, 21-81.

Bernstein, M.E. & Carroll, G.C. (1977). Internal fungi in old-growth Douglas fir foliage. Canadian Journal of Botany, *55*, 644-53.

Bose, S.R. (1947). Hereditary (seed-borne) symbiosis in *Casuarina equisetifolia*. Nature (London), *159*, 512-4.

Boursnell, J.G. (1950). The symbiotic seed-borne fungus in the Cistaceae. I. Distribution and function of the fungus in the seedling and in the tissues of the mature plant. Annals of Botany, N.S., *14*, 217-43.

Carroll, G.C. & Carroll, F.E. (1978). Studies on the incidence of coniferous needle endophytes in the Pacific Northwest. Canadian Journal of Botany, *56*, 3032-43.

Carroll, G.C. & Petrini, O. (1983). Patterns of substrate utilization by some fungal endophytes from coniferous foliage. Mycologia, *75*, 53-63.

Christiansen, H. (1972). On the development of pollen and the fertilization mechanisms of *Larix* and *Pseudotsuga menziesii*. Silvae Genetica, *21*, 166-75.

Clay, K., Hardy, T.N. & Hammond, A.M. (1985). Fungal endophytes of grasses and their effects on an insect herbivore. Oecologia, *66*, 1-5.

Claydon, N., Grove, J.F. & Pople, M. (1985). Elm bark beetle boring and feeding deterrents from *Phomopsis oblonga*. Phytochemistry, *24*, 937-43.

Condrashoff, S.F. (1961 a). Three new species of *Contarinia* Rond. (Diptera: Cecidomyiidae) in Douglas fir needles. Canadian Entomologist, *93*, 123-30.

Condrashoff, S.F. (1961 b). Description and morphology of the immature stages of three closely related species of *Contarinia* Rond. (Diptera: Cecidomyiidae) from galls on Douglas fir needles. Canadian Entomologist, *93*, 833-51.

Cubit, J.D. (1974). Interactions of seasonally changing physical factors and grazing affecting high intertidal communities on a rocky shore. Ph.D. Thesis, Eugene, University of Oregon.

Diamandis, S. (1981). *Elytroderma torres-juanii* Diamandis & Minter. A serious attack on *Pinus brutia* L. in Greece. *In* Current Research on Conifer Needle Diseases, ed. C.S. Millar, pp. 9-11. Aberdeen: Aberdeen University Press.

Edmunds, G.F., Jr. & Alstad, D.N. (1978). Coevolution in insect herbivores and conifers. Science, *199*, 941-5.

Funk, A. (1985). Foliar Fungi of Western Trees. Victoria, B.C.: Canadian Forestry Service.

Funk, C.R., Halisky, P.M., Johnson, M.C., Siegel, M.R., Stewart, A.V., Ahmad, S., Hurley, R.H. & Harvey, I.C. (1983). An endophytic fungus and resistance to sod webworms: association in *Lolium perenne* L. Bio/Technology, *1*, 189-91.

Futuyma, D.J. (1979). Evolutionary Biology. Sunderland: Sinauer Associates.

Futuyma, D.J. & Slatkin, M. eds. (1983). Coevolution. Sunderland: Sinauer Associates.

Hepperly, P.R., Kirkpatrick, B.L. & Sinclair, J.B. (1985). *Abutilon theophrasti*: wild host for three fungal parasites of soybean. Phytopathology, *70*, 307-10.

Huse, K.J. (1981). The distribution of fungi in sound-looking stems of *Picea abies* in Norway. European Journal of Forest Pathology, *11*, 1-6.

Johnson, M.C., Pirone, T.P., Siegel, M.R. & Varney, D.R. (1982). Detection of *Epichloë typhina* in tall fescue by means of enzyme-linked immunosorbent assay. Phytopathology, *72*, 647-50.

Kuć, J. & Hammerschmidt, R. (1978). Acquired resistance to bacterial and fungal infections. Annals of Applied Biology, *89*, 313-7.

Kulik, M.M. (1984). Symptomless infection, persistence, and production of pycnidia in host and non-host plants by *Phomopsis batatae*, *Phomopsis phaseoli*, and *Phomopsis sojae*, and the taxonomic implications. Mycologia, *76*, 274-91.

Luttrell, E.S. (1974). Parasitism of fungi on vascular plants. Mycologia, *66*, 1-15.

McOnie, K.C. (1967). Germination and infection of citrus by ascospores of *Guignardia citricarpa* in relation to control of black spot. Phytopathology, *57*, 743-6.

Matta, A. (1971). Microbial penetration and immunization of uncongenial host plants. Annual Review of Phytopathology, *9*, 387-410.

Merrill, W., Zang, L., Bowen, K. & Kistler, B.R. (1979). *Naemacyclus minor* needlecast of *Pinus nigra* in Pennsylvania. Plant Disease Reporter, *63*, 994.

Millar, C.S. (1980). Infection processes on conifer needles. *In* Microbial Ecology of the Phylloplane, ed. J.P. Blakeman, pp. 185-209. London: Academic Press.

Millar, C.S. & Richards, G.H. (1975). The incidence of *Lophodermium* types in attached pine needles. Mitteilungen der Bundesforschungsanstalt für Forst- und Holzwirtschaft Reinbek bei Hamburg, *108*, 57-67.

Minter, D.W. (1981). Possible biological control of *Lophodermium seditiosum*. *In* Current Research on Conifer Needle Diseases, ed. C.S. Millar, pp. 67-74. Aberdeen: Aberdeen University Press.

Minter, D.W. & Millar, C.S. (1980 a). Ecology and biology of three *Lophodermium* species on secondary needles of *Pinus sylvestris*. European Journal of Forest Pathology, *10*, 169-81.

Minter, D.W. & Millar, C.S. (1980 b). A study of three pine inhabiting *Lophodermium* species in culture. Nova Hedwigia, *32*, 361-8.

Minter, D.W. Staley, J.M. & Millar, C.S. (1978). Four species of *Lophodermium* on *Pinus sylvestris*. Transactions of the British Mycological Society, *71*,

295-301.
Moss, M.O. (1984). The mycelial habit and secondary product production. *In* The Ecology and Physiology of the Fungal Mycelium, eds D.H. Jennings & A.D.M. Rayner, pp. 127-42. Cambridge: Cambridge University Press.
Muirhead, I.F. & Deverall, B.J. (1981). Role of appressoria in latent infection of banana fruits by *Colletotrichum musae*. Physiological Plant Pathology, *19*, 77-84.
Nathaniels, N.Q.R. & Taylor, G.S. (1983). Latent infection of winter oilseed rape by *Leptosphaeria maculans*. Plant Pathology, *32*, 23-31.
Nitecki, M.H., ed. (1983). Coevolution. Chicago: University of Chicago Press.
Parker, A.K. & Reid, J. (1969). The genus *Rhabdocline* Syd. Canadian Journal of Botany, *47*, 1533-45.
Petrini, O. (1984). Zur Verbreitung und Oekologie Endophytischer Pilze. Zürich: Mikrobiologisches Institut der Eidgenössischen Technischen Hochschule.
Petrini, O. & Carroll, G. (1981). Endophytic fungi in foliage of some Cupressaceae in Oregon. Canadian Journal of Botany, *59*, 629-36.
Petrini, O., Stone, J. & Carroll, F.E. (1982). Endophytic fungi in evergreen shrubs in western Oregon: a preliminary study. Canadian Journal of Botany, *60*, 789-96.
Prager, E.M., Fowler, D.P. & Wilson, A.C. (1976). Rates of evolution in the Pinaceae. Evolution, *30*, 637-49.
Price, P.W. (1980). Evolutionary Biology of Parasites. Princeton: Princeton University Press.
Quack, W., Scholl, H. & Budzikiewicz, H. (1982). Ascocorynin, a terphenylquinone from *Ascocoryne sarcoides*. Phytochemistry, *21*, 2921-3.
Rayner, M.C. (1915). Obligate symbiosis in *Calluna vulgaris*. Annals of Botany, *29*, 96-131.
Rayner, M.C. (1929). The biology of fungus infection in the genus *Vaccinium*. Annals of Botany, *43*, 55-70.
Roll-Hansen, F. & Roll-Hansen, H. (1979 a). *Neobulgaria premnophila* sp. nov. in stems of living *Picea abies*. Norwegian Journal of Botany, *26*, 207-11.
Roll-Hansen, F. & Roll-Hansen, H. (1979 b). *Ascocoryne* species in living stems of *Picea* spp. European Journal of Forest Pathology, *5*, 275-80.
Roll-Hansen, F. & Roll-Hansen, H. (1979 c). Microflora of sound-looking wood in *Picea abies* stems. European Journal of Forest Pathology, *9*, 308-16.
Sampson, K. (1933). The systemic infection of grasses by *Epichloe typhina* (Pers.) Tul. Transactions of the British Mycological Society, *18*, 30-47.
Sampson, K. (1935). The presence and absence of an endophytic fungus in *Lolium temulentum* and *L. perenne*. Transactions of the British Mycological Society, *19*, 337-43.
Sampson, K. (1937). Further observations on the systemic infection of *Lolium*. Transactions of the British Mycological Society, *21*, 84-97.
Sampson, K. (1939). Additional notes on the systemic infection of *Lolium*. Transactions of the British Mycological Society, *23*, 316-9.
Schemske, D.W. (1983). Limits to specialization and coevolution in plant-animal mutualists. *In* Coevolution, ed. M.H. Nitecki, pp. 67-109. Chicago: University of Chicago Press.
Sherwood-Pike, M., Stone, J. & Carroll, G.C. (1986). *Rhabdocline parkeri*, a ubiquitous foliar endophyte of Douglas fir. Canadian Journal of Botany, (in press).
Spurr, S.H. & Barnes, B.V. (1980). Forest Ecology, 3rd edition. New York: John Wiley & Sons.
Stone, J.K. (1985). Foliar endophytes of *Pseudotsuga menziesii* (Mirb.) Franco. Cytology and physiology of the host-endophyte relationship. Ph.D. thesis, Eugene, University of Oregon.
Thompson, J.N. (1982). Interaction and Coevolution. New York: John Wiley & Sons.
Todd, D. (1984). A field study of the endophyte fungi in Douglas fir needles. Ph.D. thesis, Eugene, University of Oregon.
Tokunaga, Y. & Ohira, I. (1973). Latent infection of anthracnose on *Citrus* in

Japan. Reports of the Tottori Mycological Institute (Japan), *10*, 693-702.

Verhoeff, K. (1974). Latent infections by fungi. Annual Review of Phytopathology, *12*, 99-110.

Volpe, E.P. (1985). Understanding Evolution. Dubuque: William C. Brown.

Webber, J. (1981). A natural control of Dutch elm disease. Nature (London), *292*, 449-51.

Webber, J. & Gibbs, J.N. (1984). Colonization of elm bark by *Phomopsis oblonga*. Transactions of the British Mycological Society, *82*, 384-52.

Weidemann, G.J. & Boone, D.M. (1984). Development of latent infections on cranberry leaves inoculated with *Botryosphaeria vaccinii*. Phytopathology, *74*, 1041-3.

White, J.F., Jr. & Cole, G.T. (1985). Endophyte-host associations in forage grasses. III. *In vitro* inhibition of fungi by *Acremonium coenophialum*. Mycologia, *77*, 487-9.

# TOXIC METABOLITES OF EPIPHYTIC AND ENDOPHYTIC FUNGI OF CONIFER NEEDLES

J.D. Miller

*Chemistry and Biology Research Institute, Agriculture Canada, Ottawa, Ontario K1A 0C6, Canada*

## *SECONDARY METABOLITES OF CONIFERS*

There is a considerable body of literature concerning the role of secondary metabolites of plants as defense compounds. Over 12,000 compounds have been identified from plants that are secondary metabolites (400,000 are guessed to exist) and many are known to have various antobiotic activities. All groups of plants produce secondary metabolites; some species produce a few compounds and some produce many. In general, gymnosperms have a lower diversity of secondary metabolites than angiosperms (Bentley *et al.* 1982). There are two ideas that need to be considered concerning this observation. Insects and angiosperms appeared at about the same time (Lower Carboniferous) with the associated powerful selection pressures on both at a time of new habitat creation and the general exploitation of insects by angiosperms for reproduction. The second factor relates to the habit of woody plants. Long-lived plants are exposed to a variety of herbivores and pathogens and hence have a variety of general defenses against insect pests, pathogens and larger animals that are both physical and chemical in nature. Generalist defensive compounds such as terpenes in conifers and tannins in deciduous trees are more abundant than is typical of specialist compounds of angiosperms (Kingsbury 1979).

An assessment of the specific secondary metabolite defensive compounds of conifers reveals that there are surprisingly few that are active against insects that attack foliage such as the budworm (*Choristoneura fumiferana* Clem.). Needles of balsam fir (*Abies balsamea*), a host tree, contain a large variety of terpenes and these occur in greater concentrations in needles than in cortex or xylem (Hunt & Von Rudloff, 1974, 1977). These compounds are thought to play a role in protecting the needles from attack by fungal pathogens (Hunt & Von Rudloff, 1977). Further, these compounds are formed as the needles age and hence are not present in high concentrations when insects such as the spruce budworm, which prefers young needles, attack the tree. In any case, no parameter of balsam fir foliar chemistry has been consistently linked to budworm survival (Mattson *et al.* 1983). The compound pungenin from spruce (*Picea glauca*) has been reported as active against the spruce budworm but again it is in low concentration when the insects are feeding. Budworms fed with this compound will show some reduction in weight compared to controls, and in females this will result in smaller numbers of eggs (Strunz *et al.* 1986). Tannin may also play a modest role in protecting the foliage of trees attacked by the spruce budworm (G.M. Strunz, personal communication). None the less, the natural defenses of spruce and fir trees to the spruce budworm are not very effective as evidenced by the mass destruction of trees in eastern Canada (Royama 1984).

### *SECONDARY METABOLITES OF FUNGI*

Fungi are also prolific in their secondary metabolites and several thousand low molecular weight compounds have been reported (e.g. Turner 1971; Turner & Aldridge 1983). Mycologists use the term secondary metabolite in a more restrictive way than botanists. A true fungal secondary metabolite is a biochemical produced upon some nutrient limitation. Although rather a lot is known about the structures and antibiotic activities of fungal secondary metabolites, very little is understood concerning the genetics and physiology of their biosynthesis.

It is possible to identify the stages of growth of a fungus relative to secondary metabolite production. In a liquid medium that supplies all nutrients in appropriate ratios with allowance for sufficient $O_2$ and $CO_2$ transfer, growth of a fungus will be rapid and linear as measured by $CO_2$ evolution or dry weight but exponential by rate of RNA synthesis. This kind of growth is termed 'balanced' (other terms are also used). Growth in terms of dry weight production continues as long as all nutrients are available and the physical conditions are appropriate and no secondary metabolites are made. The growth state termed 'stationary phase' by bacteriologists, in which dry weight production ceases but during which the cells are metabolically active, is termed 'unbalanced'. This is induced by the exhaustion of one or more nutrients in the medium. If the limiting nutrient is, for example, a nitrogen compound, the cells will continue to transport and utilize other nutrients. Secondary metabolites are produced during unbalanced growth. Because normal growth cannot proceed, primary metabolites, e.g. acetyl CoA, malonyl CoA and various amino acids, accumulate and enzymes of secondary metabolism biosynthesize a given compound. The nature of the nutrient limitation and the associated physical conditions determine which of the secondary metabolites a particular fungus will produce from its genetically defined repertoire. Different compounds can be produced by the same fungus under different environmental conditions. A useful consideration of this topic is that of Bu'Lock (1975).

A number of ideas have been advanced concerning the purpose of secondary metabolites in fungi. Because they are not produced during balanced growth, as noted above, there is no argument that they are needed for the ordinary growth of the organism. The first idea concerning their purpose was that secondary metabolites are storage compounds. Many experiments have shown that this is not true. Indeed these compounds are energy-expensive and considerable portions of metabolizable energy are used to produce them at the expense of e.g. spore production (Rhoades 1979; Miller 1985). Another hypothesis concerns the correlation of secondary metabolite production with sporulation in a number of Deuteromycetes. Some have argued that secondary metabolites play a role in sporulation (Bird *et al.* 1981). For the cases reported, however, mutants exist that lack the ability to produce the metabolites but still sporulate normally.

At the time of the discovery of penicillin and during the considerable screening of fungi for useful secondary metabolites that followed, researchers ascribed a defensive function to secondary metabolites of fungi (Brian 1957). There followed a period during which several authors argued that there is no role for these compounds and their antibiotic properties were accidental (Turner 1971; Gottlieb 1976; Ciegler 1983). In recent years, Janzen (1977) and Wicklow (1981) have outlined compelling evidence for the defensive role of these compounds in nature, the so-called 'interference competition'. There are many examples of this phenomenon in fungi growing on a variety of substrates. In trees and in wood, there is clear evidence of interference competition. Fungi such as *Trichoderma viride* produce a variety of toxic metabolites that can restrict the growth of wood rot fungi. Successions of fungi in cankers are thought to be mediated by antifungal compounds. Experiments have demonstrated that some lignicolous marine fungi produce secondary metabolites that prevent the growth of competing fungi (Wicklow 1981; Miller 1985; Miller *et al.* 1985 a). Mycorrhizal fungi are known to protect plants from certain pathogens by a number of mechanisms; Zak (1964) included among these mechanisms the production of antibiotics.

The interference competition of fungi may involve organisms other than fungi. Aflatoxins, for example, are not particularly antifungal and are produced by *Aspergillus flavus* and *A. parasiticus* when growing on corn. Wicklow & Shotwell (1983) suggested that high concentrations of aflatoxins in conidia and sclerotia prevent predation by insects. Trichothecenes produced by *Fusarium* species growing on grains are active against animals thus 'preserving' that substrate for fungal growth (Miller *et al.* 1983).

## *FUNGAL TOXINS AND PLANTS*

Fungi and plants have evolved a variety of secondary metabolites that contribute to the survival of the organism by defending against competitors or predators. The arguments used to describe the significance of secondary metabolites in fungi and plants are essentially the same (Kingsbury 1979; Wicklow 1981). Zähner & Anke (1983) stated that the genetic material for 'useful' secondary metabolites can be transferred between organisms. Indeed there is evidence of convergence of useful secondary metabolites in fungi. For example, *Fusarium graminearum* (a cereal pathogen) produces the sesquiterpene diol culmorin (Miller & Greenhalgh 1985), as does the marine lignicolous fungus *Leptosphaeria oraemaris* (Miller 1985). This compound has antifungal activity and partially explains the fact that substrates containing the respective fungi often show little or no evidence of other fungi.

Pirozynski (1981, personal communication) argued that during evolution woody plants obtained the ability to synthesize lignin through the mediation by endotrophic fungi; in particular plants with such an association are lignin producers. Perhaps an intermediate stage in this process is the fungal endophyte associations with various grasses described by K. Clay (this volume) wherein it is clearly identified that plants gain competitive advantage by harbouring fungi that produce insecticidal and fungicidal compounds (White & Cole 1985).

As noted previously, conifers such as balsam fir apparently lack insecticidal secondary metabolites that provide foliar resistance to insects such as the spruce budworm. Fungi of various kinds have an intimate association with woody plants and wood. Recently, methods of biological control of certain plant pathogens of apple have employed toxigenic fungi that produce their toxins in the tree phylloplane (Cullen & Andrews 1984). Wright *et al.* (1982) noted that mycotoxins negatively affect insects and made the point that mycotoxins from fungi in a habitat shared with insects are likely to be most toxic to insects. It is possible to speculate that fungal toxins may be 'useful' to trees.

## *INSECTS, CONIFER NEEDLES AND FUNGI*

### *Epiphytes*

The spruce budworm is an important pest of spruce and fir in eastern Canada, and this insect has been the object of much study, especially by scientists at the Maritime Forest Research Centre in New Brunswick, Canada. Royama (1984) evaluated over 30 years of population data concerning the spruce budworm in New Brunswick and concluded that there are some unknown causes of mortalities that play a role in determining the budworm population. This led to a study of phylloplane fungi on budworm-infested and uninfested balsam fir trees described by Miller *et al.* (1985 b).

Briefly, fungi were determined weekly on and in needles using a variety of techniques from budworm-infested and uninfested trees from May to July 1983 and 1984. As expected, a diverse population of saprophytic fungi was found. The presence of the spruce budworm resulted in a doubling of fungal diversity on the needles. Included in this increased diversity were a number of potentially toxigenic fungi. Also determined were fungi from frass, larvae and gut contents. It was found that the budworms were ingesting fungi from the phylloplane.

Representative fungal isolates were chosen based on their frequency of occurrence and association with the frass and larvae. These isolates were grown in fermenters, and extracts were made. These were tested for toxicity to HeLa 229 cells, and sterile hyphae from the respective fungi were incorporated into budworm rearing diet at 1% (w/w). HeLa tests were done by adding various amounts of the extracts in pure ethanol to microplate cultures of Hela cells (Miller *et al.* 1985 b). These fungi included *Aureobasidium pullulans* as well as fungi such as *Penicillium cyclopium* and *Fusarium avenaceum*. Some produced extracts that were toxic to HeLa cells and the hyphae caused mortalities and a reduced growth rate (Table 1).

### *Endophytes*

In the above described study, needle endophytes were also

recovered and one was fermented and tested for HeLa cell toxicity; it was negative at the cut-off limit chosen (5 $\mu g\ ml^{-1}$), but was toxic at slightly higher concentrations. Fungal endophytes are well known from conifers (Carroll & Carroll 1978) but there is little information concerning endophytes from eastern Canada. G.C. Carroll reported information at the 1984 Mycology Society of America meeting relating to an endophyte that when present in Douglas fir needles was associated with mortalities of *Contarinia* sp. larvae. This organism was *Rhabdocline parkeri*, referred to as PSm-1 (see also G.C. Carroll, this volume).

As noted previously, determining the precise conditions necessary to induce the production of a secondary metabolite is time-consuming and complicated. A variety of fermentation conditions were tested that varied temperature and $pO_2$ using 2% malt extract broth. Extracts were made and tested with HeLa 229 cells. The results indicated that at low $pO_2$ and low temperature fermentations of five isolates of PSm-1 yielded extracts toxic to HeLa cells at 2 $\mu g\ ml^{-1}$. This prompted further studies of Oregon endophytes and more detailed research on PSm-1.

Fermentations were made under the conditions noted above of a variety of endophytes from Oregon and extracts were tested with HeLa cells and hyphae incorporated into the rearing diet at 1% (w/w). In some cases, extracts were incorporated into rearing diets at 1 and 10 $\mu g\ ml^{-1}$. The results (Table 2) demonstrated that some endophytes make factors toxic to the spruce budworm. This 'toxicity' was expressed as both mortalities and reduced growth rate. In some cases, the material was not toxic to HeLa cells and yet caused reduced growth rate of larvae suggesting that the chemical(s) is antifeedant in nature. Extracts of PSm-1-5 also demonstrated toxicity to the spruce budworm larvae and the activity

Table 1. Toxicity of balsam fir needle epiphytic fungi to HeLa cells and budworm larvae.

| Fungus | Isolate | HeLa toxicity (2 $\mu g\ ml^{-1}$) | Toxicity to budworm larvae | |
|---|---|---|---|---|
| | | | Mortality | Growth rate reduction |
| *Phialophora bubaki* | 628 | *[1] | +[2] | -[3] |
| *Fusarium avenaceum* | 120 | * | ++++ | ++++ |
| *Aureobasidium pullulans* | 421 | * | + | ++ |
| *Geotrichum candidum* | 564 | * | + | +++ |
| *Penicillium cyclopium* | 490 | * | + | ++ |

*: 100% of the HeLa cells killed.
+ to ++++: 20-25% to 80-100% mortality or growth rate reduction.
-: no reaction.

resulted from at least three compounds.

Research is in progress concerning the association of endophytes with needle cells. Recent electron microscope studies of algal endophytes by Porter (1985) show that the fungal hyphae do not affect the plant cells and usually are very fine (1 μm in diameter). The hyphae are usually found between plant cell walls. Where the algal cells have died, the hyphae expand to a typical diameter (4-5 μm). Preliminary studies of needle endophytes indicate that the same sort of relationship applies, i.e. healthy needles contain endophytes with fine hyphae and when needles are removed, the hyphae expand.

## *CONCLUSIONS*

From a theoretical point of view and from the limited experimentation in this area, it appears that needle fungi, both epiphytes and endophytes, may play a role in conferring resistance to insect attack of conifers. Conifers, apparently lacking strong intrinsic defenses, appear to have mechanisms that allow toxigenic fungi to grow in or on the needle. G.C. Carroll (see also this volume) expressed the view that the very long reproductive cycle of conifers does not permit a rapid genetic response to new pests gene for gene. Mechanisms that allow the growth of toxigenic fungi in foliage that can mutate at a much faster rate may provide a solution to this 'problem'. Since models exist in angiosperms and seaweeds with endophytes, this does not seem too speculative, although much research is required.

Table 2. Toxicity of tree endophytes from Oregon to HeLa cells and budworm larvae.

| Fungus | HeLa toxicity (2 μg ml$^{-1}$) | Toxicity to budworm larvae | |
|---|---|---|---|
| | | Mortality | Growth rate reduction |
| PSm-1-5 | *[1] | +++[2] | +++ |
| *Phomopsis* sp. 3 | * | - | +++ |
| *Phomopsis* sp. 2 | * | - | +++ |
| *Phyllostica* sp. 1[3] | * | - | - |
| Coelomycete x-1 | - | + | +++ |
| Coelomycete x-2 | - | ++ | ++ |

1 *: 100% of the HeLa cells killed; -: no reaction.
2 For explanation of symbols, see Table 1.
3 Some *Phyllosticta* sp. isolates induced mortalities of the spruce budworm.

Perhaps most interesting is the exact nature of the needle-endophyte association. If fungal secondary metabolites are involved, then the needle-fungus association must provide suitable conditions. This may be manifested in the oscillatory growth of wood decay fungi described by Hale & Eaton (1985) (cf. Wicklow 1981). The evidence that the biomass (diameter) of endophyte is kept under plant control is intriguing. The nature of the metabolites may provide to science new compounds of theoretical and practical interest as suggested by studies of my group and those of J.K. Grove of a xylem endophyte of elm (Claydon *et al.* 1985) and others (Fisher *et al.* 1984).

*ACKNOWLEDGEMENTS*

This article has benefited from helpful discussions and assistance from G.C. Carroll, C. Clark, J. Johnson, T. Royama, D. Strongman, G.M. Strunz, N.J. Whitney and D.T. Wicklow.

*REFERENCES*

Bentley, M.D., Leonard, D.E., Leach, S., Reynolds, E., Stoddard, W., Tomkinson, B., Tompkinson, D., Strunz, G.M. & Yatagai, M. (1982). Effects of some naturally occurring chemicals and some extracts of non-host plants on feeding by spruce budworm larvae (*Choristoneura fumiferana*). Life Sciences and Agriculture Experimental Station, University of Maine, Technical Bulletin 107.

Bird, B.A., Remaley, A.T. & Campbell, I.M. (1981). Brevianamides A and B are formed only after conidiation has begun in solid cultures of *Penicillium brevicompactum*. Applied and Environmental Microbiology, *42*, 521-5.

Brian, P.W. (1957). The ecological significance of antibiotic productions. *In* Microbial Ecology, eds R.E.O. Williams & C.C. Spicer, pp. 168-88. Cambridge: Cambridge University Press.

Bu'Lock, J.D. (1975). Secondary metabolism in fungi and its relationship to growth and development. *In* The Filamentous Fungi. Vol. 1, eds J.E. Smith & D.R. Berry, pp. 33-58. New York: John Wiley & Sons.

Carroll, G.C. & Carroll, F.E. (1978). Studies on the incidence of coniferous needle endophytes in the Pacific northwest. Canadian Journal of Botany, *56*, 3034-43.

Ciegler, A. (1983). Evolution, ecology and mycotoxins: some musings. *In* Secondary Metabolism and Differentiation in Fungi, eds J.W. Bennett & A. Ciegler, pp. 429-39. New York: Marcel Dekker.

Claydon, N., Grove, J.F. & Pople, M. (1985). Elm bark beetle boring and feeding deterrents from *Phomopsis oblonga*. Phytochemistry, *24*, 937-43.

Cullen, D. & Andrews, J.H. (1984). Evidence for the role of antibiosis in the antagonism of *Chaetomium globosum* to the apple scab pathogen, *Venturia inaequalis*. Canadian Journal of Botany, *62*, 1819-23.

Fisher, P.J., Anson, A.E. & Petrini, O. (1984). Antibiotic activity of some endophytic fungi from ericaceous plants. Botanica Helvetica, *94*, 249-53.

Gottlieb, D. (1976). The production and role of antibiotics in soil. Journal of Antibiotics, *29*, 987-1000.

Hale, M.D. & Eaton, R.A. (1985). Oscillatory growth of fungal hyphae in wood cell walls. Transactions of the British Mycological Society, *84*, 277-88.

Hunt, R.S. & Rudloff, E. von (1974). Chemosystematic studies in the genus *Abies*. I. Leaf and twig oil analysis of alpine and balsam firs. Canadian Journal of Botany, *52*, 477-87.

Hunt, R.S. & Rudloff, E. von (1977). Leaf oil terpene variation in western white pine populations of the Pacific Northwest. Forest Science, *23*, 507-16.

Janzen, D.H. (1977). Why fruits rot, seeds mold, and meat spoils. American Naturalist, *111*, 691-713.

Kingsbury, J.M. (1979). The evolutionary and ecological significance of plant toxins. *In* Herbivores: Their Interaction with Secondary Plant Metabolites, eds G.A. Rosenthal & D.H. Janzen, pp. 675-702. New York: Academic Press.

Mattson, W.J., Slocum, S.S. & Koller, C.W. (1983). Spruce budworm (*Choristoneura fumiferana*). Performance in relation to foliar chemistry. USDA Forest Service Technical Report NE-85.

Miller, J.D. (1985). Secondary metabolites in marine lignicolous fungi. *In* Proceedings of the Fourth International Marine Mycology Symposium, ed. S.T. Moss. Cambridge: Cambridge University Press (in press).

Miller, J.D. & Greenhalgh, R. (1985). Nutrient effects on the biosynthesis of trichothecenes and other metabolites by *Fusarium graminearum*. Mycologia, *77*, 130-6.

Miller, J.D., Jones, E.B.G., Moharir, Y.E. & Findlay, J.A. (1985 a). Colonization of wood blocks by marine fungi in Langstone Harbour. Botanica Marina, *28*, 251-7.

Miller, J.D., Strongman, D. & Whitney, N.J. (1985 b). Observations of fungi associated with spruce budworm-infested balsam fir needles. Canadian Journal of Forest Research, *15*, 896-901.

Miller, J.D., Young, J.C. & Trenholm, H.L. (1983). *Fusarium* toxins in field corn. I. Time course of fungal growth and production of deoxynivalenol and other mycotoxins. Canadian Journal of Botany, *61*, 3080-7.

Pirozynski, K.A. (1981). Interactions between fungi and plants through the ages. Canadian Journal of Botany, *59*, 1824-7.

Porter, D. (1985). Mycoses of marine organisms: an overview of pathogenic mechanisms. *In* Proceedings of the Fourth International Marine Mycology Symposium, ed. S.T. Moss. Cambridge: Cambridge University Press (in press).

Rhoades, D.F. (1979). Evolution of plant chemical defense against herbivores. *In* Herbivores: Their Interaction with Secondary Metabolites, eds G.A. Rosenthal & D.H. Janzen, pp. 3-54. New York: Academic Press.

Royama, T. (1984). Population dynamics of the spruce budworm *Choristoneura fumiferana* Clem.: a reinterpretation of life table studies from the Green River project. Ecological Monographs, *54*, 429-62.

Strunz, G.M., Giguère, P. & Thomas, A.W. (1986). Synthesis of pungenin, a foliar constituent of some spruce species, and an investigation of its efficiency as a feeding deterrent for the spruce budworm (*Choristoneura fumiferana* Clem.). Journal of Chemical Ecology, *12*, 251-60.

Turner, W.B. (1971). Fungal Metabolites. New York: Academic Press.

Turner, W.B. & Aldridge, D.C. (1983). Fungal Metabolites II. New York: Academic Press.

White, J.F., Jr. & Cole, G.T. (1985). Endophyte-host associations in forage grasses. III. *In vitro* inhibition of fungi by *Acremonium coenophialum*. Mycologia, *77*, 487-9.

Wicklow, D.T. (1981). Interference competition and the organization of fungal communities. *In* The Fungal Community, eds D.T. Wicklow & G.C. Carroll, pp. 351-75. New York: Marcel Dekker.

Wicklow, D.T. & Shotwell, O.L. (1983). Intrafungal distribution of aflatoxins among conidia and sclerotia of *Aspergillus flavus* and *Aspergillus parasiticus*. Canadian Journal of Microbiology, *29*, 1-5.

Wright, V.F., Vesonder, R.F. & Ciegler, A. (1982). Mycotoxins and other fungal metabolites as insecticide. *In* Microbial and Viral Pesticides, ed. E. Kurstak, pp. 559-83. New York: Marcel Dekker.

Zähner, H. & Anke, H. (1983). Evolution and secondary pathways. *In* Secondary Metabolism and Differentiation in Fungi, eds J.W. Bennett & A. Ciegler, pp. 153-71. New York: Marcel Dekker.

Zak, B. (1964). Role of mycorrhizae in root disease. Annual Review of Phytopathology, *2*, 377-92.

# SECTION IV

# PLANT-PATHOGENIC AND SAPROPHYTIC PROKARYOTES

# TEMPORAL, SPATIAL, AND GENETIC VARIABILITY OF LEAF-ASSOCIATED BACTERIAL POPULATIONS

S.S. Hirano[1] and C.D. Upper[2]

[1]*Department of Plant Pathology, University of Wisconsin-Madison, 1630 Linden Drive, Madison, WI 53706, USA*
[2]*Agricultural Research Service, United States Department of Agriculture, and Department of Plant Pathology, University of Wisconsin-Madison, 1630 Linden Drive, Madison, WI 53706, USA*

## *INTRODUCTION*

The preponderance of research on leaf-associated bacteria in the past decade has dealt with those bacteria perceived as having some economic effect on the plants whose leaves they inhabit. Thus, most studies have dealt with either phytopathogenic or ice nucleation-active (INA) bacteria, or potential antagonists of these micro-organisms. Regardless of the particular interests and questions that may motivate investigators to pursue research dealing with leaf-associated bacteria, each, undoubtedly, is or has been faced with the problem of how to best describe populations of leaf-associated bacteria *quantitatively*.

Two issues that have been adequately discussed and insufficiently well resolved to warrant additional discussion here are the methods for enumeration of bacteria on a particular sample (i.e. microscopic or cultural), and methods for determining whether bacteria are endo- or epiphytic (cf. Baker 1981; Parbery *et al.* 1981; Hirano & Upper 1983). We will assume that many of the available methods, if used with thought and care, are adequate to provide numbers approximately proportional to the actual population size. These methods can be considered reasonable for estimating relative population sizes, although they may not necessarily be accurate for determination of the absolute population values. Further, since differentiation and quantitation of surface versus internal bacterial populations for a given sample is a technically difficult problem with no readily available, completely satisfactory answer, we will refer to the bacteria in which we are interested as *leaf-associated bacteria.*

The issues that relate to quantitation of leaf-associated bacteria that will be discussed here are the sampling unit and the inherent quantitative and qualitative variability associated with the leaf ecosystem. To what extent does this variability affect estimation and conceptualization of population sizes of these bacteria? How variable are bacterial population sizes on a given plant species at a given time? How rapidly do these bacterial populations change from one time to the next? Finally, how genetically variable might we expect certain leaf-associated bacteria to be?

## *LEAF-TO-LEAF VARIABILITY IN BACTERIAL POPULATION SIZE*

Approximately 25 years have elapsed since Crosse (1959)

first demonstrated that certain plant-pathogenic bacteria can be found in large numbers on the surfaces of healthy leaves. This finding led to the development of new concepts in the epidemiology of certain foliar plant diseases caused by phytopathogenic bacteria (Leben 1963, 1965; Hirano & Upper 1983). The significance of a second finding reported by Crosse (1959) has, for the most part, been overlooked. In the process of developing a procedure to compare quantitatively the epiphytic populations of *Pseudomonas syringae* pv. *morsprunorum* on leaves of two cherry cultivars, Crosse found that numbers of the bacterial pathogen ranged from undetectable to 153,500 cells per leaf on *individual* cherry leaves. Because of this quantitative variation in bacterial pathogen population size from one individual leaf to the next, Crosse (1959) reported that accurate estimates of epiphytic *P. syringae* pv. *morsprunorum* population sizes on the two cherry cultivars could be obtained by bulked samples (i.e. wherein several leaves are combined to form a single sample) only if each bulked sample contained a large number of leaves. Each of his bulked samples contained contained 192 cherry leaves. In almost all of the research that occurred during the succeeding 20 years, bulked samples have been used. Most workers have regarded 5 to 20 leaves per sample as adequate for quantitation of these bacteria (Leben, 1963; Luisetti & Paulin 1972; Lindow *et al.* 1978 a,b; Weller & Saettler 1980; Mew & Kennedy 1982). With the exception of Crosse (1959), the basis for selection of a particular number of leaves per bulked sample is rarely, if ever, justified in published reports.

If Crosse's early work was indeed an indication that the numbers of bacteria associated with different leaves within a given canopy were quite distinct from each other at a given time, then substantial conceptual and methodological issues arise about the use of bulked samples to measure epiphytic bacterial populations. If, indeed, each individual leaf (or leaflet) is a distinct ecosystem, then population values determined from bulked samples, which may represent the population on the 'average' leaf, would not include any information about the ranges in bacterial population sizes among leaves. Further, if the distribution of bacteria among leaves followed a skewed or asymmetric distribution, then an arithmetic mean, as estimated from bulked samples, might represent population sizes quite different from the population size on the median leaf. In addition, this value will likely be quite unrepresentative of bacterial populations on *most* of the leaves (i.e. ecosystems) within the canopy.

To reexamine the extent to which leaf-associated bacterial population sizes vary from one leaf to the next within a plant canopy, and the effect of this quantitative variability on estimating and understanding bacterial populations, we determined population sizes of total leaf-associated bacteria and components thereof, including ice nucleation-active bacteria, on individual leaves or leaflets from a number of plant species, including snap bean (*Phaseolus vulgaris*), soybean (*Glycine max*), corn (*Zea mays*), tomato (*Lycopersicon esculentum*), oats (*Avena sativa*) and wheat (*Triticum aestivum*) (Hirano et al. 1982). For each crop and at each sampling time, 30 individual leaves or leaflets were processed by dilution plating of leaf washings (Fig. 1). It was not uncommon to find population sizes varying by over 1,000-fold from leaf to leaf harvested at the same sampling time. In some cases, populations of the ice nucleation-active bacterial component varied over a million-fold from leaf to leaf, in spite of the care taken to harvest leaves of visually uniform size and position in the plant canopy. For most sets of population measurements on a given canopy, population sizes of total leaf-associated bacteria and selected components, could be described by the lognormal frequency distribution (Hirano *et al.* 1982; Fig. 1). Since leaf-associated bacterial population sizes follow a lognormal distribution, bulked samples will overestimate the mean population size. The

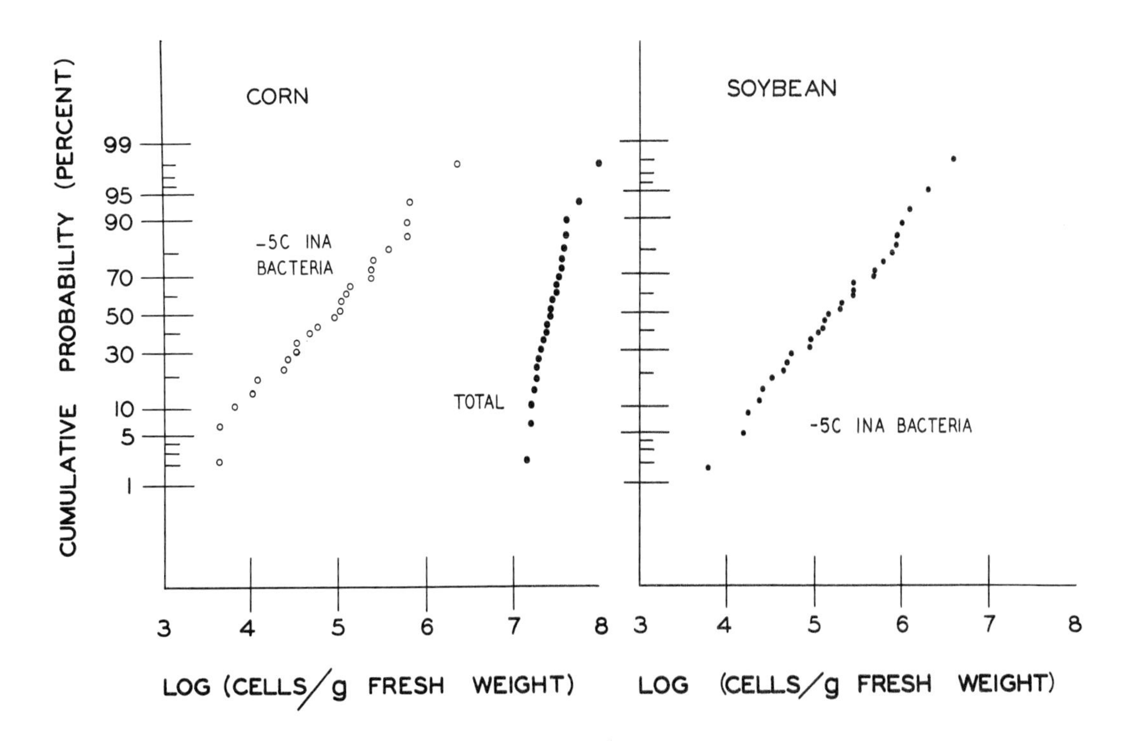


Fig. 1. Cumulative probability plots of leaf-associated bacterial populations. Populations of ice nucleation-active (INA) and total bacterial populations were determined on each of 24 individual corn leaves. INA bacterial populations were also estimated on each of 30 individual soybean leaflets. Each leaf or leaflet was washed separately and aliquots from 10-fold serial dilutions were plated onto King's medium B. INA bacteria were estimated from the dilution plates by the replica freezing method of Lindow *et al.* (1978 b). Observations from a normal distribution follow a straight line when plotted on the cumulative probability scale. Since bacterial populations have been log (base 10)-transformed, they can be said to follow a lognormal distribution.

magnitude of this overestimation is dependent on the variance of the population (Hirano *et al.* 1982).

Substantial added information is provided by knowledge of the frequency distribution (more correctly, an acceptable approximation thereof) of population sizes of bacteria among populations of leaves. From the distribution and its parameters (e.g. $\mu$ and $\sigma$) we not only know what the population is on the median leaf, but we also know the frequencies with which populations of *any given size* occur on leaves within the entire canopy! Consider the added power which this can provide in understanding the bacterial ecology of the canopy.

The finding that leaf-associated bacterial populations can be described by the lognormal distribution led to a better understanding of the quantitative relationship between population sizes of epiphytic plant-pathogenic bacteria and disease incidence. Lindemann *et al.* (1984 a,b) reported that the frequency with which populations of *P. syringae* pv. *syringae* exceeded 10,000 cells per bean leaflet was predictive of brown spot disease incidence on snap beans a week later. Rouse *et al.* (1985) supported and extended these findings with the development of a stochastic model for this system. The model integrates population sizes of epiphytic plant-pathogenic bacteria on individual leaves and the probability that a leaf with a given population size would become diseased to predict subsequent disease incidence.

On the other hand, quantitation of bacterial populations on individual leaves is certainly very time-consuming and in some situations, logistically beyond the means of some scientists. For example, an investigator faced with the determination of leaf-associated bacterial populations for each of 10 treatments, each with three replicates in a field experimental design may not find it feasible to sample and plate from 20 to 30 individual leaves from each of the replicate field plots (total = 600 to 900 individual leaves). In recognition of this problem, S.C. Chow and E.V. Nordheim (personal communication) and Chow (1985) have developed a method which will allow one to estimate the mean and variance of bacterial population measurements made on bulked samples. The number of leaves per bulked sample and the number of bulked samples to be taken per treatment depend on the precision desired and on the expected value of the population variance.

The inherent variability in bacterial population size from one leaf to the next within a plant canopy raises some rather interesting questions. What are the underlying mechanisms that generate these differences in population sizes? What roles do nutrient composition and concentration play in regulating leaf-associated bacterial population sizes? How do relatively small differences in the leaf microclimate affect population sizes? How does the composition of the microbial community on a given leaf affect the total or component bacterial population sizes on that leaf?

### *SPATIAL VARIABILITY*

Spatial distribution of micro-organisms in an apple tree canopy has been investigated by Andrews *et al.* (1980). Occasionally, population sizes of bacteria were found to be higher on leaves harvested from the interior compared to the exterior parts of apple trees. Ercolani (1979) reported that variability in population sizes of *P. syringae* pv. *savastanoi* on leaves of olive (*Olea europaea*) trees was affected more by the age of the leaves, time of leaf formation, and time of year than by the spatial (i.e. interior versus exterior) distribution of the leaves in the tree canopy.

We know from the lognormal frequency distribution, that the spatial distribution of bacteria on individual leaves or leaflets is not uniform in a canopy. Different leaves have very different leaf-associated bacterial population sizes. This is based on results obtained after random sampling and, thus, does not contain any information about the spatial placement of leaves as related to bacterial populations in a field. From the work of Lindemann *et al.* (1984 a,b) and Rouse *et al.* (1985), we know that population sizes of phytopathogenic bacteria, such as *P. syringae* pv. *syringae*, on healthy leaves are predictive of subsequent disease incidence, such as bacterial brown spot on snap beans. Many diseases of this type (e.g. halo blight and brown spot on snap beans; halo blight on oats) and probably others are reputed to be very focal in nature - that is, aggregated in space. Thus, we can expect that at some scale larger than individual leaves or leaflets and smaller than entire fields, bacterial components such as *P. syringae* are probably distributed in space in a manner in which the placement of leaves bearing populations of a given size cannot be predicted by a uniform or random distribution.

Although the role, if any, of epiphytic populations of *Xanthomonas campestris* in the epidemiology of black rot of cabbage is not clear, after symptoms have developed, cabbage plants infected with black rot follow a negative binomial distribution in commercial fields (Strandberg 1973). This may be indicative of an aggregated distribution of *X. campestris* among host plants before infection.

Beyond these studies, little is known about the spatial distribution of leaf-associated bacterial population sizes, particularly at comparable sites within a plant canopy or field. The nature of spatial distributions of leaf-associated bacterial populations presents some interesting sampling, analytical and mechanistic questions for future investigation.

### *TEMPORAL VARIABILITY*

The terms 'temporal variability' and 'population dynamics' are often used synonymously to describe changes in population sizes of leaf-associated bacteria as a function of time. How dynamic are leaf-associated bacterial populations? How rapidly do their population sizes change? Can we measure actual growth rates of these micro-organisms under field conditions? What aspects of the physical and plant environments affect changes in leaf-associated bacterial population sizes?

In many studies on changes in leaf-associated bacterial population sizes, bulked samples have been taken at weekly or monthly intervals (cf. Crosse 1959; Ercolani *et al.* 1974; Lindow *et al.* 1978 a,b; Ercolani 1979; Burkowicz 1981; Lindow 1982; Martins 1982; Mew & Kennedy 1982; Smitley & McCarter 1982; Daniel & Boher 1985). A representative example of the type of seasonal trends obtained is illustrated in Fig. 2 for total and ice nucleation-active bacterial populations on beans, soybeans, pumpkins and tomatoes (Lindow *et al.* 1978 b). Although sampling at these frequency intervals has provided some understanding of the general, long-term or seasonal trends in changes in bacterial populations, it has not fully illustrated the inherent *dynamics* of the leaf

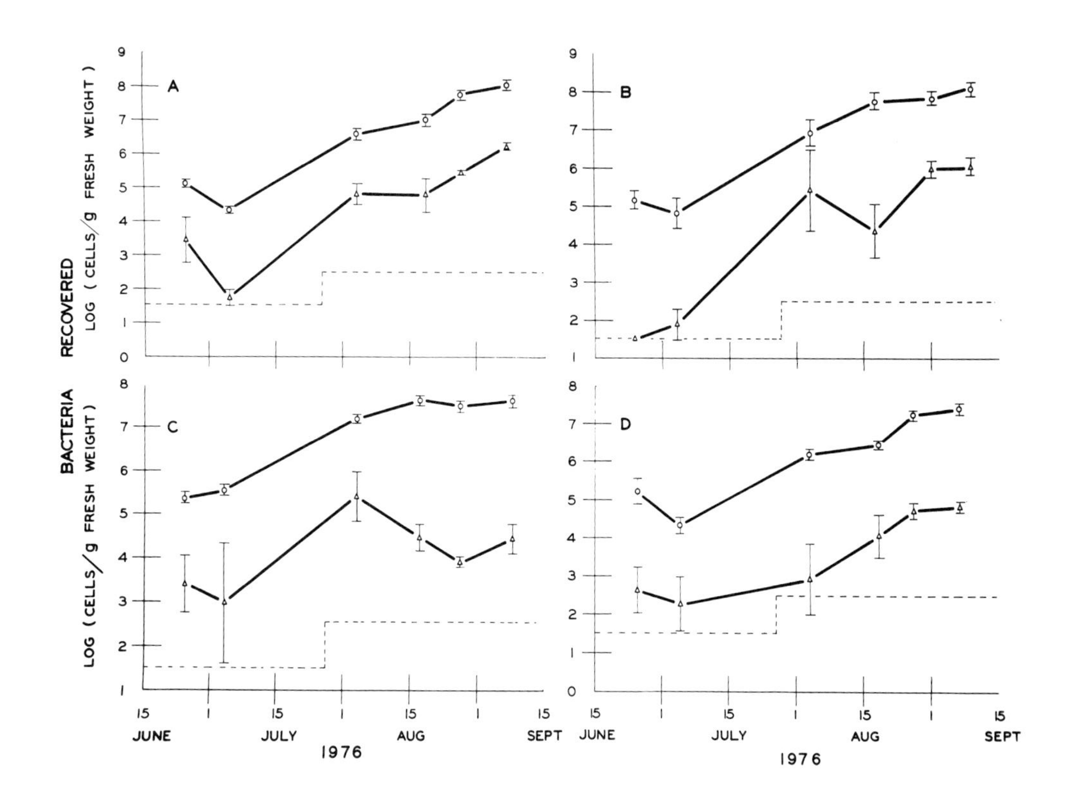


Fig. 2. Total and INA bacterial populations on normal, non-treated leaves of (A) bean, (B) soybeans, (C) pumpkins, and (D) tomatoes, sampled during the 1976 growing season. Total bacteria (o) were determined on nutrient agar, and INA bacteria (Δ) were determined on glycerol nutrient agar by dilution plating of washings of 10 to 20 g of leaves of each species from each of three replications at each date. Vertical bars represent the standard error of the mean of log populations from the three separate determinations of each sampling date. Dashed lines indicate the estimated limits of detection of INA bacteria. Reprinted from Lindow *et al.* (1978 b).

ecosystem. Anecdotal correlations from many of these seasonal studies of leaf-associated bacterial populations have produced the following generalizations:

- Bacterial populations tend to follow generally predictable seasonal trends on many crops.
- Large populations of these bacteria are frequently found after periods of relatively wet conditions.
- In the absence of rain or high humidity, bacterial populations tend to be smaller.

In many cases, relatively little change occurred from time to time. In others, large changes were found from one week, or month, to the next. For example, populations of ice nucleation-active bacteria on leaves of pear and almond trees, increased from 100 to greater than 10,000 cells $g^{-1}$ fresh weight of tissue within a month following bud break (Lindow 1982).

Weekly and monthly sampling frequencies may not allow estimates of actual *rates* of change in leaf-associated bacterial population sizes. Doubling times of one to a few hours are typical for bacteria cultured under laboratory conditions. In those cases where growth rates on plants approximate those in culture, determination of rates of changes in bacterial population sizes on or in plant tissues will have to be based on sampling frequencies of hours, not several days.

There is, however, the possibility that bacteria grow much more slowly in association with plants than in culture. Mean doubling times of 19.4 and 18.8 h were estimated for *Xanthomonas campestris* pvs *phaseoli* var. *fuscans* and *phaseoli*, respectively, on leaves of navy beans (*Phaseolus vulgaris*) (Weller & Saettler 1980). These values are based on sampling frequencies of two-to-three-day intervals following inoculation of the bean plants with rifampicin-resistant marked strains of the bacterial pathogens. Changes in population sizes of the *X. campestris* pathovars apparently followed a typical in vitro bacterial growth curve with a lag, exponential and stationary phase. In contrast to in vitro growth curves, the growth curves described by Weller & Saettler (1980) were based on data taken over a 21-day period. In the absence of more frequent sampling intervals, however, we would be cautious about these interpretations.

There is a paucity of information on growth rates of leaf-associated bacterial species, both in culture and in plants. Young *et al.* (1977) reported optimum doubling times of approximately 1.3 h at 26°C and 1.58 h at 30°C for *Pseudomonas syringae* and *Xanthomonas pruni*, respectively, in broth cultures under laboratory conditions. The rates at which bacteria multiply in association with leaves have been investigated to some extent under greenhouse or growth-chamber conditions. In some experiments, leaves of the host plant were infiltrated with the bacterium of interest, usually a phytopathogen. The plants were subsequently maintained under controlled conditions (relatively constant temperature, photoperiod). Generation times of the order of 5 to 6 h have been reported by Ercolani & Crosse (1966) for *P. syringae* pvs *syringae* and *morsprunorum* in bean and cherry leaves, respectively. Doubling times of approximately 3.5 h have been reported for strains of *P. syringae* pathogenic to maize (Gross & DeVay 1977). In other experiments, the bacterium was sprayed onto the surface of the test plants. For example, Lindow *et al.* (1983) reported a doubling time of approximately 3.8 h for an ice nucleation-active and a non-ice nucleation-active strain of *Erwinia herbicola* on mist-inoculated corn seedlings incubated in a mist chamber. In Lindow's experiments, sufficiently frequent samples were taken to assure that exponential rates of bacterial population increases would be found.

Thus, at least under controlled conditions, bacterial population sizes can increase on plants at rates of the same order of magnitude to those one might expect in culture. Are the population trends found in the field (see above) a reflection of relatively slow growth rates of leaf-associated bacteria in the field? Growth rates may be slow if, for example, substrate concentrations on leaves are low enough to seriously limit growth. On the other hand, actual short-term growth rates of bacteria on leaves may approach rates measured in the laboratory. The long-term fluctuations may represent the result of many iterations of growth, death, emigration and immigration occurring sequentially or simultaneously, at rates comparable to those expected from laboratory studies. Thus, is it possible that the long-term fluctuations in leaf-associated bacterial populations may be more an indication of changes in the carrying capacities of leaves than of population dynamics of the bacteria?

In order to appreciate these alternatives, we have begun to examine the population dynamics of leaf-associated bacteria. In our approach we regard the following as important considerations for these studies.

If bacterial generation times are as brief as one to a few hours, then sampling intervals of comparable duration are necessary to detect them. The short-term dynamics of populations of bacteria may not be paralleled on all leaves in a population of leaves. Both parameters that define the lognormal distribution (i.e. the mean and variance) should be followed as a function of time to define leaf-to-leaf trends as well as trends in the median bacterial population size. The physical environment is not constant, but continuously variable and dynamic under field conditions. Temperature, radiation, humidity and other weather parameters fluctuate diurnally. Rain (or irrigation) events of varying intensity and duration may occur. It is difficult to mimic the natural environment under controlled conditions. Under natural conditions, the microbial community associated with leaves is comprised of a range of micro-organisms - both in numbers and types. The different components probably interact to influence the overall dynamics of any given component as well as the total community. It is difficult to simulate these natural communities under controlled conditions.

Thus, our study was conducted under field conditions, and included determination of both the mean and variance of bacterial population sizes at frequent intervals. A detailed weather record was also obtained. Snap bean (*Phaseolus vulgaris*) was selected as the plant species for intense sampling of leaf-associated bacterial populations. One of the predominant bacterial species on snap bean leaflets in Wisconsin is ice nucleation-active *P. syringae*. Short-term changes in the frequency distribution (i.e. the lognormal) of bacterial populations associated with individual snap bean leaflets were determined by sampling every two hours over a 26-hour period (Hirano *et al.* 1984). Populations of *P. syringae*, presumptive *P. mesophilica* and total bacteria were estimated for each of 30 individual leaflets at each sampling time. The mean and the standard error of the mean of population sizes of *P. syringae* at each of the sampling times are shown in Fig. 3. Populations of *P. syringae* increased 28.5-fold from 0900 on 20 July to 0900 21 July. A doubling time of approximately 4.5 h was estimated from the continuous, nearly linear increase in the logarithm of population size observed throughout the initial 24 hours. The continuous growth of *P. syringae* throughout the 24-hour period was quite unexpected since free water was not visible

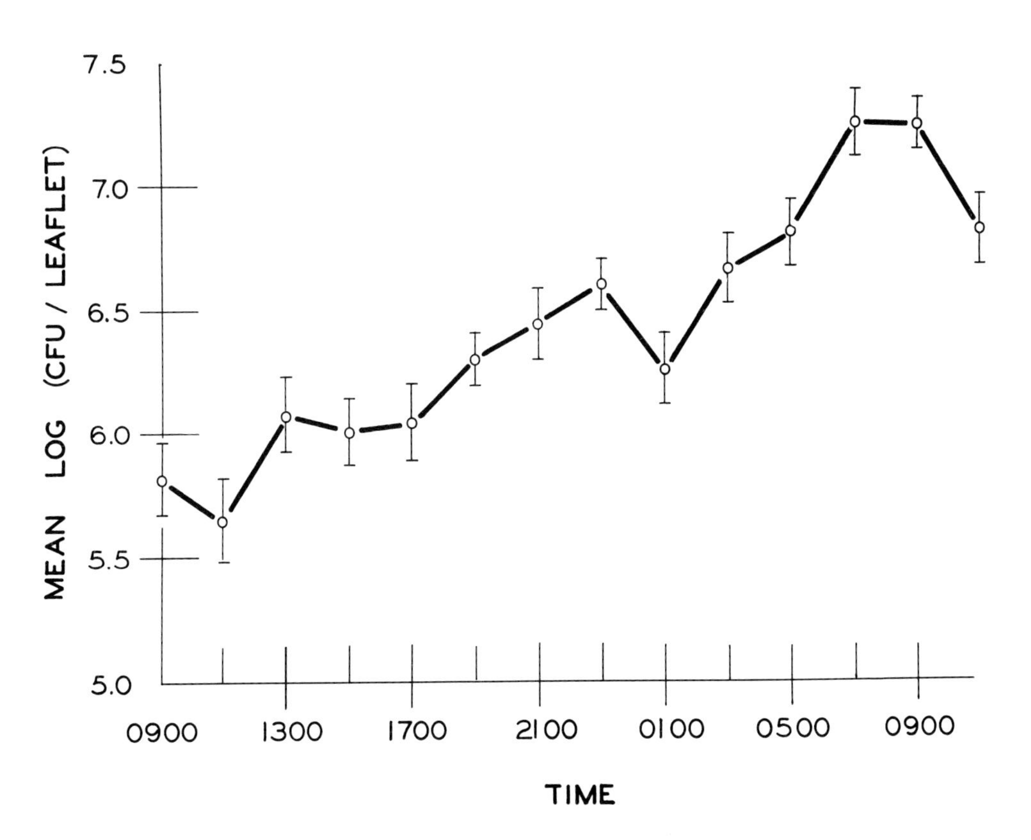


Fig. 3. Diurnal fluctuation in population sizes of *Pseudomonas syringae*. Population sizes of *P. syringae* were estimated on each of 30 individual snap bean leaflets at each sampling time. Population estimates were made from dilution plates of individual leaflet homogenates. The mean log-transformed population value and the standard error of the mean are plotted above. Sampling was conducted at two-hour intervals starting at 0900 Central Daylight Savings Time.

on the leaf surfaces during the daylight hours. The maximum and minimum temperatures for the day were approximately 33 and 20.5°C, respectively. Although no rain fell during the experiment, 15 mm of rain fell from 2002 to 2122 on 19 July, 12 to 13 hours before the start of the experiment. Population variance for *P. syringae* decreased from 0.60 to 0.37 from 0900 of day one to day two. Thus, if variance is used as the criterion of leaf-to-leaf variability, only 62% of the variability present on 20 July remained one day later. This change in population variance may be indicative of the distinctiveness of each leaf ecosystem - that is, populations of *P. syringae* did not respond uniformly on each individual leaflet under the environmental conditions that occurred on this particular day. The temporal variability in population size from leaf to leaf, harvested at a given time, is evident in Fig. 4.

Bacteria presumptively identified as *Pseudomonas mesophilica* (cf. Austin & Goodfellow 1979) were also enumerated on the same set of leaflets. The changes in population size of this pseudomonad did not parallel those of

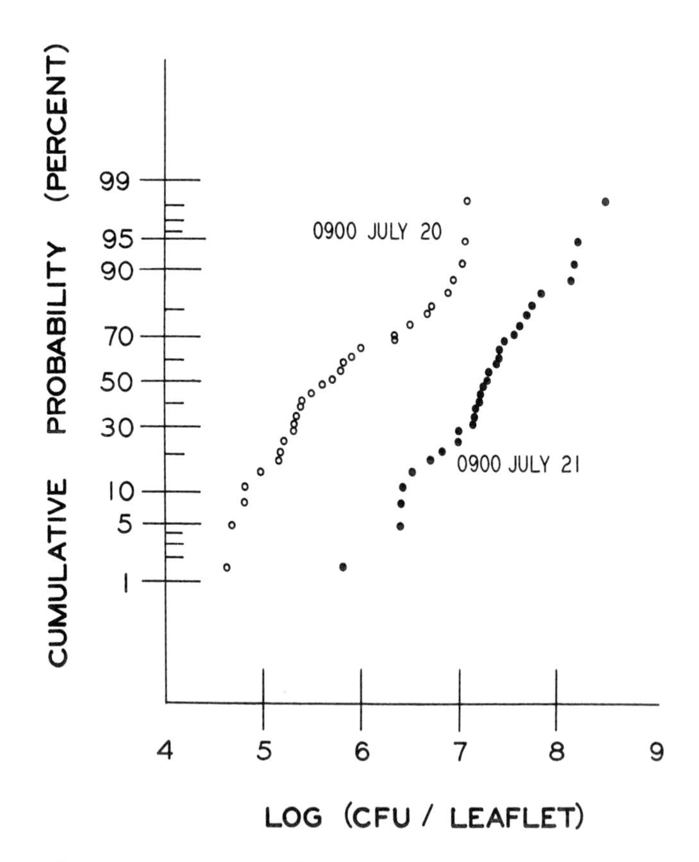


Fig. 4. Frequency distributions of population sizes of *P. syringae* on each of 30 individual snap bean leaflets harvested at 0900 July 20 and 0900 July 21. The median *P. syringae* population increased 28.5-fold during the 24-hour period.

*P. syringae*. During the daylight hours when leaves were visually dry, the median population size of this bacterium declined about three-fold. After dewfall, in the evening, the bacteria grew, so that by 0900 of the next day, the median population was the same as it had been 24 hours earlier. A comparison of the population levels of this bacterium at 0900 of day one to 0900 of day two might lead one to conclude erroneously that population sizes of *P. mesophilica* did not change during the 24-hour period. Total bacterial populations associated with the bean leaflets increased four-fold during the 24-hour period.

Under the environmental conditions of this particular 24-hour period, the population size of one bacterial species increased dramatically. Another decreased during the day, but grew during the night, so that no net change was found. Thus, each of the bacterial components of the leaf epiflora responded in its own unique way to the prevailing environmental conditions. The sum of these component changes yielded a four-fold net increase in total bacterial population size. Since the weather conditions are not constant from day to day, the observed changes discussed above, were unique to the environmental conditions of that one day. A crop such as snap beans has a growing period of approximately 65 to 70 days, or 65 to 70 individual sets of physical and biological conditions, each 24 hours in duration. Intense sampling over a 24-hour period reveals only a small fraction of the season-long trends in bacterial population sizes. However, this type of experiment provides a better understanding of the very *dynamic* nature of leaf-associated bacterial populations.

Population dynamics of *P. syringae* on snap beans have also been studied by sampling each morning between 0800 and 0830, five consecutive mornings of a week (Hirano & Upper, unpublished results). Although rates of bacterial population increase and decrease can only be inferred from these determinations (i.e. sampling frequency is too infrequent for reliable rate estimations), net changes in population sizes during many 24-hour periods can be obtained and compared to the weather and state of plant growth during that 24-hour period. From these studies during one growing season and on one snap bean cultivar, daily net changes on the order of five-fold or less in population size of *P. syringae* were more frequently observed than large changes. However, on one occasion, population size of *P. syringae* increased 155-fold from 0800 of one day to 0800 of the next. If this bacterial component increased exponentially throughout the 24 hours, then populations must have doubled, on the average, every 3.3 h. Such large increases in population sizes of *P. syringae* are usually preceded by rain.

## *GENETIC VARIABILITY OF LEAF-ASSOCIATED BACTERIA*

A few reports on the taxonomic identification of different genera of bacteria isolated from leaves of a given plant species have been published (Dickinson *et al.* 1975; Billing 1976; Ercolani 1983; Morris 1985). Representatives of the genera *Achromobacter*, *Cellulomonas*, *Corynebacterium*, *Erwinia*, *Flavobacterium*, *Oerskovia*, *Pseudomonas* and *Xanthomonas* were identified from *Lolium perenne* by Dickinson *et al.*

(1975). For the most part, however, identification of the saprophytic bacteria associated with leaves has not received much attention. Plant pathologists frequently focus on the bacterial population of interest and group most or all of the saprophytes as 'others'.

Within a single species of bacterium, genetic variability among strains is frequently illustrated by varying patterns of substrate utilization (cf. Sands *et al.* 1970; Ercolani 1983; Morris 1985; C.E. Morris and D.I. Rouse, this volume). The ability of a given bacterial strain to utilize a particular carbon or nitrogen source is scored yes or no. These scores are obtained for a number of substrates per bacterial strain and the overall patterns of substrate utilization are compared. If different strains utilize a given substrate at varying rates or at varying efficiences, the manner in which the test is conducted may substantially influence the answer.

To determine how quantitatively variable a particular phenotype might be distributed among different strains of a given species, we estimated the relative ice nucleation activity of a number of *Pseudomonas syringae* strains. Ice nucleation activity is an inherent genotypic property among bacteria that possess this activity and a bacterial strain can be scored as either having or not having the activity. However, some strains are more efficient ice nucleators than others (Maki *et al.* 1974; Lindow *et al.* 1978 a). The ratio of ice nuclei to viable bacterial cells (i.e. the nucleation frequency) is one measure of relative ice nucleation activity. The distributions of nucleation frequencies measured at -5°C among (a) 154 strains of phytopathogenic *P. syringae* from the National Collection of Plant Pathogenic Bacteria, Harpenden, England, (b) 144 strains of *P. syringae* isolated from nondiseased leaf surfaces of several plant species in Wisconsin, USA, and (c) 42 putative strains of *Erwinia herbicola* could each be described by the lognormal (Hirano *et al.* 1985) (Fig. 5). Gross *et al.* (1983) have also reported on the measured variability in nucleation frequencies among strains of ice nucleation-active bacteria from the Pacific Northwest. Their data are consistent with a lognormal distribution of nucleation frequencies among ice-nucleating bacterial isolates from fruit trees.

Genetic variability among strains of a given species can also be demonstrated by phage sensitivity patterns (Billing 1963). When approximately 20 isolates of *P. syringae* were recovered from each of seven individual bean leaflets, and tested for their sensitivity against three different phages (Table 1), two observations were made. First, not unexpectedly, the different *P. syringae* strains differed in phage sensitivity patterns. Second, the proportion of strains per phage response class differed from one leaf to the next (Table 2). This can be taken as an indication that for a given component, such as *P. syringao*, the population size of the component can vary from one leaf to the next, and moreover, the composition in terms of the genetic variability of the strains on one leaf may differ from the next.

It is frequently assumed that competition among strains of a species or among different species or genera of bacteria for substrate and/or space

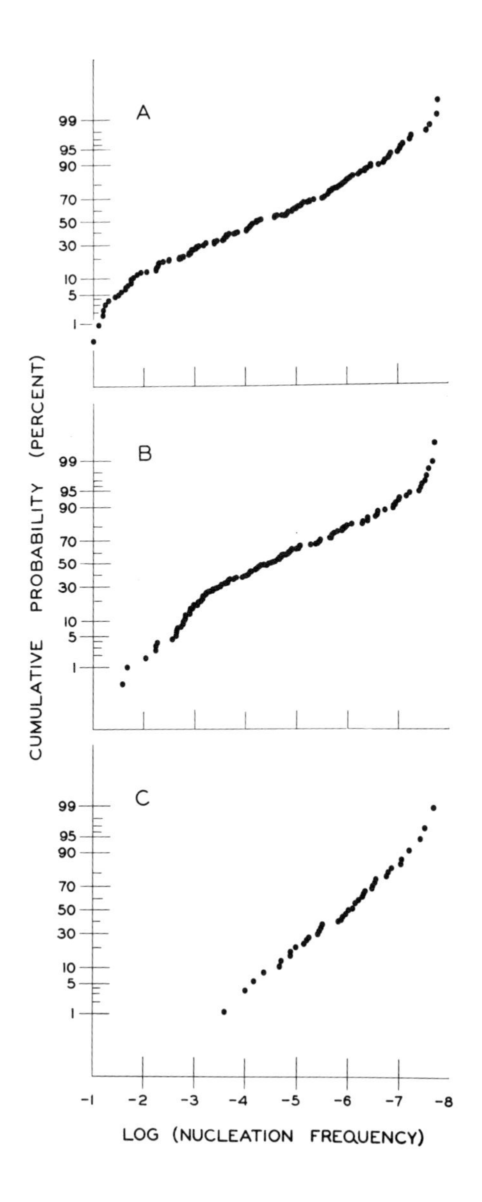


Fig. 5. Lognormal distributions of nucleation frequencies for each of three groups of INA bacteria. (A) *P. syringae* from the National Collection of Plant Pathogenic Bacteria; (B) *P. syringae* from non-diseased leaves; and (C) *Erwinia herbicola*. All strains were grown for 2 days on nutrient glycerol agar at 24°C. Nucleation frequency for each strain was calculated from the concentration of ice nuclei active at -5°C and the density (colony forming units $ml^{-1}$) of the test bacterial suspension. Reprinted from Hirano *et al.* (1985).

plays a role in regulating the overall qualitative and quantitative composition of bacterial communities associated with leaves. By measuring different phenotypic responses of leaf-associated bacteria from a statistical perspective, our understanding of the ecology of these micro-organisms may be further enhanced. It is apparent from both ice nucleation frequencies and phage sensitivities that the *P. syringae* distributed among bean leaflets constitute a substantially variable gene pool, with regard to expression of these two traits. There is every

Table 1. Response classes of *P. syringae* strains to three bacteriophages.

| Phage | Response class[1] | | | | | | | |
|---|---|---|---|---|---|---|---|---|
| | 1 | 2 | 3 | 4 | 5 | 6 | 7 | 8 |
| SO21 | + | + | + | + | - | - | - | - |
| SO30 | + | + | - | - | + | + | - | - |
| P115 | + | - | - | + | + | - | + | - |

[1] The sensitivity pattern of strains of *P. syringae* to each of three bacteriophages was classified into one of eight response classes. For example, response class 1 included all bacterial strains that were sensitive to all three phages. Response class 8 included all bacterial strains that were resistant to all three phages.

Table 2. Phenotypic variability among strains of *Pseudomonas syringae* based on phage sensitivity patterns. Values represent the proportion of isolates per response class per leaflet.

| Bean leaflet | Number of *P. syringae* tested | Response class | | | | | | | |
|---|---|---|---|---|---|---|---|---|---|
| | | 1 | 2 | 3 | 4 | 5 | 6 | 7 | 8 |
| A | 27 | 0.92 | 0 | 0 | 0 | 0 | 0 | 0 | 0.08 |
| B | 28 | 0.81 | 0.18 | 0 | 0 | 0 | 0 | 0 | 0 |
| C | 30 | 0.76 | 0.03 | 0 | 0 | 0 | 0 | 0 | 0.20 |
| D | 30 | 0.23 | 0.06 | 0 | 0 | 0 | 0 | 0 | 0.70 |
| E | 29 | 0 | 0 | 0 | 0 | 0.27 | 0 | 0 | 0.72 |
| F | 29 | 0 | 0 | 0 | 0 | 0.68 | 0.03 | 0 | 0.27 |
| G | 30 | 0 | 0 | 0 | 0 | 0 | 0 | 0 | 1.0 |

reason to believe that they are equally variable with respect to other traits. Thus, this species may be regarded as representing a broad range of genotypes, each likely to respond slightly differently to the selection pressure presented by any given set of plant and physical environmental conditions. This genetic variability should be kept in mind when the possibility of using a single genotype of any bacterial species for a potential biocontrol agent in the leaf-bacterial ecosystem is proposed.

## *CONCLUSION*

Experiments are often designed to minimize variability in the results obtained. Variability due to experimental error should indeed be minimized. However, the variability *inherent* to a given system may contain critical information about the system, and should not be ignored or discarded as a negative attribute of the system. This is clearly the case with the variability among population sizes of bacteria associated with the leaves of an individual crop canopy.

Determination of the frequency distribution of bacteria among leaves can provide additional information about the system. This information has been valuable for studying details of the population dynamics of leaf-associated bacteria. Our understanding of the relationship between population sizes of leaf-associated phytopathogenic bacteria and subsequent disease development has benefited substantially from including information on variability of bacterial population sizes among populations of leaves. It is apparent that use of the information obtained from the variability of populations among leaves will aid in studying the spatial distribution of bacteria in canopies, and the mechanisms that determine these population sizes.

Short-term changes in population sizes of leaf-associated bacteria reveal the leaf as a very dynamic ecosystem. Population sizes on the median leaf, and the leaf-to-leaf variation in bacterial population sizes can change very rapidly. Bacterial growth rates on leaves in the field approach those for the same species in cultures.

## *ACKNOWLEDGEMENTS*

We thank D.K. Schaefer for typing this manuscript and S.A. Vicen for assistance in preparation of the illustrations. Research has been supported by the Agricultural Research Service, USDA and the USDA Competitive Research Grants Office (Grant No. 84-CRCR-1-1422).

## *REFERENCES*

Andrews, J.H., Kenerley, C.M. & Nordheim, E.V. (1980). Positional variation in phylloplane microbial populations within an apple tree canopy. Microbial Ecology, *6*, 71-84.

Austin, B. & Goodfellow, M. (1979). *Pseudomonas mesophilica*, a new species of pink bacteria isolated from leaf surfaces. International Journal of Systematic Bacteriology, *29*, 373-8.

Baker, J.H. (1981). Direct observation and enumeration of microbes on plant surfaces by light microscopy. *In* Microbial Ecology of the Phylloplane, ed. J.P. Blakeman, pp. 3-14. London: Academic Press.

Billing, E. (1963). The value of phage sensitivity tests for the identification of phytopathogenic *Pseudomonas* spp. Journal of Applied Bacteriology, *26*, 193-210.

Billing, E. (1976). The taxonomy of bacteria on the aerial parts of plants. *In* Microbiology of Aerial Plant Surfaces, eds C.H. Dickinson & T.F. Preece, pp. 223-73. London: Academic Press.

Burkowicz, A. (1981). Population dynamics of epiphytic *Pseudomonas morsprunorum* on the leaf surfaces of two sweet cherry cultivars. Fruit Science Reports, *8*, 37-47.

Chow, S.C. (1985). Resampling procedures for the estimation of nonlinear functions of parameters. Ph.D. thesis, University of Wisconsin-Madison.

Crosse, J.E. (1959). Bacterial canker of stone-fruits. IV. Investigation of a method for measuring the inoculum potential of cherry trees. Annals of Applied Biology, *47*, 306-17.

Daniel, J.-F. & Boher, B. (1985). Epiphytic phase of *Xanthomonas campestris* pathovar *manihotis* on aerial parts of cassava. Agronomie, *5*, 111-5.

Dickinson, C.H., Austin, B. & Goodfellow, M. (1975). Quantitative and qualitative studies of phylloplane bacteria from *Lolium perenne*. Journal of General Microbiology, *91*, 157-66.

Ercolani, G.L. (1979). Distribuzione di *Pseudomonas savastanoi* sulle foglie dell' Olivo. Phytopathologia Mediterranea, *18*, 85-8.

Ercolani, G.L. (1983). Variability among isolates of *Pseudomonas syringae* pv. *savastanoi* from the phylloplane of the olive. Journal of General Microbiology, *129*, 901-16.

Ercolani, G.L. & Crosse, J.E. (1966). The growth of *Pseudomonas phaseolicola* and related plant pathogens in vivo. Journal of General Microbiology, *45*, 429-39.

Ercolani, G.L., Hagedorn, D.J., Kelman, A. & Rand, R.E. (1974). Epiphytic survival of *Pseudomonas syringae* on hairy vetch in relation to epidemiology of bacterial brown spot of bean in Wisconsin. Phytopathology, *64*, 1330-9.

Gross, D.C., Cody, Y.S., Proebsting, E.L., Jr., Radamaker, G.K. & Spotts, R.A. (1983). Distribution, population dynamics, and characteristics of ice nucleation-active bacteria in deciduous fruit tree orchards. Applied and Environmental Microbiology, *46*, 1370-9.

Gross, D.C. & DeVay, J.E. (1977). Population dynamics and pathogenesis of *Pseudomonas syringae* in maize and cowpea in relation to the in vitro production of syringomycin. Phytopathology, *67*, 475-83.

Hirano, S.S., Baker, L.S. & Upper, C.D. (1985). Ice nucleation temperature of individual leaves in relation to population sizes of ice nucleation active bacteria and frost injury. Plant Physiology, *77*, 259-65.

Hirano, S.S., Nordheim, E.V., Arny, D.C. & Upper, C.D. (1982). Lognormal distribution of epiphytic bacterial populations on leaf surfaces. Applied and Environmental Microbiology, *44*, 695-700.

Hirano, S.S., Rouse, D.I. & Upper, C.D. (1984). Epidemiology of bacterial brown spot and ecology of *Pseudomonas syringae* pv. *syringae* on *Phaseolus vulgaris*. *In* Proceedings of the 2nd Working Group on *Pseudomonas syringae* pathovars, eds C.G. Panagopoulos, P.G. Psallidas & A.S. Alivizatos, pp. 24-7. Athens, Greece: The Hellenic Phytopath. Soc.

Hirano, S.S. & Upper, C.D. (1983). Ecology and epidemiology of foliar bacterial plant pathogens. Annual Review of Phytopathology, *21*, 243-69.

Leben, C. (1963). Multiplication of *Xanthomonas vesicatoria* on tomato seedlings. Phytopathology, *53*, 778-81.

Leben, C. (1965). Epiphytic microorganisms in relation to plant disease. Annual Review of Phytopathology, *3*, 209-30.

Lindemann, J., Arny, D.C. & Upper, C.D. (1984 a). Epiphytic populations of *Pseudomonas syringae* pv. *syringae* on snap bean and nonhost plants and the incidence of bacterial brown spot disease in relation to cropping

patterns. Phytopathology, *74*, 1329-33.
Lindemann, J., Arny, D.C. & Upper, C.D. (1984 b). Use of an apparent infection threshold population of *Pseudomonas syringae* to predict incidence and severity of brown spot of bean. Phytopathology, *74*, 1334-9.
Lindow, S.E. (1982). Population dynamics of epiphytic ice nucleation active bacteria on frost sensitive plants and frost control by means of antagonistic bacteria. *In* Plant Cold Hardiness and Freezing Stress--Mechanisms and Crop Implications. Vol. II, eds P.H. Li & A. Sakai, pp. 395-416. New York: Academic Press.
Lindow, S.E., Arny, D.C., Barchet, W.R. & Upper, C.D. (1978 a). The role of bacterial ice nuclei in frost injury to sensitive plants. *In* Plant Cold Hardiness and Freezing Stress--Mechanisms and Crop Implications. Vol. I, eds P.H. Li & A. Sakai, pp. 249-63. New York: Academic Press.
Lindow, S.E., Arny, D.C. & Upper, C.D. (1978 b). Distribution of ice nucleation-active bacteria on plants in nature. Applied and Environmental Microbiology, *36*, 831-8.
Lindow, S.E., Arny, D.C. & Upper, C.D. (1983). Biological control of frost injury: an isolate of *Erwinia herbicola* antagonistic to ice nucleation active bacteria. Phytopathology, *73*, 1097-102.
Luisetti, J. & Paulin, J.-P. (1972). Etudes sur les bactérioses des arbres fruitiers. III. Recherche du *Pseudomonas syringae* (van Hall) à la surface des organes aériens du poirier et étude de ses variations quantitatives. Annales de Phytopathologie, *4*, 215-27.
Maki, L.R., Galyan, E.L., Chang-Chien, M. & Caldwell, D.R. (1974). Ice nucleation induced by *Pseudomonas syringae*. Applied Microbiology, *28*, 456-9.
Martins, J.M.S. (1982). Characteristics and population densities of fluorescent pseudomonads from cherry and apricot leaf surfaces in Portugal. García de Orta, Série Estudos Agronómicos, *9*, 249-54.
Mew, T.W. & Kennedy, B.W. (1982). Seasonal variation in populations of pathogenic Pseudomonads on soybean leaves. Phytopathology, *72*, 103-5.
Morris, C. (1985). Diversity of epiphytic bacteria on snap bean leaflets based on nutrient utilization abilities: biological and statistical considerations. Ph.D. thesis, University of Wisconsin-Madison.
Parbery, I.H., Brown, J.F. & Bofinger, V.J. (1981). Statistical methods in the analysis of phylloplane populations. *In* Microbial Ecology of the Phylloplane, ed. J.P. Blakeman, pp. 47-65. London: Academic Press.
Rouse, D.I., Nordheim, E.V., Hirano, S.S. & Upper, C.D. (1985). A model relating the probability of foliar disease incidence to the population frequencies of bacterial plant pathogens. Phytopathology, *75*, 505-9.
Sands, D.C., Schroth, M.N. & Hildebrand, D.C. (1970). Taxonomy of phytopathogenic Pseudomonads. Journal of Bacteriology, *101*, 9-23.
Smitley, D.R. & McCarter, S.M. (1982). Spread of *Pseudomonas syringae* pv. *tomato* and role of epiphytic populations and environmental conditions in disease development. Plant Disease, *66*, 713-7.
Strandberg, J. (1973). Spatial distribution of cabbage black rot and the estimation of diseased plant populations. Phytopathology, *63*, 998-1003.
Surico, G., Kennedy, B.W. & Ercolani, G.L. (1981). Multiplication of *Pseudomonas syringae* pv. *glycinea* on soybean primary leaves exposed to aerosolized inoculum. Phytopathology, *71*, 532-6.
Weller, D.M. & Saettler, A.W. (1980). Colonization and distribution of *Xanthomonas phaseoli* and *Xanthomonas phaseoli* var. *fuscans* in field grown navy beans. Phytopathology, *70*, 500-6.
Young, J.M., Luketina, R.C. & Marshall, A.M. (1977). The effects of temperature on growth *in vitro* of *Pseudomonas syringae* and *Xanthomonas pruni*. Journal of Applied Bacteriology, *42*, 345-54.

# EPIPHYTIC SURVIVAL OF BACTERIAL LEAF PATHOGENS

Y. Henis[1] and Y. Bashan[2]

[1]*Department of Plant Pathology and Microbiology, Faculty of Agriculture, The Hebrew University of Jerusalem, P.O. Box 12, Rehovot 76-100, Israel*
[2]*Department of Plant Genetics, The Weizmann Institute of Science, Rehovot 76-100, Israel*

## *INTRODUCTION*

Most plant-pathogenic bacteria are non-spore-forming, Gram-negative rods. Although strongly affected by adverse environmental conditions, such as drought, low relative humidity, irradiation and nutrient deficiency, these bacteria manage to multiply and spread after inflicting damage to susceptible host plants. During the last two decades, the intriguing ecological, physiological and epidemiological problems regarding plant-pathogenic bacteria were extensively reviewed (Leben 1965, 1981; Schuster & Coyne 1974; Cook & Baker 1983; Hirano & Upper 1983; Bashan 1985). The present review deals with some basic yet still open questions concerning the growth and survival of phytopathogenic bacteria on leaves. It is not intended to be a detailed up-to-date source of literature, although an attempt is made to present examples and evidence from recent relevant studies.

## *ENDOPHYTIC VERSUS EPIPHYTIC FLORA: DISTINCTION AND DYNAMICS*

Hirano & Upper (1983) defined epiphytic bacteria as "those bacteria that can be removed from above-ground plant parts by washing". This definition, however, does not distinguish between residents and casuals (Leben 1965). Epiphytic bacteria can be either pathogenic or saprophytic, whereas endophytic bacteria, defined as the bacteria inhabiting the leaf intercellular spaces or substomatal cavities, are usually pathogenic. Extensive exchanges between these populations often occur, especially through stomata. This exchange is probably dependent on the microclimatic conditions inside and outside the leaf. Techniques employed to distinguish between these phases include ultraviolet (UV) irradiation (Barnes 1965) and chemical surface disinfection to remove the epiphytic population (Henis *et al.* 1980; Sharon *et al.* 1982 b). Epiphytic populations of *Pseudomonas syringae* pv. *alboprecipitans* penetrate into sweet corn leaves through closed stomata and multiply inside the leaf (Gitaitis *et al.* 1981). Epiphytic *P. syringae* pv. *tomato* penetrate into intercellular spaces through open stomata as well as through intact or broken leaf trichomes (Schneider & Grogan 1977; Bashan *et al.* 1981) and epiphytic *Xanthomonas campestris* pv. *vesicatoria* penetrate through leaf veins (Sharon *et al.* 1982 a). The importance of stomata as infection sites, as well as locations for egress source of *X. campestris* pv. *pruni* in peach leaves, was reported by Miles *et al.* (1977). In this case, the bacterial mass was exuded from the stomata and became a new source for infections during the pre-symptomatic period.

During development of bacterial speck disease of tomato, endophytic bacteria migrated from intercellular spaces to the external leaf surface (Bashan *et al.* 1981), whereas in bacterial scab disease of pepper the pathogen multiplied either on the outer surface or inside the tissue (Sharon *et al.* 1982 b). Bonn *et al.* (1985) working with *P. syringae* pv. *tomato* reported that after inoculation there was a temporary decrease in epiphytic population, though later it increased to high levels.

The endophytic population of a resistant tomato cultivar infected with *P. syringae* pv. *tomato* was almost totally inhibited - unlike the susceptible cultivars. On the other hand, the epiphytic population was similar in the resistant and the susceptible cultivars (Fallik *et al.* 1983).

### *MULTIPLICATION*

Bacterial plant pathogens multiply before, during and after symptom development (Leben 1965, 1981; Cook & Baker 1983; Hirano & Upper 1983; Bashan 1985). Crosse (1959) was the first to demonstrate that massive epiphytic populations of *P. syringae* pv. *morsprunorum*, the causal agent of bacterial canker of stone fruit trees, were present on the surface of symptomless cherry leaves, acting perhaps as a source of inoculum for disease initiation. Similarly, English & Davis (1960) isolated unidentified pathogenic fluorescent *Pseudomonas* spp. from leaves and fruit twigs of healthy peach and almond trees, as well as from weeds in the field.

Based on the above findings and on his own, Leben (1965) suggested that the pathogen lived and multiplied on external surfaces of apparently symptomless plants. These populations are referred to as 'resident'. Under optimal conditions for disease outbreak such as appropriate temperature, availability of free water, high relative humidity and the plant's suitable physiological conditions, such an inoculum may increase in number and initiate a disease, without additional external source of inoculum. Many studies support this concept. *X. campestris* pv. *vesicatoria* was found as resident on tomato seedlings (Leben 1963) and pepper plants (Bashan *et al.* 1985 a). Monitoring of *P. syringae* pv. *syringae* in bean leaves, revealed the resident phase of the bacteria in nature (Leben *et al.* 1970). Pohronezny *et al.* (1977) regularly recovered *P. syringae* pv. *lachrymans* from within symptomless tissues. Bonn *et al.* (1985) reported that *P. syringae* pv. *tomato* was an epiphyte on symptomless tomato plants; when planted in the field, disease symptoms were developed. *P. syringae* pv. *glycinea* was recovered from healthy soybean seedlings. The pathogen could multiply in symptomless buds. Other unidentified bacteria were found to colonize soybean buds as individuals or in colonies. Thus, soybean buds may serve as a site for multiplication for a variety of bacteria (Leben *et al.* 1968 b).

Hirano & Upper (1983) presented a list of epiphytically growing pathogenic bacteria, further supporting the concept that epiphytism is a general rather than an exceptional phenomenon in nature. However, not every foliar pathogen has a resident population on its host plant. *Erwinia amylovora*, the causal agent of fire-blight in apples and pears, was detected on leaves only after fire-blight was spread all over the orchard (Thomson *et al.* 1975). Independently and concurrently, Sutton & Jones (1975) suggested that this disease broke out only from a temporary epiphytic population and not from a resident one. However, they could not monitor such small populations which might initiate this disease (Crosse *et al.* 1972).

*Compatibility*

Studies of *X. campestris* pv. *vesicatoria* in pepper leaves showed that several days after inoculation the pathogenic population was always smaller in the resistant cultivar (Stall & Cook 1966). The same trend was shown by the bacterial population of *P. syringae* pv. *tomato* (Bashan *et al.* 1981). Also, *P. syringae* pv. *phaseolicola* grew faster and to a higher total number in leaves of susceptible than in tolerant bean cultivars (Stadt & Saettler 1981). However, intensive multiplication occurred also in the resistant cultivar (Omer & Wood 1969). Working with *P. syringae* pv. *glycinea* in soybean leaves, Mew & Kennedy (1971) found a normal pattern of bacterial multiplication of three pathogenic isolates in susceptible and resistant cultivars. Similarly, both compatible (*P. syringae* pv. *phaseolicola*) and incompatible (*P. syringae* pv. *mors-prunorum*) pathogens multiplied in bean plants, though at different rates and reaching different final levels. Whereas the growth of the incompatible bacteria ceased one to three days after inoculation, the compatible pathogen continued to multiply until five days later, when visible disease symptoms appeared. The final population size of the incompatible strain was always smaller (Ercolani & Crosse 1966). Studies of multiplication on uniform tabacco callus tissue showed that the compatible *P. syringae* pv. *tabaci* multiplied rapidly and spread all over the callus within two days after inoculation, while the incompatible *P. syringae* pv. *pisi* multiplied relatively slowly and remained at the inoculation site (Huang & Van Dyke 1978).

*Environmental effects*

The effects of environmental conditions on population size, multiplication rate and survival of leaf bacteria is well known and documented for many bacterial diseases. The following major environmental factors will be considered: (a) temperature, (b) free water, (c) relative humidity (r.h.) and (d) irradiation.

*Temperature.* Pathogenic bacteria differ greatly in their temperature requirements for survival, multiplication and initiation of disease symptoms. Morton (1966) reported that *X. vesicatoria* pv. *campestris* caused maximal development of spot symptoms in excised pepper and tomato leaves at optimum temperature of 20 to 25°C, which was lower than the optimum required for the pathogen's development. The different temperature requirement of this pathogen at different geographical regions was demonstrated by Diab *et al.* (1982 a), who showed enhancement in the pathogen's growth and in disease development at high temperatures (30 to 35°C) during the growth season of pepper in Israel. The optimum temperature for *P. syringae* pv. *tomato* infection was 22 to 25°C (Schneider & Grogan 1978). However, infection could be initiated to a lesser extent over a temperature range of 18 to 30°C (Bashan *et al.* 1978; Okon *et al.* 1978), while bacteria could withstand temperatures up till 45°C (Devash *et al.* 1980). Similarly, the optimum temperature for development of bacterial blight of peas in Israel caused by *P. syringae* pv. *pisi* was 25±2°C (Bashan & Kenneth 1983).

*Relative humidity.* A high r.h., usually over 90%, is required to ensure multiplication of the bacterial population on leaf surfaces. A decrease in r.h. prevents further multiplication and may even cause a population decrease. According to Davis & Halmos (1958), when tomato plants were exposed to 100% r.h. prior to inoculation with *X. campestris* pv. *vesicatoria*, the longer the exposure, the greater was the disease severity. More precise studies indicated that *X. campestris* pv. *vesicatoria* induced symptoms only under conditions of 87 to 97% r.h. (Basu & Wallen 1966). R.h. had a significant effect on bacterial scab development in pepper: a high r.h. for long periods always favoured infection. Even when inoculated plants were exposed to a high r.h. for only a few hours during a day or two, the pathogen could cause disease symptoms. Short unfavourable periods of low r.h. following inoculation temporarily halted disease development; when high r.h. conditions were resumed symptoms continued to develop. On the other hand, long periods of low r.h. irreversibly prevented disease development, even when high r.h. was later provided (Diab *et al.* 1982 b). Similarly, when infected tomato plants were subjected to periodic mist, the number of *P. syringae* pv. *tomato* markedly increased in the rhizosphere, stems and leaves before symptom appearance (Devash *et al.* 1980). Further evidence for the major role of r.h. in disease development under field conditions was demonstrated with bacterial speck of tomato: decreasing the r.h. below 80% drastically reduced the appearance of new symptoms in new leaves (Yunis *et al.* 1980 b).

*Free water.* Free water on the leaf surface, due to fog, irrigation and dew, is probably essential for the bacteria to reach the infection site (Chet *et al.* 1973; Raymundo & Ries 1980). In the absence of free water, the pathogen's transfer requires the intervention of some external active or passive vectors (Venette 1982; Bashan 1985). Free water enhances bacterial multiplication at the onset of disease development. Opening of the stomata by free water on the leaves also favoured secondary infections of healthy plants by most bacterial pathogens under field conditions (Shaw 1935; Yunis *et al.* 1980 b). Working with *X. campestris* pv. *vesicatoria*, Cook & Stall (1977) reported that soaking of pepper leaves in water was an essential prerequisite for endophytic multiplication of the pathogen. Inoculation of pepper with *X. campestris* pv. *vesicatoria* followed by long dry periods in the absence of free water, prevented subsequent disease development (Diab *et al.* 1982 b). Subjecting tomato plants to free water prior to inoculation with *P. syringae* pv. *tomato* was an essential factor for successful infection (Bashan *et al.* 1978; Yunis *et al.* 1980 a,b).

*Irradiation.* Both visible and UV light as well as disinfectants are harmful to leaf populations of non-spore-forming bacterial pathogens. UV irradiation at doses that do not cause any damage to the plant tissue can totally eliminate the pathogen population. There is evidence that the leaf tissue serves as a protective environment for the endophytic though not for the epiphytic population: the endophytic population of *X. campestris* pv. *phaseoli* var. *sojensis* in soybean leaves was not affected by UV irradiation that was highly destructive to the bacterial cells either in water or on the leaf surface (Barnes 1965). Surface disinfection of pepper with sodium hypochlorite, which usually destroys most of *X. campestris* pv. *vesicatoria* in aqueous suspension, did not reduce the number of endophytic bacteria (Sharon *et al.* 1982 b). Only lyophilization of diseased pepper leaves, which eliminated nearly all the epiphytic microflora, including the bacterial pathogen, caused some reduction in the intercellular bacterial population (Bashan *et al.* 1982 a). This finding was recently used for the development of a method for long-term preservation of phytopathogenic bacteria without the loss of their virulence (Y. Bashan and Y. Okon, unpublished results).

## *PATHOGEN'S FATE IN NECROTIC TISSUES*

Allington & Chamberlain (1949) found that multiplication of *P. syringae* pv. *glycinea* and *X. campestris* pv. *phaseoli*, presumably in the intercellular spaces, was nearly equal in resistant and susceptible hosts. The bacterial population within the compatible host continued to increase until destruction of the tissue occurred. Leben *et al.* (1968 a,b) showed that the population of *P. syringae* pv. *glycinea*, after reaching its maximum between the 7th and the 15th day following inoculation, declined as lesions became necrotic and leaves dried. A substantial number of pathogens were present before symptom appearance, and a low inoculum concentration resulted in lower population maxima and delayed symptom expression. The latter results were confirmed for *P. syringae* pv. *tomato* and *X. campestris* pv. *vesicatoria* (Bashan *et al.* 1978; Diab *et al.* 1982 a).

C. Leben and his associates speculated that in older drying leaves and in necrotic lesions, bacteria are 'glued down' and serve as a slow release reservoir for future disease outbreak. *P. syringae* pv. *tomato* and *X. campestris* pv. *vesicatoria* were followed on living leaves of tomato and pepper, respectively, using scanning electron microscopy (Figs 1 and 2). After development of necrosis, bacteria were not found in the many necrotic sites which were carefully examined, but detected only in apparently healthy infected tissue surrounding the necrotic area. Thus, bacteria presumably died or lysed with the plant cells in the lesion itself (Bashan *et al.* 1981; Sharon *et al.* 1982 a). Thus, the fate of phytopathogenic bacteria in the necrotic tissue still remains open.

## *LEAF AGE*

Leaf ageing, through its association with increasing resistance, affects both epiphytic and endophytic bacterial multiplication: the more resistant the leaf, the smaller is the bacterial population.

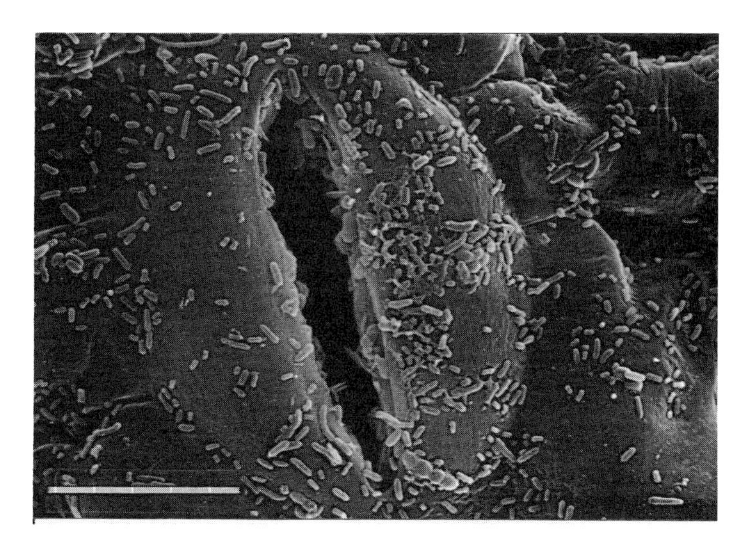

Fig. 1. Scanning electron micrographs showing *Pseudomonas syringae* pv. *tomato* around a stoma of a symptomless susceptible tomato leaf, 100 h after inoculation with $10^6$ CFU $ml^{-1}$ and incubation under partial mist conditions at 24±2°C (Y. Bashan, Y. Okon and Y. Henis, unpublished results). Scale bar equals 10 µm.

Fig. 2. Scanning electron micrograph showing *Xanthomonas campestris* pv. *vesicatoria* on epidermal cells above a vein of a symptomless susceptible pepper leaf, 100 h after inoculation with $10^6$ CFU $ml^{-1}$ and incubation under partial mist conditions at 30±2°C (Y. Bashan, Y. Okon and Y. Henis, unpublished results). Scale bar equals 10 µm.

*X. campestris* pv. *vesicatoria* is more likely to attack young pepper leaves which develop to full size after infection (Diab *et al.* 1982 a). Davis & Halmos (1958) and later Nayudu & Walker (1960) and Leben (1963) claimed that the younger the tomato leaves, the higher their susceptibility to *X. campestris* pv. *vesicatoria*, whereas the older leaves did not develop lesions even if plants were exposed to conditions favourable for disease development. Studies with *P. syringae* pv. *tomato* indicated that the pathogen could infect tomato plants at all stages of plant development (Bashan *et al.* 1978), but the yield loss was higher when young plants were infected (Yunis *et al.* 1980 b). However, the epiphytic pathogenic population was not measured in these studies.

### *PRE-INOCULATION FACTORS*

Pre-inoculation treatment of tomato leaves with carborundum, spraying leaves with diluted wax solvents or pre-incubating plants under mist for 24 h prior to inoculation increased significantly the number of lesions per leaflet (Bashan *et al.* 1978). Scanning electron micrographs of leaves taken immediately after pretreatment and before inoculation showed open stomata in the mist and 40°C treatments. In the case of *X. campestris* pv. *vesicatoria* the pathogen was capable of infecting pepper plants without any mechanical or chemical pretreatments, providing that the plants were later subjected to periodic mist for several days (Diab *et al.* 1982 a).

Agricultural practices affect bacterial development and disease severity through their influence on microclimatic conditions. Rapid multiplication of many leaf pathogens occurs in plants growing out of season in the permanently humid conditions prevailing under plastic tunnels. The mode of irrigation also influences disease severity. Sprinkle irrigation caused temporarily humid conditions inside dense foliage, enabling diseases such as bacterial scab of pepper to develop in the middle of the dry, hot Israeli summer. Drip irrigation which kept the upper plant surface dry for most of the growth season reduced bacterial multiplication and spread as well as severity and incidence of several vegetable bacterial diseases (Volcani 1962, 1964, 1969; Yunis *et al.* 1980 a,b; Palti 1981; Bashan *et al.* 1985 a).

### *MECHANICAL BARRIERS*

It is generally agreed that natural openings in the external surface of the plant, such as stomata, nectaries, hydathodes, lenticels and broken glandular hairs are the main sites of entry for many species of phytopathogenic leaf bacteria (Rolfs 1915; Lewis & Goodman 1965; Burki 1972). Schneider & Grogan (1977) suggested that the tomato leaf trichomes are the source of resident inoculum and serve as sites of penetration by *P. syringae* pv. *tomato*. Bashan *et al.* (1981) confirmed these findings and located the primary site of penetration to open stomata and trichome bases. Open stomata at high r.h. were implicated as a factor affecting disease severity in tomato plants infected with this pathogen (Bashan *et al.* 1978). Analysis of the relationship between the morphology of leaf surfaces of tomato cultivars and leaf invasion by

*P. syringae* pv. *tomato* revealed a highly significant correlation between availability of natural sites of infection and disease. However, the role of natural sites for bacterial invasion is probably limited since even in highly resistant cultivars there are enough natural openings to enable successful bacterial penetration (Bashan *et al.* 1985 b).

### *EPIPHYTIC POPULATION AND DISEASE*

There is no doubt that the epiphytic population of a pathogen can eventually lead to disease outbreak. Yet, the question remains whether this state is essential for disease development. Addressing this issue, three experimental prerequisites were proposed by Hirano & Upper (1983):

(a) epiphytic populations should be measured during the experiment;
(b) different levels of the epiphytic populations should be investigated;
(c) quantitative disease assessments should be carried out.

Few examples of leaf diseases caused by epiphytic bacteria fulfill the above requirements: ice nucleation-active bacteria (Lindow 1983), brown spot disease of snap beans caused by *P. syringae* pv. *syringae* (Lindemann *et al.* 1984) and halo blight of oats caused by *P. syringae* pv. *coronafaciens* (Hirano *et al.* 1982). It appears that every disease has its threshold level of the pathogen's population below which no visible symptoms develop during the host's life span. This may explain the sporadic appearance of a disease in the field, i.e. few infections in some growing seasons and heavy epidemics in others, probably depending on environmental factors.

Weller & Saettler (1980) showed a correlation between disease severity and bacterial number per leaflet. They also found that under field conditions, disease symptoms appeared when the population level of *X. campestris* pv. *phaseoli* or of *X. campestris* pv. *phaseoli* var. *fuscans* reached $5 \times 10^6$ CFU per leaflet. Data showing a relationship between the population size of *P. syringae* pv. *tomato* and bacterial speck development in tomato plants were presented by Smitley & McCarter (1982). The data were valid only under environmental conditions favourable for disease development.

A statistical model designed to explain the quantitative relationships between the epiphytic population of pathogenic bacteria and disease incidence was suggested (Rouse *et al.* 1981). This model takes into account host susceptibility, environmental factors, different rates of bacterial multiplication in the field and bacteria number. However, some bacterial diseases do not follow the description for epiphytic bacterial multiplication in plants. Under field conditions, no positive correlation was found between the number of bacteria in pear flowers infected with *E. amylovora* and disease incidence (Miller & Schroth 1972); no relationship was found between *X. campestris* pv. *vesicatoria* number in pepper plants and symptom expression (Bashan *et al.* 1985 a). Therefore, this model should be treated at present with some caution.

Epiphytic leaf pathogens must compete for their survival with the surrounding saprophytic micro-organisms (Schuster & Coyne 1974). Competition between pathogens is possible too. *P. syringae* pv. *syringae* was frequently recovered from mixed infections with *X. campestris* pv. *vesicatoria* on tomato plants but was never found with *P. syringae* pv. *tomato* (Gitaitis *et al.* 1985), a probable reflection of the competition between the two pathogens.

*SYMPTOMLESS PLANTS*

Kennedy (1969) used an enrichment leaf culture to detect *P. syringae* pv. *glycinea* in soybean seeds in the presence of saprophytes. Henis *et al.* (1980) and Sharon *et al.* (1982 b) compared the multiplication of *X. campestris* pv. *vesicatoria* and *P. syringae* pv. *tomato* in the presence and absence of *P. fluorescens* on pepper and tomato plants, respectively. The saprophyte multiplied mainly on the leaf surface and surpassed the pathogen, whereas the latter multiplied only within the leaf. This finding was used to develop a highly sensitive method, based on surface sterilization, for detection of a very small number of pathogenic bacteria in seeds and in symptomless leaves. This method was adopted by some industrial seed quality control units in Israel (Bashan & Assouline 1983).

Leben (1965, 1981) and Cook & Baker (1983) suggested the existence of commensalism between plants and their related pathogens in the epiphytic growth phase. This approach assumes nutritional relationships between the two organisms in which the pathogen could obtain nutrients, as leaf exudates, without causing any damage to the plant ('biotrophic parasitism'). Under conditions suitable for disease development the pathogen causes a partial or total damage to its host and multiplies in the dead tissue ('necrotrophic parasitism').

The endophytic population is not always correlated with visible disease symptoms. Monitoring the endophytic population of *X. campestris* pv. *vesicatoria* in susceptible pepper leaves under field conditions showed no correlation between disease severity, inoculum concentration and endophytic population. A massive endophytic population developed, with very few visible disease symptoms on the leaves (Bashan *et al.* 1985 a). Mew & Kennedy (1982) reported that pathogenic bacteria could be washed away in great numbers from symptomless soybean leaves. Tomato plants which were symptomless from sowing to flowering stage had 20 to 30% less foliage when grown in soil infested with *P. syringae* pv. *tomato* compared to plants growing in uninfested soil (Bashan & Okon 1981).

Pepper seeds collected from symptomless plants grown in seed production fields were found to be contaminated with virulent isolates. The seeds were germinated in the greenhouse and transferred to mist chambers. No visible symptoms developed under these conditions. However, the endophytic pathogenic population of *X. campestris* pv. *vesicatoria* was $10^5$ to $10^6$ CFU $g^{-1}$ leaf. When these seedlings were later transferred into the field for an additional two months, no symptoms appeared but the endophytic pathogenic bacterial population maintained its high level throughout the growing season (Bashan *et al.* 1982 a).

### *SURVIVAL IN DETACHED LEAVES, ROOTS AND WOODY TISSUES*

Epiphytic bacteria can survive on their host leaves throughout the season and in plant debris from one season to the next. There is evidence to support the idea that most residential epiphytic bacteria of leaves of temperate zones, including the leaf pathogens, die concomitantly with the leaf death and only the endophytic bacteria survive (Leben 1981). Plant debris comprises an important site of survival for many pathogens.

*X. campestris* pv. *vesicatoria* survived and retained its virulence on dry tomato and pepper leaves for at least 12 months. After 4 months the pathogen reached a steady level of $10^4$ CFU $g^{-1}$ dried leaves (Bashan *et al.* 1982 a) and was capable of surviving in the rhizosphere of dead plants (Peterson 1963). Crossan & Morehart (1964) showed that this pathogen became established within various internal tissues of pepper plants exhibiting external symptoms of bacterial spot. The pathogen enters readily through stomata and veins (Sharon *et al.* 1982 a) destroying cells adjacent to the vascular system. Survival in such woody tissues may explain occasional disease outbreaks. Moisture of the debris is a crucial factor for the pathogen's survival. Under moist conditions, the debris decomposes, including the survived bacterial cells (Leben 1981), as in the case of the bacterial scab pathogen (Lewis & Brown 1961). *P. syringae* pv. *tomato* can survive in plant debris either above or below the soil surface for 30 weeks (Chambers & Merriman 1975). Virulent *X. campestris* pv. *alfalfae* were recovered from naturally infested alfalfa debris for a period of a year. Dried alfalfa shoots kept in the laboratory contained the pathogen eight years after collection (Claflin & Stuteville 1973).

### *ROLE OF SEEDS, NON-HOSTS AND WEEDS IN PATHOGEN SURVIVAL*

*X. campestris* pv. *phaseoli* var. *fuscans* retained in bean seeds its viability, virulence and physiological properties for three years over a range of temperatures from -20 to 35°C. Maintaining the bacteria at 25 to 35°C on agar media reduced their survival and virulence; nevertheless, viability was lost only after nine months (Basu & Wallen 1966). *X. campestris* pv. *phaseoli* survived on bean seeds for 3 to 24 years (Zaumeyer & Thomas 1957; Basu & Wallen 1966). All surviving isolates were pathogenic to their respective hosts at the seedling stage. *X. campestris* pv. *vesicatoria* and *P. syringae* pv. *tomato* were found in commercial seed lots, packed for 10 and 20 years, respectively (Bashan *et al.* 1982 b). *P. syringae* pv. *tomato* found on host seeds as cell aggregates inside holes and cavities (Devash *et al.* 1980) could be eliminated from seeds by pulp fermentation (Chambers & Merriman 1975). No heat-tolerant strain was found. Thermal death point was as low as 48°C and the bacterial population was already affected at a relatively low temperature of 35°C. However, few surviving cells were capable of causing speck symptoms on young seedlings (Devash *et al.* 1980). Similarly to *P. syringae* pv. *pisi* in pea seedlings (Bashan & Kenneth 1983), other members of *P. syringae* pathovars, such as *P. syringae* pv. *syringae* pathogenic to wheat, can also survive on most seed lots (Otta 1977).

Pepper and tomato seeds infested with *X. campestris* pv. *vesicatoria* are known as a major source of inoculum for bacterial scab of pepper (Gardner & Kendrick 1921; Lewis & Brown 1961; Crossan & Morehart 1964; Shekhawat & Chakravarti 1976). The number of pathogens on the pepper seeds decreased with time, reaching a steady level five months after seed infestation. Seeds also became infested in apparently healthy pepper fields grown for seed production (Bashan *et al.* 1982 a). The number of surviving epiphytic pathogenic cells on dry seeds was very small. Furthermore, this small population was masked by a large population of saprophytic bacteria, thus complicating the detection of the pathogen (Schaad 1982; Sharon *et al.* 1982 b; Bashan & Assouline 1983). Disease initiation is dependent on environmental conditions. Very few bacterial cells among the vast amount of seeds may produce epidemics, but under favourable conditions even heavily infested seeds may give rise to healthy plants.

A low level of *X. campestris* pv. *vesicatoria* was detected in field soil during the winter season. This suggests that the soil was continuously supplied with the pathogen from standing stalks of dead tomato plants, since the pathogen disappeared in non-sterile soil held at 25°C within two weeks (Peterson 1963). He further suggested that the pathogen could not survive for more than two weeks in natural soil and for two months in sterile soil. However, Bashan *et al.* (1982 a) demonstrated that even when the pathogenic population in the soil decreases markedly with time, it was possible to monitor a few surviving cells in the field the next season, 18 months after the field was cropped with diseased pepper plants.

Survival of pathogenic bacteria in the host and non-host rhizosphere is very important for the pathogen's overwintering (Schuster & Coyne 1974). A population of *X. campestris* pv. *vesicatoria* developed in the rhizosphere of apparently symptomless pepper plants grown under mist, but not under dry conditions. When seeds of sorghum, pea, cucumber, bean, wheat, soybean and tomato were infested, the pathogen could multiply and survive for at least three months in non-host rhizosphere but not in their foliage. Except for tomato seeds, the pathogen was naturally eliminated from all other non-host seeds within 30 days (Diachun & Valleau 1946; Bashan *et al.* 1982 a). *P. syringae* pv. *tomato* survived for at least several days on many non-host cultivated plants (Bonn & MacNeill 1983). Kritzman & Zutra (1983) demonstrated the survival of *P. syringae* pv. *lachrymans* in soil, plant debris, rhizosphere of non-host plants and in the vascular system of cucumber plants.

Survival on weeds adjacent to or in the field is an important factor, as pathogens cannot be totally eliminated, and contaminated weeds may cause epidemics in the cultivated crop. A highly virulent strain of *P. solanacearum* had a resident phase on hairy vetch, a common weed in bean plots in Wisconsin. The pathogen was the main component of the Gram-negative epiphytic microflora throughout the year (Ercolani *et al.* 1974). Gitai-

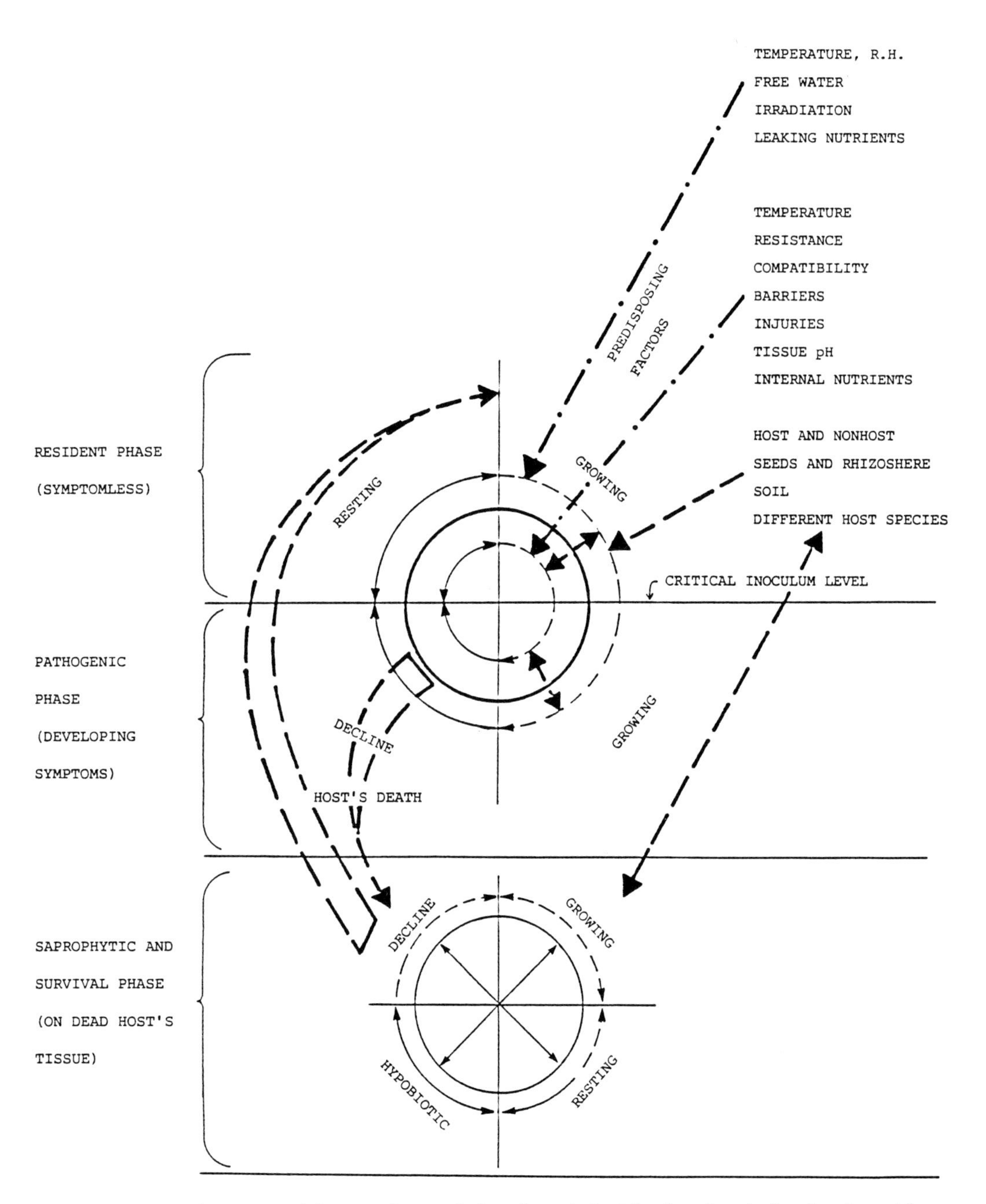


Fig. 3. Life cycle model of epiphytic bacterial plant pathogens (after Leben 1981).
—·—▶ Predisposing factors, ▬▬▶ Bacterial spread, ⋙ Bacterial spread in one direction, ◀——▶ Change in bacterial growth phase, – – –▶ Bacterial growth, ▭ Leaf tissue.

tis *et al.* (1985) isolated *P. syringae* pv. *syringae* from leaf washings of symptomless wild cherry and rye leaves adjacent to or within tomato transplant fields. McCarter *et al.* (1983) and Schneider & Grogan (1977) showed that *P. syringae* pv. *tomato* was present on native weeds and transmitted by them to tomato fields.

*SUMMARY*

The main sites which determine the survival and activity of epiphytic pathogens are the host's leaf, the dead host's tissue, seeds, rhizosphere and the soil (Fig. 3). The pathogen can stay on the leaf surface in a resting, growing or declining phase. The pathogen may survive in dead or detached host tissues in growing, resting, hypobiotic or decline phases. With regard to disease development three main phases can be distinguished in the pathogen's life cycle (Leben 1981): resident (symptomless host), pathogenic (developing visible disease symptoms) and saprophytic phase. It seems that most bacteria, either pathogenic or saprophytic, can multiply on the surface of plant leaves, providing that available nutrients are supplied externally or by the host plant. Compatible pathogenic bacteria can multiply freely both outside and inside the leaf, whereas saprophytic bacteria are usually restricted to the external surface only. The population of pathogenic bacteria usually increases massively during disease development. The endophytic population can develop in the intercellular leaf spaces only if these are filled up with water. A better understanding of the physiological basis of the behaviour of epiphytic pathogens in their natural environment may improve the control of diseases caused by these bacteria.

*ACKNOWLEDGEMENTS*

Y. Bashan is a recipient of a Sir Charles Clore fellowship and participated in this review in memory of the late Mr Avner Bashan.

*REFERENCES*

Allington, W.B. & Chamberlain, D.W. (1949). Trends in the population of pathogenic bacteria within leaf tissues of susceptible and immune plant species. Phytopathology, *39*, 656-60.

Barnes, E.H. (1965). Bacteria on leaf surfaces and in intercellular leaf spaces. Science, *147*, 1151-2.

Bashan, Y. (1985). Are bacteria-plant cell interactions specific? *In* Recent Topics in Experimental and Conceptual Plant Pathology, ed. R.S. Singh (in press).

Bashan, Y. & Assouline, I. (1983). Complementary bacterial enrichment techniques for the detection of *Pseudomonas syringae* pv. *tomato* and *Xanthomonas campestris* pv. *vesicatoria* in infested tomato and pepper seeds. Phytoparasitica, *11*, 187-93.

Bashan, Y., Azaizeh, M., Diab, S., Yunis, H. & Okon, Y. (1985 a). Crop loss of pepper plants artificially infected with *Xanthomonas campestris* pv. *vesicatoria* in relation to symptom expression. Crop Protection, *4*, 71-84.

Bashan, Y., Diab, S. & Okon, Y. (1982 a). Survival of *Xanthomonas campestris* pv. *vesicatoria* in pepper seeds in symptomless and dry leaves and in soil.

Plant and Soil, *68*, 161-70.
Bashan, Y. & Kenneth, R. (1983). The occurrence of bacterial blight of peas in Israel. Phytoparasitica, *11*, 113-5.
Bashan, Y. & Okon, Y. (1981). Inhibition of seed germination and development of tomato plants in soil infested with *Pseudomonas tomato*. Annals of Applied Biology, *98*, 413-7.
Bashan, Y., Okon, Y. & Henis, Y. (1978). Infection studies of *Pseudomonas tomato*, causal agent of bacterial speck of tomato. Phytoparasitica, *6*, 135-44.
Bashan, Y., Okon, Y. & Henis, Y. (1982 b). Long-term survival of *Pseudomonas syringae* pv. *tomato* and *Xanthomonas campestris* pv. *vesicatoria* in tomato and pepper seeds. Phytopathology, *72*, 1143-4.
Bashan, Y., Okon, Y. & Henis, Y. (1985 b). Morphology of leaf surfaces of tomato cultivars in relation to possible invasion into the leaf by *Pseudomonas syringae* pv. *tomato*. Annals of Botany, *55*, 803-9.
Bashan, Y., Sharon, E., Okon, Y. & Henis, Y. (1981). Scanning electron and light microscopy of infection and symptom development in tomato leaves infected with *Pseudomonas tomato*. Physiological Plant Pathology, *19*, 139-44.
Basu, P.K. & Wallen, V.R. (1966). Influence of temperature on the viability, virulence, and physiologic characteristics of *Xanthomonas phaseoli* var. *fuscans* in vivo and in vitro. Canadian Journal of Botany, *44*, 1239-45.
Bonn, W.G., Gitaitis, R.D. & MacNeill, B.H. (1985). Epiphytic survival of *Pseudomonas syringae* pv. *tomato* on tomato transplants shipped from Georgia. Plant Disease, *69*, 58-60.
Bonn, W.G. & MacNeill, B.H. (1983). Establishment and survival of a double-marked strain of *Pseudomonas syringae* pv. *tomato* on tomato and other crops. Phytopathology, *73*, 363 (Abstr.).
Burki, T. (1972). *Pseudomonas tomato* (Okabe) Alstatt, Erreger einer für die Schweiz neuen Tomatenbakteriose. Schweizerische Landwirtschaftliche Forschung, *11*, 97-107.
Chambers, S.C. & Merriman, P.R. (1975). Perennation and control of *Pseudomonas tomato* in Victoria. Australian Journal of Agricultural Research, *26*, 657-63.
Chet, I., Zilberstein, Y. & Henis, Y. (1973). Chemotaxis of *Pseudomonas lachrymans* to plant extracts and to water droplets collected from the leaf surfaces of resistant and susceptible plants. Physiological Plant Pathology, *3*, 473-9.
Claflin, L.E. & Stuteville, D.L. (1973). Survival of *Xanthomonas alfalfae* in alfalfa debris and soil. Plant Disease Reporter, *57*, 52-3.
Cook, A.A. & Stall, R.E. (1977). Effects of watersoaking on response to *Xanthomonas vesicatoria* in pepper leaves. Phytopathology, *67*, 1101-3.
Cook, R.J. & Baker, K.F. (1983). The Nature and Practice of Biological Control of Plant Pathogens, pp. 62-3. St. Paul: Am. Phytopath. Soc.
Crossan, D.F. & Morehart, A.L. (1964). Isolation of *Xanthomonas vesicatoria* from tissues of *Capsicum annuum*. Phytopathology, *54*, 358-9.
Crosse, J.E. (1959). Bacterial canker of stone-fruits. IV. Investigation of a method for measuring the inoculum potential of cherry trees. Annals of Applied Biology, *47*, 306-17.
Crosse, J.E., Goodman, R.N. & Shaffer, W.H., Jr. (1972). Leaf damage as a predisposing factor in the infection of apple shoots by *Erwinia amylovora*. Phytopathology, *62*, 176-82.
Davis, D. & Halmos, S. (1958). The effect of air moisture on the predisposition of tomato to bacterial spot. Plant Disease Reporter, *42*, 110-1.
Devash, Y., Okon, Y. & Henis, Y. (1980). Survival of *Pseudomonas tomato* in soil and seeds. Phytopathologische Zeitschrift, *99*, 175-85.
Diab, S., Bashan, Y. & Okon, Y. (1982 a). Studies of infection with *Xanthomonas campestris* pv. *vesicatoria*, causal agent of bacterial scab of pepper in Israel. Phytoparasitica, *10*, 183-91.
Diab, S., Bashan, Y., Okon, Y. & Henis, Y. (1982 b). Effect of relative humidity on

bacterial scab caused by *Xanthomonas campestris* pv. *vesicatoria* on pepper. Phytopathology, *72*, 1257-60.
Diachun, S. & Valleau, W.D. (1946). Growth and overwintering of *Xanthomonas vesicatoria* in association with wheat roots. Phytopathology, *36*, 277-80.
English, H. & Davis, J.R. (1960). The source of inoculum for bacterial canker and blast of stone fruit trees. Phytopathology, *50*, 634 (Abstr.).
Ercolani, G.L. & Crosse, J.E. (1966). The growth of *Pseudomonas phaseolicola* and related plant pathogens *in vivo*. Journal of General Microbiology, *45*, 429-39.
Ercolani, G.L., Hagedorn, D.J., Kelman, A. & Rand, R.E. (1974). Epiphytic survival of *Pseudomonas syringae* on hairy vetch in relation to epidemiology of bacterial brown spot of bean in Wisconsin. Phytopathology, *64*, 1330-9.
Fallik, E., Bashan, Y., Okon, Y., Cahaner, A. & Kedar, N. (1983). Inheritance and sources of resistance to bacterial speck of tomato caused by *Pseudomonas syringae* pv. *tomato*. Annals of Applied Biology, *102*, 365-71.
Gardner, M.W. & Kendrick, J.B. (1921). Bacterial spot of tomato. Journal of Agricultural Research, *21*, 123-56.
Gitaitis, R.D., Jones, J.B., Jaworski, C.A. & Phatak, S.C. (1985). Incidence and development of *Pseudomonas syringae* pv. *syringae* on tomato transplants in Georgia. Plant Disease, *69*, 32-5.
Gitaitis, R.D., Samuelson, D.A. & Strandberg, J.O. (1981). Scanning electron microscopy of the ingress and establishment of *Pseudomonas alboprecipitans* in sweet corn leaves. Phytopathology, *71*, 171-5.
Henis, Y., Okon, Y., Sharon, E. & Bashan, Y. (1980). Detection of small numbers of phytopathogenic bacteria using the host as an enrichment medium. Journal of Applied Bacteriology, *49*, unnumbered page (Abstr.).
Hirano, S.S., Rouse, D.I. & Upper, C.D. (1982). Frequency of bacterial ice nuclei on snap bean (*Phaseolus vulgaris* L.) leaflets as a predictor of bacterial brown spot. Phytopathology, *72*, 1006 (Abstr.).
Hirano, S.S. & Upper, C.D. (1983). Ecology and epidemiology of foliar bacterial plant pathogens. Annual Review of Phytopathology, *21*, 243-69.
Huang, J.S. & Van Dyke, C.G. (1978). Interaction of tobacco callus tissue and *Pseudomonas tabaci*, *P. pisi* and *P. fluorescens*. Physiological Plant Pathology *13*, 65-72.
Kennedy, B.W. (1969). Detection and distribution of *Pseudomonas glycinea* in soybean. Phytopathology, *59*, 1618-9.
Kritzman, G. & Zutra, D. (1983). Survival of *Pseudomonas syringae* pv. *lachrymans* in soil, plant debris and the rhizosphere of non-host plants. Phytoparasitica, *11*, 99-108.
Leben, C. (1963). Multiplication of *Xanthomonas vesicatoria* on tomato seedlings. Phytopathology, *53*, 778-81.
Leben, C. (1965). Epiphytic microorganisms in relation to plant disease. Annual Review of Phytopathology, *3*, 209-30.
Leben, C. (1981). How plant-pathogenic bacteria survive. Plant Disease, *65*, 633-7.
Leben, C., Daft, G.C. & Schmitthenner, A.F. (1968 a). Bacterial blight of soybeans: population levels of *Pseudomonas glycinea* in relation to symptom development. Phytopathology, *58*, 1143-6.
Leben, C., Rusch, V. & Schmitthenner, A.F. (1968 b). The colonization of soybean buds by *Pseudomonas glycinea* and other bacteria. Phytopathology, *58*, 1677-81.
Leben, C., Schroth, M.N. & Hildebrand, D.C. (1970). Colonization and movement of *Pseudomonas syringae* on healthy bean seedlings. Phytopathology, *60*, 677-80.
Lewis, G.D. & Brown, D.H. (1961). Studies on the overwintering of *Xanthomonas vesicatoria* in New Jersey. Phytopathology, *51*, 577 (Abstr.).
Lewis, S. & Goodman, R.N. (1965). Mode of penetration and movement of fire blight bacteria in apple leaf and stem tissue. Phytopathology, *55*, 719-23.
Lindemann, J., Arny, D.C. & Upper, C.D. (1984). Use of an apparent infection threshold population of *Pseudomonas syringae* to predict incidence and

severity of brown spot of bean. Phytopathology, *74*, 1334-9.
Lindow, S.E. (1983). The role of bacterial ice nucleation in frost injury to plants. Annual Review of Phytopathology, *21*, 363-84.
McCarter, S.M., Jones, J.B., Gitaitis, R.D. & Smitley, D.R. (1983). Survival of *Pseudomonas syringae* pv. *tomato* in association with tomato seed, soil, host tissue, and epiphytic weed host in Georgia. Phytopathology, *73*, 1393-8.
Mew, T.W. & Kennedy, B.W. (1971). Growth of *Pseudomonas glycinea* on the surface of soybean leaves. Phytopathology, *61*, 715-6.
Mew, T.W. & Kennedy, B.W. (1982). Seasonal variation in populations of pathogenic *Pseudomonas* on soybean leaves. Phytopathology, *72*, 103-5.
Miles, W.G., Daines, R.H. & Rue, J.W. (1977). Presymptomatic egress of *Xanthomonas pruni* from infected peach leaves. Phytopathology, *67*, 895-7.
Miller, T.D. & Schroth, M.N. (1972). Monitoring the epiphytic population of *Erwinia amylovora* on pear with a selective medium. Phytopathology, *62*, 1175-82.
Morton, D.J. (1966). Bacterial spot development in excised pepper and tomato leaves at several temperatures. Phytopathology, *56*, 1194-5.
Nayudu, M.V. & Walker, J.C. (1960). Bacterial spot of tomato as influenced by temperature and by the age and nutrition of the host. Phytopathology, *50*, 360-4.
Okon, Y., Bashan, Y. & Henis, Y. (1978). Studies of bacterial speck of tomato caused by *Pseudomonas tomato*. Proceedings of the Fourth International Conference on Plant Pathogenic Bacteria, ed. Station de Pathologie Végétale et Phytobacteriologie, pp. 699-702. Beaucouzé: Inst. Nat. Rech. Agron.
Omer, M.E.H. & Wood, R.K.S. (1969). Growth of *Pseudomonas phaseolicola* in susceptible and resistant bean plants. Annals of Applied Biology, *63*, 103-16.
Otta, J.D. (1977). Occurrence and characteristics of isolates of *Pseudomonas syringae* on winter wheat. Phytopathology, *67*, 22-6.
Palti, J. (1981). Cultural Practices and Infectious Crop Diseases. Advanced Series in Agricultural Sciences No. 9. New York: Springer-Verlag.
Peterson, G.H. (1963). Survival of *Xanthomonas vesicatoria* in soil and diseased tomato plants. Phytopathology, *53*, 765-7.
Pohronezny, K., Leben, C. & Larsen, P.O. (1977). Systemic invasion of cucumber by *Pseudomonas lachrymans*. Phytopathology, *67*, 730-4.
Raymundo, A.K. & Ries, S.M. (1980). Chemotaxis of *Erwinia amylovora*. Phytopathology, *70*, 1066-9.
Rolfs, F.M. (1915). A bacterial disease of stone fruits. Cornell University Agricultural Experiment Station Memoir No. 8.
Rouse, D.I., Hirano, S.S., Lindemann, J., Nordheim, E.V. & Upper, C.D. (1981). A conceptual model for those diseases with pathogen population multiplication independent of disease development. Phytopathology, *71*, 901 (Abstr.).
Schaad, N.W. (1982). Detection of seedborne bacterial plant pathogens. Plant Disease, *66*, 885-90.
Schneider, R.W. & Grogan, R.G. (1977). Tomato leaf trichomes, a habitat for resident populations of *Pseudomonas tomato*. Phytopathology, *67*, 898-902.
Schneider, R.W. & Grogan, R.G. (1978). Influence of temperature on bacterial speck of tomato. Phytopathology News, *12*, 204 (Abstr.).
Schuster, M.L. & Coyne, D.P. (1974). Survival mechanisms of phytopathogenic bacteria. Annual Review of Phytopathology, *12*, 199-221.
Sharon, E., Bashan, Y., Okon, Y. & Henis, Y. (1982 a). Presymptomatic multiplication of *Xanthomonas campestris* pv. *vesicatoria* on the surface of pepper leaves. Canadian Journal of Botany, *60*, 1041-5.
Sharon, E., Okon, Y., Bashan, Y. & Henis, Y. (1982 b). Detached leaf enrichment: a method for detecting *Pseudomonas syringae* pv. *tomato* and *Xanthomonas campestris* pv. *vesicatoria* in seeds and symptomless leaves of tomato

and pepper. Journal of Applied Bacteriology, *53*, 371-7.
Shaw, L. (1935). Intercellular humidity in relation to fire-blight susceptibility in apple and pear. Cornell University Agricultural Experiment Station Memoir No. 181.
Shekhawat, P.S. & Chakravarti, B.P. (1976). Factors affecting development of bacterial leaf spot of chillies caused by *Xanthomonas vesicatoria*. Indian Phytopathology, *29*, 392-7.
Smitley, D.R. & McCarter, S.M. (1982). Spread of *Pseudomonas syringae* pv. *tomato* and role of epiphytic populations and environmental conditions in disease development. Plant Disease, *66*, 713-4.
Stadt, S.J. & Saettler, A.W. (1981). Effect of host genotype on multiplication of *Pseudomonas phaseolicola*. Phytopathology, *71*, 1307-10.
Stall, R.E. & Cook, A.A. (1966). Multiplication of *Xanthomonas vesicatoria* and lesion development in resistant and susceptible pepper. Phytopathology, *56*, 1152-4.
Sutton, T.B. & Jones, A.L. (1975). Monitoring of *Erwinia amylovora* populations on apple in relation to disease incidence. Phytopathology, *65*, 1009-12.
Thomson, S.V., Schroth, M.N., Moller, W.J. & Reil, W.O. (1975). Occurrence of fire blight of pears in relation to weather and epiphyte populations of *Erwinia amylovora*. Phytopathology, *65*, 353-8.
Venetto, J.R. (1982). How bacteria find their hosts. *In* Phytopathogenic Prokaryotes. Vol. 2, eds M.S. Mount & G.H. Lacy, pp. 4-30. New York: Academic Press.
Volcani, Z. (1962). Bacterial spot disease of tomatoes and peppers in Israel. Plant Disease Reporter, *46*, 175.
Volcani, Z. (1964). Development of angular leaf spot of cucumbers grown under plastic in Israel. Plant Disease Reporter, *48*, 256-7.
Volcani, Z. (1969). The effect of mode of irrigation and wind direction on disease severity caused by *Xanthomonas vesicatoria* on tomato in Israel. Plant Disease Reporter, *53*, 459-61.
Weller, D.M. & Saettler, A.W. (1980). Colonization and distribution of *Xanthomonas phaseoli* and *Xanthomonas phaseoli* var. *fuscans* in field-grown navy beans. Phytopathology, *70*, 500-6.
Yunis, H., Bashan, Y., Okon, Y. & Henis, Y. (1980 a). Two sources of resistance to bacterial speck of tomato caused by *Pseudomonas tomato*. Plant Disease, *64*, 851-2.
Yunis, H., Bashan, Y., Okon, Y. & Henis, Y. (1980 b). Weather dependence, yield losses and control of bacterial speck of tomato caused by *Pseudomonas tomato*. Plant Disease, *64*, 937-9.
Zaumeyer, W.J. & Thomas, H.R. (1957). A Monographic Study of Bean Diseases and Methods for Their Control. USDA Technical Bulletin No. 868.

# EPIPHYTIC COLONIZATION OF HOST AND NON-HOST PLANTS BY PHYTOPATHOGENIC BACTERIA

T.W. Mew and C.M. Vera Cruz

*Department of Plant Pathology, The International Rice Research Institute, Los Baños, Laguna, Philippines*

## *INTRODUCTION*

Two and a half decades have elapsed since the first report by Crosse (1963) that large numbers of *Pseudomonas syringae* pv. *mors-prunorum* were detected on the leaf surfaces of cherry. Since then, there is little doubt that most if not all foliar pathogenic bacteria are natural inhabitants of the aerial parts of the host plant. The recognition of an epiphytic resident phase (Leben 1965) in the life of these bacterial pathogens is significant. It has been demonstrated that epiphytic bacterial pathogens can occur in two distinct phases, viz. a non-pathogenic and a pathogenic one. As non-pathogens, they serve as potential inoculum on leaves where they reside or for other plant parts. Conducive conditions on the aerial surfaces can lead to the normal pathogenic state.

The recognition of the epiphytic phase of bacterial pathogens has further extended researches on their ecology and their roles as potential sources of inoculum of plant diseases they cause (see Y. Henis and Y. Bashan, this volume). Information on bacterial brown spot of bean caused by *P. syringae* pv. *syringae* has clearly demonstrated the potential sources of inoculum to be more diverse than had been indicated earlier (Ercolani *et al.* 1974; Lindemann *et al.* 1984). Classically, scientists had attributed brown spot disease epidemics entirely to infected seed sources or to debris from infected plants (Hagedorn & Patel 1965; Hoitink *et al.* 1968). More recent studies show that the pathogen occurred not only on the leaf surface of the host plant, snap bean, but also on the leaves of non-hosts, such as hairy vetch, oak, black locust, rye and sow thistle, growing near commercial bean fields (Ercolani *et al.* 1974; Lindemann *et al.* 1984). All these plants supported epiphytic populations of *P. syringae* pv. *syringae*.

A pathogen's epiphytic stage needs to be considered in studies on the epidemiology and nature of the diseases they cause. Recent reviews by Hirano & Upper (1983) and Y. Henis and Y. Bashan (this volume) have given a very comprehensive account on all aspects of epiphytic pathogenic bacteria. Our present review is therefore limited to examine only information relative to colonization of leaf surfaces of resistant and susceptible genotypes of host plants and non-host plants. We are especially interested in the fate and behaviour of the epiphytic bacterial pathogens on resistant and susceptible host genotypes. We can consider this stage as a pre-infection phase of the disease, i.e. a transition from non-pathogenic to pathogenic activity and also as a stage of the response of the host plants. Despite voluminous researches focused on epiphytic bacterial pathogens, there is little work being carried out on this area of research. We therefore must rely largely on the information from our own studies. We will try to demonstrate that there are distinct host responses to epiphytic bacterial pathogens after they arrive at the sites of entry. Distinct phenomena in relation to such pre-infection activity are postulated.

## *EPIPHYTIC POPULATION ON HOST AND NON-HOST PLANTS*

In their review, Hirano & Upper (1983) concluded that resistant and susceptible host genotypes and non-hosts are potential reservoirs in the dissemination of epiphytic plant-pathogenic bacteria. They also concluded that "field resistance of certain cultivars to epipabs (epiphytic phytopathogenic bacteria) may be a function of the inability of that epipab to colonize and establish large populations on resistant compared to susceptible cultivars". The questions then arise whether there are differential responses of host genotypes to epipab colonization, and whether epipabs are inhabitants or invaders (sensu Garrett 1970) on non-host or resistant host genotypes. Equally, we can ask whether the arrival of an epipab on a receptive site initiates infection, and whether or not the epipab on a leaf surface can be merely a transient inoculum from other neighbouring plants. To resolve these issues, the influence of resistant host genotypes and non-host plants on epipab activity needs further examination.

### *Non-host plants*

Phytopathogenic bacteria multiply only slightly in the tissues of non-host plants for a relatively short duration immediately after artificial introduction. The response of non-host plant tissues is termed as hypersensitive reaction (Klement & Goodman 1967). It is not known, however, whether or not the bacteria would invade non-host plants naturally.

Foliar bacterial pathogens are reported to become epiphytes on leaf surfaces or aerial plant parts of non-host plants. They are likely to be epiphytes of many non-host plants. On these plants, we are not sure if they are invaders or inhabitants, but they could also serve as inoculum for nearby host plants.

Lindow *et al.* (1978) found that *P. syringae* was widely distributed as an epiphyte on apparently healthy plants of 74 out of 95 species sampled. Only conifers as a group did not harbour epiphytic populations of *P. syringae*. None of the isolates from other plant species was tested to determine if they were potentially pathogenic to their plant associates. Furthermore, the epiphytic populations were greater on symptomless bean leaflets and corn leaves in bean-growing areas than in areas where there was no commercial snap bean production. The pathogen was also detected by Lindemann *et al.* (1984) on different non-host plants near commercial snap bean fields. It was not determined if these bacteria were on the non-host plants only as invaders. It is important whether they 'multiplied' on the leaf surfaces of these plants or remained 'static' in a small population.

When bush bean (*Phaseolus vulgaris* L.) was planted near a soybean field, *P. syringae* pv. *glycinea* was detected on the surfaces of bush bean leaves throughout the growing season but always at a low population density (Fig. 1). The bacteria were evidently on the bush bean leaf surfaces only as invaders and not as inhabitants.

*Host plants*

It is advantageous for the foliar plant-pathogenic bacteria to adapt and reside on leaf surfaces of host plants. Residency is only the first step leading to the invasion of the host. The role of the epiphytic populations of these bacterial pathogens needs to be considered not only as potential inoculum, as has been reviewed recently (Hirano & Upper 1983), but also as part of the infection process. A successful infection is possible only if the bacterial cells are firmly established and reach the infection court. If this happens, the host genotype is likely to respond to the colonization. A degree of host specificity relative to host genotypes and pathogen virulence has been demonstrated (Crosse 1963; Mew & Kennedy 1971, 1982; Daub & Hagedorn 1981).

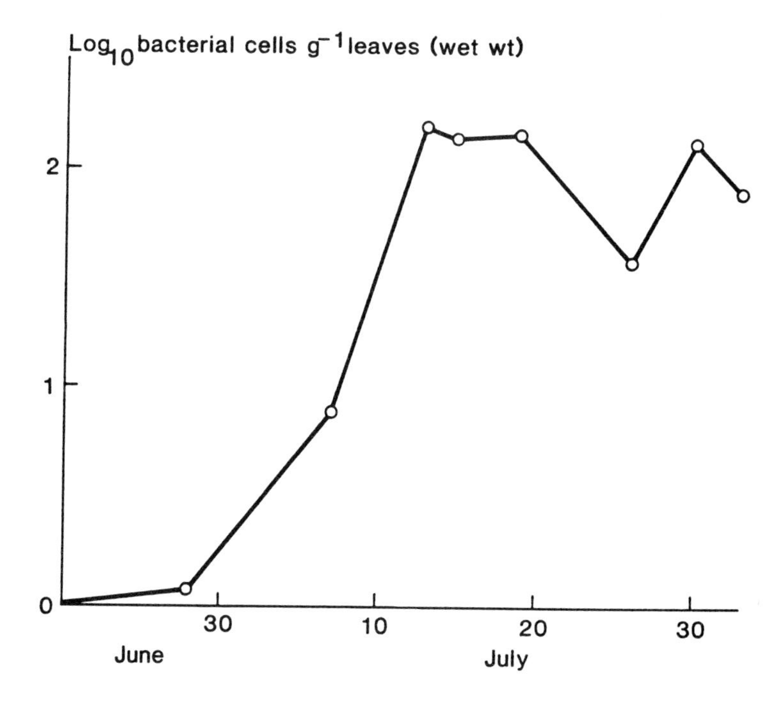


Fig. 1. Population of plant-pathogenic fluorescent pseudomonads detected in the phyllosphere of garden bean (cultivar Bush Blue Lake) grown in rows adjacent to soybean. Each point represents a mean of four replicates (20 leaves each).

Some of the foliar epiphytic bacterial plant pathogens such as *P. syringae* gain recognition by becoming inoculum for a more severe infection of other parts of the plant or of other hosts. Others, such as *Xanthomonas campestris* pv. *oryzae* (Mew *et al.* 1984) and probably those causing foliar diseases are known to enter the infection courts on the leaf surfaces where they have established their residency. The role of epiphytes relies much on the nature of the specific diseases and on the manner and site of entry to the host. There is little doubt that foliar bacterial pathogens multiply on the leaf surfaces. It is questionable, however, whether they multiply on the leaf surfaces of resistant as well as susceptible host genotypes. Limited information has suggested that they do not, and also cannot do equally well on both host genotypes. The part played by the host genotypes on epiphytic colonization is not yet fully understood. The issue, therefore, that should be resolved is whether or not there are specific leaf surface niches for colonization.

*Host genotypes.* The widespread occurrence of foliar bacterial pathogens as epiphytes on symptomless and/or healthy leaves of both resistant and susceptible host genotypes has been well documented. In investigating a method for measuring the inoculum potential of *P. syringae* pv. *morsprunorum* on cherry trees, Crosse (1959) found that the numbers of bacteria varied on healthy cherry leaf surfaces. He showed that the bacterial pathogen was abundant on the leaf surfaces in the autumn. Moreover, the results suggested that the numbers present were related to the inherent resistance of the varieties. Later, in six separate comparisons in different seasons and localities consistently higher leaf surface population densities were found on the susceptible cv. Napoleon than on the resistant cv. Roundel. In another survey, however, leaf surface population densities on cv. Roundel were found to be significantly higher than on cv. Napoleon. This reversal in the ability of the two cultivars to support epiphytic populations of *P. syringae* pv. *morsprunorum* was found to be associated with a variant of the causal organism which had overcome the resistance of cv. Roundel. The variant, designated as race 2, appeared to establish a population on cv. Roundel in early May about 1,000 times greater than the populations on cv. Napoleon.

The information indeed suggested preferential colonization ability of a bacterial plant pathogen and differences in the infective potential of the population related to host genotype leading to larger or smaller populations. Evidence for preferential colonization of leaf surfaces by foliar bacterial pathogens was found in the ability of *P. syringae* pv. *morsprunorum* and *P. syringae* pv. *syringae* to establish themselves on host cherry and pear, respectively, whereas these bacterial pathogens failed to colonize the leaf surfaces of the non-host plants (Ercolani 1969). This suggests that the ability of pathogenic bacteria to survive epiphytically may be a fairly specific phenomenon. The influence of host genotypes cannot be neglected.

Monitoring the differences in epiphytic populations of bacterial pathogens on leaf surfaces of resistant and susceptible host genotypes of three sour cherry cultivars, Sobiczewski (1978) found that pathogenic pseudomonads causing bacterial canker were present on the susceptible cv. Nefris. On a less susceptible cultivar, North Star, the pathogenic pseudomonads were present sporadically. On the resistant cv. Schathenmorelle pathogenic pseudomonads were rarely present in one year and were sporadically present in another year. Pathogenic pseudomonads were never dominant in either the less susceptible or the resistant cultivar. When *Xanthomonas populi* was sprayed onto very susceptible and resistant poplar clones, a large number of bacterial cells were detected on leaf surfaces on the susceptible clones, but none on those of the resistant poplar (Ridé *et al.* 1978). *Pseudomonas syringae* pv. *glycinea* is inherently capable of multiplying on soybean leaf surfaces (Mew & Kennedy 1971) and also can colonize surfaces of soybean buds (Leben *et al.* 1968). In field-grown soybean in Minnesota, the bacteria could be detected on soybean leaves approximately 6 weeks after planting. It was also noted that early in the growing season pathogenic races within the epiphytic population of *P. syringae* pv. *glycinea* tended to be specific to the cultivar from which they were isolated (Mew & Kennedy 1982). Late in the season a population virulent to a number of cultivars increased while cultivar-specific components of the total decreased. Large differences in epiphytic population of *P. syringae* pv. *syringae* on the leaves of resistant snap bean cultivar WBR 13 and susceptible cultivar Eagle were recorded (Daub & Hagedorn 1981). On intermediately disease-resistant breeding lines densities of epiphytic populations of the bacterium were also intermediate.

Several reports indicate that the bacterial pathogens can be detected on symptomless leaves of both resistant and susceptible cultivars. *Xanthomonas campestris* pv. *manihotis* occurred on leaf surfaces of both susceptible and resistant cassava varieties (Persley 1978). *P. syringae*, the causal bacterium of wheat leaf necrosis, moved from inoculated wheat seeds to the seedlings and survived as an epiphyte on the leaves of both resistant and susceptible genotypes (Fryda & Otta 1978). Apparently, resistance is not related to the inhibition of moving bacterial inoculum from seed to seedlings. On the other hand, significantly larger populations of *P. syringae* pv. *glycinea* developed on germinating seeds of a susceptible soybean genotype compared to a resistant one (Laurence & Kennedy 1974). It is assumed that the establishment of the bacteria on surfaces of soybean buds may be related to this population on germinating seeds.

The above-described experiments indicate that there is a differential host response to bacterial movement from one plant part to another. Movement is possibly related to 'physical force' rather than a selective stress expressed by the host genotypes. Moist conditions favoured epiphytic establishment (Fryda & Otta 1978). Ridé *et al.* (1978) found that during dry periods and at high temperatures susceptible and resistant poplar clones maintained equal populations of *X. populi*.

*Site specificity.* Although the foliar bacterial plant pathogens establish themselves as epiphytes, it is not quite certain how they are distributed on the leaf surfaces. This can be of importance in relation to the entry through wounds or other openings. Leben *et al.* (1970) indicated that *P. syringae* pv. *syringae* was recovered from terminal buds and from newly unfurled leaves and stipules of healthy bean seedlings 9 to 22 days after buds had been inoculated. They also showed by leaf prints that *P. syringae* pv. *syringae* spread over the wet leaf lamina after the lamina had been inoculated at one point near the petiole junction.

More detailed site colonization was demonstrated recently (Mew *et al.* 1984). On rice leaves, the differential multiplication of virulent and avirulent *Xanthomonas campestris* pv. *oryzae* strains to rice cultivars was noted. The bacterial pathogen causes bacterial blight of rice and invades the host plants either through wounds or hydathodes. Hydathodes, 1-5 mm in length, are normally located near the edge of the rice leaf and appear as 'white streaks' to the naked eyes. These hydathodes predominantly occur toward the tip of a leaf and are densely distributed along the edge of the upper surface of leaf blades (Zee 1981; Mew *et al.* 1984). Mew *et al.* (1984) indicated that they did not observe marked differences in numbers of bacterial cells 1 h after inoculation on all cultivar-strain combinations used. Twenty-four to 48 h later, the virulent strain multiplied immediately outside the water pores and some bacteria gained entrance through these pores, while the avirulent strain did not increase in number significantly during the time. They observed the bacterial cells only on the water pores and not on other surface areas nor the stomata. The data indicate that *X. campestris* pv. *oryzae* was not distributed randomly on the rice leaf surface thus suggesting specific sites for epiphytic colonization of foliar plant pathogens.

To further confirm if foliar bacterial plant pathogens colonize only specific sites on leaf surfaces, we studied the epiphytic behaviour of *X. campestris* pv. *oryzicola*, the causal bacterium of rice leaf streak. The bacteria invade the parenchymatous tissues. They were found on and in the vicinity of the stomata, including the papillae of the guard cells, trichome bases and other parts of the leaf surface covered with wax particles (Fig. 2). The bacteria did not, however, colonize the water pores which are free from wax particles. Morphogenetically, water pores are transformed from stomata in rice (Xu & Xu 1982). There are water pores in which one of the two giant guard cells still retain the features of the guard cells of a stoma, including the papillae and wax particles (Fig. 3). Interestingly, *X. campestris* pv. *oryzicola* also colonizes those with wax particles. The observation has suggested specific niches of the rice leaf surface for colonization by the two bacterial pathogens of rice (Fig. 4).

*Mechanisms for preferential colonization.* There is evidence that host genotypes influence the preferential colonization of foliar bacterial plant pathogens on leaf surfaces. There are sites in which one bacterial pathogen will and another will not colonize, thus, development of specific niches on plant surfaces undoubtedly is an evolutionary process. Lindemann *et al.* (1984) studied *P. syringae* pv. *syringae* on snap bean and suggested that in areas where the host plants were widely and continuously cultivated the epiphytic population should have some selective advantage over other bacteria on host and neighbouring non-host plants when they are competing for bean leaves. The same hypothesis can suggest that virulent bacterial strains are more likely to predominate on susceptible than on resistant host genotypes. The preferential establisment of race 2 of *P. syringae* pv. *morsprunorum* on cherry cultivar Roundel and that of race 1 (the typical race) on cv. Napoleon (Freigoun & Crosse 1975) points to this effect of the host on the epiphytic population of the pathogen. Similar evidence was provided by the studies on the differential colonization of *P. syringae* pv. *glycinea* on soybean cultivars (Mew & Kennedy 1971, 1982), *X. campestris* pv. *oryzae* races on rice (Mew *et al.* 1984), *X. populi* on susceptible and resistant host genotypes of poplars (Ridé *et al.* 1978), *P. syringae* pv. *morsprunorum* and pv. *syringae* on cherry cultivars (Crosse 1963; Sobiczewski 1978) and cherry and pear (Ercolani 1969), *P. syringae* pv. *syringae* on snap bean (Daub & Hagedorn 1981), and *P. phaseolicola* on bean (Stadt & Saettler 1981).

Our recent studies with *Xanthomonas campestris* pv. *oryzae* and *X. campestris* pv. *oryzicola* have demonstrated that there are specific niches on leaf surfaces of rice for colonization by the two bacteria (Table 1). Site specificity appears related to the nature of the pathogenesis and the natural openings that the bacterial pathogens can enter if wounds are absent. *X. campestris* pv. *oryzae* seems to lack the ability to colonize rice leaf areas covered with wax particles and cannot establish full residency on 'non-receptive' sites. In contrast, *X. campestris* pv. *oryzicola*, a parenchymatous pathogen, is found everywhere except on the

Fig. 2. *Xanthomonas campestris* pv. *oryzicola* colonizing a stoma and its vicinity including the papillae of the guard cells (A) and a trichome base (B) on leaf surfaces of rice cultivar IR58. Bars represent 10 μm.

water pores and the guard cells free from wax.

Crosse (1963) speculated that the fluctuations in epiphytic populations of *P. syringae* pv. *morsprunorum* on resistant cherry cultivars may be due to the nutritional status of the leaf surfaces and the occurrence of toxic substances and changeable microbial components residing on the

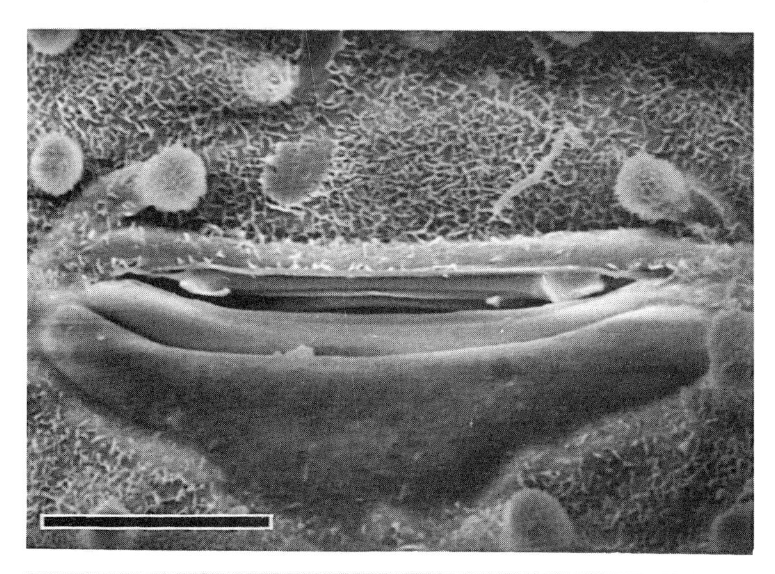

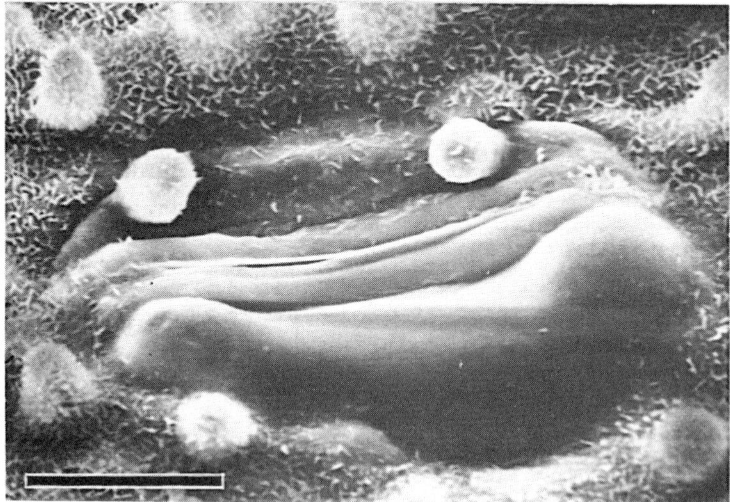

Fig. 3. Water pores along hydathodes on the leaf surface of rice cv. TR58 showing two giant guard cells. One of them is still covered with wax particles thus retaining the features of the guard cells of a stoma including the papillae and wax particles. Bars represent 10 µm.

leaf surfaces. Ridé *et al.* (1978) showed that the level of non-pathogenic bacteria on the poplar leaf surfaces was higher on resistant clones than on susceptible clones. They also showed that a substance excreted

Table 1. Colonization of receptive sites on the leaf surface of the susceptible rice cultivars CAS 209 and IR58 by *Xanthomonas campestris* pvs *oryzae* and *oryzicola*, respectively.

| Receptive site | Pv. *oryzae* | Pv. *oryzicola* |
|---|---|---|
| Water pore | +[1] | - |
| Stoma | - | + |
| Glandular trichome | - | + |
| Prickle hair (large and small) | - | + |
| Papilla | - | + |
| Cork cell | - | + |

[1] +: presence of cells at the site; -: absence of cells at the site.

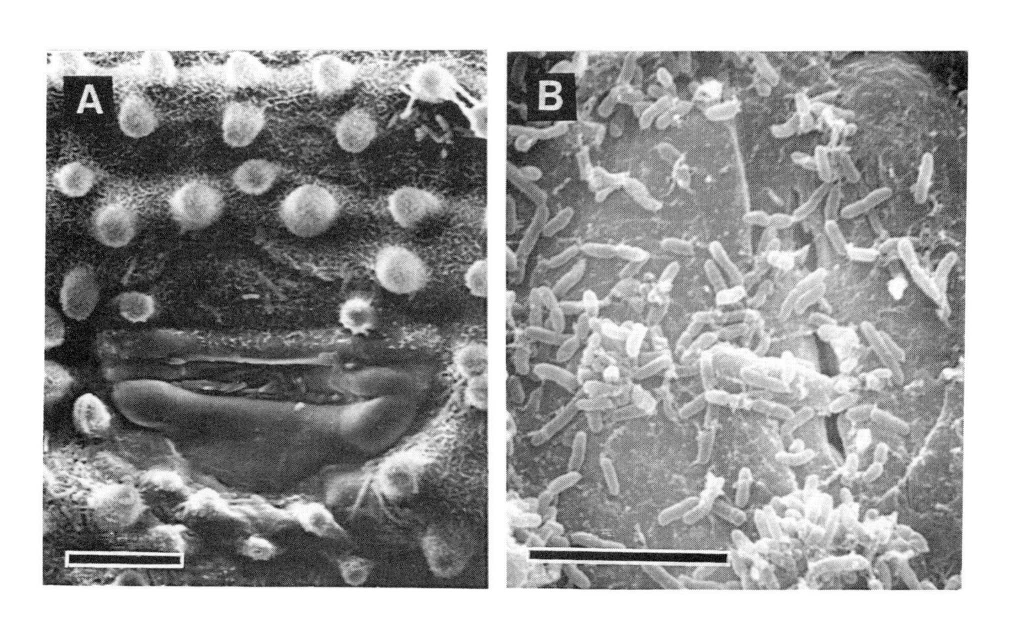


Fig. 4. A. On the leaf surface of rice cv. IR58, *Xanthomonas campestris* pv. *oryzicola* colonized one of the guard cells of the water pore covered with wax particles (bar equals 10 µm). B. *X. campestris* pv. *oryzae* colonized and multiplied at the water pore of the hydathode of rice cv. CAS 209 (bar: 5 µm).

from epidermal cells of the stipules of the resistant clone could exert an antibiotic activity against the canker bacterium, *X. populi*. Sobiczewski (1978) showed that the density of the pathogenic pseudomonads on cherry leaves was completely independent from the associated epiphytic bacterial populations on the cherry cultivars. Other mechanisms are likely to be involved.

Studying a biotrophic fungal system, Niks (1981, 1983 a,b) observed that on resistant barley genotypes and on wheat, *Puccinia hordei* may form appressoria, but that early abortion of the colony immediately after penetration occurred. On another non-host plant, lettuce, *P. hordei* would not form appressoria. He postulated a lack of the stimuli required to direct the germ tubes toward the stomata, the site of penetration. This may be a common defense mechanism of non-host plants to most biotrophic fungal pathogens penetrating through stomata.

In the case of foliar bacterial plant pathogens, the receptive site on the leaf surfaces of susceptible host genotypes may secrete substances that stimulate and direct the bacterial multiplication. Recently, Takahashi & Doke (1984) indicated that unwashed bacterial cells of *Xanthomonas campestris* pv. *citri* were found to adhere preferentially to a wounded portion of citrus leaf tissues rather than to intact leaf surfaces. The washed or unwashed cells of an extracellular polysaccharide-(EPS)-non-producing mutant could not do so on wounded tissues. They proposed that an EPS-agglutinin interaction may cause the bacterial adhesion by reacting to some substances exposed in the wounded area of host leaf tissues.

On a resistant host-parasite system, exudates from water pores of rice may immobilize the bacterial cells by embedding them, as a defense mechanism, to prevent the cells from entering the infection courts (Mew *et al.* 1984). Immobilization was not observed on a susceptible host-parasite system. The exudates therefore appear to determine incompatibility to colonization at the receptive site. When *X. campestris* pv. *oryzicola* colonized the leaf surface of a susceptible host genotype, the bacterial cells were observed to colonize the base of trichomes, the stomata and their vicinity. It appears that the bacteria could colonize the rice leaf surface covered with wax particles except the water pores. There are three kinds of trichomes on a rice leaf, viz. the glandular trichomes and two types of prickle hairs, one small and one large hair located along the cork cells. As they are easily broken off, secretion was often observed, especially in the glandular trichomes. Cells of *X. campestris* pv. *oryzicola* often colonized the secretion and formed small colonies. These secretions appeared to have a stimulating effect on virulent bacteria but on a resistant host genotype the bacterial cells appear to become embedded in the wax (Fig. 5).

The preferential colonization by a foliar bacterial plant pathogen on the leaf surface of a resistant and susceptible host genotype could be due to agglutinants such as those shown on the wounded leaf tissues of citrus to *X. campestris* pv. *citri*. The degree of inhibition of the virulent and avirulent strains of the bacterial pathogen may be related to the degree of resistance of the host. The substances responsible for such agglutination are likely components of the secretion from trichomes, water pores or epidermal cells. The relationship of receptive and non-receptive sites on leaf surfaces and the mechanisms for prefer-

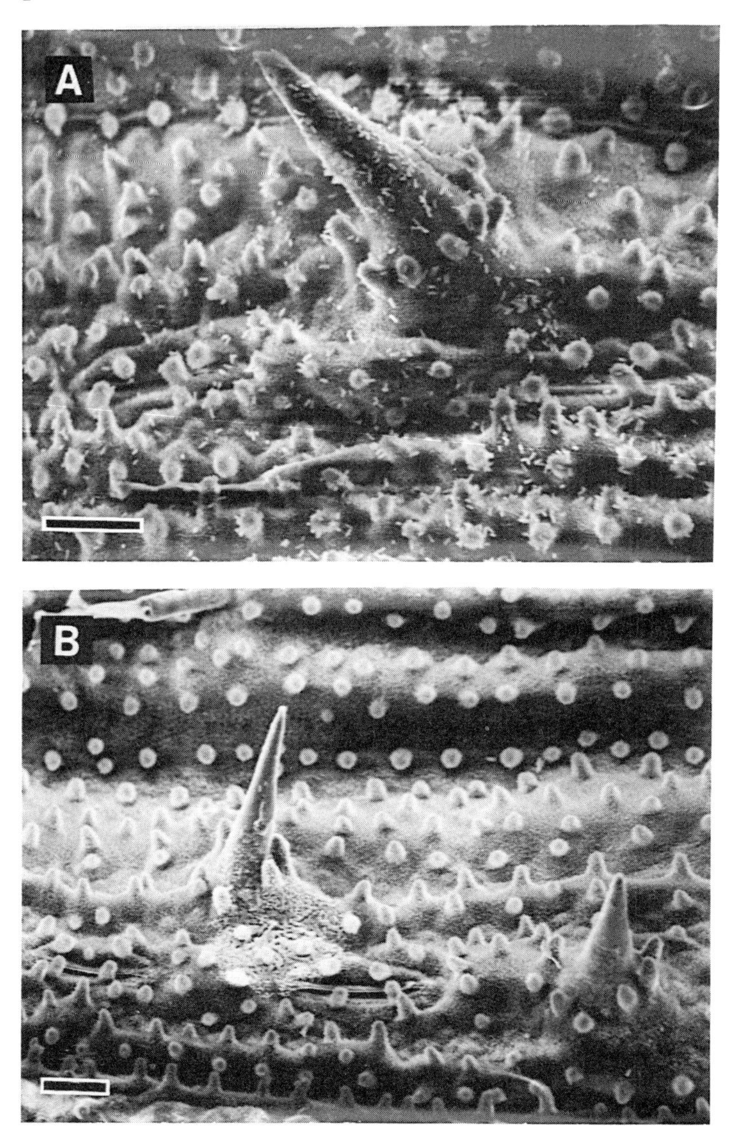


Fig. 5. Scanning electron micrographs of the leaf surface of rice cultivar IR58 inoculated with *Xanthomonas campestris* pv. *oryzicola*. A. Distribution of the bacterial cells of virulent strain BLS 127. B. The weakly virulent strain BLS 117 appeared to be localized and embedded in the wax particles of the leaf surface. Bars represent 10 μm.

ential colonization is postulated in Table 2.

*CONCLUSION*

It is well established that some plant-pathogenic bacteria can colonize the phyllosphere as inhabitants before infection takes place. Bacteria multiply on the plant surface and form a potential inoculum; very often this multiplication may be independent of disease development. Some bacterial species, especially *P. syringae*, are widely distributed as epiphytes on many host and non-host plants. However, no studies have been conducted to clearly establish if pathogenic bacteria on non-host plants behave as inhabitants or as invaders.

Host genotype influences the colonization of plant-pathogenic bacteria at specific site(s) on leaf surfaces. A degree of host specificity is indicated in some cases, but a few exceptions have also been reported. On rice, virulent strains of *X. campestris* pv. *oryzae* appear more able than avirulent strains to colonize and multiply on a compatible host genotype. *X. campestris* pv. *oryzae* colonizes specifically the water pores on the rice leaf surface prior to infection. Similar studies with *X. campestris* pv. *oryzicola*, a pathogen of parenchymatous tissue of rice leaves, indicate that there are specific leaf surface niches for colonization in comparison to *X. campestris* pv. *oryzae* as epiphytes before infection occurs. *X. campestris* pv. *oryzae* colonizes only the water pores or the vicinity of the water pores. *X. campestris* pv. oryzicola colonizes the stomata or the vicinity of the stomata including the papillae of the guard cells of the stoma, trichome bases, and other parts of the leaf surface covered with wax particles. *X. campestris* pv. *oryzicola* does not colonize the water pores, where wax particles are not found.

Table 2. Postulated relationship between foliar bacterial plant pathogens and colonization of receptive sites of resistant and susceptible host genotypes[1].

| Host genotype | Receptive sites | |
|---|---|---|
| | Compatible | Incompatible |
| Resistant | -[2] | - |
| Susceptible | + | - |

[1] Foliar bacterial pathogens may also colonize non-host plants but they may occupy the sites temporarily as invaders.

[2] -: Inhibitory effect on colonization; +: stimulating effect on colonization.

On the non-receptive site of a susceptible host genotype, the bacterial cells appear to be aggregated and immobilized. The avirulent strain was also immobilized on the receptive site of a resistant host genotype. Phytopathogenic bacteria on host plants are stimulated on receptive sites (or infection courts) on leaf surfaces while there may be an inhibitory mechanism on non-host or on incompatible host genotypes.

## *REFERENCES*


Crosse, J.E. (1959). Bacterial canker of stone-fruits. IV. Investigation of a method for measuring the inoculum potential of cherry trees. Annals of Applied Biology, *47*, 306-17.

Crosse, J.E. (1963). Bacterial canker of stone-fruits. V. A comparison of leaf-surface populations of *Pseudomonas mors-prunorum* in autumn on two cherry varieties. Annals of Applied Biology, *52*, 97-104.

Daub, M.E. & Hagedorn, D.J. (1981). Epiphytic populations of *Pseudomonas syringae* on susceptible and resistant bean lines. Phytopathology, *71*, 547-50.

Ercolani, G.L. (1969). Sopravvivenza epifitica di popolazioni di *Pseudomonas mors-prunorum* Wormald da Ciliego e di *P. syringae* van Hall da Pero sulla pianta ospite di provenienza e sull'altra pianta. Phytopathologia Mediterranea, *8*, 197-206.

Ercolani, G.L., Hagedorn, D.J., Kelman, A. & Rand, R.E. (1974). Epiphytic survival of *Pseudomonas syringae* on hairy vetch in relation to epidemiology of bacterial brown spot of bean in Wisconsin. Phytopathology, *64*, 1330-9.

Freigoun, S.O. & Crosse, J.E. (1975). Host relations and distribution of a physiological and pathological variant of *Pseudomonas morsprunorum*. Annals of Applied Biology, *81*, 317-30.

Fryda, S.J. & Otta, J.D. (1978). Epiphytic movement and survival of *Pseudomonas syringae* on spring wheat. Phytopathology, *68*, 1064-7.

Garrett, S.D. (1970). Pathogenic Root-infecting Fungi. Cambridge: Cambridge University Press.

Hagedorn, D.J. & Patel, P.N. (1965). Halo blight and bacterial brown spot of bean in Wisconsin in 1964. Plant Disease Reporter, *49*, 591-5.

Hirano, S.S. & Upper, C.D. (1983). Ecology and epidemiology of foliar bacterial plant pathogens. Annual Review of Phytopathology, *21*, 243-69.

Hoitink, H.A.J., Hagedorn, D.J. & McCoy, E. (1968). Survival, transmission, and taxonomy of *Pseudomonas syringae* van Hall, the causal organism of bacterial brown spot of bean (*Phaseolus vulgaris* L.). Canadian Journal of Microbiology, *14*, 437-41.

Klement, Z. & Goodman, R.N. (1967). The hypersensitive reaction to infection by bacterial plant pathogens. Annual Review of Phytopathology, *5*, 17-44.

Laurence, J.A. & Kennedy, B.W. (1974). Population changes of *Pseudomonas glycinea* on germinating soybean seeds. Phytopathology, *64*, 1470-1.

Leben, C. (1965). Epiphytic microorganisms in relation to plant disease. Annual Review of Phytopathology, *3*, 209-30.

Leben, C., Rusch, V. & Schmitthenner, A.F. (1968). The colonization of soybean buds by *Pseudomonas glycinea* and other bacteria. Phytopathology, *58*, 1677-81.

Leben, C., Schroth, M.N. & Hildebrand, D.C. (1970). Colonization and movement of *Pseudomonas syringae* on healthy bean seedlings. Phytopathology, *60*, 677-80.

Lindemann, J., Arny, D.C. & Upper, C.D. (1984). Epiphytic populations of *Pseudomonas syringae* pv. *syringae* on snap bean and nonhost plants and the incidence of bacterial brown spot disease in relation to cropping patterns. Phytopathology, *74*, 1329-33.

Lindow, S.E., Arny, D.C. & Upper, C.D. (1978). Distribution of ice nucleation-active bacteria on plants in nature. Applied and Environmental Microbiology, *36*, 831-8.

Mew, T.W. & Kennedy, B.W. (1971). Growth of *Pseudomonas glycinea* on the surface of soybean leaves. Phytopathology, *61*, 715-6.

Mew, T.W. & Kennedy, B.W. (1982). Seasonal variation in populations of pathogenic pseudomonads on soybean leaves. Phytopathology, *72*, 103-5.

Mew, T.W., Mew, I.C. & Huang, J.S. (1984). Scanning electron microscopy of virulent and avirulent strains of *Xanthomonas campestris* pv. *oryzae* on rice leaves. Phytopathology, *74*, 635-41.

Niks, R.E. (1981). Appressorium formation of *Puccinia hordei* on partially resistant barley and two non-host species. Netherlands Journal of Plant Pathology, *87*, 201-7.

Niks, R.E. (1983 a). Comparative histology of partial resistance and the nonhost reaction to leaf rust pathogens in barley and wheat seedlings. Phytopathology, *73*, 60-4.

Niks, R.E. (1983 b). Haustorium formation by *Puccinia hordei* in leaves of hypersensitive, partially resistant, and non-host plant genotypes. Phytopathology, *73*, 64-6.

Persley, G.J. (1978). Epiphytic survival of *Xanthomonas manihotis* in relation to the disease cycle of cassava bacterial blight. Proceedings of the 4th International Conference on Plant Pathogenic Bacteria, ed. Stat. Path. Végét. Phytobacteriologie, pp. 773-7. Beaucouzé: Inst. Nat. Rech. Agron.

Ridé, M., Ridé, S. & Poutier, J.-C. (1978). Factors affecting the survival of *Xanthomonas populi* on aerial structures of poplar. Proceedings of the 4th International Conference on Plant Pathogenic Bacteria, ed. Stat. Path. Végét. Phytobacteriologie, pp. 803-14. Beaucouzé: Inst. Nat. Rech. Agron.

Sobiczewski, P. (1978). Epiphytic populations of pathogenic pseudomonads on sour cherry leaves. Proceedings of the 4th International Conference on Plant Pathogenic Bacteria, ed. Stat. Path. Végét. Phytobacteriologie, pp. 753-62. Beaucouzé: Inst. Nat. Rech. Agron.

Stadt, S.J. & Saettler, A.W. (1981). Effect of host genotype on multiplication of *Pseudomonas phaseolicola*. Phytopathology, *71*, 1307-10.

Takahashi, T. & Doke, N. (1984). A role of extracellular polysaccharides of *Xanthomonas campestris* pv. *citri* in bacterial adhesion to citrus leaf tissues in preinfectious stage. Annals of the Phytopathological Society of Japan, *50*, 565-73.

Xu, S.H. & Xu, X.P. (1982). [The Morphology and Anatomy of Rice (in Chinese).] Beijing: Agriculture Publication.

Zee, S.Y. (1981). Wheat and Rice Plants. A Scanning Electron Microscope Survey. Hongkong: Cosmos Book Ltd.

# MYCOPLASMAS ON PLANT SURFACES

R.E. McCoy

*Champlain Isle Agro Associates, East Shore Road, Isle La Motte, VT 05463, USA*

## *INTRODUCTION*

Bacteria, fungi and higher organisms have traditionally been associated with the plant surface; however, it is only recently that mycoplasmas have been found occupying this niche. Mycoplasmas, prokaryotic organisms of the division Tenericutes (Table 1), have largely been overlooked because of their small size, exacting nutritional requirements, and relatively recent introduction to plant microbiologists.

The initial reports of plant-mycoplasma associations were largely supported by ultrastructural evidence and were always related to the induction of disease by the mycoplasma or mycoplasma-like organism (MLO). The habitat of these potential pathogens was apparently limited to living sieve tube elements of the phloem of angiosperms (McCoy 1983, 1984). Attempts to isolate and culture plant-pathogenic mycoplasmas have been successful only for the helical forms of the genus *Spiroplasma* (Whitcomb 1981). Taxonomic characterization of the non-helical MLOs has not been possible in the absence of cultural confirmation of their identity; all attempts to isolate these organisms have ended in failure. Although a number of early mycoplasma isolates were obtained from diseased plants (Hampton *et al.* 1969; Lin *et al.* 1970; Daniels & Meddins 1971; Giannotti & Vago 1971; Kleinhempel *et al.* 1972), characterization demonstrated them to be known animal-inhabiting species of the genera *Acholeplasma* or *Mycoplasma*. None of the isolates was demonstrated to be plant-pathogenic. In retrospect, it seems likely that some of the isolates may have represented organisms present on the plant surface.

The first definitive isolation of mycoplasmas from plant surfaces was reported by Davis (1978) when he cultured both helical and non-helical forms from tulip tree (*Liriodendron tulipifera*) flowers. Surface sterilization prevented such isolations. This initial report was followed by a host of collaborative findings from other regions and using different plants and plant parts. The discovery of these largely heretofore unknown mycoplasmas has opened a new chapter in mycoplasmal taxonomy, and has posed some intriguing new questions pertaining to the ecological relationships of these organisms.

## *THE MYCOPLASMAS*

By way of introduction, the mycoplasmas are the smallest, simplest cellular organisms known. They are characterized by the lack of a cell wall and are bounded only by a single membrane. Mycoplasmas are extraordinarily small and generally polymorphic. They are visible with the light microscope at only the highest magnifications using phase contrast or dark field optics. Due to cellular plasticity most mycoplasma cells can pass through 0.22-μm pore diameter filters, and, therefore, they pose severe problems as contaminants in certain sterile filtration operations. As they lack a cell wall the mycoplasmas are generally insensitive to penicillin and other cell wall-inhibiting antibiotics. This also renders them extremely labile in solutions of low osmolarity. Mycoplasma cells contain chromatin and ribosomes and do not characteristically have any other inclusions. Recent reviews on the biology of the plant and insect-inhabiting mycoplasmas have been given by Whitcomb (1981), McCoy (1981, 1984) and Bové (1984).

### *ISOLATION AND IDENTIFICATION*

Surface mycoplasmas may be isolated from flowers, extrafloral nectaries or other plant parts by the following simple procedure (Davis 1978; McCoy *et al.* 1979). Samples are collected in plastic bags without touching by hand. The samples are then rinsed briefly in liquid growth medium, and the medium is passed through a 0.45-μm pore diameter filter to remove bacteria and larger contaminating organisms. Aliquots of 2 to 3 drops of filtrate are then added to tubes containing 2 to 5 ml sterile medium. The inoculated tubes may then be incubated at *ca* 30°C

Table 1. Phylogenetic relationships of the mycoplasmas to other prokaryotes. From Murray (1984) and Tully (1985).

| Classification | Characteristics |
|---|---|
| Kingdom: Prokaryotae | |
| Division I: Gracilicutes | the Gram-negative bacteria |
| Division II: Firmicutes | the Gram-positive bacteria |
| Division III: Tenericutes | the mycoplasmas[1] |
| Order: Mycoplasmatales | require cholesterol |
| Family: Mycoplasmataceae | polymorphic |
| Genus: *Mycoplasma* | |
| Genus: *Ureaplasma* | require urea |
| Genus: *Spiroplasma* | helical morphology |
| Order: Acholeplasmatales | no cholesterol requirement |
| Family: Acholeplasmataceae | |
| Genus: *Acholeplasma* | |
| Genus of uncertain affiliation: *Anaeroplasma* | |
| Division IV: Mendosicutes | the archaeobacteria |

[1] The trivial term mycoplasma is commonly used to refer to all members of the Tenericutes.

until growth is observed, usually for 2 to 15 days for primary isolations. Due to their small size, most mycoplasmas do not produce turbidity in broth. Therefore, phenol red (phenolsulfonphthalein) is often added to broth as an indicator of growth. Broth which has changed colour should be observed at 1000 x with phase contrast optics to determine if the culture contains bacteria or helical or non-helical mycoplasmas.

The mycoplasmas isolated from plant surfaces are generally less nutritionally exacting than those which are pathogenic to animals or plants. However, the use of a more complex medium will allow a greater number of strains to be isolated and growth will be more rapid (McCoy *et al.* 1979; Bové *et al.* 1983). MC broth is an inexpensive medium that will support growth of a number of plant surface-associated mycoplasmas (McCoy *et al.* 1979). This medium contains Difco (Detroit, MI) PPLO broth plus 20% horse serum, 2% fresh yeast extract, 0.12 M sucrose, and 0.28 mM phenol red. The horse serum and yeast extract should be added aseptically after autoclaving the base medium. Prior heat treatment of the serum for 1 hour at 50°C will eradicate any potentially contaminating animal mycoplasmas. Mycoplasma media should never be pipetted by mouth as this may introduce another source of contaminating mycoplasmas. A second, more complex, medium that will yield superior results in primary isolations of plant surface mycoplasmas is SP-4 medium (Tully *et al.* 1977).

The first step in identifying a wall-free prokaryote is to observe cellular morphology with phase contrast optics at 1000x to determine if the organism is helical and therefore belongs to the genus *Spiroplasma*. Identification beyond this stage is primarily serological or biochemical and the reader is referred to Razin & Tully (1983) for an excellent book on mycoplasma characterization.

## *MYCOPLASMAS FROM PLANT SURFACES*

The initial isolation of mycoplasmas from plant surfaces came about following the discovery that a newly discovered pathogen of honey-bees was a spiroplasma (Clark 1977). It was postulated that this organism, *Spiroplasma melliferum* (Clark *et al.* 1985), might occur on flowers visited by bees. Davis (1978) then made systematic isolations from tulip tree flowers to search for this spiroplasma. His results were positive in that spiroplasmas were isolated, but the honey-bee spiroplasma was not obtained. Instead, spiroplasmas of two previously undescribed serogroups were isolated (Muniyappa & Davis 1980). These organisms have since been named *S. floricola* (Davis 1981), and *S. apis* (Mouches *et al.* 1983). Later isolations from Florida did, however, finally turn up the elusive *S. melliferum* on flower surfaces (Davis 1981).

The first corroborating report of flower surface mycoplasmas came from Florida (McCoy *et al.* 1979). The Florida isolates included a spiroplasma related to *S. apis* from *Calliandra haematocephalus* flowers (Tully *et al.* 1980; McCoy *et al.* 1982), a new species, *Acholeplasma florum* (Stephens *et al.* 1981; McCoy *et al.* 1984) from flowers of *Citrus* sp., and an as yet unnamed *Mycoplasma* sp. also from *Citrus* flowers. Additional flower isolates were reported from Europe (Junca *et al.* 1980; Vignault *et al.* 1980), and central and eastern United States (Whitcomb *et al.* 1982). These included *S. apis*, *S. floricola*, and a new spiroplasma placed in subgroup I-6 and termed Maryland flower spiroplasma (MFS). Also, in the United States a non-helical strain reported by Whitcomb *et al.* (1982) was determined to be related to *A. florum* by serological techniques. Davis (1978) also mentioned that several non-helical mycoplasmas had been isolated from tulip tree flowers. Two of these tested by the growth inhibition serological assay were also identified as *A. florum* (McCoy, unpublished results).

In addition to the isolations from flower surfaces, a number of mycoplasma strains were isolated from a variety of plants and plant parts since 1979. Eden-Green & Tully (1979) reported the isolation of strains of *A. axanthum* and *A. oculi* from senescent or dead bud tissues of coconut palms (*Cocos nucifera*) affected by lethal yellowing, a disease associated with MLOs in the vascular system. These organisms normally exist as mammalian or avian parasites. It was suggested that they occurred in rotting palm tissues as epiphytes or saprophytes. Further isolations from healthy and diseased coconut palm bud tissues yielded a spiroplasma distinct from other spiroplasma strains (Eden-Green & Waters 1981; Eden-Green *et al.* 1983). The occurrence of this spiroplasma in both healthy and diseased palms would tend to preclude its designation as a pathogen.

In another series of isolations from plants, *A. laidlawii*, *A. axanthum*, *A. oculi* and an untyped acholeplasma were isolated from broccoli (*Brassica oleracea* var. *italica*) and kale (*B. oleracea* var. *acephala*) (Somerson *et al.* 1982). In addition, *M. verecundrum* was isolated from kale. All of these species are normally considered to be animal parasites and their presence on plant tissue was surprising.

## *ISOLATION OF PLANT SURFACE MYCOPLASMAS FROM INSECTS*

While the discovery of *S. melliferum*, causal agent of honey-bee spiroplasmosis, initiated the search for mycoplasmas on flower surfaces, it also stimulated research into the role played by insects in the ecology of these organisms. It has been known for some time that leafhoppers served as alternate hosts and vectors of the apparently plant-pathogenic MLOs. Now it appears that insects may play a similar if not dominant role in the ecology of the mycoplasmas found on plant surfaces.

In reverse order, *S. apis* was first found on flower surfaces, and later was found to be the pathogen of May disease of honey-bees in France (Mouches *et al.* 1982, 1983). Similarly, the MFS, isolated from 10% of *Eupatorium maculatum* flowers examined in Maryland, was found in 12 to 38 tiger swallowtail butterflies (*Pterourus glaucus*) examined (Hackett *et al.* 1984). McCoy *et al.* (1981) found a range of spiroplasmas to multiply in hemolymph of the wax moth (*Galleria mellonella*). These included the plant pathogens *S. citri* and the corn stunt spiroplasma, the tick spiroplasma, *S. mirum*, and the flower spiroplasmas *S. apis* and *S. floricola*. Dowell *et al.* (1981) reported that of the above spiroplasmas, only the flower forms were strongly pathogenic to the wax moth.

In an effort to look into the relationships of the flower surface mycoplasmas to visiting insects, J.H. Tsai and R.E. McCoy (unpublished results) made isolations from insects visiting *Melaleucca* sp. flowers in Florida. Of 100 insects examined 30 isolations were made. These included 14 isolations of *S. melliferum* from 23 honey-bees collected. This species was also isolated once from an anthophorine bee (*Xylocopa micans*). *X. micans* also yielded *S. floricola* and *Mycoplasma* sp. strain M1. The majority of *S. floricola* and *S. apis* isolates came from the bee *Agapostemon splendeus*. Surprisingly, all *A. florum* isolates came from flies (*Cochliomyia macellaria*). This grouping indicates a strong host preference for these mycoplasma strains.

Clark (1984) classified mycoplasma infections of insects into four groupings:
Type A: temporary gut infection or contamination. He suggested that the majority of insect-spiroplasma associations were of this type, especially among those insects which visit flowers.
Type B: permanent gut infection such as he found for *S. floricola* in empidid flies (*Trichina* sp.) and for a spiroplasma in Colorado potato beetle (Clark *et al.* 1982).
Type C: gut infection with hemolymph invasion such as found with *S. apis* and *S. melliferum* in honey-bee, and for the plant MLOs in leafhoppers.
Type D: hemolymph infection with non-cultivable mycoplasmas in the sex ratio spiroplasma of *Drosophila willistoni*.

## *ECOLOGY OF PLANT SURFACE MYCOPLASMAS*

The mycoplasmas isolated from plant surfaces appear to have an intimate relationship with the visiting insect fauna. The ready isolation of many mycoplasma types from flower surfaces is consistent with this hypothesis, especially in view of the concurrent isolation of many of the same strains from flower-visiting insects. The isolation of acholeplasmas and a spiroplasma in rotting tissues of coconut palm may be due to the common colonization of these tissues by dipteran larvae. The isolation of strains normally found as parasites or pathogens of mammals or birds from coconut palm and vegetables was unexpected and opens several new avenues of investigation. The relationship of mycoplasmas to insects has been known for some time for the plant-pathogenic spiroplasmas and MLOs. In retrospect, it is not surprising to find other insect mycoplasmas on the surface of plants. The exact ecological roles played by these organisms remain an open question.

## *REFERENCES*

Bové, J.M. (1984). Wall-less prokaryotes of plants. Annual Review of Phytopathology, *22*, 361-96.

Bové, J.M., Whitcomb, R.F. & McCoy, R.E. (1983). Culture techniques for spiroplasmas from plants. *In* Methods in Mycoplasmology. Vol. 2. Diagnostic Mycoplasmology, eds J.G. Tully & S. Razin, pp. 225-34. New York: Academic Press.

Clark, T.B. (1977). *Spiroplasma* sp., a new pathogen in honey bees. Journal of Invertebrate Pathology, *29*, 112-3.

Clark, T.B. (1984). Diversity of spiroplasma host-parasite relationships. Israel Journal of Medical Sciences, *20*, 995-7.

Clark, T.B., Whitcomb, R.F. & Tully, J.G. (1982). Spiroplasmas from coleopterous insects: new ecological dimensions. Microbial Ecology, *8*, 401-9.

Clark, T.B., Whitcomb, R.F., Tully, J.G., Mouches, C., Saillard, C., Bové, J.M., Wróblewski, H., Carle, P., Rose, D.L., Henegar, R.B. & Williamson, D.L. (1985). *Spiroplasma melliferum*, a new species from the honeybee (*Apis mellifera*). International Journal of Systematic Bacteriology, *35*, 296-308.

Daniels, M.J. & Meddins, B.M. (1971). Isolation and cultivation of plant mycoplasmas. John Innes Institute, Annual Report, *62*, 99-101.

Davis, R.E. (1978). Spiroplasma associated with flowers of the tulip tree (*Liriodendron tulipifera* L.). Canadian Journal of Microbiology, *24*, 954-9.

Davis, R.E. (1981). The enigma of the flower spiroplasmas. *In* Mycoplasma Diseases of Trees and Shrubs, eds K. Maramorosch & S.P. Raychaudhuri, pp. 259-79. New York: Academic Press.

Dowell, R.V., Basham, H.G. & McCoy, R.E. (1981). Influence of five spiroplasma strains on growth rate and survival of *Galleria mellonella* (Lepidoptera: Pyralidae) larvae. Journal of Invertebrate Pathology, *37*, 231-5.

Eden-Green, S.J., Archer, D.B., Tully, J.G. & Waters, H. (1983). Further studies on a spiroplasma isolated from coconut palms in Jamaica. Annals of Applied Biology, *102*, 127-34.

Eden-Green, S.J. & Tully, J.G. (1979). Isolation of *Acholeplasma* spp. from coconut palms affected by lethal yellowing disease in Jamaica. Current Microbiology, *2*, 311-6.

Eden-Green, S.J. & Waters, H. (1981). Isolation and preliminary characterization of a spiroplasma from coconut palms in Jamaica. Journal of General Microbiology, *124*, 263-70.

Giannotti, J. & Vago, C. (1971). Rôle des mycoplasmes dans l'étiologie de la phyllodie du trèfle: culture et transmission expérimentale de la maladie. Physiologie Végétale, *9*, 541-53.

Hackett, K., Clark, T.B., Hicks, A., Whitcomb, R.F., Lowry, E. & Batra, S.W.T. (1984). Occurrence and frequency of subgroup I-6 spiroplasma in arthropods associated with old fields in Maryland and Virginia. Israel Journal of Medical Sciences, *20*, 1006-8.

Hampton, R.O., Stevens, J.O. & Allen, T.C. (1969). Mechanically transmissible mycoplasma from naturally infected peas. Plant Disease Reporter, *53*, 499-503.

Junca, P., Saillard, C., Tully, J., Garcia-Jurado, O., Degorce-Dumas, J.-R., Mouches, C., Vignault, J.-C., Vogel, R., McCoy, R., Whitcomb, R., Williamson, D., Latrille, J. & Bové, J.M. (1980). Caractérisation de spiroplasmes isolés d'insectes et de fleurs de France continental, de Corse et du Maroc. Proposition pour une classification des spiroplasmes. Comptes Rendus Hebdomadaires des Séances de l'Académie des Sciences, Série D, *290*, 1209-12.

Kleinhempel, H., Müller, H.M. & Spaar, D. (1972). Isolierung und Kultivierung von Mycoplasmatales aus Weissklee mit Blutenvergrünungssymptomen. Archiv für Pflanzenschutz, *8*, 361-70.

Lin, S.C., Lee, C.S. & Chiu, R.J. (1970). Isolation and cultivation of, and inoculation with, a mycoplasma causing white leaf disease of sugarcane. Phytopathology, *60*, 795-7.

McCoy, R.E. (1981). Wall-free prokaryotes of plants and invertebrates. *In* The Prokaryotes. A Handbook on Habitats, Isolation, and Identification of Bacteria, eds M.P. Starr, H. Stolp, H.G. Trüper, A. Balows & H.G. Schlegel, pp. 2238-46. Berlin: Springer-Verlag.

McCoy, R.E. (1983). Proposed subgroups of spiroplasmas of high guanine plus cytosine content, group IV. Yale Journal of Biology and Medicine, *56*, 593

-8.
McCoy, R.E. (1984). Mycoplasma-like organisms of plants and invertebrates. *In* Bergey's Manual of Systematic Bacteriology. Vol. 1, 9th edition, eds N.R. Krieg & J.G. Holt, pp. 792-3. Baltimore: Williams & Wilkins.

McCoy, R.E., Basham, H.G. & Davis, R.E. (1982). Powder puff spiroplasma: a new epiphytic mycoplasma. Microbial Ecology, *8*, 169-80.

McCoy, R.E., Basham, H.G., Tully, J.G., Rose, D.L., Carle, P. & Bové, J.M. (1984). *Acholeplasma florum*, a new species isolated from plants. International Journal of Systematic Bacteriology, *34*, 11-5.

McCoy, R.E., Davis, M.J. & Dowell, R.V. (1981). In vivo cultivation of spiroplasmas in larvae of the greater wax moth. Phytopathology, *71*, 408-11.

McCoy, R.E., Williams, D.S. & Thomas, D.L. (1979). Isolation of mycoplasmas from flowers. *In* Proceedings of Republic of China-United States Cooperative Science Seminar on Mycoplasma Diseases of Plants, eds H.J. Su & R.E. McCoy, pp. 75-81. Taipei: National Science Council.

Mouches, C., Bové, J.M., Albisetti, J., Clark, T.B. & Tully, J.G. (1982). A spiroplasma of serogroup IV causes a May-disease-like disorder of honeybees in southwestern France. Microbial Ecology, *8*, 387-99.

Mouches, C., Bové, J.M., Tully, J.G., Rose, D.L., McCoy, R.E., Carle-Junca, P., Garnier, M. & Saillard, C. (1983). *Spiroplasma apis*, a new species from the honey bee (*Apis mellifera*). Annales de Microbiologie, *A134*, 383-97.

Muniyappa, V. & Davis, R.E. (1980). Occurrence of spiroplasmas of two serogroups on flowers of the tulip tree (*Liriodendron tulipifera* L.) in Maryland. Current Science, *49*, 58-60.

Murray, R.G.E. (1984). Kingdom Procaryotae Murray, 1968, $252^{AL}$. *In* Bergey's Manual of Systematic Bacteriology. Vol. 1, 9th edition, eds N.R. Krieg & J.G. Holt, pp. 35-6. Baltimore: Williams & Wilkins.

Razin, S. & Tully, J.G., eds (1983). Methods in Mycoplasmology. Vol. 1. Mycoplasma Characterization. New York: Academic Press.

Somerson, N.L., Kocka, J.P., Rose, D. & Del Guidice, R.A. (1982). Isolation of acheloplasmas and a mycoplasma from vegetables. Applied and Environmental Microbiology, *43*, 412-7.

Stephens, E.B., Aulakh, G.S., McCoy, R.E., Rose, D.L., Tully, J.G. & Barile, M.F. (1981). Lack of genetic relatedness among some animal and plant acholeplasmas by nucleic acid hybridization. Current Microbiology, *5*, 367-70.

Tully, J.G. (1985). International Committee on Systematic Bacteriology Subcommittee on the Taxonomy of *Mollicutes*. Minutes of the Interim Meeting. International Journal of Systematic Bacteriology, *35*, 378-81.

Tully, J.G., Rose, D.L., Garcia-Jurado, O., Vignault, J.-C., Saillard, C., Bové, J.M., McCoy, R.E. & Williamson, D.L. (1980). Serological analysis of a new group of spiroplasmas. Current Microbiology, *3*, 369-72.

Tully, J.G., Whitcomb, R.F., Clark, H.F. & Williamson, D.L. (1977). Pathogenic mycoplasmas: cultivation and vertebrate pathogenicity of a new spiroplasma. Science, *195*, 892-4.

Vignault, J.-C., Bové, J.M., Saillard, C., Vogel, R., Farro, A., Venegas, L., Stemmer, W., Aoki, S., McCoy, R., Al-Beldawi, A.S., Larue, M., Tuzcu, O., Ozsan, M., Nhami, A., Abassi, M., Bonfils, J., Moutous, G., Fos, A., Poutiers, F. & Viennot-Bourgin, G. (1980). Mise en culture de spiroplasmes à partir de matériel végétal et d'insectes provenant de pays circum-méditeranéens et du Proche Orient. Comptes Rendus Hebdomadaires des Séances de l'Académie des Sciences, Série D, *290*, 775-8.

Whitcomb, R.F. (1981). The biology of spiroplasmas. Annual Review of Entomology, *26*, 397-425.

Whitcomb, R.F., Tully, J.G., Rose, D.L., Stephens, E.B., Smith, A., McCoy, R.E. & Barile, M.F. (1982). Wall-less prokaryotes from fall flowers in central United States and Maryland. Current Microbiology, *7*, 285-90.

## SECTION V

# BIOLOGICAL CONTROL ON AERIAL PLANT SURFACES

# STRATEGIES AND PRACTICE OF BIOLOGICAL CONTROL OF ICE NUCLEATION-ACTIVE BACTERIA ON PLANTS

S.E. Lindow

*Department of Plant Pathology, University of California, 147 Hilgard Hall, Berkeley, CA 94720, USA*

## *INTRODUCTION*

This review chapter will discuss mechanisms of bacterial interactions on leaf surfaces while emphasizing those applicable to the biological control of plant frost injury. Although biological control of target organisms important in plant frost injury has only recently been attempted, this model system has many features which have allowed the elucidation of important concepts regarding the manipulation of the microbial communities on leaf surfaces. The practice of manipulation of epiphytic microbial communities to achieve biological control of plant frost injury will be illustrated with recent results of studies of population dynamics of bacteria on treated and untreated almond trees. Important, unresolved questions pertaining to manipulations of leaf surface microbial communities will also be addressed.

## *BACTERIAL ICE NUCLEATION AND PLANT FROST INJURY*

Frost injury is a serious abiotic disease of frost-sensitive plants. The fruits of many subtropical plant species, and the flowers, foliage and fruits of many annual and perennial temperate plants are already damaged at temperatures between 0 and -5°C. Losses in agricultural production in the United States alone are estimated at nearly one billion ($10^9$) dollars annually (White & Haas 1975). Frost-sensitive plants, such as almond, are distinguished from frost-hardy plants by their relative inability to tolerate ice formation within their tissues (Chandler 1958; Mayland & Cary 1970; Burke *et al.* 1976). Ice formed on the surface of frost-sensitive plants spreads rapidly in continuous moisture films to intercellular locations within the plant (Mayland & Cary 1970). Subsequent ice propagation throughout all intercellular locations ultimately leads to the formation of intracellular ice which causes mechanical disruption of plant cell membranes (Burke *et al.* 1976). Frost-sensitive plants have no significant mechanism of frost tolerance, i.e. the restriction of ice propagation only to intercellular locations, and must avoid ice formation to avoid frost injury (Levitt 1972).

Five different bacterial species have the ability to serve as catalysts for ice formation both in vitro as well as on plant surfaces. Most isolates of approximately half of the described pathovars (Dye *et al.* 1980) of *Pseudomonas syringae* van Hall exhibit ice nucleation activity in vitro at temperatures above -5°C (Maki *et al.* 1974; Arny *et al.* 1976;

Hirano *et al.* 1978; Paulin & Luisetti 1978). Most isolates of *Pseudomonas viridiflava* (Burkholder) Paulin & Luisetti 1978 and *Xanthomonas campestris* pv. *translucens* (Jones, Johnson & Reddy) Dye also exhibit ice nucleation activity at temperatures above -5°C (S. Kim, D.C. Sands, C.S. Orser and S.E. Lindow, unpublished results). Some isolates (probably less than 10% of all strains) of *Pseudomonas fluorescens* Migula and *Erwinia herbicola* (Löhnis) Dye also possess ice nucleation activity at temperatures above -5°C (Lindow *et al.* 1978 b; Maki & Willoughby 1978; Yankofsky *et al.* 1981; Makino 1982).

Ice nucleation-active (INA) bacteria have a widespread distribution as epiphytes on plants in temperate zones throughout the world. One or more species of INA bacteria have been found on nearly all non-coniferous plant species investigated. While the population dynamics of INA bacteria on plants have been investigated only in a few cases (see also S.S. Hirano and C.D. Upper, this volume), the population size of INA bacteria measured on plant surfaces at a given time varies widely (Lindow *et al.* 1978 a,b,c, 1982 a, 1983 a,b; Lindow 1982 a,b, 1983 a,b; Makino 1982; Gross *et al.* 1983; Lindow & Connell 1984). Populations of INA bacteria varied from very low levels (undetectable to less than approximately 100 cells $g^{-1}$ fresh weight) to in excess of $10^7$ cells $g^{-1}$ fresh weight on certain tree crops such as pear and almond (Lindow 1982 b, 1983 a; Gross *et al.* 1983; Lindow & Connell 1984) as determined from leaf washings. The population size of INA bacteria on a given plant species has also been reported to vary greatly spatially (Lindemann *et al.* 1984; Andersen & Lindow 1985). Airborne inoculum of INA bacterial species originating on nearby plants was suggested as the determinant for such spatial variations in epiphytic bacterial population size (Lindemann *et al.* 1984; Andersen & Lindow 1985).

The population size of epiphytic INA bacteria on plants determines the likelihood of freezing injury to plants at temperatures above -5 to -6°C. Nearly all plant species investigated do not contain intrinsic ice nuclei active at temperatures above -5 to -6°C (Kaku 1964, 1975; Arny *et al.* 1976; Marcellos & Single 1976, 1979; Lindow *et al.* 1978 a, 1982 a; Rajashekar *et al.* 1983). The difficulty of obtaining axenically grown woody plant tissues makes the origin of ice nuclei reported in field and greenhouse-grown woody tissues, such as peach, more difficult to determine (Ashworth & Davis 1984; Gross *et al.* 1984). The majority of plant species tested, however, has been shown to readily supercool to below -5°C. Exogenous ice-nucleating agents are required to initiate damaging ice formation in frost-sensitive plant species at above -5°C. Since all INA bacterial species, particularly *P. syringae*, are efficient producers of ice nuclei active at temperatures above -5°C, they have been shown to influence the supercooling point of plant tissues (Arny *et al.* 1976; Lindow *et al.* 1978 a,b, 1982 a,b; Anderson *et al.* 1982; Gross *et al.* 1984). The incidence of frost damage to several plant species, both in greenhouse and in field conditions, is related directly with the logarithm of population size of INA bacteria on plants (Lindow *et al.* 1978 a,b, 1982 a,b; Lindow & Connell 1984). However, not every bacterial cell within a strain of INA bacteria is active as an ice nucleus at a given temperature at any given time, either in vitro or on plants. Although

INA bacteria can catalyze ice formation in water supercooled to temperatures as high as -1°C, not every cell in a population of cells is active at a given time (Lindow *et al.* 1978 a,b, 1982 a,b). For each strain of INA bacteria, each cell has a finite, but very low, probability of being active in ice nucleation at a given temperature. The frequency of ice nucleation activity at a given temperature for any one bacterial species, such as *P. syringae*, varies widely. The nucleation frequency for *P. syringae* varies from as little as $10^{-8}$ per cell at -5°C for some strains to over $10^{-1}$ per cell for more active strains in vitro (Hirano *et al.* 1978; Lindow *et al.* 1982 b). The average frequency of ice nucleation activity on plants appears to be approximately $10^{-3}$ per cell (Lindow 1982 b; Gross *et al.* 1983; Lindow & Connell 1984). Since not every bacterial cell is active in ice nucleation, the incidence of frost damage to plants (incidence of ice-nucleating events on plants) is related directly with the logarithm of the number of bacterial ice nuclei per unit mass (Lindow *et al.* 1978 a).

Large reductions in the population size of INA bacteria on leaf surfaces are necessary to achieve a substantial reduction in the incidence of frost damage to plants at a given temperature. Since the incidence of frost damage to plants is related directly with the logarithm of bacterial population size (with a relatively small value of the regression coefficient (Lindow *et al.* 1978 a)), small reductions in numbers of epiphytic bacteria, i.e. less than five-fold, do not substantially reduce the incidence of frost damage to plants. Epiphytic populations of INA bacteria must be reduced over 50-fold to achieve a 50% reduction in the incidence of frost damage at a given temperature.

The supercooling point of individual plant parts, the lowest temperature that can be achieved before ice nucleation and therefore frost injury occurs, is related directly with the logarithm of bacterial population size. At temperatures above -5°C, the probability of a given bacterial cell being active as an ice nucleus decreases logarithmically with increasing temperature. Therefore, as the number of INA bacteria on an experimental sample, such as a leaf, increases, the probability of that sample containing a higher temperature ice nucleus will also increase. The percentage of leaves that freeze at a given temperature is an estimate of the frequency with which leaves contain a sufficiently large population of INA bacteria to contribute an ice nucleus (Hirano *et al.* 1985). One bacterial ice nucleus per leaf is sufficient to cause ice formation in the entire leaf and cause frost damage to that leaf. As shown in Fig. 1 for *P. syringae* strain Cit 7 (rifampicin-resistant), a significant loglinear relationship between the freezing temperature of individual leaves and the population size of strain Cit 7 is observed. As the mean log bacterial population size associated with a given leaf decreases, the freezing temperature decreases. It is also clear from Fig. 1, however, that large reductions in the epiphytic population size of INA bacteria are required to achieve significant (greater than 1°C) reductions in freezing temperatures of leaves. Measurement of the supercooling point of leaves by a 'tube nucleation assay' is also predictive of plant frost sensitivity under field conditions (Hirano *et al.* 1985).

*BIOLOGICAL CONTROL OF INA BACTERIA*

Modification of epiphytic bacterial communities by application of individual bacterial strains ('antagonists') appears to be due to an exclusion of subsequent colonists ('challengers') by the applied strain rather than a displacement of pre-existing colonists. The population size of INA bacteria and subsequent frost damage to corn seedlings following challenge inoculations of such plants with INA bacteria were reduced by applications of certain INA bacteria only when such antagonistic bacteria were applied to plants prior to, or at the same time as, challenge inoculations (Lindow *et al.* 1983 a). If antagonistic bacteria were applied to plants after the INA bacteria, there was no reduction in INA bacterial populations compared to plants treated only with INA bacteria (Lindow *et al.* 1983 a). Thus, the relative time of inoculation of plants with INA and non-INA bacteria greatly determines the relative population size of INA bacteria applied as challenge inoculum. Naturally

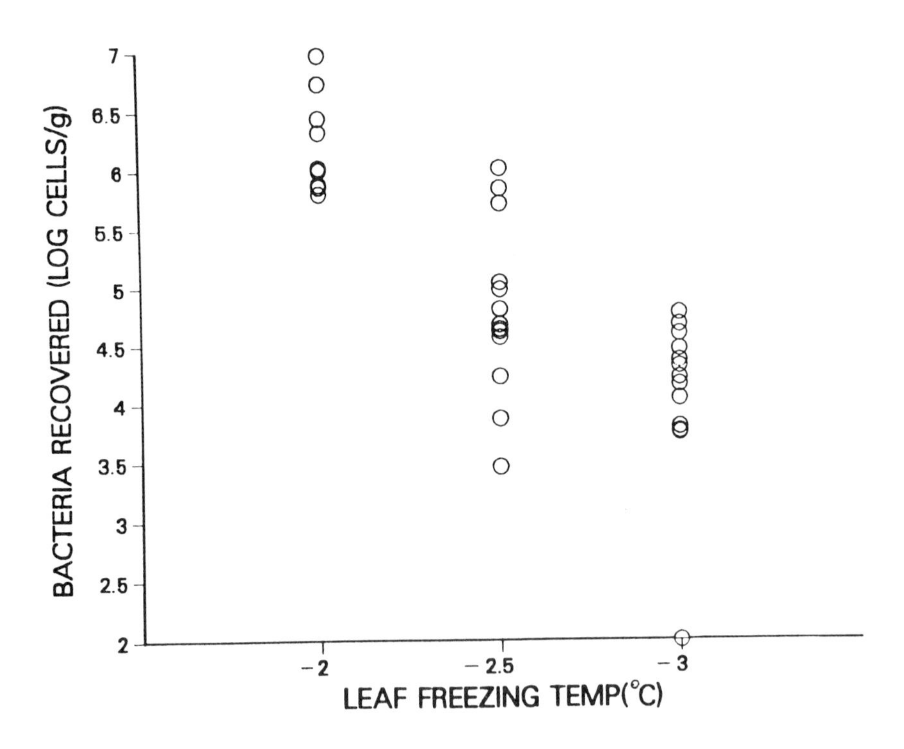


Fig. 1. Relationship between the population size of *P. syringae* strain Cit 7 on individual bean (*Phaseolus vulgaris*) leaves and the supercooling point of these leaves. Individual bean leaves were immersed in a test tube containing 10 ml of ice nuclei-free water and cooled slowly. Leaves that froze at each of the temperatures shown on the abscissa were warmed to 20°C, sonicated in an ultrasonic cleaning bath at 20°C for 9 minutes, and the population size of *P. syringae* strain Cit 7 determined by dilution plating of leaf washings on King's medium B containing 100 μg rifampicin $ml^{-1}$.

occurring strains of *P. fluorescens* and *P. putida* applied to pear trees under field conditions reduced the population size of naturally occurring INA bacteria compared to untreated trees only when applied to trees prior to the occurrence of large population sizes of INA bacteria (Lindow 1982 a). The population size of INA bacteria on treated as well as untreated pear trees was similar for several days following inoculation of trees with non-INA bacteria (Lindow 1982 b, 1983 a, 1985 a). However, the population size of INA bacteria on untreated trees increased relative to those on treated trees with increasing time after inoculation for up to 30 days (Lindow 1982 b, 1983 a, 1985 a). When antagonistic bacteria were applied to pear trees only after large population sizes of *P. syringae* had developed, no subsequent differences in population size of *P. syringae* between treated and untreated trees were observed.

The effectiveness of selected non-INA bacteria to exclude INA bacteria during subsequent challenge inoculations increased with increasing relative population size of the antagonistic bacteria. The maximum epiphytic bacterial population size observed on symptomless plants is often approximately $10^7$ cells $g^{-1}$ fresh weight ($10^6$ cells $cm^{-2}$) (Crosse 1965; Lindow *et al.* 1978 c; Gross *et al.* 1983; Hirano & Upper 1983). The application of very large numbers of bacteria onto plant surfaces generally does not result in an epiphytic population size greater than $10^7$ cells $g^{-1}$ when epiphytic populations are measured one or more days after inoculation (Crosse 1971; Lindow 1982 b, 1985 a; Lindow *et al.* 1983 b). One or more physical or chemical factors probably limit the total number of bacteria on plant surfaces. An inverse relationship between the population size of antagonistic bacteria and that of co-existing applied INA bacteria has been observed both under greenhouse conditions (Lindow *et al.* 1983 b) as well as in field conditions (Lindow 1982 b, 1985 a; Lindow & Connell 1984). When a non-INA *E. herbicola* isolate was applied to plants prior to INA strains of this species, it grew rapidly and comprised over 90% of the total bacteria found on plants (Lindow *et al.* 1983 a). Similarly, when an INA strain of *E. herbicola* was inoculated onto plants prior to a non-INA strain of this species, it grew rapidly and comprised the majority of bacterial cells recoverable from plant surfaces (Lindow *et al.* 1983 a). As the population size of a non-INA *E. herbicola* strain increased on plants, the population size of an INA strain of *E. herbicola* decreased, as did frost damage to plants (Lindow *et al.* 1983 a). Similarly, frost damage to plants treated with a single application of an INA *E. herbicola* strain decreased as the concentration of non-INA *E. herbicola* cells applied prior to the INA strain increased up to $10^8$ cells $ml^{-1}$ (Lindow *et al.* 1983 a). Therefore, it appears that modification of the leaf surface microflora is by pre-emptive exclusion rather than displacement mechanisms and requires the development of a sufficiently large population size of the antagonistic bacterium on the plant surface. Thus, if both antagonistic and target organisms have similar growth rates on plant surfaces, the arrival of antagonistic organisms on plants must precede that of target organisms in order that a large population size be achieved prior to any multiplication of target organisms. Early and rapid epiphytic colonization of leaves appears to be a prerequisite for biological control of frost injury and probably for modifications of leaf surface bacterial populations in general without the use of other perturbants.

*MECHANISMS OF ANTAGONISM OF BACTERIA ON LEAF SURFACES*

Antibiosis is a commonly assumed mechanism for the interaction of bacterial species on leaf surfaces as in other habitats (Teliz-Ortiz & Burkholder 1964; Goodman 1965; Leben 1965; Leben & Daft 1965; Chakravarti *et al.* 1972; Thomson *et al.* 1976; Beer *et al.* 1980; Spurr 1981). Many bacterial strains which exhibit biological control of either fungal or bacterial plant pathogens exhibit in vitro antibiosis. Although some investigators have noted a poor correlation between antibiosis in vitro and in vivo antagonism (Leben 1965; Hsu & Lockwood 1969; Lundborg & Unestam 1980; Andrews 1985; Lindow 1985 a), few critical examinations of the importance of antibiosis on interactions of bacteria on leaf surfaces have been made. Several types of antibiotics including bacteriocin-like compounds and siderophores have been suggested as inhibitory agents responsible for biological control of certain foliar diseases. While in certain cases these compounds could explain the biological control by the producing agents (Leben 1965; Beer *et al.* 1980), tests showing their necessity were lacking. Some studies indicate that bacterial siderophores are not produced on leaf surfaces (Lindow *et al.* 1984).

Antibiosis was found not to be necessary for the pre-emptive exclusion of INA bacteria by non-INA bacteria on plant surfaces (Lindow 1985 a). Only 58% of 88 bacteria antagonistic to *P. syringae* on leaf surfaces produced compounds inhibitory to *P. syringae* on any of several culture media evaluated (Lindow 1985 a). Although it could not be conclusively demonstrated that all strains did not produce an antibiotic on some culture medium, the bulk of the evidence suggested that many strains inhibitory to *P. syringae* in vivo were not inhibitory in vitro. Furthermore, antibiosis in antibiotic-producing strains was shown not be necessary for exclusion of *P. syringae* on leaf surfaces using antibiosis-deficient mutants. Frost injury to corn seedlings pre-inoculated with antibiosis-deficient mutants of 24 of 25 non-INA antibiosis-positive bacterial antagonists was significantly smaller than on plants treated only with water before inoculation with *P. syringae* strains. *P. syringae* populations on plants pretreated with antibiosis-deficient mutants or with antibiosis-positive antagonistic bacteria did not differ at the time of assay (Lindow 1985 a). Therefore, antibiosis appears not to be necessary for successful antagonism among bacteria on leaf surfaces either when antagonistic bacteria are allowed to colonize plants prior to challenge inoculations with INA bacteria or when co-inoculated with challenge strains.

Competition for limiting environmental resources appears sufficient to account for antagonism among bacterial species on leaf surfaces. Experiments using both near-isogenic (chemically induced mutations) or fully isogenic (site-directed in vitro mutagenesis) bacterial strains have shown this to be likely. Near-isogenic strains of *P. syringae*, deficient in expression of ice nucleation activity, have been obtained following mutagenesis with ethyl methanesulfonate and other mutagens (Lindow &

Staskawicz 1981). Treatment of leaves with Ice$^{-}$ mutants of *P. syringae* prior to challenge inoculations with Ice$^{+}$ parental strains of these bacteria, reduced both the population size of Ice$^{+}$ *P. syringae* strains, and frost damage to leaves at -5°C compared to plants treated only with Ice$^{+}$ *P. syringae* strains (Lindow 1985 a). The population size of Ice$^{-}$ mutants of *P. syringae* did not differ from that of Ice$^{+}$ parental strains when either was inoculated singly onto the surfaces of corn or bean plants. Thus, not only is antibiosis not necessary for the successful antagonism of bacteria on leaf surfaces, it is unable to explain the successful antagonism of near-isogenic strains of bacteria on the same leaf.

Fully isogenic Ice$^{+}$ and Ice$^{-}$ strains of *P. syringae* have been constructed to further test the sufficiency of competition on leaf surfaces as a mechanism of biological control of frost injury. Genes conferring ice nucleation activity in both *P. syringae* and *E. herbicola* have been cloned and express the Ice phenotype in recipient *Escherichia coli* cells (Orser *et al.* 1983, 1984, 1985; Kozloff *et al.* 1984; Warren *et al.* 1985). The gene determining ice nucleation in *P. syringae* is rather large (approximately 4.2 kb) and is under the control of its own cis-acting regulatory region (Orser *et al.* 1985). Deletions within this *ice* region have been constructed in vitro by molecular genetic techniques (Orser *et al.* 1984, 1985). Reciprocal exchange of the deletion-containing *ice* gene constructed in vitro (carried on a plasmid such as pBR325 which does not replicate within *P. syringae*) for the homologous functional *ice* gene in recipient *P. syringae* strains occurs following conjugal transfer of such a plasmid to recipient strains (Orser *et al.* 1984). The single genetic change, only within the *ice* region, resulting from this procedure can be verified by Southern blot analysis (Orser *et al.* 1984). Ice$^{-}$ deletion mutants of *P. syringae* constructed in vitro do not differ from Ice$^{+}$ parental strains in their growth rate on plant surfaces, their ability to establish a particular population size following inoculation of any of 25 different plant species, their survival during repeated freeze/thaw cycles, their survival in soil, or in any of over 100 biochemical and physiological tests (Lindemann & Suslow 1985; Lindow 1985 b). Ice$^{-}$ deletion mutants of *P. syringae* constructed in vitro when applied to corn, bean, tomato, potato or strawberry plants reduced the population size of Ice$^{+}$ *P. syringae* strains applied as challenge inoculations compared to plants not treated with Ice$^{-}$ bacteria prior to such challenge inoculations (Lindemann *et al.* 1985; Lindow 1985 a). Reductions in populations of Ice$^{+}$ *P. syringae* strains on plants pre-inoculated with Ice$^{-}$ strains were inversely proportional to the population size of Ice$^{-}$ *P. syringae* strains at the time of challenge inoculations (Lindemann *et al.* 1985; Lindow 1985 b). A similar reciprocal reduction in population size of isogenic strains of *P. tomato* was observed when either of two isogenic strains was used alternatively as an 'antagonist' or 'challenge' strain (Lindemann 1985). Thus, competition among isogenic strains on plants appears to be a common mechanism of interaction and one which can lead to large changes in the balance of microbial population sizes on leaves when modified by artificial inoculations of 'antagonist' strains.

The limiting resources for which competition may occur on leaf surfaces can include specific localized sites that are habitable on leaf surfaces as well as limitations of nutrients and other factors that may occur at such sites. Several investigators have noted the non-random distribution of bacterial cells on leaf surfaces (Leben & Daft 1966; Leben *et al.* 1970; Hirano *et al.* 1982). Although bacteria may be more likely to occur at certain sites such as near veins or abaxial epidermal cell junctions, not all such localities are colonized by bacteria at a given time (Leben & Daft 1966). The distribution of bacterial population sizes among individual leaves of a given plant species has been shown to be lognormal (Hirano *et al.* 1982; Haefele & Lindow 1984; see also S.S. Hirano and C.D. Upper, this volume). The lognormal distribution of population sizes of bacteria is also observed even if the area within an examined leaf is decreased (C.D. Upper, personal communication). Thus, no matter at what scale population size is being described, certain leaves, certain portions of leaves, or even collections of a few epidermal cells, support the growth and/or survival of many more bacteria than do the average leaf or subsection of a leaf. Such lognormal distributions of bacterial population sizes cannot be explained by a non-random deposition of initial bacterial inoculum. Uniformly inoculated leaves at first have a normal distribution of bacterial population sizes among treated leaves, but quickly develop a lognormal distribution of population sizes (Haefele & Lindow 1984). It is attractive to hypothesize that the environment, either chemical or physical, at a few such localized sites may be far more conducive for the growth and/or survival of bacterial cells on the leaf surface than most other sites. Most competition among bacterial cells may, therefore, be expected to occur at such localized areas (Fig. 2).

A lack of significant increases in bacterial population sizes following application of nutrients to leaf surfaces (Lindow *et al.* 1983 b; Morris & Rouse 1985) may be a reflection of the relative lack of habitable sites on the leaf surface. Nutrients applied to uninhabitable sites on leaves if not mobile on the leaf surface, would not become available to resident bacteria. Similarly, a lack of habitable sites would restrict population sizes itself. A clearer understanding of factors limiting bacterial population sizes on leaf surfaces should lead not only to a better understanding of the mechanisms of interaction between bacteria on leaf surfaces, but also to selection of more efficient antagonistic bacteria.

## *IMPLEMENTING BIOLOGICAL CONTROL ON ALMOND AND OTHER CROPS*

Because antagonistic bacteria are effective only in exclusion of INA bacteria, a knowledge of the population dynamics of target bacterial species on plants is required to determine times at which exclusion of target bacteria would be optimized. The population size of INA bacteria on untreated field-grown potato, citrus, pear and tomato plants varies greatly during the growing season (Lindow *et al.* 1978 b; Lindow 1982 b; Lindow & Connell 1984). Population size of INA bacteria on potato and tomato was smallest shortly after emergence of these annual plants from seeds (Lindow 1982 b). Population size of INA bacteria increased by a factor of over $10^3$ and $10^4$ within 2 to 5 weeks after emergence of potato and tomato, respectively. The population size of INA bacteria on pear was smallest shortly after the emergence of leaves and flowers from overwintering buds in early spring (Lindow 1982 b, 1983 a, 1985 a; Gross *et al.* 1983). The INA bacterial population size increased by a factor of over $10^4$ within 4 weeks after the appearance of young vegetative tissues during spring. The population size of INA bacteria declined approximately 10 weeks after emergence of vegetative tissue on pear with the onset of hot, dry weather typical of the Mediterranean climate of California (Lindow 1982 b, 1983 a, 1985 a). No INA bacteria are found on various *Citrus* species, including navel orange in California during the summer months of April through September. INA bacteria are detected on navel orange only starting in October when maximum air temperatures begin to decline and free moisture first appears on leaves due to rain, and increasingly heavy dew at night (Lindow 1982 b). The total population size of INA bacteria on navel orange reaches a maximum in early January, coinciding with the period of maxi-

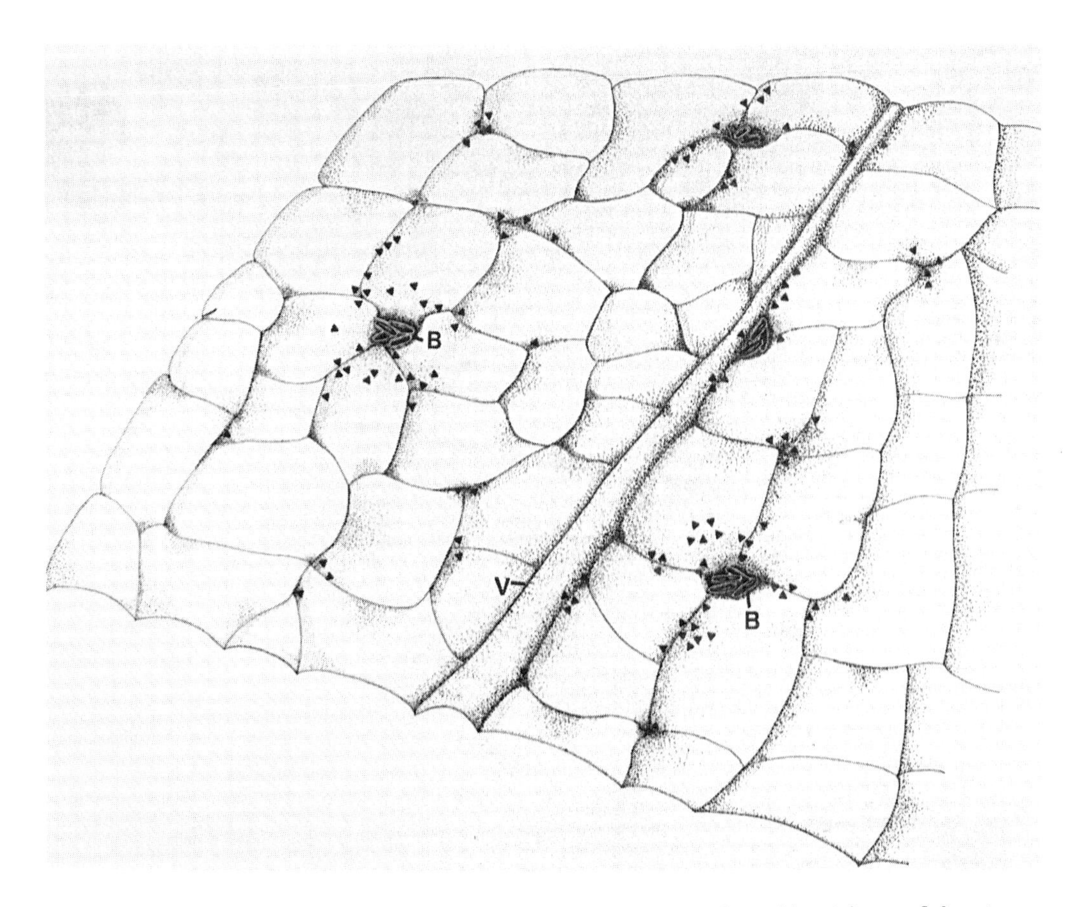


Fig. 2. Depiction of the non-random localization of bacterial cells (B) on a leaf surface including veins (V) and competition for irregularly occurring nutrients (▲).

mum frost hazard to this crop (Lindow 1982 b). Antagonistic bacteria could be applied to navel oranges upon the onset of favourable environmental conditions in October, and prior to the colonization of foliage by INA bacteria. For other crops, favourable conditions for bacterial growth occur when natural populations of INA bacteria are small following seedling or bud emergence. Thus, early spring periods exist in which environmental conditions are favourable for the development of epiphytic populations of INA bacteria, and in which the colonization of young, emerging tissues of potato, tomato and pear is incomplete. Applications of antagonistic bacteria to these crops at this time have prevented subsequent increases in population size of INA bacteria (Lindow 1982 b, 1983 a, 1985 a; Lindow & Connell 1984).

Large seasonal variations in the population size of INA bacteria on untreated almond trees were observed during the years 1982 to 1985. Population dynamics of INA bacteria on almond were similar to those seen on pear. Relatively small populations of INA bacteria were seen on young, vegetative or reproductive tissues in early spring (Fig. 3). Large increases in population size of INA bacteria were seen in the years 1982 to 1984, and more modest increases in 1985, within 2 to 4 weeks following flowering of trees. The maximum population size of INA bacteria on almond trees during these years was dependent on the frequency and abundance of rainfall during the months of February, March and April when newly expanded almond leaf and flower tissue is available for colonization by INA bacteria (Fig. 3). Small total populations of INA bacteria in 1985 reflect an almost total absence of rainfall during the early spring of that year. Antagonistic bacteria, used as competitors to exclude subsequent increases in INA bacterial population sizes on almond trees, must be inoculated onto trees shortly after flowering in these trees, to precede the development of populations of the target species.

Antagonistic bacteria applied to plants with the purpose of biological control of INA bacteria became the dominant bacterial species recovered over a period from several days to more than two months. The absolute population size of antagonistic bacteria, genetically marked by spontaneous mutations conferring resistance to the antibiotic rifampicin to allow their unambiguous identification and enumeration, varied between different plant species to which they were applied (Lindow 1982 b, 1983 a, 1985 a; Lindow *et al.* 1983 b; Lindow & Connell 1984). However, their population sizes as a proportion of the total population size of bacteria recoverable on King's medium B (King *et al.* 1954) was often larger than 90%. The population size of antagonistic bacteria applied to pear trees increased for 2 to 3 weeks following application and constituted between 90 to 95% of the total bacteria recoverable from leaf surfaces of treated pear trees (Lindow 1982 b, 1983 a, 1985 a). Reductions in populations of INA bacteria on treated plants were highest when the antagonistic bacteria comprised a large percentage of the total viable recoverable bacteria on treated plants. An inverse relationship between the population size of antagonistic bacteria applied to plants and the population size of INA bacteria was seen under field conditions on treated almond trees (Lindow & Connell 1984).

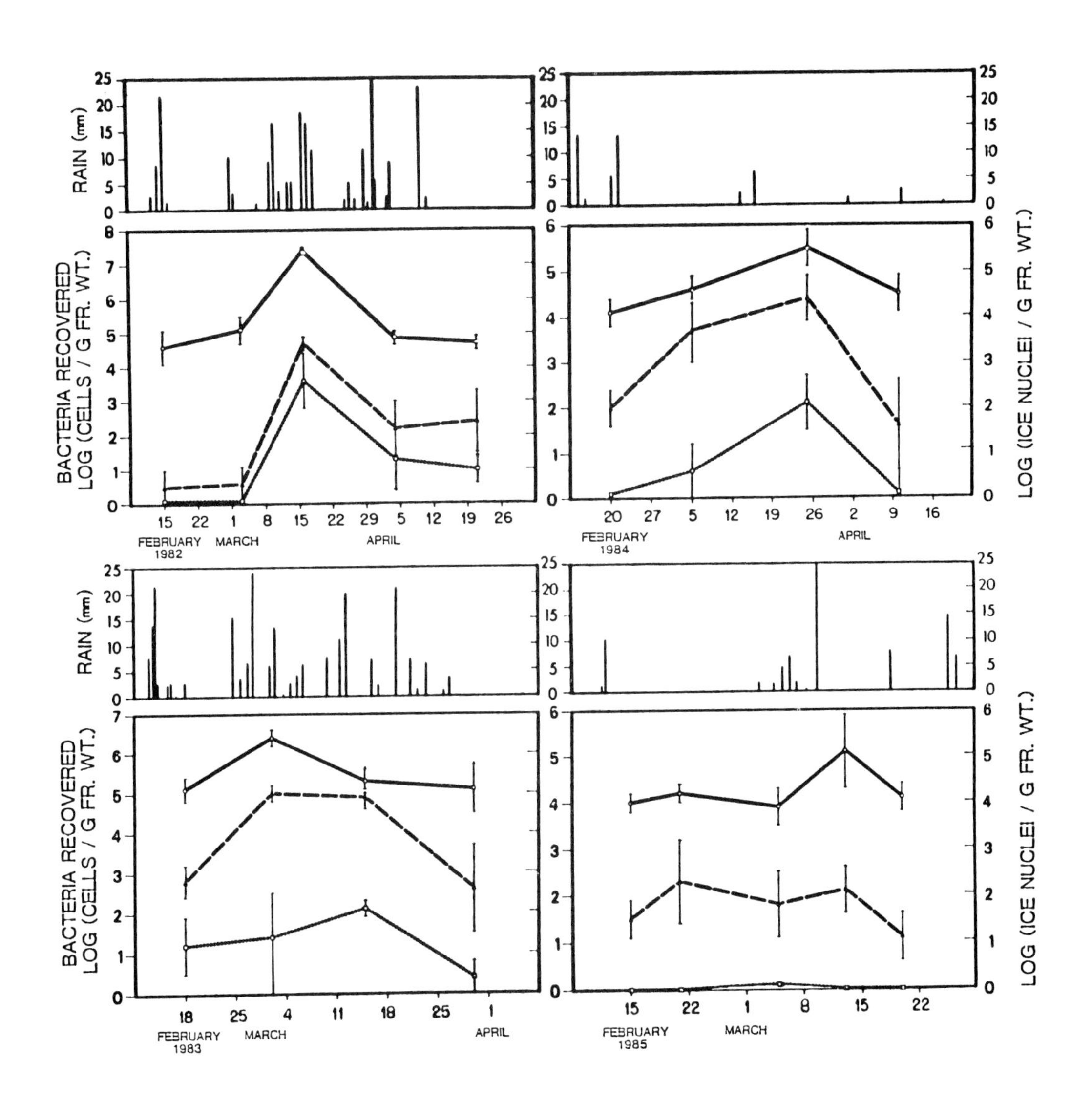


Fig. 3. Total (o——o) and INA bacteria, principally *P. syringae* (▲---▲) and ice nuclei active at -5°C (□······□) on leaves and flowers of untreated almond (*Prunus dulcis* cv. Ne Plus) trees during the 1982 to 1985 growing seasons grown near Modesto, California. The vertical bars represent the standard error of the mean of the log of bacterial populations or ice nuclei. Bacterial populations were determined from leaf washings on King's medium B. Numbers of ice nuclei on plants were determined by a droplet freezing assay (Lindow *et al.* 1978 b) of leaf washings as described previously (Lindow & Connell 1984).

The population dynamics of different bacterial strains applied to plants such as almond under field conditions may differ considerably. The population size of *P. syringae* strain Cit 13-12, a chemically derived Ice$^-$ mutant of *P. syringae* strain Cit 13, was large on treated almond trees and comprised over 90% of the total viable bacteria recovered from almond for over 4 weeks after a single application of this isolate (Fig. 4). In contrast, the population size of the naturally occurring Ice$^-$ strain, A506, of *P. fluorescens* was approximately 10-fold smaller than that of *P. syringae* strain Cit 13-12 and generally constituted less than 90% of the total bacteria recovered from treated almond trees (Fig. 4). In particular, population size of strain A506 dropped dramatically approximately 4 weeks following inoculation and constituted less than 0.01% of the total bacteria recovered from treated trees after this time. The population size of INA bacteria was reduced from 10- to over 100-fold on trees treated with strains Cit 13-12 or A506 compared to untreated trees (Figs 3 and 4), particularly, when measured between 1 and 3 weeks following a single application of these strains to trees. Although the population size of INA bacteria remained small for over 6 weeks on almond trees treated with strain Cit 13-12, it increased after 3 to 4 weeks on trees treated with A506 when the population size of this antagonistic bacterium decreased (Fig. 4). The number of ice nuclei on almond trees treated with either strain Cit 13-12 or A506 was reduced from 10- to 100-fold compared to untreated trees (Figs 3 and 4). The ratio of the number of ice nuclei to the number of bacterial cells capable of ice nucleation (nucleation frequency) on treated trees was the same on both treated and untreated trees.

The average supercooling point of almond trees treated with antagonistic bacteria is generally lower than that of untreated trees for up to several weeks following a single application of antagonistic bacteria. The supercooling point of almonds treated with one of several antagonistic bacteria was reduced from 0.5 to over 1.5°C (Fig. 5). The reduction of the supercooling point of almond fruiting spurs by antagonistic bacteria was nearly as large as reductions following repeated applications of bactericides such as cupric hydroxide (Fig. 5). The reduction of the supercooling point of almond spurs treated with antagonistic bacteria was predictable based on reductions of the population size of INA bacteria (Figs 3 and 4), and the relationship between population size of INA bacteria and the supercooling point of individual plant parts (Fig. 1). Thus, it appears that some antagonistic bacteria, by reducing the population size of antagonistic bacteria over 10- to 100-fold, can cause significant reductions in the nucleating temperatures of treated plants and frost damage. The most effective antagonistic bacteria, therefore, appear sufficiently efficacious for their consideration as practical biological control agents of plant frost injury. Due to rapid changes in the population size of INA bacteria on untreated plants, significant differences in the effectiveness of applied antagonistic bacteria in reducing the population size of INA bacteria might be expected if even small changes in their time of application had been made. Further work is necessary to determine the duration of 'windows' of time, during which different plant species can support populations of applied antagonistic bacteria, but have not yet been colonized by INA bacteria.

*UNINVESTIGATED FACTORS IMPORTANT TO THE APPLICATION OF BIOLOGICAL CONTROL IN THE PHYLLOPLANE*

The results of the many experiments summarized in this review indicate that individual non-INA bacteria applied to plants under field conditions can achieve large, and relatively stable, population sizes on plants, and therefore reduce the population size of INA bacteria on these plants. However, predicting the efficacy of this procedure in different times and locations will require better understanding of several different aspects of leaf surface microbial ecology. The specificity of interactions among bacteria on leaf surfaces is largely unknown. Although antibiosis can be eliminated as an important mechanism of antagonism among those strains that have been studied in detail, the importance of antibiotics in interactions of at least some bacteria on

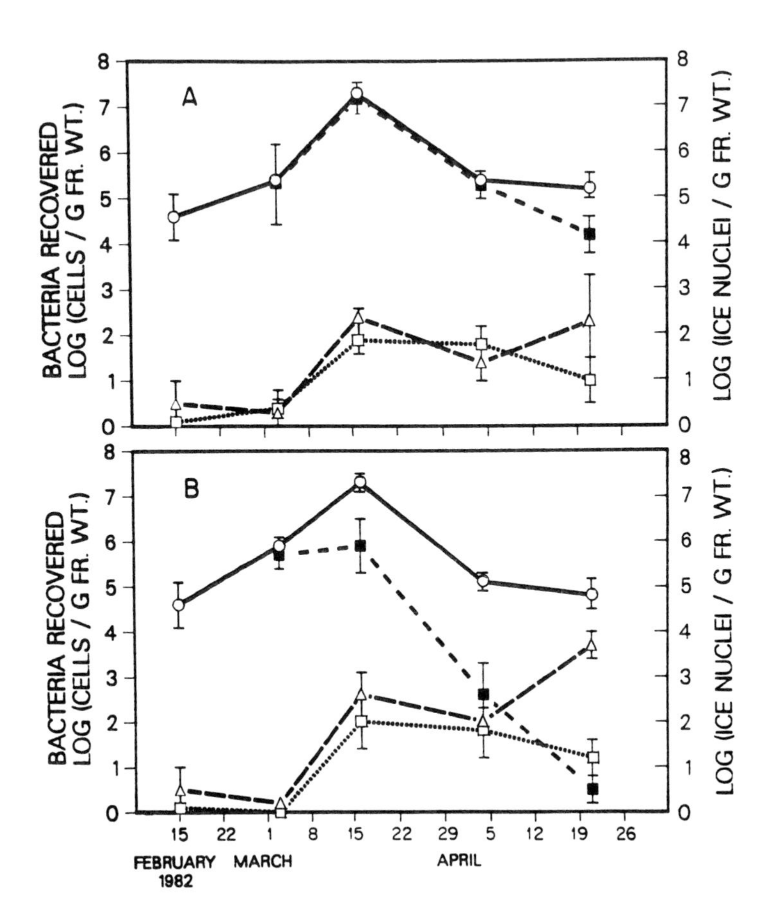


Fig. 4. Total (o) and INA bacteria (Δ) and ice nuclei active at -5°C (□) on almond trees treated once with *ca* $10^7$ cells $ml^{-1}$ of *P. syringae* strain Cit 13-12 (A, ■) or *P. fluorescens* strain A506 (B, ■) on 16 February, 1982 near Modesto, California. The vertical bars represent the standard error of the mean log populations.

leaf surfaces cannot be excluded. Although competition for limiting environmental resources appears to be a more general mechanism of antagonism on leaf surfaces, little is known of the specificity of competition on leaves. Even though the population size of a number of different bacterial species applied as antagonistic bacteria to corn was similar (and large), these strains differed considerably in the reduction of the population size of INA bacteria that they achieved (Lindow 1985 a). Thus, a fair degree of specificity of interspecies competition may exist on plants. Such differences in competition between different species may represent differences in biochemical and genetic traits, thus differences in ecological niches. Therefore, different environmental resources may limit different bacterial species on leaf surfaces. A high degree of genetic similarity, and thus similarity in ecological niche, should increase competition between bacterial strains. Insufficient evidence exists at this time to define the specificity of intraspecific interactions of bacteria on leaf surfaces. Such specificity can be tested,

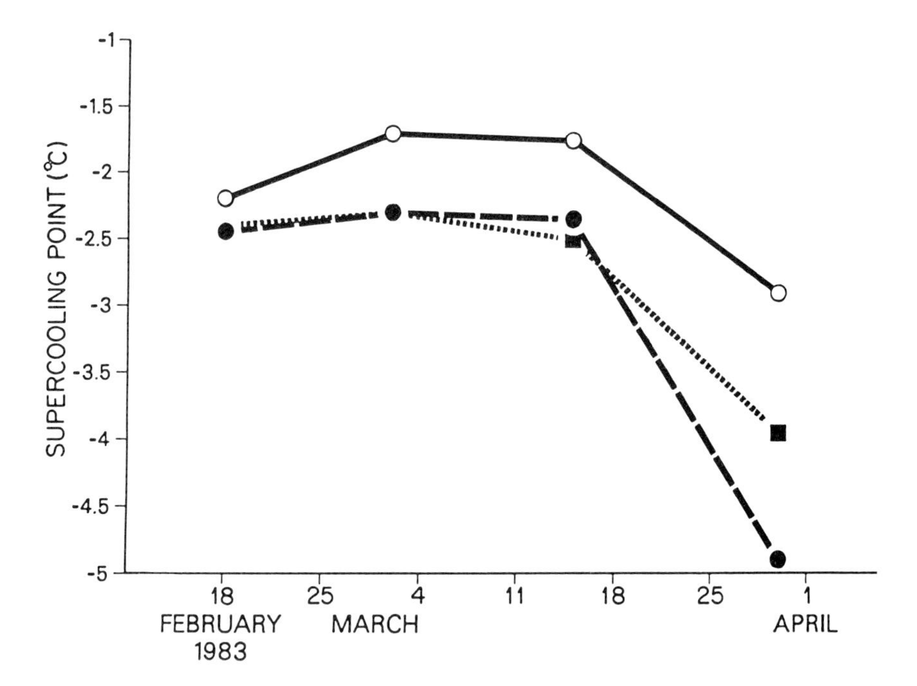


Fig. 5. Supercooling point of fruiting spurs of almond grown near Modesto, California, untreated (o) or treated on 18 February with *ca* $10^7$ cells of *P. fluorescens* strain A506 $ml^{-1}$ (■) or treated weekly starting on 18 February with cupric hydroxide (Kocide 101, Kocide Chemical Company, 1.2 g $l^{-1}$) (•). Eighty fruiting spurs from each treatment were collected at the times indicated on the abscissa and the supercooling point determined in a tube nucleation assay similar to that reported by Hirano *et al.* (1985) and as described in the legend to Fig. 1.

however, by the use of isogenic and non-isogenic strains of the same bacterial species. Molecular biological techniques now exist for the construction of isogenic bacterial strains (Ruvkun & Ausubel 1981), e.g. site-directed Ice$^-$ deletion mutants of *P. syringae* (Orser *et al.* 1983, 1984, 1985). The results of Morris & Rouse (1985, this volume) indicate that considerable physiological diversity exists between different strains of the same bacterial species, even on individual leaves. If the specificity of competition on leaf surfaces within a bacterial species is great, biological control of INA bacteria or of other epiphytic bacterial species on leaves harbouring a genetically diverse target species may be poor. Similarly, S.S. Hirano and C.D. Upper (this volume) have presented evidence that the population size of individual bacterial species on leaf surfaces undergoes considerable temporal variability. It will be important to know, however, whether all strains of a genetically diverse bacterial species undergo similar temporal responses to environmental fluctuations or whether different strains respond differently to environmental variables determining such temporal variations in population size. Should the genetic diversity evident among certain bacterial species such as *P. syringae* be reflected in a significantly different response to environmental stress, the application of individual antagonistic bacteria to plants which harbour a genetically diverse target species would prove unsuccessful or less successful than on plants harbouring a less diverse target species. Not only might genetic differences in the target bacterial species affect biological control, but different varieties of a given plant species or different plant species may affect the population size of antagonistic bacteria and target bacterial species differently. Environmental resources for which an antagonistic bacterium competes successfully on one host species may not be limiting on a different variety or a different plant species. Thus, some interaction of the host plant and the antagonistic bacterium in determining the efficiency of biological control might be expected. No data relating to this question are now available however.

The lack of effective and economical bactericides is a strong stimulus for the evaluation of biological control agents, not only against INA bacteria, but also against other plant-pathogenic bacteria. A consistent and effective biological control of populations of target bacteria on plant surfaces will require a better understanding of the interactions of micro-organisms on leaf surfaces. That effective biological control of INA bacteria and certain plant diseases can be achieved is a strong stimulus to pursue a better understanding of important features of leaf surface microbial ecology.

*ACKNOWLEDGEMENTS*

Some of the work reported here was supported by the Almond Board of California. I wish to thank W. Asai, Cooperative Extension, University of California, Modesto, California, for assistance in almond field trials, M. Owen, J. Papp and D. Okamoto for valuable technical assistance in laboratory measurements of bacterial populations and plant freezing points, and K. Callan for preparation of some of the figures and illustrations.

*REFERENCES*

Andersen, G.L. & Lindow, S.E. (1985). Local differences in epiphytic bacterial population size and supercooling point of citrus correlated with type of surrounding vegetation and rate of bacterial immigration. Phytopathology, *75*, 1321 (Abstr.).

Anderson, J.A., Buchanan, D.W., Stall, R.E. & Hall, C.B. (1982). Frost injury of tender plants increased by *Pseudomonas syringae* van Hall. Journal of the American Society for Horticultural Science, *107*, 123-5.

Andrews, J.H. (1985). Strategies for selecting antagonistic microorganisms from the phylloplane. *In* Biological Control on the Phylloplane, eds C.E. Windels & S.E. Lindow, pp. 31-44. St. Paul: Am. Phytopath. Soc.

Arny, D.C., Lindow, S.E. & Upper, C.D. (1976). Frost sensitivity of *Zea mays* increased by application of *Pseudomonas syringae*. Nature (London), *262*, 282-4.

Ashworth, E.N. & Davis, G.A. (1984). Ice nucleation within peach trees. Journal of the American Society for Horticultural Science, *109*, 198-201.

Beer, S.V., Norelli, J.L., Rundle, J.R., Hodges, S.S., Palmer, J.R., Stein, J.I. & Aldwinckle, H.S. (1980). Control of fireblight by non-pathogenic bacteria. Phytopathology, *70*, 459 (Abstr.).

Burke, M.J., Gusta, L.V., Quamme, H.A., Weiser, C.J. & Li, P.H. (1976). Freezing and injury in plants. Annual Review of Plant Physiology, *27*, 507-28.

Chakravarti, B.P., Leben, C. & Daft, G.C. (1972). Numbers and antagonistic properties of bacteria from buds of field-grown soybean plants. Canadian Journal of Microbiology, *18*, 696-8.

Chandler, W.H. (1958). Cold resistance in horticultural plants: a review. Proceedings of the American Society for Horticultural Science, *64*, 552-72.

Crosse, J.E. (1965). Bacterial canker of stone fruits. VI. Inhibition of leaf-scar infection of cherry by a saprophytic bacterium from the leaf surfaces. Annals of Applied Biology, *56*, 149-60.

Crosse, J.E. (1971). Interactions between saprophytic and pathogenic bacteria in plant disease. *In* Ecology of Leaf Surface Micro-organisms, eds T.F. Preece & C.H. Dickinson, pp. 283-90. London: Academic Press.

Dye, D.W., Bradbury, J.F., Goto, M., Hayward, A.C., Lelliott, R.A. & Schroth, M.N. (1980). International standards for naming pathovars of phytopathogenic bacteria and a list of pathovar names and pathotype strains. Review of Plant Pathology, *59*, 153-68.

Goodman, R.N. (1965). In vitro and in vivo interactions between components of mixed bacterial cultures isolated from apple buds. Phytopathology, *55*, 217-21.

Gross, D.C., Cody, Y.S., Proebsting, E.L., Radamaker, G.K. & Spotts, R.A. (1983). Distribution, population dynamics, and characteristics of ice nucleation-active bacteria in deciduous fruit tree orchards. Applied and Environmental Microbiology, *46*, 1370-9.

Gross, D.C., Proebsting, E.L., Jr. & Andrews, P.K. (1984). The effects of ice nucleation-active bacteria on temperatures of ice nucleation and freeze injury of *Prunus* flower buds at various stages of development. Journal of the American Society for Horticultural Science, *109*, 375-80.

Haefele, D.M. & Lindow, S.E. (1984). Changes in leaf surface characteristics influence the mean, variance, and nucleation frequency of epiphytic ice nucleation active bacterial populations. Phytopathology, *74*, 882 (Abstr.).

Hirano, S.S., Baker, L.S. & Upper, C.D. (1985). Ice nucleation temperature of individual leaves in relation to population sizes of ice nucleation active bacteria and frost injury. Plant Physiology, *77*, 259-65.

Hirano, S.S., Maher, E.A., Kelman, A. & Upper, C.D. (1978). Ice nucleation activity of fluorescent plant pathogenic pseudomonads. Proceedings of the 4th International Conference on Plant Pathogenic Bacteria, ed. Stat. Path.

Végét. Phytobacteriologie, pp. 717-24. Beaucouzé: Inst. Nat. Rech. Agron.

Hirano, S.S., Nordheim, E.V., Arny, D.C. & Upper, C.D. (1982). Lognormal distribution of epiphytic bacterial populations on leaf surfaces. Applied and Environmental Microbiology, *44*, 695-700.

Hirano, S.S. & Upper, C.D. (1983). Ecology and epidemiology of foliar bacterial plant pathogens. Annual Review of Phytopathology, *21*, 243-69.

Hsu, S.C. & Lockwood, J.L. (1969). Mechanisms of inhibition of fungi in agar by Streptomycetes. Journal of General Microbiology, *57*, 149-58.

Kaku, S. (1964). Undercooling points and frost resistance in mature and immature leaf tissues of some evergreen plants. Botanical Magazine Tokyo, *77*, 283-9.

Kaku, S. (1975). Analysis of freezing temperature distribution in plants. Cryobiology, *12*, 154-9.

King, E.O., Ward, M.K. & Raney, D.E. (1954). Two simple media for the demonstration of pyocyanin and fluorescin. Journal of Laboratory and Clinical Medicine, *44*, 301-7.

Kozloff, L., Lute, M. & Westaway, D. (1984). Phosphatidylinositol as a component of the ice nucleating site of *Pseudomonas syringae* and *Erwinia herbicola*. Science, *226*, 845-6.

Leben, C. (1965). Epiphytic microorganisms in relation to plant disease. Annual Review of Phytopathology, *3*, 209-30.

Leben, C. & Daft, G.C. (1965). Influence of an epiphytic bacterium on cucumber anthracnose, early blight of tomato, and northern leaf blight of corn. Phytopathology, *55*, 760-2.

Leben, C. & Daft, G.C. (1966). Migration of bacteria on seedling plants. Canadian Journal of Microbiology, *12*, 1119-23.

Leben, C., Schroth, M.N. & Hildebrand, D.C. (1970). Colonization and movement of *Pseudomonas syringae* on healthy bean seedlings. Phytopathology, *60*, 677-80.

Levitt, J. (1972). Responses of Plants to Environmental Stress. New York: Academic Press.

Lindemann, J. (1985). Genetic manipulation of microorganisms for biological control. *In* Biological Control on the Phylloplane, eds C.E. Windels & S.E. Lindow, pp. 116-30. St. Paul: Am. Phytopath. Soc.

Lindemann, J., Arny, D.C. & Upper, C.D. (1984). Epiphytic populations of *Pseudomonas syringae* pv. *syringae* on snap bean and nonhost plants and the incidence of bacterial brown spot disease in relation to cropping patterns. Phytopathology, *74*, 1329-33.

Lindemann, J., Joe, L. & Moayeri, A. (1985). Reciprocal competition between INA$^+$ wild-type and INA$^-$ deletion mutant strains of *Pseudomonas* on strawberry blossoms. Phytopathology, *75*, 1361 (Abstr.).

Lindemann, J. & Suslow, T.V. (1985). Characteristics relevant to the question of environmental fate of genetically engineered INA$^-$ deletion mutant strains of *Pseudomonas*. *In* Proceedings of the 6th International Conference on Plant Pathogenic Bacteria, ed. E. Civerolo. College Park, MD: U. S. Department of Agriculture (in press).

Lindow, S.E. (1982 a). Epiphytic ice nucleation-active bacteria. *In* Phytopathogenic Prokaryotes. Vol. 1, eds M.S. Mount & G.H. Lacy, pp. 335-62. New York: Academic Press.

Lindow, S.E. (1982 b). Population dynamics of epiphytic ice nucleation active bacteria on frost sensitive plants and frost control by means of antagonistic bacteria. *In* Plant Cold Hardiness, and Freezing Stress - Mechanisms and Crop Implications, eds P.H. Li & A. Sakai, pp. 395-416. New York: Academic Press.

Lindow, S.E. (1983 a). Methods of preventing frost injury caused by epiphytic ice-nucleation-active bacteria. Plant Disease, *67*, 327-33.

Lindow, S.E. (1983 b). The role of bacterial ice nucleation in frost injury to plants. Annual Review of Phytopathology, *21*, 363-84.

Lindow, S.E. (1985 a). Integrated control and role of antibiosis in biological control of fireblight and frost injury. *In* Biological Control on the Phylloplane, eds C.E. Windels & S.E. Lindow, pp. 83-115. St. Paul: Am. Phytopath. Soc.

Lindow, S.E. (1985 b). Ecology of *Pseudomonas syringae* relevant to the field use of Ice$^-$ deletion mutants constructed in vitro for plant frost control. *In* Engineered Organisms in the Environment: Scientific Issues, eds H.O. Halvorson, D. Pramer & M. Rogul, pp. 23-35. Washington, DC: Am. Soc. Microbiol.

Lindow, S.E., Arny, D.C., Barchet, W.R. & Upper, C.D. (1978 a). The role of bacterial ice nuclei in frost injury to sensitive plants. *In* Plant Cold Hardiness and Freezing Stress - Mechanisms and Crop Implications, eds P.H. Li & A. Sakai, pp. 249-63. New York: Academic Press.

Lindow, S.E., Arny, D.C. & Upper, C.D. (1978 b). *Erwinia herbicola*: a bacterial ice nucleus active in increasing frost injury to corn. Phytopathology, *68*, 523-7.

Lindow, S.E., Arny, D.C. & Upper, C.D. (1978 c). Distribution of ice nucleation-active bacteria on plants in nature. Applied and Environmental Microbiology, *36*, 831-8.

Lindow, S.E., Arny, D.C. & Upper, C.D. (1982 a). Bacterial ice nucleation: a factor in frost injury to plants. Plant Physiology, *70*, 1084-9.

Lindow, S.E., Arny, D.C. & Upper, C.D. (1983 a). Biological control of frost injury: an isolate of *Erwinia herbicola* antagonistic to ice nucleation active bacteria. Phytopathology, *73*, 1097-102.

Lindow, S.E., Arny, D.C. & Upper, C.D. (1983 b). Biological control of frost injury: establishment and effects of an isolate of *Erwinia herbicola* antagonistic to ice nucleation active bacteria on corn in the field. Phytopathology, *73*, 1102-6.

Lindow, S.E. & Connell, J.H. (1984). Reduction of frost injury to almond by control of ice nucleation active bacteria. Journal of the American Society for Horticultural Science, *109*, 48-53.

Lindow, S.E., Hirano, S.S., Arny, D.C. & Upper, C.D. (1982 b). Relationship between ice nucleation frequency of bacteria and frost injury. Plant Physiology, *70*, 1090-3.

Lindow, S.E., Loper, J.E. & Schroth, M.N. (1984). Lack of evidence for *in situ* fluorescent pigment production by *P. s. syringae* on leaf surfaces. Phytopathology, *74*, 825-6 (Abstr.).

Lindow, S.E. & Staskawicz, B.J. (1981). Isolation of ice nucleation deficient mutants of *Pseudomonas syringae* and *Erwinia herbicola* and their transformation with plasmid DNA. Phytopathology, *71*, 237 (Abstr.).

Lundborg, A. & Unestam, T. (1980). Antagonism against *Fomes annosus*. Comparison between different test methods *in vitro* and *in vivo*. Mycopathologia, *70*, 107-15.

Maki, L.R., Galyan, E.L., Chang-Chien, M.-M. & Caldwell, D.R. (1974). Ice nucleation induced by *Pseudomonas syringae*. Applied Microbiology, *28*, 456-9.

Maki, L.R. & Willoughby, K.J. (1978). Bacteria as biogenic sources of freezing nuclei. Journal of Applied Meteorology, *17*, 1049-53.

Makino, T. (1982). Micropipette method: a new technique for detecting ice-nucleation activity of bacteria and its application. Annals of the Phytopathological Society of Japan, *48*, 452-7.

Marcellos, H. & Single, W.V. (1976). Ice nucleation on wheat. Agricultural Meteorology, *16*, 125-9.

Marcellos, H. & Single, W.V. (1979). Supercooling and heterogeneous nucleation of freezing of tissue of tender plants. Cryobiology, *16*, 74-7.

Mayland, H.F. & Cary, J.W. (1970). Chilling and freezing injury to growing plants. Advances in Agronomy, *22*, 203-34.

Morris, C.E. & Rouse, D.I. (1985). Role of nutrients in regulating epiphytic bacterial populations. *In* Biological Control on the Phylloplane, eds C.E. Windels & S.E. Lindow, pp. 63-82. St. Paul: Am. Phytopath. Soc.

Orser, C.S., Lotstein, R., Lahue, E., Willis, D.K., Panopoulos, N.J. & Lindow, S.E. (1984). Structural and functional analysis of the *Pseudomonas syringae* pv. *syringae ice* region and construction of *ice*$^-$ deletion mutants. Phytopathology, *74*, 798 (Abstr.).

Orser, C., Staskawicz, B.J., Loper, J., Panopoulos, N.J., Dahlbeck, D., Lindow, S.E. & Schroth, M.N. (1983). Cloning of genes involved in bacterial ice nucleation and fluorescent pigment/siderophore production. *In* Molecular Genetics of the Bacteria-Plant Interaction, ed. A. Pühler, pp. 353-61. Berlin: Springer-Verlag.

Orser, C.S., Staskawicz, B.J., Panopoulos, N.J., Dahlbeck, D. & Lindow, S.E. (1985). Cloning and expression of bacterial ice nucleation genes in *Escherichia coli*. Journal of Bacteriology, *164*, 359-66.

Paulin, J.-P. & Luisetti, J. (1978). Ice nucleation activity among phytopathogenic bacteria. Proceedings of the 4th International Conference on Plant Pathogenic Bacteria, ed. Stat. Path. Végét. Phytobacteriologie, pp. 725-31. Beaucouzé: Inst. Nat. Rech. Agron.

Rajashekar, C.B., Li, P.H. & Carter, J.V. (1983). Frost injury and heterogeneous ice nucleation in leaves of tuber-bearing *Solanum* species. Plant Physiology, *71*, 749-55.

Ruvkun, G.B. & Ausubel, F.M. (1981). A general method for site-directed mutagenesis in prokaryotes. Nature (London), *289*, 85-8.

Spurr, H.W. (1981). Experiments on foliar disease control using bacterial antagonists. *In* Microbial Ecology of the Phylloplane, ed. J.P. Blakeman, pp. 369-81. London: Academic Press.

Teliz-Ortiz, M. & Burkholder, W.H. (1960). A strain of *Pseudomonas fluorescens* antagonistic to *Pseudomonas phaseolicola* and other bacterial plant pathogens. Phytopathology, *50*, 119-23.

Thomson, S.V., Schroth, M.N., Moller, W.J. & Reil, W.O. (1976). Efficacy of bactericides and saprophytic bacteria in reducing colonization and infection of pear flowers by *Erwinia amylovora*. Phytopathology, *66*, 1457-9.

Warren, G., Wolber, P. & Green, R. (1985). Functional significance of oligonucleotide repeats in a bacterial ice nucleation gene. *In* Proceedings of the 6th International Conference on Plant Pathogenic Bacteria, ed. E. Civerolo. College Park, MD: U. S. Department of Agriculture (in press).

White, G.F. & Haas, J.E. (1975). Assessment of Research on Natural Hazards (pp. 304-12). Cambridge, MA: MIT Press.

Yankofsky, S.A., Levin, Z. & Moshe, A. (1981). Association with citrus of ice-nucleating bacteria and their possible role as causative agents of frost damage. Current Microbiology, *5*, 213-7.

# TACTICS AND FEASIBILITY OF GENETIC ENGINEERING OF BIOCONTROL AGENTS

N.J. Panopoulos

*Department of Plant Pathology, University of California, 147 Hilgard Hall, Berkeley, CA 94720, USA*

## *INTRODUCTION*

Micro-organisms interact with plants in a variety of ways, both direct and indirect. Typical examples of direct interactions are those involving classical pathogenesis, symbiosis and production of plant growth-stimulating, growth-retarding or other substances directly affecting the development of plants. Indirect interactions are those in which antagonistic micro-organisms influence the biology of others which directly affect plant growth, nutrition or health. Similar interactions occur between insects, nematodes or other plant pests and weeds which, along with pathogens, are frequently major limiting factors in commercial agriculture and forestry.

The vast array of direct and indirect interactions involving micro-organisms as partners provide broad opportunities for genetic engineering of microbial biocontrol agents directed at plant pathogens or other injurious micro-organisms as well as against weeds and pests. Several biocontrol agents have been successfully employed for such purposes in experimental or commercial agriculture. However, their selection has been largely empirical and any further improvements and development were based exclusively on classical mutation-selection techniques. The advent of recombinant DNA technology and the wide array of tools now available for in vitro and in vivo manipulation of microbial, plant and arthropod genomes have raised hopes that tailor-made biocontrol agents can be constructed.

Extensive discussions and/or documentation on biological control principles, practices, underlying mechanisms and various strategies for improving the efficacy of biological control agents have been extensively discussed (e.g. Baker & Cook 1974; Vidaver 1976, 1982, 1983; Moore & Warren 1979; Kerr 1980; Papavizas & Lumsden 1980; Papavizas 1981, 1986; Schroth & Hancock 1981; Cook & Baker 1983; Lindow 1983, this volume; Schroth *et al.* 1984; Lindemann 1985; Napoli & Staskawicz 1985). Technical accounts on the theory and practice of genetic engineering, in vivo and in vitro mutagenesis, can be found elsewhere (e.g. Bernard & Helinski 1980; Maniatis *et al.* 1982; Williamson 1982; Szostak & Rothstein 1984; Woo 1984). This article will focus on selected aspects of these subjects that are pertinent to the development of microbial biocontrol agents for plant diseases.

### *GENETIC TOOLS AND STRATEGIES*

The use of classical genetic techniques of mutation-selection in biological control agent research and development thus far has been limited to a few properties, model agents and systems. Our limited knowledge of the mechanisms and components which determine an organism's biocontrol potential and the lack of prior genetic work with most candidate agents and the significant amount of effort genetically needed to adapt the classical genetic tools to new organisms has generally discouraged researchers in this area. The development of genetic engineering during the past decade has expanded enormously the range of genetic modifications that may be used in the development of biological control agents, but has also reduced drastically the need for developing many of the necessary tools in each potential agent. Many modifications of genetic material can either be performed in vitro or in well characterized living host cells, such as *Escherichia coli* and yeast. However, many phenotypes of interest to biocontrol agent research cannot be easily recognized in such uncongenial hosts. Empirical clues coupled with the use of improved methods of mutagenesis, cloning and complementation of appropriate mutants will continue to be the tactical basis for identifying and manipulating genes that determine biological performance of a given agent in the biocontrol context.

#### *General strategies and tools for cloning genes*

The cloning, analysis and modification of DNA molecules are accomplished with a simple set of techniques which are well described in the literature (e.g. Maniatis *et al.* 1982; Woo 1984) and are constantly becoming improved as well as simplified. In addition to the numerous *E. coli* plasmid and phage vectors, broad host range as well as shuttle vectors suitable for the propagation of recombinant molecules in a variety of Gram-negative and Gram-positive bacteria, yeasts and other fungi with biocontrol potential are available (e.g. Kahn *et al.* 1979; Bernard & Helinski 1980; Ditta *et al.* 1980; Bagdasarian *et al.* 1981; Bagdasarian & Timmis 1982; Friedman *et al.* 1982; Knauf & Nester 1982; Leemans *et al.* 1982; Maniatis *et al.* 1982; Frey *et al.* 1983; Tait *et al.* 1983; Priefer *et al.* 1985; Turgeon *et al.* 1985; Yoder *et al.* 1985).

Two general strategies are usually followed in gene cloning. One involves the initial construction of what might be called a transposon insertion library, i.e. the isolation of a large number of Tn-generated mutants (see next section). For bacterial genomes (*ca* 4 x $10^3$ kilobase pairs or 4 to 5 x $10^3$ genes), the screening of 5 to 10 x $10^3$ survivors of random Tn mutagenesis can in theory insure the isolation of mutants in virtually any of the organism's non-essential genes. The mutants may be examined individually by applying tests appropriate to the phenotype of interest. These tests can be simple or cumbersome depending on the assay methods available. Once a mutant affected in the phenotype of interest is identified the Tn target fragment can be readily cloned in a plasmid vector by taking advantage of the selectable phenotype of the Tn element or in a phage vector by screening the library through plaque hybridization with a Tn probe. The cloning of the wild-type fragment can usually proceed as discussed below. Certain suicide plasmids induce mutagenesis by vector insertion which permits direct cloning of the target segments without addition of an exogenous vector (Donald *et al.* 1985). Proof that the Tn target fragment or wild-type equivalent indeed controls the phenotype of interest may be obtained by marker exchange (homogenotization) experiments.

Another strategy involves the construction of a genomic library, preferably from randomly generated fragments of total DNA and usually on a plasmid or cosmid vector. Cosmid vectors directly select for cloning of large inserts by virtue of the packaging size-limit of phages. Phage λ vectors do not allow for subsequent complementation experiments in non-hosts of the phage but may be used when the products or functions of the genes of interest are known or when molecular, immunological, enzymatic or other means for identifying the clones of interest are available. For technical reasons, most libraries are constructed with mobilizable cosmid vectors in *E. coli* K12 hosts, preferably deficient in general recombination (*recA*) and DNA restriction (*hsdR* or *hsdS*) but able to modify DNA (*hsdM*$^{+}$) to improve recombinant yield, maintain clonal stability and facilitate interstrain transfer of recombinant plasmids. The capacity of cosmid vectors is such that any bacterial gene or DNA sequence non-lethal to the host or the vector may be represented once or twice in 200 to 500 recombinant colonies.

Generally, insert frequency (i.e. the percentage of clones carrying recombinant molecules) can be determined by small-scale extractions and electrophoresis of plasmid DNA after digestion with a restriction endonuclease (miniscreens). Several vectors allow for insertional inactivation of an antibiotic resistance gene by the cloned DNA so that insert frequency may be determined by replica plating on antibiotic-containing medium. Some vectors allow for the direct selection for recombinant molecules (direct selection vectors), while certain procedures and cloning protocols disallow the circularization of the vector and, thus, reduce the proportion of non-recombinant molecules in the final population. With certain plasmids, such as the pUC series and pLAFR3 (B.J. Staskawicz, personal communication), and hosts (JM series), insert frequency can be readily determined by including a chromogenic substrate (X-gal) for β-galactosidase in the medium (see Maniatis *et al.* 1982).

Tests for gene representation in a genomic library can be based on any readily assayable gene/phenotype. One test that is applicable with any plasmid vector is to determine the frequency of clones that are complemented for an auxotrophic requirement(s) (most bacterial genes and several fungal genes for amino acid biosynthesis are expressed sufficiently well in *E. coli* to enable it to grow without the required supplement). In my laboratory we have also used successfully the '*recA* test' by replicating a few hundred colonies on mutagen-containing media (MMS or mitomycin C) or on regular complex media and exposing the plates to a dose of ultraviolet (UV) light that is lethal to *rec*$^{-}$ strains (2.5 W m$^{-2}$ for *E. coli* HB101). Since most libraries are made in *recA* hosts and since a number of *recA*-like genes have been shown to restore UV and mutagen resistance to *E. coli* (e.g. Better & Helinski 1983; Keener *et al.* 1984; M.J. Hickman, C.S. Orser, D.K. Willis and N.J. Panopoulos, unpublished results), this test may be generally applicable in cloning work.

Once a library is made, it may be screened for the gene of interest in a number of ways. As with Tn insertion libraries, this can be cumbersome when an in vitro test is not available. In practice, one proceeds either through complementation of available mutants or via Southern blot hybridization with Tn target fragments previously cloned. Any type of mutant is useful at this step. A cautionary note concerns the need to verify that cloned fragments of interest actually originate in the DNA of the strain used and not from unsuspected culture contaminants or DNA adhering to glass container walls. This is readily accomplished by showing that the cloned fragment is colinear with its homologue in the source-organism's genome.

In addition to the methods described above DNA segments from a variety of Gram-negative bacteria may be cloned by in vivo methods that do not require an in vitro step. Plasmid pULB113 (RP4-mini-Mu) is a derivative of the broad host range IncP1 plasmid RP4 carrying the mini-Mu transposon Mu3A which is devoid of all of the lethal functions but retains the transposition-related properties of phage Mu. During propagation in a bacterial host this plasmid generates R-primes which carry chromosomal segments up to 80 kb long (Van Gijsegem & Toussaint 1982; Lejeune *et al.* 1983; Schoonejans & Toussaint 1983; Chatterjee *et al.* 1985). A variety of other tools for in vivo genetic engineering have also been developed (see Kleckner *et al.* 1977; Simon *et al.* 1983).

*Transposon mutagenesis*

Gene inactivation by chemical or radiation mutagenesis has several shortcomings, such as the uncontrolled nature of the mutations and the low frequency of mutant recovery for any given locus. The use of high doses of chemical mutagens or radiation to increase mutant yields compounds the problem of multiple silent mutations scattered throughout the genome or clustered in regions which were undergoing replication at the time of treatment, or in various other 'hot spots' (see Miller 1972). To overcome these problems a variety of methods for localized, directed or controlled mutagenesis have been developed (see Itakura *et al.* 1984; Botstein & Shortle 1985).

A commonly used method for obtaining mutants is by using transposable elements (Tn) specifying resistance to an antibiotic. Such elements are usually transferred to a recipient cell via a carrier replicon, either a suicide plasmid or a defective phage. Transposition involves the production of a new copy of the Tn element by specialized mechanisms encoded by the transposon. One copy is retained in the carrier molecule while the other becomes inserted at a new site. Since Tn elements do not normally contain replication origins the copy which transposes can be maintained in the recipient cell only as an integral part of a resident genomic element, for example its chromosome. Some transposons, such as Tn*5*, insert fairly randomly while others (e.g. Tn*7*) have preferred sites. A good test for insertion randomness is to determine the frequency of Tn-induced auxotrophs (usually 0.5 to 2%) or to determine the sizes of Tn target fragments in a randomly selected sample of mutants by blot hybridization with a Tn-specific probe.

Transposition frequencies may vary with the element and cell type and many strains are non-receptive to attempted introduction of the suicide element, due to DNA restriction or other barriers. A common practice which often works is to screen several strains and select those which yield progeny at frequencies well above the frequency with which spontaneous mutations leading to resistance to the counterselecting antibiotics occur. Alternatively, a different donor species may be tried. When one is forced to work with a non-receptive but biologically interesting strain it is possible to select receptive mutants by chemical mutagenesis and selection for rare recombinants acquiring a plasmid (Mindich *et al.* 1976; N.J. Panopoulos, unpublished results). Such mutants may also be more efficient donors because they would 'modify' carrier DNA and thus provide protection against restriction in the isogenic recipient (M. Rella, personal communication).

Several Tn-carrier plasmids have been successfully used to deliver transposons to plant-pathogenic, symbiotic, saprophytic and other bacteria (see Panopoulos 1981; Berg & Berg 1983; De Bruijn & Lupski 1984; Mills 1985). Those most commonly employed have the broad conjugational host range of P or W group plasmids and suicide properties provided either by a narrow host range replication system or by the presence of phage Mu1. Phage λ carriers are less frequently used for Tn mutagenesis in these bacteria which presumably lack suitable adsorption sites. Extension of phage λ to *Rhizobium* sp. was accomplished by prior introduction of a recombinant plasmid expressing the λ receptor protein (De Vries *et al.* 1984).

Selection of transposon insertion mutants is accomplished by plating on antibiotic-supplemented plates. High basal levels of antibiotic tolerance and high mutation rates to the counter-selecting antibiotics are often encountered among *Pseudomonas* spp. and other plant-associated bacteria. These problems may be overcome by choosing a different transposon. Several derivatives of Tn*5* carrying an alternate marker (e.g. Tet, *lac*) are also available (e.g. Bellofatto *et al.* 1984; Kroos & Kaizer 1984). With proper adjustment of antibiotic levels every cell that survives suicide Tn-mutagenesis gives rise to a mutant colony usually carrying one inserted Tn copy and, presumably, one mutation. Exceptions to this have been noted, for example, when an inverted repeat of Tn*5* (IS50) transposes independently of or in addition to the element, when phage Mu insertions occur concurrently with Tn*5* on Mu-containing plasmids, or when endogenous insertion sequences in the recipient have also transposed to a different site (see Berg & Berg 1983; Leong 1984; Mills 1985).

The suitability of Tn mutagenesis in genetic analysis has been further extended by the construction of several modified Tn*5*, Tn*3* and Mu derivatives containing a promoterless gene (Bellofatto *et al.* 1984; Castilho *et al.* 1984; Kroos & Kaizer 1984; Stachel *et al.* 1985; Perkins & Youngman 1986). Transcriptional or translational fusions between these genes and those located at the Tn-target site facilitate analysis of gene regulation of the transcriptional or translational level. Translational fusions yield hybrid proteins which facilitate the identification and purification of unknown gene products (Silhavy & Beckwith 1985).

Although Tn insertion usually inactivates gene function, activation of genes located on either side of Tn*5* (see Berg & Berg 1983; De Bruijn & Lupski 1984) and Tn*10* (Chiampi *et al.* 1982; Simons *et al.* 1984) has been observed. This is due to the presence of outward reading promoters and/or the creation of fortuitous promoters of the Tn::junction site.

## *GENETIC MODIFICATIONS*

Broadly speaking, genetic modifications of micro-organisms may involve the addition of new genetic material, which may include chemically synthesized genes or oligonucleotides, the deletion of resident genes or the restructuring of parts of the organism's own genome. Only some of these modifications can be performed with conventional genetic techniques, such as undirected in vivo mutagenesis, selection and DNA transfers through standard gene exchange and recombination processes operating in each organism. Nonconventional methods involve one or more in vitro steps catalyzed by enzymes acting on DNA (or RNA) substrates. Certain genetic elements, such as plasmids, phages or other viruses and transposons serve as vehicles for the introduction, stabilization, replacement, deletion or amplification of DNA segments in the organism of interest.

### *Gene deletions*

The construction of ice nucleation deletion mutants of *P. syringae* as biocontrol agents of frost injury illustrates a generally applicable method for deleting undesirable genes or replacing alleles in bacteria (Fig. 1). The nucleation gene (*ice*) was initially cloned by cosmid-cloning procedures (Orser *et al.* 1985). The extent of the coding regions was then determined by progressive subcloning of DNA fragments, deleting overlapping segments of the minimal-size insert and isolating and scoring the Ice phenotype of Tn*5* insertions throughout the cloned DNA segment. Based on this information portions internal to the coding region bracketed by suitable restriction sites were excised in vitro and confirmed not to confer an Ice$^+$ phenotype to *E. coli* or *P. syringae*. The fact that the phenotype was expressed in *E. coli* and could be easily scored greatly expedited the initial cloning and many of the above steps. The vector was plasmid pBR325, which replicates normally in *E. coli* but not in *P. syringae*. Because the plasmid is not self-transferable the deleted derivatives were introduced into *P. syringae* by triparental matings using another self-transmissible plasmid as a helper element. An alternate method, which was not tried, would have been transformation with purified plasmid DNA. Antibiotic-resistant colonies (Tc$^r$) were selected in which the vector integrated into *P. syringae* chromosome presumably by a single recombination event between one of the remaining *ice* region segments and its chromosomal homologue. These strains were phenotypically Ice$^+$, and contained a tandem duplication of the part of their chromosome (*cis*-merodiploids) consisting of the parts of the *ice* region remaining on the deletion plasmid and the chromosomal wild-type allele separated by the vector DNA. They were subsequently cycled several times in antibiotic-free broth, and several derivatives were obtained which had lost the vector markers and sequences as well as the ability to nucleate ice and cause frost injury to plants in growth chamber tests. Loss of these features presumably occurred by a second cross-over between the homologous duplicate sequence regions lying on the opposite side of the deletion relative to the site of the first cross-over event. Unique restriction sites within and outside the *ice* region were used as reference points to confirm the predicted structure of the ice region in these derivatives (Ice$^-$ excision haploids) through Southern blot hybridization with appropriate DNA segments as probes (C.S. Orser, R. Lotskin, D.K. Willis, N.J. Panopoulos and S.E. Lindow, unpublished results).

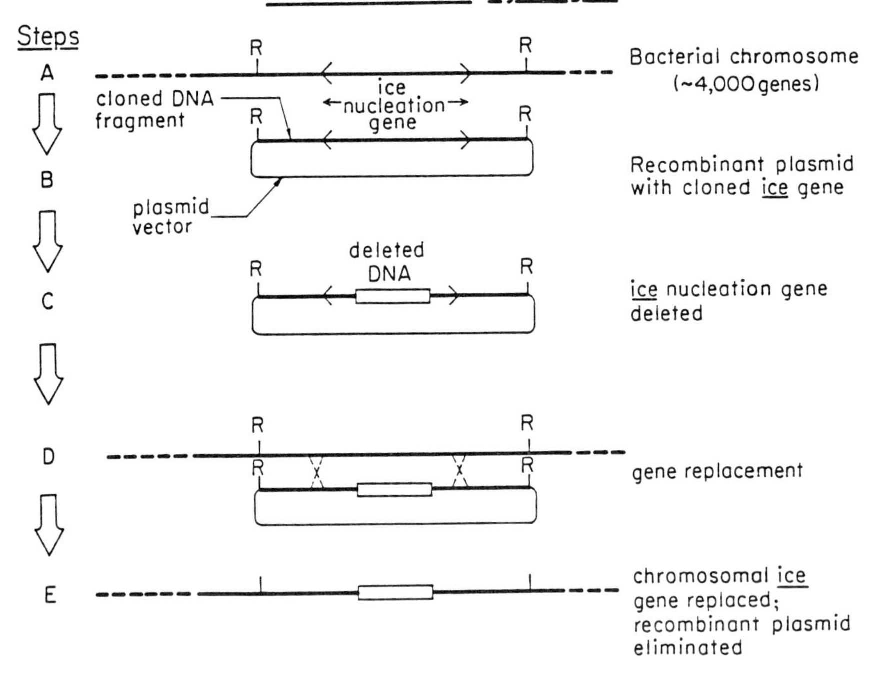


Fig. 1. Replacement mutagenesis in *Pseudomonas syringae* using in vitro-constructed *ice* deletion alleles. A deletion removing a segment internal to the *ice* gene is shown. R designates restriction endonuclease cleavage sites at the ends of the cloned fragment. An *E. coli* host was used in steps B and C and *P. syringae* in A, D and E (Orser *et al.* 1985).

Variations of the above method include the use of *lacZ*-containing vectors to permit visual screening of haploidized strains (G. Warren, personal communication) and the use of replicative vectors derived from broad host-range plasmids which are either lost spontaneously in antibiotic-free medium (P.B. Lindgren and N.J. Panopoulos, unpublished results; B.J. Staskawicz, personal communication) or eliminated by subsequent introduction of an incompatible plasmid (Ruvkun & Ausubel 1981). Other modified forms of allele replacement are described by Botstein & Shortle (1985).

*Bacillus* spp. are frequently mentioned in the literature on biological control (e.g. Baker & Cook 1974). A method to generate deletions in *Bacillus subtilis* has been developed (Niandet *et al.* 1985). Linear molecules composed of two regions homologous with the chromosome flanking a 'central' non-homologous segment can integrate into the chromosome by two cross-overs provided that the flanking regions are in the same relative orientation. If no essential genes are present between the flanking regions and if the recombinant molecules do not replicate in the cell, the central region of the integrating molecule replaces its chromosomal homologue (if essential genes are present merodiploid strains can be produced).

### *Gene additions*

There are two basic requirements that must be fulfilled in order to accomplish gene addition in any organism, namely, insuring the gene's replication between cell divisions and its proper partition into the daughter cells at each division. DNA replication in vivo is a complex process with a strict requirement for a starting point, the origin of replication (*ori*, *rep*) as well as a terminus. Prokaryotic chromosomes, phages and most plasmids have single replication origins so that most DNA fragments derived from them are not proper substrates for the cellular DNA polymerases to initiate DNA replication. Eukaryotic DNAs have a higher density of replication initiation points, the autonomous replicating sequences (*ars*), which can function as such in different eukaryotic cell types (Petes 1980; Stinchcomb *et al.* 1980). Partition functions are normally provided for in all bacterial chromosomes and plasmids; eukaryotic chromosome partition is controlled by the centromeres.

Gene addition may, therefore, be accomplished either by inserting the gene in an existing chromosome or constructing a new chromosome. Cloning in plasmid vectors is an example of the latter approach. More complicated is the construction of new chromosomes not involving a plasmid module (e.g. eukaryotic chromosomes) which requires elements with centromer and telomere sequences. However, this has been accomplished, for example, in yeast (Murray & Szostak 1983). Insertion of genes into ordinary chromosomes utilizes indirect methods because of the inherent difficulty in isolating and restructuring entire chromosomes in vitro. One commonly used method involves the use of accessory genetic elements, such as transposons or insertion sequences (IS), which are unable to replicate independently but empowered with ability to insert themselves into DNA genomes in vivo. The gene is initially inserted into these elements by in vitro techniques in a region that is not essential for transposition and the composite element is then transferred into the recipient cell where it is rescued by becoming inserted (transposing) into a resident genomic element. Another method exploits the inherent ability of cells to recombine DNA molecules that are perfectly homologous, or nearly so, with their own cellular DNA by a process known as homologous recombination. Molecules having homologous regions on one or both sides of a foreign gene can be assimilated into the resident genomic elements by either one or two cross-over events depending on whether the element is circular or linear, respectively. In bacteria, yeasts and sometimes *Aspergillus nidulans*, this process guides the recombining molecules to the genomic sites where the homologous resident sequences are normally located. Any piece of DNA cloned from the recipient organism can be used for this purpose. In this, as in Tn- or IS-mediated transfers, disruption of gene function at the insertion point may occur. This can be avoided by selecting insertions in non-essential intergenic (non-coding, non-regulatory) regions or by providing a duplicate copy of the interrupted gene (or the entire transcriptional unit if the gene belongs to an operon).

### *New methods for blocking gene expression*

A new method for preventing gene expression without altering the target gene exploits the ability of non-translated RNA molecules complementary to coding RNAs (antisense copy) to form duplex RNA molecules in vivo, thereby preventing mRNA translation. With some, but few, known exceptions (reviewed by Normak *et al.* 1983) cellular mRNAs represent complementary sequences of only one strand of duplex DNA. However, cells can be programmed to transcribe the opposite strand by cloning the DNA segment in a 'flipped' orientation relative to its native promoter region or by providing a suitable promoter 5' upstream from the antisense region. Duplex formation between sense and antisense RNAs is highly specific. Antisense molecules complementary to various parts of coding regions or to as little as 52 nucleotides including the 5' upstream region of the sense mRNA inhibit the expression of targeted endogenous genes (Izant & Weintraub 1985). Although not yet applied to biocontrol agent research, this method for blocking gene expression is potentially useful especially with diploid, multinucleate or heterokaryotic fungi where more than one gene copy must be targeted by other mutational methods, or whenever allele replacement cannot be directed to their homologous region, or when the target gene is present in more than one copy per haploid genome.

## *GENES AND MICRO-ORGANISMS WITH BIOCONTROL POTENTIAL*

Biological efficacy of a biocontrol agent is undoubtedly determined by a constellation of factors, mostly undefined. However, like any other property they can be genetically dissected by mutation and complementation analysis. Such analysis may exploit naturally existing variation between competent and non-competent strains or variation induced by artificial mutagenesis. The potential utility of heterospecific DNA transfers, now made routine by the recombinant DNA technology for genetic analysis purposes as well as for strain construction and improvement, remains to be explored.

The properties which may be desired in a microbial biocontrol agent will naturally differ according to the task the agent is called to perform. Ability to multiply and establish itself in the intended habitat in sufficient numbers long enough to exert an effect upon the target organism is one basic prerequisite to efficacy. However, there are few clearcut or documented examples of specific traits that have been successfully introduced into micro-organisms from another source or of traits that occur naturally in the candidate agent but were eliminated with a resultant improvement in habitat colonization under field conditions. There is a strong need to address these and other pertinent questions at a fundamental level. A molecular genetic approach would seem worthwhile as well as necessary, if habitat colonization and competition factors are to be elucidated. At present, such research in most fungi with biocontrol potential would be seriously limited by current technology. Although available tools could presumably be adapted to such organisms, several years of effort would be needed. For most bacterial agents, on the other hand, technological limitations are not serious and the smaller genome size compared to fungi should further simplify the task. The logistics that pertain to such analysis (i.e. testing of several hundreds or a few thousands of strains) should be no more discouraging than, for instance, the analysis of bacterial symbiosis or pathogenicity which have been successfully accomplished in a number of cases. However, simplified test systems need to be developed, which shorten the time required for performance tests to a minimum, still retain the physical and biological parameters, and provide some degree of containment for experiments involving the use of transposable elements as mutagens or recombinant plasmids as cloning vehicles.

A number of bacterial phytopathogens are specially adapted for epiphytic growth and often constitute a dominant component of microbial flora on foliage or other above-ground plant surface habitats (Hirano & Upper 1983). In theory, stable genetically engineered mutations leading to loss of pathogenicity but not affecting adversely the organism's epiphytic colonization potential would provide a good example of genetic manipulations for the development of biocontrol agents directed against the pathogen itself. Theoretically, such mutants would be expected to have maximum habitat overlap and, therefore, competitive exclusion potential against their natural counterpart. Recent work on the genetics of pathogenicity makes the above hypothesis easily testable. Some aspects of this work are briefly reviewed below.

Bacterial pathogenicity on plants is determined by several genes, the number of which is not precisely known, and their exact role in pathogenesis varies. Certain genes encode the production of enzymes involved in the biosynthesis of phytotoxins, growth hormones or enzymes capable of degrading plant cell wall or other constituents. Since these genes are positively needed for pathogenesis, their inactivation (e.g. through deletion replacement techniques) would destroy the organism's pathogenic potential or reduce its virulence depending on whether the metabolite or enzyme is a pathogenicity or a virulence factor (see Yoder 1982). In some instances such a modification will have no effect if the bacterium has alternate functions for pathogenicity or if duplicate or multiple genes are present (e.g. genes encoding for pectate lyase isoenzymes in soft rot Erwiniae; D. Coplin, personal communication).

Recent studies (Mills 1985; Niepold *et al.* 1985; Panopoulos & Peet 1985; Panopoulos *et al.* 1985; P.B. Lindgren, R.C. Peet and N.J. Panopoulos, unpublished results) revealed other classes of genes that are also needed in a positive sense for pathogenicity and, although their products and functions are still unknown at the biochemical level, they appear to be distinct from the genes mentioned above. These genes appear to be conserved at the DNA sequence level in several phytopathogens and in some cases they are interchangeable. Thus, they may be easily cloned from other organisms of interest by screening a genomic library with segments already cloned or possibly through complementation tests with existing mutants. In *P. syringae* pv. *phaseolicola* many of these genes control pathogenicity on the homologous host and the elicitation of the hypersensitive reaction (HR) on non-host plants while others control pathogenicity only. The fact that HR-controlling sequences also control pathogenicity on the organism's homologous host and that HR-inducing ability is a common characteristic of phytopathogenic *Pseudomonas* spp. and several other plant pathogens suggests that certain genes may be absolutely required for these bacteria to be pathogenic on any plants. If so, phytopathogens from which such genes are deleted would be expected to be universally phytosafe. The effects of such deletions on epiphytic colonization potential of these organisms have not yet been assessed. A Tn*5*-induced non-pathogenic mutant of a brown spot strain of *P. syringae* pv. *syringae* (which still elicited HR) constructed in this laboratory was shown in greenhouse tests to colonize bean foliage, compete very effectively against its parent and to reduce substantially disease incidence (D.K. Willis, S.E. Lindow and N.J. Panopoulos, unpublished results).

Besides positively-acting genes for pathogenicity or virulence, pathogens often have genes which act in a negative sense upon host range. Such genes determine race-cultivar specificity, for example in *P. syringae* pv. *glycinea* (Staskawicz *et al.* 1984) and *Xanthomonas campestris* pv. *malvacearum* (Gabriel & Lazo 1985), by restricting the host range of individual races to only those cultivars of soybean lacking genes for race-specific resistance (R genes). Furthermore, when such genes are inactivated by mutation the host range of a particular race is extended to one or more resistant cultivars which carry functionally correspondent R genes. These genes, therefore, act in accordance with the gene-for-gene hypothesis (Flor 1955) which operates in many plant host-pathogen and pest systems. It is conceivable that deletion of one or more such genes from weed pathogens would extend their host range and thus make them more effective weed biocontrol agents.

*BIOSAFETY AND CONTAINMENT*

The range of habitat which a given biocontrol agent may successfully colonize can be assumed to be determined primarily by the agent's natural adaptation and fitness characteristics. The possibility that these properties may be altered intentionally or unintentionally by genetic modification in undesirable ways is a cause for concern. Some modifications (e.g. ice nucleation, toxin or pathogenicity gene deletions) would appear to be benign in and of themselves because they do not represent addition of genetic material from outside the organism's natural gene pool, do not alter the transmission potential of the genes involved and would seriously reduce or eliminate the agent's pathogenic or injurious potential, at least on plants. Furthermore, these or analogous genetic events almost certainly occur spontaneously in the organism and can be readily produced by ordinary chemical or radiation mutagenesis. For other types of modifications this may not be the case. The regularly repeated consensus sequence structure of *ice* genes (Green & Warren 1985; M. Mindrinos, L.G. Rahme, N.J. Panopoulos and S.E. Lindow, unpublished results) makes them especially prone to spontaneous deletion formation. Small-scale field trials with such agents on plants would be unlikely to cause a serious problem. However, this may not be a safe assumption for deployment on a large scale since the target organism may perform other, useful functions in the ecosystem, a possibility which although conjectural, would be imprudent to totally dismiss. This illustrates a basic contradiction in many biological control strategies, whether through natural or engineered agents.

It is possible to envision genetic modifications that would enable candidate biocontrol agents to attack only the intended target organism, or would make the agent habitat-dependent. Such possibilities would have considerable scientific and ecological merit. However, we presently lack knowledge on what determines habitat specificity and fitness. The likely, although untested possibility that such modifications reduce the organism's field performance, which would make it commercially unacceptable, strongly discourages initiatives in this direction. In theory, habitat dependency may be imparted upon an organism in a number of ways. For instance, modifications can be envisioned whereby one of the organism's 'household' genes is placed under the genetic control of a gene which is specifically induced by a factor(s) encountered only in its intended habitat. Although such factors are largely unknown at present they are likely to exist for many types of habitats.

A different type of containment would aim at reducing gene mobility under natural conditions. Genetic exchanges are more likely between organisms that are genetically closely related. Genetic elements such as plasmids, transposons or insertion sequences are generally more mobile than ordinary chromosomal genes. Transmission to the target organism of genetic information encoding immunity functions that normally protect the biocontrol agent producing an antimicrobial compound (e.g. a bacteriocin) would be clearly undesirable from the standpoint of an agent's commercial durability. Restriction or containment of genes within the organism in which they were originally engineered is desirable from an environmental, regulatory and legal standpoint. A range of genetic techniques could be used to minimize this problem or to prevent expression of engineered genes in other organisms, but detailed discussions would be outside the scope of this article.

*MONITORING OF RECOMBINANT MICRO-ORGANISMS IN ENVIRONMENTAL RELEASE PROGRAMS*

The debate over biosafety of small-scale field trials of

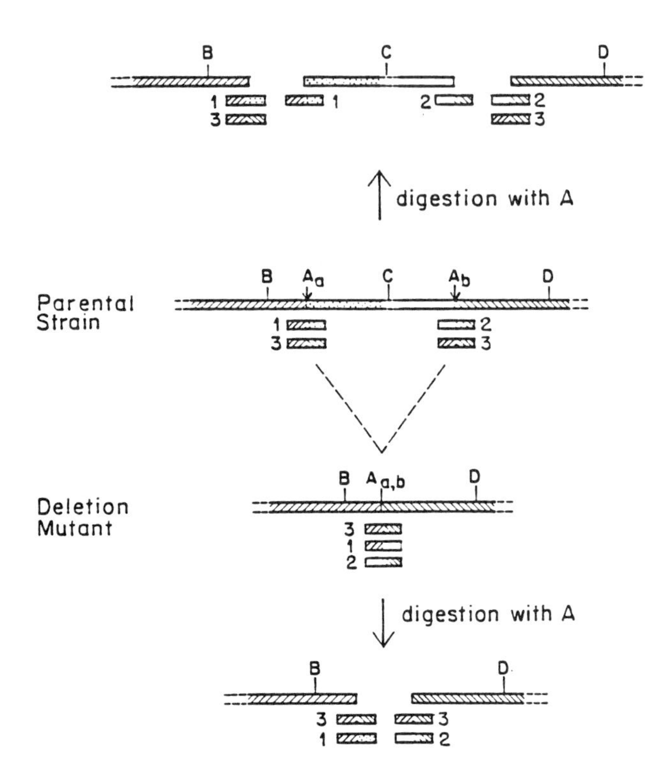


Fig. 2. Detection and identification of hypothetical wild-type and deletion mutants by hybridization with synthetic oligonucleotide probes. $A_a$, $A_b$, B, C and D designate restriction endonuclease cleavage sites. Three different probes are depicted. Probes 1 and 2 match the nucleotide sequence at the ends of the $A_a$-$A_b$ segment which is removed by the deletion. Similarly shaded areas represent matching sequences on the probes and chromosomes. Each probe is shown underneath each point of full or partial homology for ease of observation. Only full matches will produce a hybridization signal. Thus, probes 1 and 2 will specifically detect the wild-type strain (hybridization at points $A_a$ and $A_b$, respectively) while probe 3 will only detect the mutant strain. Susceptibility of the hybridization signal prior to digestion of DNA with enzyme A provides further positive proof of strain identity.

Ice⁻ deletion mutants of *P. syringae* has brought into focus legal and regulatory aspects of environmental release programs involving genetically engineered organisms. These issues are likely to have a long-term impact on the future of biotechnology worldwide. One important question related to this subject is whether microbial strains can be readily and unambiguously identified once they are released. For most micro-organ-

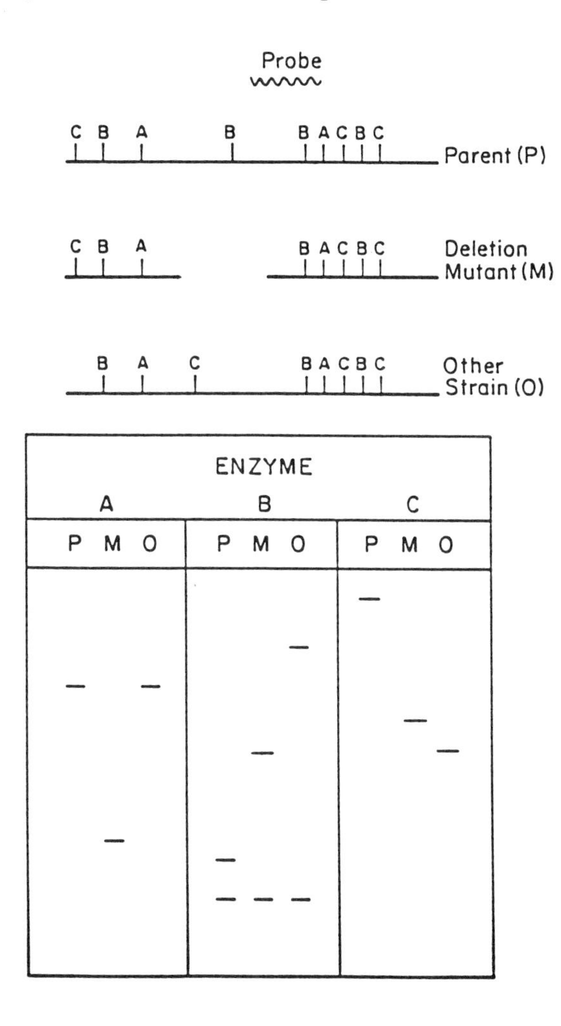


Fig. 3. Differentiation between parental (P), deletion mutant (M) and other closely related strains (O) based on restriction fragment length polymorphisms (RFLPs). An appropriate nucleic acid probe (shown at the top of the diagram) will hybridize to unique segments in each of the strains (fragment sizes are only indicative). B and C designate restriction sites in the genomic DNA. A is invariant in all three strains; B is within the deleted fragment and, therefore, absent from the mutant; C is a variant site among closely related wild-type strains.

isms present taxonomic information is either inadequate or difficult to adapt to routine use when large numbers of specimens must be analyzed. Selective or differential media are lacking for many microbial species. When such media are available, they may be useless in situations other than those for which they were developed (e.g. when samples from a different or a more complex habitat must be examined). Nutritional or other identification tests (e.g. bacteriocin or phage typing, pathogenicity), although adequate for certain micro-organisms or taxonomic groups, are not available for many others or too laborious for large-scale monitoring programs. For genetically modified micro-organisms, the identification requirements are considerably more stringent than those of ordinary taxonomic or population studies. Risk assessment studies as well as patent enforcement will require unambiguous identification of strains released in the environment. Detection methods must be able to differentiate a mutant strain from all closely related strains including its genetic parent; they must be adaptable to routine screening of large numbers of colonies with a minimum number of tests at the lowest possible cost.

When a new gene or synthetic DNA has been added to the organism in question, the same molecule can be used as a hybridization probe to screen any number of colonies or samples through dot-blot procedures. Similar possibilities also exist for other types of genetic modification which

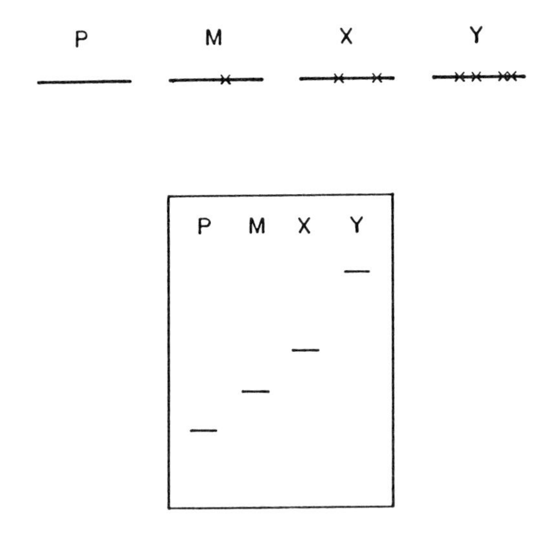


Fig. 4. Differentiation between parental (P), mutant (M) and closely related strains (X, Y) based on differential melting of DNA hybrids in polyacrylamide formamide-urea gradient gels (e.g. Myers *et al.* 1985). Movement of the fragment stops when they reach a point in the gel where denaturant concentration is sufficient to melt the duplex. Single base pair mismatches (indicated by ×) are sufficient for separation.

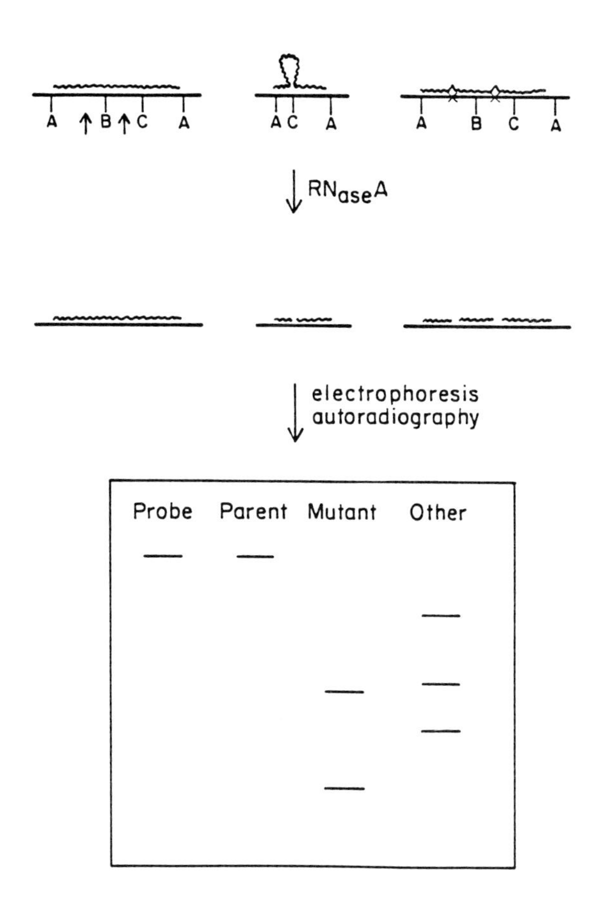


Fig. 5. Differentiation among closely related strains based on detection of nucleotide sequence polymorphisms by RNase A cleavage of RNA:DNA hybrids. Labelled RNA probes prepared by in vitro transcription of DNA fragments cloned adjacent to an SP6 or other similar promoter are hybridized to DNA that has been cleaved at sites (A) outside the region complementary to the probe. RNA:DNA duplexes are then cleaved with RNase A and separated by denaturing gel electrophoresis (Myers *et al.* 1985). The number of RNA fragments detected equals the number of mismatches in the region covered by the probe plus one. If the sum of the fragment lengths differs from that of the probe deletions or insertions relative to the probe are present in the strains. The points of mismatches can be mapped by cleavage with an asymmetrically located site (C) within the region homologous to the probe. An alternative method would involve cleavage of DNA:DNA hybrid duplexes with the single-strand-specific nuclease.

do not involve the addition of new genetic material. For example, when a DNA segment is deleted, nucleotide sequences that were previously separated are brought adjacent to each other, forming a novel joint (Fig. 2). Similarly, inversions, or additions of new pieces of DNA (synthetic or cloned from another cell), as well as single nucleotide substitutions have a similar effect.

Synthetic oligonucleotides complementary to the sequence created at the novel joint provide one way to positively detect such modifications. The template specificity of oligonucleotide probes is such that they will hybridize only if perfectly base-paired. For example, the absolute specificity of nonadecanucleotides (19-mers) permitted the distinction between mutant alleles of the β-globin gene in humans carrying the sickle-cell trait and the diagnosis of α-thalassemias and α-trypsin deficiency resulting from single base pair alterations in the respective structural genes (see Itakura *et al.* 1984 for a review). The use of synthetic oligonucleotides to positively identify and distinguish deletion mutants from their genetic parent is illustrated in Fig. 2. Restriction fragment length polymorphisms (RFLPs) are also increasingly being used in prenatal diagnosis of human diseases (see Jordan 1985 for a review). These and other methods, capable of detecting and differentiating specific strains from closely related counterparts are illustrated in Figs 3, 4 and 5. Details of the methods may be found in Itakura *et al.* (1984), Myers *et al.* (1985) and references cited therein.

## *CONCLUSIONS AND PROSPECTS*

After more than a decade since their original discovery recombinant DNA techniques are rapidly entering a stage of broad use in microbial biocontrol agent genetic research and development. The lack of genetic knowledge about these organisms is no longer a barrier to their genetic modification. The ability to clone, alter and introduce DNA sequences in a given organism offers exciting prospects for analyzing microbial fitness and competition-determining traits and for redesigning biocontrol agents to perform desirable functions. A critical factor in this effort is biosafety as it relates to potentially adverse short and long-term biological impact on target and non-target organisms, evolving governmental regulations and public acceptance of the prospect of releasing genetically modified organisms in the environment. Although in many cases adverse effects on non-target organisms or the environment would be unlikely, if at all measurable, on a small scale, this cannot be persuasively argued for large-scale deployment. Innovative approaches to biosafe agent design and long-term risk assessment methods constitute an important new challenge in biological control and genetic research in general.

## *ACKNOWLEDGEMENTS*

Work in the author's laboratory reported here was supported in part by a National Science Foundation Grant PCM 84-09723. I thank P.B. Lindgren, D.K. Willis, R.C. Peet, M.J. Hickman, M. Mindrinos, B.J. Staskawicz and S.E. Lindow for sharing unpublished data.

*REFERENCES*

Bagdasarian, M., Lurz, R., Rückert, B., Franklin, F.C.H., Bagsadarian, M.M., Fey, J. & Timmis, K.N. (1981). Specific-purpose plasmid cloning vectors. II: Broad host range, high copy number, RSF1010-derived vectors, and a host-vector system for gene cloning in *Pseudomonas*. Gene, *16*, 237-47.

Bagdasarian, M. & Timmis, K.N. (1982). Host vector systems for gene cloning in *Pseudomonas*. Current Topics in Microbiology and Immunology, *96*, 46-67.

Baker, K.F. & Cook, R.J. (1974). Biological Control of Plant Pathogens. San Francisco: W.H. Freeman & Co.

Bellofatto, V., Shapiro, L. & Hodgson, D.A. (1984). Generation of a Tn*5* promoter probe and its use in the study of gene expression in *Caulobacter cresentus*. Proceedings of the National Academy of Sciences of the U.S.A., *81*, 1035-9.

Berg, D.E. & Berg, C.M. (1983). The prokaryotic transposable element Tn*5*. Bio/Technology, *1*, 417-35.

Bernard, H.U. & Helinski, D.R. (1980). Bacterial plasmid cloning vehicles. *In* Genetic Engineering - Principles and Methods, eds J. Settow & A. Hollaender, pp. 113-67. New York: Plenum Press.

Better, M. & Helinski, D.R. (1983). Isolation and characterization of the *recA* gene of *Rhizobium meliloti*. Journal of Bacteriology, *155*, 311-6.

Botstein, D. & Shortle, D. (1985). Strategies and applications of in vitro mutagenesis. Science, *229*, 1193-201.

Castilho, B.A., Olfson, P. & Casadaban, M.J. (1984). Plasmid insertion mutagenesis and *lac* gene fusion with mini-Mu bacteriophage transposons. Journal of Bacteriology, *158*, 488-95.

Chatterjee, A.K., Ross, L.M., McEvoy, J.L. & Thurn, K.K. (1985). pULB113, an RP4::mini-Mu plasmid, mediates chromosomal mobilization and R-prime formation in *Erwinia amylovora*, *Erwinia chrysanthemi* and subspecies of *Erwinia carotovora*. Applied and Environmental Microbiology, *50*, 1-9.

Chiampi, M.S., Schmid, M.B. & Roth, J.R. (1982). Transposon Tn*10* provides a promotor for transcription of adjacent sequences. Proceedings of the National Academy of Sciences of the U.S.A., *79*, 5016-20.

Cook, R.J. & Baker, K.F. (1983). The Nature and Practice of Biological Control of Plant Pathogens. St. Paul: Am. Phytopath. Soc.

De Bruijn, F.J. & Lupski, J.R. (1984). The use of transposon Tn*5* mutagenesis in the rapid generation of correlated physical and genetic maps of DNA segments cloned into multicopy plasmids - a review. Gene, *27*, 131-49.

De Vries, G., Raymond, C.K. & Ludwig, R.A. (1984). Extension of bacteriophage λ host range: selection, cloning and characterization of a constitutive λ receptor gene. Proceedings of the National Academy of Sciences of the U.S.A., *81*, 6080-4.

Ditta, G., Stanfield, S., Corbin, D. & Helinski, D.R. (1980). Broad host range DNA cloning system for Gram-negative bacteria: construction of a gene bank of *Rhizobium meliloti*. Proceedings of the National Academy of Sciences of the U.S.A., *77*, 7347-51.

Donald, G.K., Raymond, C.K. & Ludwig, R.A. (1985). Vector insertion mutagenesis of *Rhizobium* strain ORS571: direct cloning of mutagenized sequences. Journal of Bacteriology, *162*, 317-23.

Flor, H.H. (1955). Host-parasite interaction in flax rust - its genetics and other implications. Phytopathology, *45*, 680-5.

Frey, J., Bagdasarian, M., Feiss, D., Franklin, C.H. & Deshusses, J. (1983). Stable cosmid vectors that enable the introduction of cloned fragments into a wide range of Gram-negative bacteria. Gene, *24*, 299-308.

Friedman, A.M., Long, S.R., Brown, S.E., Buikema, W.J. & Ausubel, F.M. (1982). Construction of a broad host range cosmid cloning vector and its use in the genetic analysis of *Rhizobium meliloti*. Gene, *18*, 289-96.

Gabriel, D. & Lazo, G.K. (1985). Specific avirulence genes from *Xanthomonas campestris* pv. *malvacearum*. *In* Plant Cell/Cell Interactions, eds I. Sussex,

A. Ellingboe, M. Crouch & R. Mulmberg, pp. 103-7. Cold Spring Harbor: Cold Spring Harbor Laboratory.

Green, R.L. & Warren, G.J. (1985). Physical and functional repetition in a bacterial ice nucleation gene. Nature (London), *317*, 645-8.

Hirano, S.S. & Upper, C.D. (1983). Ecology and epidemiology of foliar bacterial plant pathogens. Annual Review of Phytopathology, *21*, 243-69.

Itakura, K., Rossi, J.J. & Wallace, R.B. (1984). Synthesis and use of synthetic oligonucleotides. Annual Review of Biochemistry, *53*, 323-56.

Izant, J.G. & Weintraub, H. (1985). Constitutive and conditional suppression of exogenous and endogenous genes by anti-sense RNA. Science, *229*, 345-52.

Jordan, B.R. (1985). Antenatal diagnosis by DNA analysis: current status, future developments and a few unanswered questions. BioEssays, *2*, 196-201.

Kahn, M., Kolter, R., Thomas, C., Figurski, D., Meyer, R., Remaut, E. & Helinski, D.R. (1979). Plasmid cloning vehicles derived from plasmids ColE1, R6K, and RK2. Methods in Enzymology, *68*, 268-80.

Keener, S.L., McNamee, K.P. & McEntee, K. (1984). Cloning and characterization of *recA* genes from *Proteus vulgaris*, *Erwinia carotovora*, *Shigella flexneri* and *Escherichia coli* B/r. Journal of Bacteriology, *160*, 153-60.

Kerr, A. (1980). Biological control of crown gall through production of agrocin 84. Plant Disease, *64*, 25-30.

Kleckner, N., Roth, J. & Botstein, D. (1977). Genetic engineering *in vivo* using translocatable drug resistance elements: new methods in bacterial genetics. Journal of Molecular Biology, *116*, 125-59.

Knauf, V.C. & Nester, E.W. (1982). Wide host range cloning vectors: a cosmid clone bank of an *Agrobacterium* Ti plasmid. Plasmid, *8*, 45-54.

Kroos, L. & Kaizer, D. (1984). Construction of Tn*5lac*, a transposon that fuses *lacZ* expression to exogenous promoters and its introduction into *Myxococcus xanthus*. Proceedings of the National Academy of Sciences of the U.S.A., *81*, 5816-20.

Leemans, J., Langenakens, J., DeGreeve, H., Deblaere, R., Van Montagu, M. & Schell, J. (1982). Broad host range vectors derived from the W plasmid pSa. Gene, *19*, 361-4.

Lejeune, P., Margaey, M., Van Gijsegem, F., Faelen, M., Gerits, J. & Toussaint, A. (1983). Chromosome transfer and R-prime plasmid formation mediated by plasmid pULB113 (RP4::Mini-Mu) in *Alkaligenes eutrophus* CH34 and *Pseudomonas fluorescens* 6.2. Journal of Bacteriology, *155*, 1015-26.

Leong, S. (1984). Gene transfer in phytopathogenic prokaryotes. *In* Advances in Molecular Genetics of the Bacteria-Plant Interaction, eds A.A. Szalay & R.P. Legocki, pp. 207-11. Ithaca, N.Y.: Media Services, Cornell University Publishers.

Lindemann, J. (1985). Genetic manipulation of microorganisms for biological control. *In* Biological Control on the Phylloplane, eds C.E. Windels & S.E. Lindow, pp. 116-30. St. Paul: Am. Phytopath. Soc.

Lindow, S.E. (1983). The role of bacterial ice nucleation in frost injury to plants. Annual Review of Phytopathology, *21*, 363-84.

Maniatis, T., Fritsch, E.F. & Sambrook, J. (1982). Molecular Cloning: A Laboratory Manual. Cold Spring Harbor: Cold Spring Harbor Laboratory.

Miller, J.H. (1972). Experiments in Molecular Genetics. Cold Spring Harbor: Cold Spring Harbor Laboratory.

Mills, D. (1985). Transposon mutagenesis and its potential for studying virulence genes in plant pathogens. Annual Review of Phytopathology, *23*, 297-320.

Mindich, L., Cohen, J. & Weisburd, M. (1976). Isolation of nonsense suppressor mutants in *Pseudomonas*. Journal of Bacteriology, *126*, 177-82.

Moore, L.W. & Warren, G. (1979). *Agrobacterium radiobacter* strain 84 and biological control of crown gall. Annual Review of Phytopathology, *17*, 163-79.

Murray, A.W. & Szostak, J.W. (1983). Construction of artificial chromosomes in yeast. Nature (London), *305*, 189-93.

Myers, R.M., Larin, Z. & Maniatis, T. (1985). Detection of single base substitutions by ribonuclease cleavage at mismatches in RNA-DNA duplexes. Science, *230*, 1242-6.

Napoli, C. & Staskawicz, B.J. (1985). Molecular genetics of biological control agents of plant pathogens: status and prospects. *In* Biological Control in Agricultural Integrated Pest Management Systems, eds M.A. Hoy & D.C. Herzog, pp. 455-63. New York: Academic Press.

Niandet, B., Janniere, L. & Ehrlich, S.D. (1985). Integration of linear, heterologous DNA molecules into the *Bacillus subtilis* chromosome: mechanism and use in induction of predictable rearrangements. Journal of Bacteriology, *163*, 111-20.

Niepold, F., Anderson, D. & Mills, D. (1985). Cloning determinants of pathogenesis from *Pseudomonas syringae* pv. *syringae*. Proceedings of the National Academy of Sciences of the U.S.A., *82*, 406-10.

Normak, S., Berstrom, S., Edlund, T., Grundstrom, T., Jaurin, B., Lindberg, F.P. & Olson, O. (1983). Overlapping genes. Annual Review of Genetics, *17*, 495-525.

Orser, C.S., Staskawicz, B.J., Panopoulos, N.J., Dahlbeck, D. & Lindow, S.E. (1985). Cloning and expression of bacterial ice nucleation genes in *Escherichia coli*. Journal of Bacteriology, *164*, 359-66.

Panopoulos, N.J. (1981). Emerging tools for *in vitro* and *in vivo* manipulation of phytopathogenic Pseudomonads and other non-enteric gram-negative bacteria. *In* Genetic Engineering in the Plant Sciences, ed. N.J. Panopoulos, pp. 163-86. New York: Praeger.

Panopoulos, N.J., Lindgren, P.B., Willis, D.K. & Peet, R.C. (1985). Clustering and conservation of genes controlling the interactions of *Pseudomonas syringae* pathovars with plants. *In* Plant Cell/Cell Interactions, eds I. Sussex, A. Ellingboe, M. Crouch & R. Mulmberg, pp. 69-75. Cold Spring Harbor: Cold Spring Harbor Laboratory.

Panopoulos, N.J. & Peet, R.C. (1985). The molecular genetics of plant pathogenic bacteria and their plasmids. Annual Review of Phytopathology, *23*, 381-419.

Papavizas, G.C., ed. (1981). Biological Control in Crop Production. Totowa, NJ: Allanheld, Osmun & Co.

Papavizas, G.C. (1986). Genetic manipulation to improve effectiveness of biocontrol fungi for plant disease control. *In* Non-conventional Approaches to Plant Disease Control, ed. I. Chet. New York: John Wiley & Sons, Inc. (in press).

Papavizas, G.C. & Lumsden, R.D. (1980). Biological control of soil-borne fungal propagules. Annual Review of Phytopathology, *18*, 389-413.

Perkins, J.B. & Youngman, P.J. (1986). Construction and properties of Tn*917-lac*, a transposon derivative that mediates transcriptional gene fusions in *Bacillus subtilis*. Proceedings of the National Academy of Sciences of the U.S.A., *83*, 140-4.

Petes, T.D. (1980). Molecular genetics of yeast. Annual Review of Biochemistry, *49*, 845-76.

Priefer, U.B., Simon, R. & Pühler, A. (1985). Extension of the host range of *Escherichia coli* vectors by incorporation of RSF1010 replication and mobilization functions. Journal of Bacteriology, *163*, 325-30.

Ruvkun, G.B. & Ausubel, F.M. (1981). A general method for site-directed mutagenesis in prokaryotes. Nature (London), *289*, 85-8.

Schoonejans, E. & Toussaint, A. (1983). Utilization of plasmid pULB113 (RP4::Mini-Mu) to construct a linkage map of *Erwinia carotovora* subsp. *chrysanthemi*. Journal of Bacteriology, *154*, 1489-92.

Schroth, M.N. & Hancock, J.G. (1981). Selected topics in biological control. Annual Review of Microbiology, *35*, 453-76.

Schroth, M.N., Loper, J.E. & Hildebrand, D.C. (1984). Bacteria as biocontrol agents of plant disease. *In* Current Perspectives in Microbial Ecology, eds M.J. Klug & C.A. Reddy, pp. 362-9. Washington, DC: Am. Soc. Microbiol.

Silhavy, T.J. & Beckwith, J.R. (1985). Uses of *lac* fusions for the study of biological problems. Microbiological Reviews, *49*, 398-418.

Simon, R., Priefer, U. & Pühler, A. (1983). A broad host range mobilization system for *in vivo* genetic engineering: transposon mutagenesis in Gram-negative bacteria. Bio/Technology, *1*, 784-90.

Simons, R.W., Hoopes, B.C., McClure, W.R. & Kleckner, N. (1984). Three promoters near the termini of IS*10*: pIN, pOUT and pIII. Cell, *34*, 673-82.

Stachel, S.E., Gynheung, A., Flores, C. & Nester, E.W. (1985). A Tn*3* *lacz* transposon for the random generation of β-galactosidase gene fusions: application to the analysis of gene expression in *Agrobacterium*. EMBO Journal, *4*, 891-8.

Staskawicz, B.J., Dahlbeck, D. & Keen, N.T. (1984). Cloned avirulence gene of *Pseudomonas syringae* pv. *glycinea* determines race-specific incompatibility on *Glycine max* (L.) Merr. Proceedings of the National Academy of Sciences of the U.S.A., *81*, 6024-8.

Stinchcomb, D.T., Thomas, M., Kelly, J., Selker, E. & Davis, R.W. (1980). Eucaryotic DNA segments capable of autonomous replication in yeast. Proceedings of the National Academy of Sciences of the U.S.A., *77*, 4558-63.

Szostak, J.W. & Rothstein, R.J. (1984). Theory and practice of genetic engineering. *In* Plant-Microbe Interactions: Molecular and Genetic Perspectives. Vol. 1, eds T. Kosuge & E.W. Wester, pp. 125-45. New York: Macmillan.

Tait, R.C., Close, T.J., Lundquist, R.C., Hagiya, M., Rodriguez, R.L. & Kado, C.I. (1983). Construction and characterization of a versatile broad-host range DNA cloning system for gram-negative bacteria. Bio/Technology, *1*, 269-75.

Turgeon, B.G., Garber, R.C. & Yoder, D.C. (1985). Transformation of the fungal maize pathogen *Cochliobolus heterostrophus* using the *Aspergillus nidulans amdS* gene. Molecular and General Genetics, *201*, 450-3.

Van Gijsegem, F. & Toussaint, A. (1982). Chromosome transfer and R-prime formation by an RP4-mini-Mu derivative in *Escherichia coli*, *Salmonella typhimurium*, *Klebsiella pneumoniae* and *Proteus mirabilis*. Plasmid, *6*, 30-44.

Vidaver, A.K. (1976). Prospects for control of phytopathogenic bacteria by bacteriophages and bacteriocins. Annual Review of Phytopathology, *14*, 451-65.

Vidaver, A.K. (1982). Biological control of plant pathogens with prokaryotes. *In* Phytopathogenic Prokaryotes. Vol. 2, eds M.S. Mount & G.H. Lacy, pp. 387-97. New York: Academic Press.

Vidaver, A.K. (1983). Bacteriocins: the lure and the reality. Plant Disease, *67*, 471-5.

Williamson, R. (1982). Genetic Engineering 3. New York: Academic Press.

Woo, S. (1984). DNA Recombinant Technology. Boca Raton: CRC Press.

Yoder, C.O. (1982). Use of pathogen produced toxins in genetic engineering of plants and pathogens. *In* Genetic Engineering of Plants - An Agricultural Perspective, eds T. Kosuge, C.P. Meredith & A. Hollaender, pp. 335-53. New York: Plenum Press.

Yoder, C.O., Turgeon, B.G. & Garber, R.C. (1985). Molecular technology for studying fungus/plant interactions. *In* Plant Cell/Cell Interactions, eds I. Sussex, A. Ellingboe, M. Crouch & R. Mulmberg, pp. 131-5. Cold Spring Harbor: Cold Spring Harbor Laboratory.

# USE OF HYPERPARASITES IN BIOLOGICAL CONTROL OF BIOTROPHIC PLANT PATHOGENS

L. Sundheim

*Division of Plant Pathology, Norwegian Plant Protection Institute, P.O. Box 70, N-1432 Ås-NLH, Norway*

## *INTRODUCTION*

Mycoparasitism, the parasitism of one fungus on another fungus, is common in all major groups of fungi from simple chytrids to higher basidiomycetes (Lumsden 1981). Biotrophic plant-pathogenic fungi also have their mycoparasites. The term hyperparasite is used for an organism parasitizing a primary parasite. Some of the hyperparasites recorded on biotrophs are potentially useful in biological control of these important plant pathogens.

Most of the literature on hyperparasites of biotrophic plant pathogens consists of reports about new parasites on powdery mildews and rusts. Several authors have studied the host-parasite relationships in considerable detail. A number of inoculation experiments have been performed to distinguish true parasites from fungicolous organisms merely growing on plant-parasitic fungi. However, few have attempted to use hyperparasites in disease control under practical conditions in greenhouses and fields.

Several recent reviews deal with hyperparasites of biotrophic plant pathogens (e.g. Hawksworth 1980; Kuhlman 1980; Kranz 1981; Lumsden 1981; Blakeman & Fokkema 1982; Jarvis 1983; Mendgen 1983; Upadhyay & Rai 1983). The present review focuses on those hyperparasites which are potentially useful in biological control of biotrophic plant pathogens.

## *HYPERPARASITES FOR CONTROL OF POWDERY MILDEWS*

Relatively few hyperparasites have been described on the powdery mildews, considering the large number of economically important plant diseases caused by members of the Erysiphaceae (Hijwegen & Buchenauer 1984).

### *Ampelomyces quisqualis Ces.*

The common coelomycete *Ampelomyces quisqualis* was first described as a parasite on powdery mildews in the middle of the nineteenth century. For a number of years the validity of this description was questioned by mycologists, who believed that the pycnidia on the conidiophores, hyphae and cleistothecia represented an additional spore form of the powdery mildews.

Hashioka & Nakai (1980) investigated the ultrastructure of growth and development of the pycnidia. *A. quisqualis* penetrates from cell to cell by forcing its hyphae through the septal pores of the powdery mildew and continues normal growth during the gradual degeneration of the infected cells. Beuther *et al.* (1981) found no evidence of toxin production by *A. quisqualis*.

Sundheim & Krekling (1982) studied the infection process of *A. quisqualis* on the cucumber powdery mildew *Sphaerotheca fuliginea* (Fr.) Poll. by scanning electron microscopy. The hyphae of the hyperparasite are only about one fifth the width of those of the powdery mildew. Within 24 h after inoculation the hyperparasite had germinated and the germ tubes had developed appressorium-like structures at the point of contact with the powdery mildew host. Within 5 days the hyperparasite had developed pycnidia with conidia on the powdery mildew hyphae and conidiophores (Fig. 1).

Yarwood (1939) sprayed conidial suspensions of *A. quisqualis* on clover leaves infected with powdery mildew (*Erysiphe trifolii* Grev.) and left the leaves to overwinter outdoors. In the next spring, he found numerous dark-brown pycnidia embedded in the dead leaf tissue and, by reisolation, he proved that the hyperparasite survives as a saprophyte in the dead leaf tissue.

*A. quisqualis* is widely distributed on many Erysiphaceae in both tropical and temperate climates. Inoculation experiments with isolates from powdery mildews on several plant families have not given any indication of host specificity (Philipp & Crüger 1979). Outside the Erysiphaceae there are few records of host fungi for *A. quisqualis*. Jarvis & Slingsby (1977) presented evidence that *A. quisqualis* is parasitic on *Botrytis cinerea* Pers.: Fr., *Alternaria solani* Sor., *Colletotrichum coccodes* (Wallr.) Hughes and *Cladosporium cucumerinum* Ell. & Arth. Philipp & Crüger (1979) obtained infection of two *Mucor* spp., *Rhizopus stolonifer* (Ehrenb.) Link and a *Pilobolus* sp., but they were unable to infect 14 other plant-pathogenic fungi including *Bremia lactucae* Regel and *Peronospora parasitica* (Pers.: Fr.) Fr.

The first attempt to use *A. quisqualis* in control of powdery mildews was made by Yarwood (1932), who used conidial suspensions to control *E. trifolii* on red clover. He reported that production of powdery mildew conidia ceased about 8 days after inoculation with the hyperparasite. Odintsova (1975) reported on control of apple powdery mildew (*Podosphaera leucotricha* (Ell. & Ev.) Salmon) by application of conidial suspensions of the hyperparasite. Powdery mildews on field crops are parasitized by *A. quisqualis* late in the growing season (Cvjetković & Mandić 1980).

Jarvis & Slingsby (1977) obtained good control of the cucumber powdery mildew *S. fuliginea*, when conidial suspensions of the hyperparasite were applied together with regular water sprays. Also without these water sprays, *A. quisqualis* reduced the powdery mildew attack and increased cucumber yield. They also noted small angular leafspots and sunken

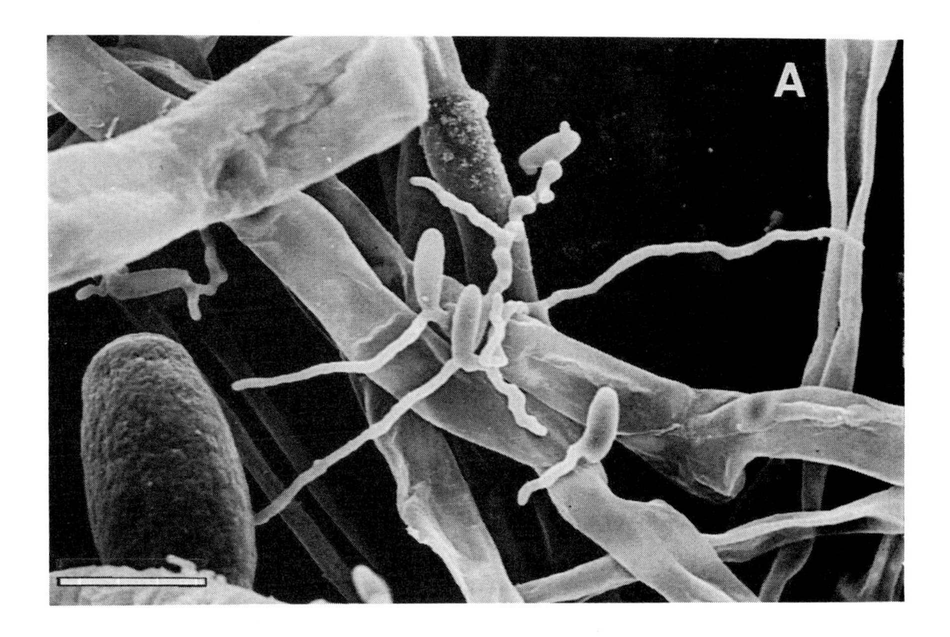


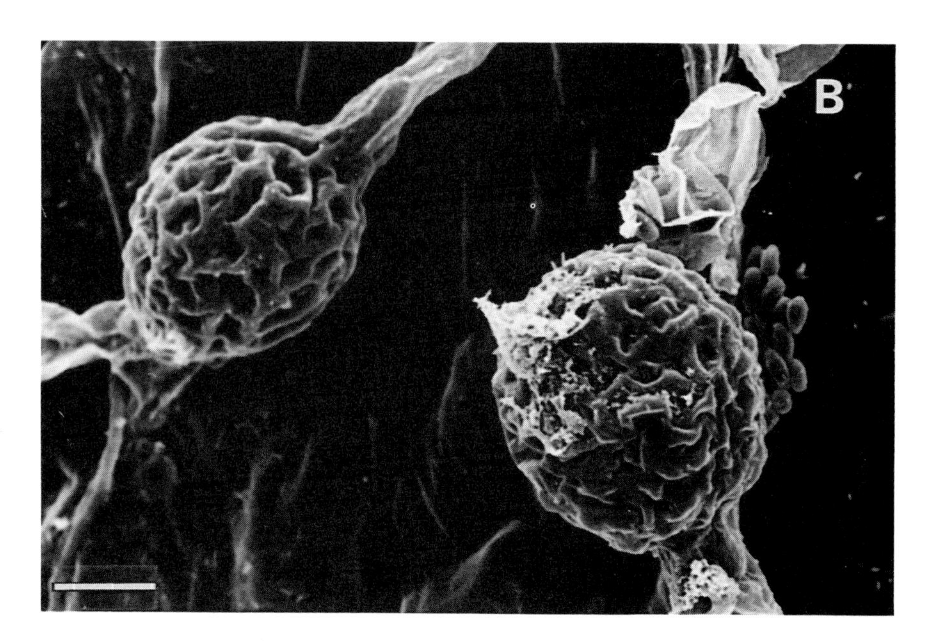


Fig. 1. Scanning electron microscopy of the hyperparasite *Ampelomyces quisqualis* and the powdery mildew, *Sphaerotheca fuliginea*, on greenhouse cucumber. A. The hyperparasite has germinated and in some places penetrated the host hyphae, 48 h after inoculation. B. Five days after inoculation *A. quisqualis* has produced pycnidia and conidia. Bars represent 10 μm.

lesions on mature cucumber fruits sprayed with the hyperparasite.

Sztejnberg (1979) sprayed *A. quisqualis* at 10-day intervals in greenhouse experiments, and obtained good control of *S. fuliginea* on cucumber and watermelon, and of powdery mildews on some other crops. Under conditions of high humidity, Philipp & Crüger (1979) obtained infection by the hyperparasite on cucumber powdery mildews on plants in protected cultivation and on crops in the field. Weekly sprays with spore suspensions controlled the powdery mildew development during periods with humid weather in field-grown crops. In dry periods, *A. quisqualis* had less effect.

Sundheim (1982) did experiments on the control of *S. fuliginea* on greenhouse cucumber. The yield increases over an untreated control were similar with weekly applications of conidial suspensions of the hyperparasite and when the fungicide triforine was used every two weeks. In experiments on cucumber infected with another powdery mildew, *Erysiphe cichoracearum* DC., in commercial greenhouses, *A. quisqualis* extensively parasitized this mildew. The cucumber yield was comparable to the yield from quinomethionate-sprayed plants.

Puzanova (1984) tested the effect of ampelomycin, a preparation produced from a strain of *A. quisqualis*, isolated from *E. cichoracearum*. Ampelomycin gave effective control of powdery mildews on several host plants.

*A. quisqualis* tolerates several fungicides used for powdery mildew control (Philipp & Crüger 1979; Philipp *et al.* 1982; Sundheim & Amundsen 1982; Philipp & Kirchhoff 1983). Thus, the hyperparasite should be well adapted to integrated control programs. Sundheim (1982) combined the application of the hyperparasite with reduced rates of fungicides. Philipp *et al.* (1984) reported that most of the acaricides, insecticides and fungicides tested by them had only a slight or moderate inhibitory effect, when applied at normal rates. In disease control experiments, the hyperparasite was applied with regular intervals to cover new growth of the host plant and to protect against the rapid dissemination of the powdery mildews. Philipp *et al.* (1984) maintained that passive transport of the hyperparasite within infected powdery mildew conidia plays an important role in dissemination of the hyperparasite. Parasitism of powdery mildew on unsprayed control plots also indicates spread of airborne inoculum (Sundheim 1982).

Neither suspensions of freeze-dried spores nor agar cultures of *A. quisqualis* had any toxic effects when subjected to a standard test for acute oral toxicity in rats. An eye irritation test in rabbits was also negative (Sundheim 1983, 1984).

### *Tilletiopsis spp.*

*Tilletiopsis* is a genus of ballistosporic yeasts of the family Sporobolomycetaceae. *Tilletiopsis* spp. are common phylloplane fungi. Hoch & Provvidenti (1979) detected antagonism between a *Tilletiopsis* sp. and the cucumber powdery mildew *Sphaerotheca fuliginea*.

Inoculation with a spore suspension of *Tilletiopsis* sp. on detached, mildew-infected cucumber leaves eliminated the superficial hyphae and conidial inoculum of the powdery mildew. Development of *Tilletiopsis* sp. on the *S. fuliginea* colonies could be observed 48 h after inoculation. Within 5 days the powdery mildew was eliminated. When the hyperparasite was applied at lower spore concentrations, the time needed to eradicate the powdery mildew increased by several days. Treatment of cucumber leaves with *Tilletiopsis* sp. spores up to 8 days before inoculation with *S. fuliginea* prevented powdery mildew development. The isolate used to control *S. fuliginea* also controlled apple powdery mildew *Podosphaera leucotricha* and the powdery mildew of grape, *Uncinula necator* (Schw.) Burr (Hoch & Provvidenti 1979). Hartmann *et al.* (1984) tested the parasitism of *Tilletiopsis* spp. on *S. fuliginea*. All suppressed the powdery mildew colonies compared to the untreated control.

*Verticillium lecanii (Zimm.) Viegas*

The host range of *Verticillium lecanii* includes arthropods, rust fungi, powdery mildews and many other fungi (Hall 1981). In inoculation experiments on *S. fuliginea*-infected greenhouse cucumber, Spencer & Ebben (1983) noted that the hyperparasite did not survive for very long on the cucumber leaves, but suspending the conidia in a solution with 2% glycerol and 1% gelatin improved the survival. In a yield trial, cucumber plants sprayed with *V. lecanii* had less mildew than the unsprayed control, but there was no significant increase in the cucumber yield on plants sprayed with the hyperparasite.

*Cladosporium spp.*

*Cladosporium* spp. have been reported to parasitize several powdery mildews. Mathur & Mukerji (1980) sprayed a spore suspension of *C. spongiosum* onto leaves of *Morus alba* infected with *Phyllactinia guttata* (Wallr.) Lev. and obtained some disease control. *C. spongiosum* also parasitizes and inhibits conidial germination of *P. dalbergiae* Pirozynski on *Dalbergia sissoo* (Mathur & Mukerji 1981), while *C. cladosporioides* (Fres.) de Vries parasitizes *Erysiphe cichoracearum* on *Xanthium strumarium*.

*Acremonium alternatum*

Malathrakis (1985) found *Acremonium alternatum* Linc: Fr. to parasitize cucurbits powdery mildew (*S. fuliginea*). Optimum temperature for spore germination and infection with the hyperparasite is 27°C. At this temperature the thallus of *S. fuliginea* is parasitized within 3 days.

*HYPERPARASITES FOR CONTROL OF RUST FUNGI*

A number of fungi have been described as parasites of Uredinales on herbaceous and woody plants. Some of the hyperparasites have been used in biological control experiments.

*Eudarluca caricis (Fr.) O. Eriks.*

The anamorph *Sphaerellopsis filum* (Biv.-Bern.: F.) Sutton of *Eudarluca caricis* parasitizes many species of rust fungi. It is frequently found in the urediniosori of economically important species, including the cereal rusts *Puccinia coronata* Corda, *P. graminis* Pers., *P. recondita* Rob., *P. sorghi* Schw. and *P. striiformis* Westend. (Von Schroeder & Hassebrauk 1957; Eriksson 1966; Kranz 1973; Chaǐka 1981; Kranz & Brandenburger 1981). Kuhlman & Matthews (1976) extended the host range of *E. caricis* to include the important pine rusts *Cronartium fusiforme* Hedgc. & Hunt. and *C. strobilinum* (Arthur) Hedgc. & Hahn. In a newly revised host list, Kranz & Brandenburger (1981) enumerated 369 rust species as hosts of *E. caricis*. The hyperparasite is most common in tropical and subtropical countries (Kranz 1969; Morelet & Pinon 1973; González Avila & Castellanos 1978; Kala & Gaur 1983).

The conidial state, *S. filum*, is easily observed under low magnification as shiny, black, spherical pycnidia in clumps between the urediniospores. It is also reported from spermogonia, aecia and telia of its hosts.

There are some disagreements as to whether *E. caricis* is able to penetrate the sub-basal hyphae of the urediniosori. Carling *et al.* (1976) examined the infection process by transmission and scanning electron microscopy and concluded that unspecialized hyphae penetrate the urediniospores. They characterized the *E. caricis*-*P. graminis* relation as a destructive biotrophic relationship and showed that penetration of the spore wall is due to a combination of mechanical and enzymatic processes. *E. caricis* is able to germinate on wheat leaves without the presence of its rust host, but urediniospores of *P. recondita* f. sp. *tritici* stimulate the germination of the hyperparasite (Ståhle & Kranz 1984).

Swendsrud & Calpouzos (1970) found that *E. caricis* affected the germination of wheat leaf rust (*P. recondita* f. sp. *tritici*) urediniospores, but Ståhle & Kranz (1984) were unable to find any significant differences between the percentages of urediniospore germination, when wheat plants inoculated with both *P. recondita* f. sp. *tritici* and the hyperparasite were compared to plants inoculated with the rust only. Swendsrud & Calpouzos (1972) compared the effects of different inoculation sequences on the infection of *P. recondita* f. sp. *tritici* by *E. caricis*. Application of the hyperparasite to wheat leaves 3 days prior to rust inoculation resulted in more severe rust attacks and fewer uredinia infected, than when both organisms were applied simultaneously. Application of the hyperparasite 3 days after inoculation with the rust had the same effect as simultaneous inoculation. These results suggest that the *E. caricis* conidia have a limited ability to survive on the wheat leaves in the absence of the rust host. Hau & Kranz (1978) developed a model to assess the effectiveness of *E. caricis* in limiting the development of *P. recondita* f. sp. *tritici*. They concluded that the hyperparasite will reduce the final severity of leaf rust by 60 to 80%, when the pycnidia occupy 40 to 60% of the total rust pustule surface.

Kuhlman *et al.* (1978) evaluated the potential of *E. caricis* for biological control of *Cronartium fusiforme*, causal agent of southern fusiform rust of pine, and *C. strobilinum*. Surveys indicated that natural infection by the hyperparasite reduced the percentage of sori which contained telia. In inoculation experiments with *C. fusiforme* and *E. caricis* on oak, the rust sori became infected by the hyperparasite, but this did not reduce the number of telia formed in the sori. Because of the short, irregular cycle of *C. fusiforme* on oak, its alternate host, Kuhlman *et al.* (1978) concluded that biological control of *C. fusiforme* by *E. caricis* is not practical.

*Verticillium spp.*

As discussed above, *Verticillium lecanii* can parasitize both arthropods and fungi. It is parasitic on scale insects, aphids and other pests (Hall 1980). Allen (1982) inoculated rust-infected bean plants with a *V. lecanii* isolate from the aphid *Brachyocaudus helichrysi* (Kltb.). The aphid isolate colonized the uredinia of the bean rust *Uromyces appendiculatus* (Pers.) Unger. Grabski & Mendgen (1985) used *V. lecanii* successfully in biological control of bean rust in greenhouse experiments, but they failed to prevent spread of the bean rust in the field.

The host range of *V. lecanii* includes several cereal rusts. McKenzie & Hudson (1976) reported it to be common on *Puccinia graminis* f. sp. *tritici*. Among other rust hosts of this catholic hyperparasite are *P. chrysanthemi* Roze (Kotthoff 1937), *Hemileia vastatrix* (Locci *et al.* 1971) and *Uromyces dianthi* (Pers.) Niessl. (Spencer 1980). A synonym is *V. hemileiae*, known as hyperparasite of *H. vastatrix* (Gams 1971).

Mendgen (1981) inoculated stripe rust (*P. striiformis*)-infected wheat leaves with *V. lecanii* conidia labelled with fluorescent antibodies. He demonstrated that the urediniospore wall became dissolved following infection. The hyperparasite requires air humidities above 80% for growth in the stripe rust pustule, but optimal conditions are relative air humidities in the range of 95 to 100%. Temperatures between 15 and 18°C allow good development of the hyperparasite on stripe rust, and high light intensities have a positive effect on parasitism.

Spencer (1980) found *V. lecanii* on *Uromyces dianthi* in a carnation nursery. In laboratory experiments *V. lecanii* either prevented development of carnation rust or arrested formation of urediniospores. He also obtained significant control of carnation rust in a greenhouse experiment using the hyperparasite. Spencer & Atkey (1981) compared the development of carnation rust on plants inoculated with the rust only and plants inoculated with both *U. dianthi* and *V. lecanii*. The numbers of uredinia per plant were reduced by 84 to 90% when the hyperparasite was applied together with the rust urediniospores.

Lim & Nik (1983) identified *V. psalliotae* Treschow as the white hyperparasite growing on uredinia of coffee rust (*H. vastatrix*) in Malaysia. Some uredinia were completely covered by the hyperparasite, which prevented further development of the rust. The hyperparasite penetrated living urediniospores and filled the spore with hyphae. No penetration of the vegetative rust hyphae was observed.

Hall (1980) did infection experiments with *V. lecanii* isolated from *E. graminis* and four rust species on the aphid *Macrosiphoniella sanborni* (Gillette). Some of the isolates of fungal origin were found to be highly pathogenic also on the aphid, and Hall (1980) concluded that this indicates the possibility of using a single *V. lecanii* strain in biological control of both pests and diseases. On the other hand, the broad spectrum of efficiency implies risks of eliminating desirable, parasitic arthropods.

### *Tuberculina spp.*

The hyphomycete genus *Tuberculina* includes several hyperparasites on aecia and uredinia of rust fungi. Sundaram (1962) observed *T. costaricana* Syd. in uredinia of *Puccinia penniseti* Zimm. and some other *Puccinia* and *Uromyces* species in India. Reis (1982) reported on a *Tuberculina* sp. on groundnut rust (*P. arachidis* Speg.) in Mozambique. Sharma *et al.* (1977) did inoculation experiments with *T. costaricana* on *P. arachidis* and obtained heavy parasitism of the rust urediniospores.

*T. maxima* Rostrup was described from cankers of white pine blister rust (*Cronartium ribicola* J. C. Fischer) on *Pinus strobus*. In western North America *T. maxima* is commonly found on a number of pine rusts (Powell 1971). Kuhlman & Miller (1976) found that between 15 and 20% of *C. quercuum* (Berk.) Miyabe ex Shirai f. sp. *fusiforme* galls on *Pinus taeda* in North and South Carolina were colonized by *T. maxima*. The purple mould *T. maxima* has been considered an hyperparasite of the rusts. Wicker (1981), reviewing the research on *T. maxima*, pointed out that there is no experimental evidence of a direct, nutritional relationship between *T. maxima* and the rust. Wicker & Woo (1973) studied the histology of the purple mould-rust association and found that *C. ribicola* does not degrade pine cells. However, when *T. maxima* is present in rust-infected tissue a rapid degradation occurs. Walls, cytoplasm and nuclei in infected pine cells are destroyed following penetration by *T. maxima* hyphae, but the hyperparasite is not able to attack rust-free pine tissue. Wicker & Woo (1973) concluded that *T. maxima* destroys the nutritional base of the rust and suppresses production of both spermogonia and aecia. Thus, it reduces the inoculum production and delays rust damage, but *T. maxima* does not control the disease (Wicker 1981).

### *Scytalidium uredinicola Kuhlman et al.*

Kuhlman *et al.* (1976) described *Scytalidium uredinicola* as a new parasite of *Cronartium quercuum* f. sp. *fusiforme* on loblolly pine (*Pinus taeda*) and slash pine (*P. elliottii* var. *elliottii*). Based on observations in three pine plantations during four seasons, Kuhlman (1981 a,b) found *S. uredinicola* to be the most common hyperparasite on these two pines in North and South Carolina.

Hiratsuka *et al.* (1979) reported *S. uredinicola* to be common on the western gall rust *Endocronartium harknessii* (J. P. Moore) Y. Hiratsuka in Canada, and subsequent investigations showed that more than 80% of the galls were infected in some localities (Tsuneda *et al.* 1980). *S. uredinicola* can parasitize the rust spores without penetrating and causes a degeneration of the spore wall and the cell contents. Tsuneda *et al.* (1980) found that the hyphae of the parasite develop in the rust sori and in the rust hyphae within the pine tissue. *S. uredinicola* produces numerous arthrospores by fragmentation of the vegetative hyphae. The hyperparasite develops slowly and inactivation of one rust gall is usually not completed within the year of infection. In the end the whole rust gall is replaced by the numerous arthrospores of the parasite. J.E. Cunningham and M.A. Pickard (unpublished results) identified a metabolite of *S. uredinicola* which in low concentration inhibits germination of *E. harknessii* teliospores. Tsuneda *et al.* (1980) considered *S. uredinicola* a promising biological control agent against *E. harknessii* because it is a successful parasite of the rust fungus in nature and kills the rust galls. As *E. harknessii* is an autoecious rust, spores produced on the pine reinfect pine directly. Thus, a reduction of spore production gives a directly proportional decrease in inoculum.

### *Aphanocladium album (Preuss) W. Gams*

The hyphomycetous fungus *Aphanocladium album* is a weak hyperparasite of Myxomycetes and many other fungi (Gams 1971). It was found to induce telia formation in cereal rusts (Biali *et al.* 1972). At high humidities it produces a fine, white mycelium over and around the uredinia. Forrer (1977) found evidence that metabolic products from *A. album* induce formation of telia in several rust fungi. Application of a purified, cell-free extract caused precocious telia production in *Puccinia graminis* f. sp. *tritici*, *P. sorghi* and *P. recondita*.

Koç *et al.* (1983) established that *A. album* is a necrotrophic hyperparasite of rusts. Urediniospores of *P. graminis* f. sp. *tritici* are parasitized by penetration. In penetrated spores the cytoplasma disintegrated and disappeared. Within 3 to 4 days after application of a conidial suspension, the rust sori became completely covered by the cotton-like, white mycelium of the hyperparasite. Srivastava *et al.* (1985 a) demonstrated that *A. album* produces chitinase on a chitin-containing medium.

Koç & Défago (1983) inoculated 14 rust species with a conidial suspension of *A. album* and all were heavily parasitized within one week after inoculation. Also the teliospores of three species of Ustilaginales were slightly parasitized. Srivastava *et al.* (1985 b) obtained 90 to 95% infection of teliospores of chrysanthemum white rust (*Puccinia horiana* P. Henn.) and a lesser frequency of infection in other microcyclic rusts.

### *Cladosporium spp.*

Several *Cladosporium* spp. have been reported as hyperparasites of rust fungi (Sharma & Heather 1981). Steyaert (1930) described *C. hemileiae* Steyaert as a parasite on the coffee rust fungus *Hemileia vastatrix* in Zaire. Ullasa (1968) reported on a *Cladosporium* sp. on *Puccinia solmsii* P. Henn. in India.

Tsuneda & Hiratsuka (1979) described the penetration of *C. gallicola* Sutton on the western gall rust *Endocronartium harknessii*. The hyperparasite grows very rapidly and its mycelium covers *E. harknessii* within a few days after the bark ruptures and exposes the rust spores. The mycoparasite penetrates viable teliospores. *C. gallicola* is usually restricted to the outer rust spore layers, and it does not affect the basal region of galls and the rust hyphae in the pine tissue. Tsuneda & Hiratsuka (1979) found evidence for an enzymatic action in this contact parasitism. The parasite is not known to produce antibiotics.

Traquair *et al.* (1984) studied the parasitism of *C. uredinicola* Speg. on *Puccinia violae* (Schum.) DC., a common rust on garden violet. The hyperparasite affects the urediniospores before penetration and invades the spores from small swellings on the hyphal tips. Intercellular hyphae of *C. uredinicola* develop in the parasitized spores.

Omar & Heather (1979) experimented with a *Cladosporium* sp. on the poplar rust *Melampsora larici-populina* Kleb. growing on leaf discs or germinating on cover slips. When conidia of this *Cladosporium* sp. were applied 1 h before or together with the urediniospores, germination of *M. larici-populina* was significantly reduced and the hyperparasite caused lysis of the urediniospores. Srivastava *et al.* (1985 b) compared the hyperparasites *C. sphaerospermum* Penzig, *C. uredinicola*, *Aphanocladium album* and *Verticillium lecanii* in inoculation experiments with *Puccinia horiana* and three other microcyclic rusts. The two *Cladosporium* spp. were less effective than *V. lecanii* and *A. album* against the rust species tested.

## *CONCLUSION*

Hyperparasites present an attractive alternative to fungicides in control of biotrophic plant pathogens. There are probably no environmental hazards involved in using these widespread enemies of powdery mildews and rusts in reducing disease losses. However, as pointed out by Jarvis (1983), the hyperparasitic principle may appear simple, but it is difficult to exploit in practical disease control. Three different organisms are involved, viz. the host plant, the pathogen and the hyperparasitic fungus. Each are affected by the environment, cultural practices and pest control programs.

In several countries fungal preparations have recently been registered as pesticides for the control of aphids, whiteflies and other pests. In a total disease and pest management program, disease control by fungicides is poorly compatible with biocontrol agents. Thus, more emphasis should be given to the development of hyperparasites of important plant pathogens on crops in protected cultivation, where biological pest control is practiced.

Tsuneda & Hiratsuka (1981) maintain that pine stem rusts have characteristics which should facilitate biological control by hyperparasitic fungi. The rust cankers are perennial, and there are abundant supplies of rust spores available as a food source for the parasites. The western gall rust, *Endocronartium harknessii*, does not have an alternate host and its inoculum potential is directly proportional to the spore production on the pine.

The research on hyperparasites should be expanded. Wide differences in virulence between naturally occuring strains have been noted for several hyperparasites. Selection of efficient strains should be combined with efforts to improve virulence by genetic engineering.

## *REFERENCES*


Allen, D.J. (1982). *Verticillium lecanii* on the bean rust fungus, *Uromyces appendiculatus*. Transactions of the British Mycological Society, *79*, 362-4.

Beuther, E., Philipp, W.-D. & Grossmann, F. (1981). Untersuchungen zum Hyperparasitismus von *Ampelomyces quisqualis* auf Gurkenmehltau (*Sphaerotheca fuliginea*). Phytopathologische Zeitschrift, *101*, 265-70.

Biali, M., Dinoor, A., Eshed, N. & Kenneth, R. (1972). *Aphanocladium album*, a fungus inducing teliospore production in rusts. Annals of Applied Biology, *72*, 37-42.

Blakeman, J.P. & Fokkema, N.J. (1982). Potential for biological control of plant diseases on the phylloplane. Annual Review of Phytopathology, *20*, 167-92.

Carling, D.E., Brown, M.F. & Millikan, D.F. (1976). Ultrastructural examination of the *Puccinia graminis-Darluca filum* host-parasite relationship. Phytopathology, *66*, 419-22.

Chaǐka, M.N. (1981). [Electron microscopic examination of the interactions between *Darluca filum* (Fr.) Cast. and *Puccinia graminis* Pers. f. sp. *tritici* Eriks. & E. Henn.] Mikologiya i Fitopatologiya, *15*, 105-7.

Cvjetkovič, B. & Mandič, R. (1980). [*Erysiphe cruciferarum*, the powdery mildew pathogen of cabbage, and its hyperparasite *Ampelomyces quisqualis* Ces.] Zaštita Bilja, *31*, 373-7.

Eriksson, O. (1966). On *Eudarluca caricis* (Fr.) O. Eriks., comb. nov., a cosmopolitan uredinicolous Pyrenomycete. Botaniska Notiser, *119*, 33-69.

Forrer, H.R. (1977). Der Einfluss von Stoffwechselprodukten des Mycoparasiten *Aphanocladium album* auf die Teleutosporenbildung von Rostpilzen. Phytopathologische Zeitschrift, *88*, 306-11.

Gams, W. (1971). *Cephalosporium*-artige Schimmelpilze (Hyphomycetes). Stuttgart: G. Fischer Verlag.

González Avila, M. & Castellanos, J.J. (1978). Presencia del micoparásito *Darluca filum* sobre uredosoros de *Uromyces phaseoli* var. *typica*. Ciencias de la Agricultura, *3*, 119-25.

Grabski, G.C. & Mendgen, K. (1985). Einsatz von *V. lecanii* als biologisches Schädlingsbekämpfungsmittel gegen den Bohnenrostpilz *U. appendiculatus* var. *appendiculatus* im Feld und im Gewächshaus. Phytopathologische Zeitschrift, *113*, 243-51.

Hall, R.A. (1980). Laboratory infection of insects by *Verticillium lecanii* strains isolated from phytopathogenic fungi. Transactions of the British Mycological Society, *74*, 445-6.

Hall, R.A. (1981). The fungus *Verticillium lecanii* as a microbial insecticide against aphids and scales. *In* Microbial control of pests and plant diseases 1970 - 1980, ed. H.D. Burges, pp. 438-98. New York: Academic Press.

Hartmann, H., Riggs, W.A. & Hall, J.W. (1984). Screening for biological control agents of powdery mildew (*Sphaerotheca fuliginea*) on cucumbers. Phyto-

pathology, *74*, 864 (Abstr.).
Hashioka, Y. & Nakai, Y. (1980). Ultrastructure of pycnidial development and mycoparasitism of *Ampelomyces quisqualis* parasitic on Erysiphales. Transactions of the Mycological Society of Japan, *21*, 329-38.
Hau, B. & Kranz, J. (1978). Modelrechnungen zur Wirkung des Hyperparasiten *Eudarluca caricis* auf Rostepidemien. Zeitschrift für Pflanzenkrankheiten und Pflanzenschutz, *85*, 131-41.
Hawksworth, D.L. (1981). A survey of the fungicolous conidial fungi. *In* Biology of Conidial Fungi. Vol. 1, eds G.T. Cole & B. Kendrick, pp. 171-244. New York: Academic Press.
Hijwegen, T. & Buchenauer, H. (1984). Isolation and identification of hyperparasitic fungi associated with Erysiphaceae. Netherlands Journal of Plant Pathology, *90*, 79-83.
Hiratsuka, Y., Tsuneda, A. & Sigler, L. (1979). Occurrence of *Scytalidium uredinicola* on *Endocronartium harknessii* in Alberta, Canada. Plant Disease Reporter, *63*, 512-3.
Hoch, H.C. & Provvidenti, R. (1979). Mycoparasitic relationships: cytology of the *Sphaerotheca fuliginea-Tilletiopsis* sp. interaction. Phytopathology, *69*, 359-62.
Jarvis, W.R. (1983). Progress in the biological control of plant diseases. *In* Proceedings 10th International Congress of Plant Protection 1983, pp. 1095-105. Croydon: The British Crop Protection Council.
Jarvis, W.R. & Slingsby, K. (1977). The control of powdery mildew of greenhouse cucumber by water sprays and *Ampelomyces quisqualis*. Plant Disease Reporter, *61*, 728-30.
Kala, S.P. & Gaur, R.D. (1983). New host records for *Eudarluca caricis* from India. Indian Phytopathology, *36*, 408-9.
Koç, N.K. & Défago, G. (1983). Studies on the host range of the hyperparasite *Aphanocladium album*. Phytopathologische Zeitschrift, *107*, 214-8.
Koç, N.K., Forrer, H.R. & Défago, G. (1983). Hyperparasitism of *Aphanocladium album* on aecidiospores and teliospores of *Puccinia graminis* f. sp. *tritici*. Phytopathologische Zeitschrift, *107*, 219-23.
Kotthoff, P. (1937). *Verticillium coccorum* (Petch) Westerdijk als Parasit auf *Puccinia chrysanthemi* Roze. Angewandte Botanik, *19*, 127-30.
Kranz, J. (1969). Zur natürlichen Verbreitung des Rostparasiten *Eudarluca caricis* (Fr.) O. Eriks. Phytopathologische Zeitschrift, *65*, 43-53.
Kranz, J. (1973). A host list of the rust parasite *Eudarluca caricis* (Fr.) O. Eriks. Nova Hedwigia, *24*, 169-80.
Kranz, J. (1981). Hyperparasitism of biotrophic fungi. *In* Microbial Ecology of the Phylloplane, ed. J.P. Blakeman, pp. 327-52. London: Academic Press.
Kranz, J. & Brandenburger, W. (1981). An amended host list of the rust parasite *Eudarluca caricis*. Zeitschrift für Pflanzenkrankheiten and Pflanzenschutz, *88*, 682-702.
Kuhlman, E.G. (1980). Hypovirulence and hyperparasitism. *In* Plant Disease. An Advanced Treatise. Vol. V, eds J.G. Horsfall & E.B. Cowling, pp. 363-80. New York: Academic Press.
Kuhlman, E.G. (1981 a). Mycoparasitic effects of *Scytalidium uredinicola* on aeciospore production and germination of *Cronartium quercuum* f. sp. *fusiforme*. Phytopathology, *71*, 186-8.
Kuhlman, E.G. (1981 b). Parasite interaction with sporulation by *Cronartium quercuum* f. sp. *fusiforme* on loblolly and slash pine. Phytopathology, *71*, 348-50.
Kuhlman, E.G., Carmichael, J.W. & Miller, T. (1976). *Scytalidium uredinicola*, a new mycoparasite of *Cronartium fusiforme* on *Pinus*. Mycologia, *68*, 1188-94.
Kuhlman, E.G. & Matthews, F.R. (1976). Occurrence of *Darluca filum* on *Cronartium strobilinum* and *Cronartium fusiforme* infecting oak. Phytopathology, *66*, 1195-7.
Kuhlman, E.G., Matthews, F.R. & Tillerson, H.P. (1978). Efficacy of *Darluca filum* for biological control of *Cronartium fusiforme* and *C. strobilinum*.

Phytopathology, *68*, 507-11.
Kuhlman, E.G. & Miller, T. (1976). Occurrence of *Tuberculina maxima* on fusiform rust galls in the southeastern United States. Plant Disease Reporter, *60*, 627-9.
Lim, T.K. & Nik, W.Z. (1983). Mycoparasitism of the coffee rust pathogen, *Hemileia vastatrix*, by *Verticillium psalliotae* in Malaysia. Pertanika, *6*, 23-5.
Locci, R., Minervini Ferrante, G. & Rodrigues, C.J. (1971). Studies by transmission and scanning electron microscopy on the *Hemileia vastatrix-Verticillium hemileiae* association. Rivista di Patologia Vegetale, Serie IV, *7*, 127-40.
Lumsden, R.D. (1981). Ecology of mycoparasitism. *In* The Fungal Community. Its Organization and Role in the Ecosystem, eds D.T. Wicklow & G.C. Carroll, pp. 295-318. New York and Basel: Marcel Dekker Inc.
McKenzie, E.H.C. & Hudson, H.J. (1976). Mycoflora of rust-infected and non-infected plant material during decay. Transactions of the British Mycological Society, *66*, 223-38.
Malathrakis, N.E. (1985). The fungus *Acremonium alternatum* Linc: Fr., a hyperparasite of the cucurbits powdery mildew pathogen *Sphaerotheca fuliginea*. Zeitschrift für Pflanzenkrankheiten und Pflanzenschutz, *92*, 509-15.
Mathur, M. & Mukerji, K.G. (1980). Parasitism of *Phyllactinia corylea* by *Cladosporium spongiosum*. Experientia, *36*, 89-90.
Mathur, M. & Mukerji, K.G. (1981). Antagonistic behaviour of *Cladosporium spongiosum* against *Phyllactinia dalbergiae* on *Dalbergia sissoo*. Angewandte Botanik, *55*, 75-7.
Mendgen, K. (1981). Growth of *Verticillium lecanii* in pustules of stripe rust (*Puccinia striiformis*). Phytopathologische Zeitschrift, *102*, 301-9.
Mendgen, K. (1983). Alternativen beim Pflanzenschutz? Naturwissenschaften, *70*, 235-40.
Morelet, M. & Pinon, J. (1973). *Darluca filum* hyperparasite du genre *Melampsora* sur peuplier et saule. Revue Forestière Française, *25*, 378-9.
Odintsova, O.V. (1975). [Role of a hyperparasite, *Cicinnobolus cesatii* D. By., in suppressing powdery mildew on apple trees.] Mikologiya i Fitopatologiya, *9*, 337-9.
Omar, M. & Heather, W.A. (1979). Effect of saprophytic phylloplane fungi on germination and development of *Melampsora larici-populina*. Transactions of the British Mycological Society, *72*, 225-31.
Philipp, W.-D., Beuther, E. & Grossmann, F. (1982). Untersuchungen über den Einfluss von Fungiziden auf *Ampelomyces quisqualis* im Hinblick auf eine integrierte Bekämpfung von Gurkenmehltau unter Glas. Zeitschrift für Pflanzenkrankheiten und Pflanzenschutz, *89*, 575-81.
Philipp, W.-D. & Crüger, G. (1979). Parasitismus von *Ampelomyces quisqualis* auf Echten Mehltaupilzen an Gurken und anderen Gemüsearten. Zeitschrift für Pflanzenkrankheiten und Pflanzenschutz, *86*, 129-42.
Philipp, W.-D., Grauer, U. & Grossmann, F. (1984). Ergänzende Untersuchungen zur biologischen und integrierten Bekämpfung von Gurkenmehltau under Glas durch *Ampelomyces quisqualis*. Zeitschrift für Pflanzenkrankheiten und Pflanzenschutz, *91*, 438-43.
Philipp, W.-D. & Kirchhoff, J. (1983). Wechselwirkungen zwischen Triadimefon und dem Mehltauhyperparasiten *Ampelomyces quisqualis* in vitro. Zeitschrift für Pflanzenkrankheiten und Pflanzenschutz, *90*, 68-72.
Powell, J.M. (1971). Incidence and effect of *Tuberculina maxima* on cankers of the pine stem rust, *Cronartium comandrae*. Phytoprotection, *52*, 104-11.
Puzanova, L.A. (1984). [Hyperparasites of the genus *Ampelomyces* Ces. ex Schlecht. and their possible application to biological control of powdery mildew]. Mikologiya i Fitopatologiya, *18*, 333-8.
Reis, L.G.L. (1982). Observations on groundnut rust (*Puccinia arachidis*) in Mozambique. Garcia de Orta, Série Estudos Agronomicos, *9*, 61-70.
Schroeder, H. von & Hassebrauk, K. (1957). Beiträge zur Biologie von *Darluca filum* (Biv.) Cast. und einigen anderen auf Uredineen beobachteten Pilzen.

Zentralblatt für Bakteriologie. II. Abteilung, *110*, 676-96.
Sharma, I.K. & Heather, W.A. (1981). Hyperparasitism of *Melampsora larici-populina* by *Cladosporium herbarum* and *Cladosporium tenuissimum*. Indian Phytopathology, *34*, 395-7.
Sharma, N.D., Vyas, S.C. & Jain, A.C. (1977). *Tuberculina costaricana* Syd.: a new hyperparasite on groundnut rust (*Puccinia arachidis* Speg.). Current Science, *46*, 311-2.
Spencer, D.M. (1980). Parasitism of carnation rust *Uromyces dianthi* by *Verticillium lecanii*. Transactions of the British Mycological Society, *74*, 191-4.
Spencer, D.M. & Atkey, P.T. (1981). Parasitic effects of *Verticillium lecanii* on two rust fungi. Transactions of the British Mycological Society, *77*, 535-42.
Spencer, D.M. & Ebben, M.H. (1983). Biological control of cucumber powdery mildew. Annual Report Glasshouse Crops Research Institute 1981, 128-9.
Srivastava, A.K., Défago, G. & Boller, T. (1985 a). Secretion of chitinase by *Aphanocladium album*, a hyperparasite of wheat rust. Experientia, *41*, 1612-3.
Srivastava, A.K., Défago, G. & Kern, H. (1985 b). Hyperparasitism of *Puccinia horiana* and other microcyclic rusts. Phytopathologische Zeitschrift, *114*, 73-8.
Stähle, U. & Kranz, J. (1984). Interactions between *Puccinia recondita* and *Eudarluca caricis* during germination. Transactions of the British Mycological Society, *82*, 562-3.
Steyaert, R.L. (1930). *Cladosporium hemileiae* n. sp. Un parasite de l'*Hemileia vastatrix* Berk. et Br. Bulletin de la Société Royale de Botanique de Belgique, *63*, 46-7.
Sundaram, N.V. (1962). Studies on parasites of the rusts. Indian Journal of Agricultural Science, *32*, 266-71.
Sundheim, L. (1982). Control of cucumber powdery mildew by the hyperparasite *Ampelomyces quisqualis* and fungicides. Plant Pathology, *31*, 209-14.
Sundheim, L. (1983). The hyperparasite *Ampelomyces quisqualis* in biological control of cucumber powdery mildew. *In* Proceedings 10th International Congress of Plant Protection 1983, p. 1110 (Abstr.). Croydon: The British Crop Protection Council.
Sundheim, L. (1984). L'hyperparasite *Ampelomyces quisqualis* dans la lutte contre l'oïdium du concombre. Les Colloques de l'INRA, *18*, 145-54.
Sundheim, L. & Amundsen, T. (1982). Fungicide tolerance in the hyperparasite *Ampelomyces quisqualis* and integrated control of cucumber powdery mildew. Acta Agriculturae Scandinavica, *32*, 349-55.
Sundheim, L. & Krekling, T. (1982). Host-parasite relationships of the hyperparasite *Ampelomyces quisqualis* and its powdery mildew host *Sphaerotheca fuliginea*. I. Scanning electron microscopy. Phytopathologische Zeitschrift, *104*, 202-10.
Swendsrud, D.P. & Calpouzos, L. (1970). Rust uredospores increase the germination of pycnidiospores of *Darluca filum*. Phytopathology, *60*, 1445-7.
Swendsrud, D.P. & Calpouzos, L. (1972). Effect of inoculation sequence and humidity on infection of *Puccinia recondita* by the mycoparasite *Darluca filum*. Phytopathology, *62*, 931-2.
Sztejnberg, A. (1979). Biological control of powdery mildews by *Ampelomyces quisqualis*. Phytopathology, *69*, 1047 (Abstr.).
Traquair, J.A., Meloche, R.B., Jarvis, W.R. & Baker, K.W. (1984). Hyperparasitism of *Puccinia violae* by *Cladosporium uredinicola*. Canadian Journal of Botany, *62*, 181-4.
Tsuneda, A. & Hiratsuka, Y. (1979). Mode of parasitism of a mycoparasite, *Cladosporium gallicola*, on western gall rust, *Endocronartium harknessii*. Canadian Journal of Plant Pathology, *1*, 31-6.
Tsuneda, A. & Hiratsuka, Y. (1981). Biological control of pine stem rusts by mycoparasites. Proceedings of the Japan Academy, Series B, Physical and Biological Sciences, *57*, 337-41.

Tsuneda, A. Hiratsuka, Y. & Maruyama, P.J. (1980). Hyperparasitism of *Scytalidium uredinicola* on western gall rust, *Endocronartium harknessii*. Canadian Journal of Botany, *58*, 1154-9.

Ullasa, B.A. (1968). *Cladosporium*: a new mycoparasite on rust. Current Science, *37*, 505.

Upadhyay, R.S. & Rai, B. (1983). Mycoparasitism with reference to biological control of plant diseases. *In* Recent Advances in Plant Pathology, ed. A. Husain, pp. 48-72. Lucknow: Print House.

Wicker, E.F. (1981). Natural control of white pine blister rust by *Tuberculina maxima*. Phytopathology, *71*, 997-1000.

Wicker, E.F. & Woo, J.Y. (1973). Histology of blister rust cankers parasitized by *Tuberculina maxima*. Phytopathologische Zeitschrift, *76*, 356-66.

Yarwood, C.E. (1932). *Ampelomyces quisqualis* on clover mildew. Phytopathology, *22*, 31 (Abstr.).

Yarwood, C.E. (1939). An overwintering pycnidial stage of *Cicinnobolus*. Mycologia, *31*, 420-2.

# USE OF TRICHODERMA SPP. IN BIOLOGICAL CONTROL OF NECROTROPHIC PATHOGENS

A. Tronsmo

*Department of Microbiology, Agricultural University of Norway, P.O. Box 40, N-1432 Ås-NLH, Norway*

## *INTRODUCTION*

Isolates of the fungal genus *Trichoderma* are among the most used biocontrol agents. *Trichoderma* spp. are frequently isolated from soil, and most biocontrol trials have been done against soil-borne plant pathogens (Papavizas 1985). However, pathogens on aerial plant surfaces have also been successfully controlled by *Trichoderma* spp., and these attempts will be reviewed in this paper.

### *Why Trichoderma spp. in biological control?*

Wood & Tveit (1985) considered using antagonists against pathogens on aerial plant parts. The authors reasoned that antagonists need to have a high reproductive capacity, to be able to survive unfavourable environmental conditions, and to be very aggressive or antagonistic.

*Trichoderma* spp. in many ways fulfil these demands as a biocontrol agent. They are fast-growing organisms (radial growth of 15 mm $day^{-1}$ on solid media is not unusual) and they have simple nutrient requirements. A mineral solution with one of several carbohydrates is sufficient for fast growth. Most isolates tested exhibit one or more of the following mechanisms of antagonism: production of non-volatile or volatile antibiotics, competition for nutrients, and hyphal interactions (Dennis & Webster 1971 a,b,c; Tronsmo & Dennis 1978).

Spores of *Trichoderma* spp. are used for most biocontrol purposes. Most isolates produce masses of spores, either conidia on solid surfaces or chlamydospores in liquid cultures. Because biocontrol trials against a pathogen in a field situation must be performed together with chemical control of other diseases, tolerance to agrochemicals is of importance. An advantage of *Trichoderma* spp. in this respect is that they readily acquire fungicide resistance after being subjected to fungicides (Abd-El Moity *et al.* 1982; Tronsmo 1985 a), and that it is also possible to induce fungicide resistance with ultraviolet light (UV) (Papavizas *et al.* 1982; Gullino & Garibaldi 1983; Papavizas & Lewis 1983).

### *Production of spores for biocontrol trials*

For small-scale experiments conidia of *Trichoderma* spp. have been produced on solid media in petri dishes or in bottles (Tronsmo &

Dennis 1977). However, as this method of producing spores is labour-demanding and expensive, for field trials a more effective way of producing spores is needed. Up till now only conidia have been used for biocontrol purposes on aerial plant surfaces. Production of conidia on plant materials such as barley grains (Tronsmo 1983) is very effective, but it is not known if this method will be suitable for commercial production of conidia. However, Durand (1983) has described a solid-state fermentation system that could be used for large-scale production of conidia.

Use of chlamydospores has been found more suitable for biocontrol purposes against soil-borne pathogens than conidia (J.A. Lewis, personal communication). Papavizas *et al.* (1984) have described a simple and inexpensive method for production of chlamydospores in liquid culture, but their effectiveness on aerial plant surfaces remains to be tested.

## *MECHANISMS OF MYCOPARASITISM*

### *Production of antibiotics*

*Trichoderma* spp. are well known as antagonistic fungi and have been the subject of intensive study ever since Weindling (1934) reported that culture filtrates of *T. lignorum* were toxic to *Rhizoctonia solani* and other fungi, even at high dilutions. Unfortunately, there has been some confusion about the nomenclature of the fungus used for antibiotic production. Webster & Lomas (1964) cleared up the taxonomic confusion by showing that Weindling's gliotoxin-producing isolates and the two isolates from which Brian & Hemming (1945) and Brian *et al.* (1946) obtained gliotoxin and viridin were not *Trichoderma viride*, but *Gliocladium virens*. The report of Webster & Lomas (1964) casts doubt on the identity of some isolates of *Trichoderma* spp. used by workers investigating antagonism by this fungus. However, it seems likely that most reports on antagonism by *Trichoderma* spp. relate to this genus (Dennis & Webster 1971 a).

Because of confusion about the nomenclature, a taxonomic revision was needed, and Rifai (1969) published a revision, which is now generally used. Even though Webster & Lomas (1964) have shown that the two antibiotics, gliotoxin and viridin, are not produced by *Trichoderma* spp., many other antibiotics have been reported from this genus (Dennis & Webster 1971 a).

Apart from the diffusible inhibitors, volatile inhibitors are also produced by several isolates of *Trichoderma* spp. Bilaĭ (1956) reported the presence of volatile metabolites with both antifungal and antibacterial properties. Dennis & Webster (1971 b) concluded that isolates of *Trichoderma* spp. probably produce more than one active volatile metabolite, and tentatively identified acetaldehyde as one of the inhibitory metabolites of *T. viride*. Tamimi & Hutchinson (1975) identified ethanol, acetaldehyde, ethylene, acetone and $CO_2$ in the head space above cultures of *Trichoderma* spp., and they concluded that variations in inhibition of test fungi could be accounted for by differences in the rate of $CO_2$ production. However, Tronsmo & Dennis (1978) showed that $CO_2$ did not account for all inhibition by volatile components from *Trichoderma* spp.

Several compounds have since then been detected by GC-mass spectroscopy (Tronsmo, unpublished results), among these alcohols, ketones, sesquiterpenes and 6-pentyl-α-pyrone, the characteristic coconut smell produced by several isolates of *Trichoderma* spp. The identity of the inhibitor(s) is not yet known. However, I have indications that some of these volatile metabolites may have stimulatory effects at low concentration, and inhibitory effects at higher concentrations.

On aerial plant surfaces, production of volatile metabolites is probably not an important antagonistic mechanism. It has been suggested, however, as a possible mechanism for control of fungal decay in creosoted transmission poles (Bruce *et al.* 1984).

### *Hyphal interaction*

Hyphal interaction by *Trichoderma* species is another antagonistic mechanism. Of 80 isolates of *Trichoderma* spp. tested by Dennis & Webster (1971 c) in dual culture experiments, only 10 did not show coiling. Most isolates coiled around hyphae of all six test fungi, but some did not interact with hyphae of *Fusarium oxysporum*. Hyphal interaction was studied at pH 4.0 and 6.5 and no difference was found. This was in contrast to the findings of Aytoun (1953) who observed more intense coiling at pH 3.4 than at pH 5.1 or 7.0. Different temperatures in the range 5 to 20°C had no effect on the hyphal interactions (Tronsmo & Dennis 1978).

Even though hyphal interactions are very frequent among isolates of *Trichoderma* spp., penetration of hyphae by *Trichoderma* spp. are seldom observed (Dennis & Webster 1971 c). However, scanning and transmission electron microscopy have clearly shown the attachment of coils, hooks or appressoria to hyphae, and lysed sites and penetration holes in hyphae of plant-pathogenic fungi (Elad *et al.* 1983 b,c).

### *Production of lytic enzymes*

The ability of *Trichoderma* spp. to grow on fungal cell walls, excrete cell wall-degrading enzymes, and parasitize plant-pathogenic fungi has been intensively studied. Jones & Watson (1969) showed that *T. viride* produced β-1,3-glucanase and chitinase that were able to solubilize hyphae of *Sclerotinia sclerotiorum*. Hadar *et al.* (1979) showed that *T. harzianum* produces the same enzymes when attacking *Rhizoctonia solani* cell walls. As they could not detect any antibiotic production by this isolate, they claimed that the main mechanism involved in the antagonism of *T. harzianum* against *R. solani* and *Sclerotium rolfsii* is the release of these lytic enzymes (Hadar *et al.* 1979; Elad *et al.* 1980). Isolates of *T. harzianum* which were found to differ in their ability to attack *S. rolfsii*, *R. solani* and *Pythium aphanidermatum* also differed in their levels of hydrolytic enzymes produced. This phenomenon was found to be correlated with the ability of each of these isolates to control soil-borne diseases (Elad *et al.* 1982).

### *Recognition*

How can the antagonist detect its prey? Light microscopy has revealed that hyphae of *T. harzianum* show a chemotropic response by branching and growing towards *R. solani* (Chet & Elad 1983). After the contact between the two fungi has been established, the antagonist either grows parallel to its host or coils around it. Is this caused by a tactic response? It is probably so, because when Dennis & Webster (1971 c) studied the interaction of *Trichoderma* spp. with plastic threads with diameters similar to those of fungal hyphae, they did not observe any coiling. How the mycoparasite *Trichoderma* sp. 'recognizes' its host and what determines the binding specificity, if any, in this fungus-fungus relationship is not known, but the involvement of a lectin has been proposed (Elad *et al.* 1983 a; Barak *et al.* 1985).

### *Competition*

Competition for nutrients has been well documented as an antagonistic mechanism for bacteria in the phyllosphere (Blakeman 1985). For *Trichoderma* spp. less is known about the importance of this mechanism. However, when *Trichoderma* spp. are the primary wood colonizers, they will deplete the more accessible nutrients, and thus they may hinder the normal rapid process of colonization by wood-rotting organisms (Hulme & Shields 1970, 1972).

### *Triggering of the host defence reactions*

Do *Trichoderma* spp. protect the plant by inducing plant resistance mechanisms? Probably not, because Kuč (1982) stated that to obtain immunization in the field situation, the inducer must cause lesions. As *Trichoderma* spp. are not able to penetrate a healthy leaf, it is unlikely that they will be able to induce resistance.

### *Inhibition caused by ungerminated spores of Trichoderma spp.*

Growth of several plant-pathogenic fungi was inhibited when they were inoculated on agar containing more than $10^8$ heat-killed spores of *T. viride* per plate (Ale-Agha *et al.* 1974). However, in a liquid culture with *T. harzianum* ($10^6$ conidia $ml^{-1}$) no effect was seen on germination of *Botrytis cinerea* spores before the *T. harzianum* conidia started to germinate (Stople 1982). It is, therefore, unlikely that ungerminated spores will have any biocontrol effect in the field. After germination of the *T. harzianum* spores, however, nearly 100% inhibition of the germination of *B. cinerea* was observed both at 12 and 17°C (Stople 1982).

## *APPLICATIONS OF BIOLOGICAL CONTROL*

### *Protection of wounds on trees*

Wounds on trees are easily invaded by a number of pathogens. Once established within the plant tissue, a pathogen is often protected from externally applied chemicals and antagonistic micro-organisms. This means that it is most important to protect the fresh wounds. Protection of man-made branch wounds has been attempted with fungicidal paints, but the results were mostly poor (Shigo & Wilson 1971; Mercer *et al.* 1983). The cut surface must be uncontaminated when the protective cover is applied, and breaks in the surface due to weathering allow the entry of wood-rotting fungi, which then develop in the tissue under particularly favourable conditions (Corke & Rishbeth 1981).

The first report of protection of pruning cuts is from 1657 when Austin (1657) reported that treatment of the wounds with a wash containing cow dung and urine protected apple trees against apple canker. Austin's method was later investigated by Grosclaude (1970), who concluded that its effectiveness was due to an antagonistic micro-organism within the applied soil-dung mixture.

Biological protection of pine stumps against *Heterobasidion annosum* (*Fomes annosus*), during routine thinning operations in forest plantations, by painting them immediately after cutting with a spore suspension of *Peniophora gigantea* (Rishbeth 1963) has been used commercially for a long time in many countries (Greig 1976).

Many workers have attempted to find an equally efficient organism for biological control of *H. annosum* on other genera of conifers (Seaby 1977). Among many others, isolates of *Trichoderma* spp. have been used. Seaby (1977) found that *T. viride* was able to reduce colonization of *Picea sitchensis* and *P. contorta* by *H. annosum* after artifical inoculation as effectively as 20% urea, the standard stump protection treatment used in Northern Ireland. The method was effective, both when applied in the mild Irish winter and in spring. Kallio & Hallaksela (1979) found that *T. viride* showed a considerable inhibitory effect on natural infection by *H. annosum* spores in Norway spruce (*Picea abies*) in the summer. But even if the *T. viride* strain was collected from a region about 700 km north of the test area, it did not protect the stumps during the cold season of the year. *P. gigantea*, however, gave complete protection against the natural airborne infection in both winter and summer.

Pottle *et al.* (1977) investigated the effect of inoculating *T. harzianum* into wounds of *Acer rubrum* trees. They found that inoculation of summer wounds with *T. harzianum* delayed the natural infection of wood decay fungi (Hymenomycetes) for at least 21 months, but that it had no effect on winter-inoculated wounds. Spore concentrations in the range from $10^3$ to $10^9$ spores $ml^{-1}$ had no differential effect (Smith *et al.* 1979). No indication of hyperparasitism of *T. harzianum* on the wood decay fungi was detected. Smith *et al.* (1981) postulated that replacement of the pioneer fungi, which reduce the phenolic content of the tree, with *T. harzianum*, which more slowly decreases the phenolic content, made the resistance response of the tree more effective. The decay fungi were therefore inhibited for an extended period. However, this single mechanism could not completely explain the effect of the *T. harzianum* treatment (Smith *et al.* 1981).

Unprotected cuts on plum trees have been shown to be rapidly colonized by micro-organisms and this can prevent infection by the silver leaf fungus *Chondrostereum purpureum* (Corke & Rishbeth 1981). Grosclaude

(1970) found that *T. viride* which was antagonistic in vitro against *C. purpureum*, was also highly inhibitory in vivo. This was further tested and Grosclaude *et al.* (1973) showed that plum trees were completely protected if the newly cut surface was inoculated with a conidial suspension 48 hours before inoculation with the parasite, whereas simultaneous inoculation did not protect the trees. The same protection was obtained if *T. viride* was applied as a spray or with specially designed pruning shears, which deliver a small amount of *T. viride* spores to the surface when the pruning is done (Grosclaude *et al.* 1973). Further development of pruning shears has since been described (Jones *et al.* 1975; Seaby & Swinburne 1976; Carter & Mullett 1978). In a recent trial good protection against natural infection was obtained with *T. viride* either alone or in an integrated program with the fungicide triadimefon (Woodgate-Jones & Hunter 1983).

In addition to protecting pruning wounds against new infection by *C. purpureum*, inoculation of trees with *T. viride*, particularly young trees, showing symptoms of silver leaf, has met with considerable success (Corke & Rishbeth 1981). Also, reduction in silvering of leaves of peach and nectarine has been obtained by treatment with *T. viride* (Dubos & Ricard 1974). Several inoculation methods have been tried and the most convenient method, which is now commercially adapted, is the method of Ricard *et al.* (1969) slightly modified by Corke (1974, 1978). By this technique sharpened dowels, impregnated with *T. viride*, are forced into holes drilled in the trunk of infected trees.

Dutch elm disease caused by *Ophiostoma (Ceratocystis) ulmi* is difficult to control by chemicals, and biological control of the disease with several biocontrol agents has been tried. *Trichoderma* spp. have also been used in field trials, but their effect has been variable (Gear 1985). It is therefore too early to tell if treatment by *Trichoderma* spp. can be used commercially against Dutch elm disease in the future.

### *Biological control of Botrytis cinerea on vine*

Grey mould (*Botrytis cinerea*) is one of the most serious diseases of vine and control of the disease is, therefore, very important. Biological control of this disease with *Trichoderma* spp. has been tested, with success, in France (Dubos *et al.* 1978, 1982 a,b, 1983) and Italy (Gullino & Garibaldi 1983; Bisiach *et al.* 1985; Gullino *et al.* 1985). Results from these trials will be used to illustrate some important aspects in biological control.

The effect of the concentration of the conidia in the spray solution has been investigated. Lantero *et al.* (1982) have tested spore concentrations in the range $5 \times 10^5$ to $2 \times 10^7$ $ml^{-1}$ of several isolates of *Trichoderma* spp. on disease development on detached grapes in the laboratory. They found that the highest spore concentration ($2 \times 10^7$ conidia $ml^{-1}$) was most effective. Dubos (1985) on the other hand tested the effect of the concentrations $10^6$, $10^7$, $10^8$ and $10^9$ conidia $ml^{-1}$ in a field situation and found $10^8$ spores $ml^{-1}$ to be optimal.

Time of spraying is an other important factor in application of biocontrol agents. B. Dubos and coworkers have investigated this over several years. The standard treatment they have used is one spray at each of the following phenological stages: A: flowering, B: closing of bunches, C: changing of colour, D: 3 weeks before harvest. If they compared this treatment with spraying in the flowering period only, they obtained in some years as good a control with this one spray as with four sprays, or they obtained at least 57% of the protection (Dubos *et al.* 1982 a). This clearly shows the importance of protecting the flowers. Adding the antagonist when the vine starts blooming gives rise to a colonization of *Trichoderma* spp. on the senescent floral parts. This colonization by *Trichoderma* spp. prevents the saprophytic establishment of *B. cinerea* on the vine, and delays the initial development of the disease in the vineyard (Dubos & Bulit 1981). Due to the hyperparasitic ability of *Trichoderma* spp., their establishment on the floral parts further limits the development of the grey mould on the bunches (Dubos *et al.* 1982 a).

Dubos *et al.* (1982 a) have further investigated the effect of treatment by *Trichoderma* spp. on the build-up of infection in the field. They found that addition of *Trichoderma* spp. late in the season, when the vine ceases growing, led to a decrease of the number of sclerotia on the branches. No delay in fructification of the sclerotia was found, but the number of conidia was reduced. These different mechanisms may reduce the inoculum present in the vineyard in early spring and be responsible for slower development of the disease (Dubos *et al.* 1982 a).

The importance of a biological treatment in the flowering period is well documented (Dubos *et al.* 1982 a,b), but also spraying at later stages gives better protection in most years. However, the time of spraying after the flowering period is of importance. Strizyk (1982) has investigated the epidemiology of *B. cinerea* on vine. By using his model, which in 1982 meant a delay of the treatment for one week at stage C, an increase in the effectivity of the treatment over the standard procedure from 72.5 to 86% was obtained (Dubos *et al.* 1983). This shows that the epidemiology of the disease must be taken into consideration, in order to obtain the maximal effect of the biocontrol treatment.

### *Control of Phomopsis viticola on vine*

*Phomopsis viticola*, the cause of dead arm disease, has been controlled in six successive years in the Bordeaux region by spraying the vine at the dormant bud stage with *T. harzianum*. Control was comparable to the pre-budding treatment with sodium arsenite or two post-budding treatments with mancozeb (Bugaret *et al.* 1983).

### *Control of B. cinerea on strawberries*

*B. cinerea* causes an important fruit rot, grey mould, wherever strawberries are grown. The fungus may, under favourable disease conditions, attack the berries and other parts of the plant. The disease is favoured by high humidity and low temperature (Coley-Smith *et al.* 1980). The first attempt of biological control of this disease was reported by Bhatt & Vaughan (1962), who obtained a yield increase by spraying with *Cladosporium herbarum* conidia. However, attempts to repeat these results have been unsuccessful (Blakeman & Fokkema 1982).

Also, field trials with *Trichoderma* spp. as the antagonist have been performed. Spraying the strawberries with a conidial suspension ($10^7$ conidia $ml^{-1}$) of *T. viride* and *T. harzianum* three times during the flowering period, the last spray being applied 14 days before the first harvest, reduced natural infection of *B. cinerea* from 21 to 12%. This was as effective as a chemical treatment with dichlofluanid (Tronsmo & Dennis 1977). Further tests with *T. harzianum* have been performed, but with less success than previously reported (Tronsmo 1986).

Trials under controlled conditions in the glasshouse have produced variable results. D'Ercole (1985) obtained satisfactory control with *T. viride*, whereas Gullino *et al.* (1985) were unable to control the disease with an isolate of *Trichoderma* spp. that was effective against *B. cinerea* on grapes.

### *Control of B. cinerea on apple*

*B. cinerea* may attack apple flowers under humid conditions and cause dry eye rot (Tronsmo & Raa 1977 a). Wilted petals are probably an important nutrient base for *B. cinerea* (Tronsmo 1983) and the pathogen is able to penetrate from the petals into the sepals where rot development may start two months later (Tronsmo *et al.* 1977; Tronsmo & Raa 1978).

The first attempt at biological control of this disease showed that *T. pseudokoningii* was able to reduce artificial inoculation but not the natural infections with *B. cinerea* (Tronsmo & Raa 1977 b). The isolate used, however, was unable to grow below 9°C. Because the temperature during the flowering period was often below 9°C, new trials were performed with an isolate selected for antagonistic properties at low temperature (Tronsmo & Dennis 1978). With this *T. harzianum* isolate the natural infection by *B. cinerea* was significantly reduced (Tronsmo & Ystaas 1980; Tronsmo 1983). Several nutrients have been tested as additives to the conidia and the best results have been obtained with soluble cellulose (Tronsmo 1983). Carboxymethylcellulose acts as a nutrient for *Trichoderma* spp. and its sticky nature helps the conidia to stay on the plants. Unfortunately, it is also a nutrient for *B. cinerea*, and without the antagonist it stimulates the disease (Tronsmo 1983). Therefore, a better additive to the spore solution than carboxymethylcellulose is needed.

### *Control of post-harvest diseases of fruits and vegetables*

Biological control of post-harvest diseases should be easier than field protection of the crop, because at the post-harvest stage one is better able to control the climate around the produce. However, few attempts at biological control of post-harvest diseases have been described so far (Cook & Baker 1983; Wilson & Pusey 1985).

Post-harvest rot of strawberries caused by *B. cinerea* and *Mucor piriformis* has been reduced by spraying in the flowering period with *T. viride* and *T. harzianum* (Tronsmo & Dennis 1977; Tronsmo 1986). The biocontrol treatment was more effective against *M. piriformis* than against *B.cinerea* (Tronsmo 1986). De Matos (1983) showed that green mould (*Penicillium digitatum*) could be reduced from 35 to 8% by inoculating *T. viride* with the pathogen into the lemon peel. Rot of apples caused by *B. cinerea* was completely controlled by simultaneous inoculation with *T. pseudokoningii* when the apples were stored at 22°C, but not when stored at 4°C (Tronsmo & Raa 1977 b). This effect of temperature on the antagonistic action of *T. pseudokoningii* can be explained by the different temperature requirements of the two fungi. *B. cinerea* is able to grow below 0°C whereas the *T. pseudokoningii* isolate was unable to grow below 9°C.

When protecting fruits and vegetables in cold storage one therefore needs to take the temperature demand of the antagonist and the parasite into account. Tronsmo & Dennis (1978) were able to select several isolates of *Trichoderma* spp. capable of growth and antagonism at low temperature.

One of these isolates has been used on carrots as a post-harvest dip. The carrots were stored in large bins covered with polyethylene at 0°C for up to 9 months (Tronsmo & Hoftun 1984). The biological treatment reduced the rots caused by *B. cinerea*, *Mycocentrospora acerina* and *Rhizoctonia carotae*, the main spoilage organisms in cold storage in Norway, in two successive years (Tronsmo 1985 b). This experiment shows that it is possible to use *Trichoderma* spp. as a post-harvest treatment, even if the produce is stored close to the minimum growth temperature of the antagonist.

## *INTEGRATED CONTROL*

Integrated control with fungal antagonists and reduced dosages of chemicals has received increased attention during the last few years. Carter & Prince (1974) were the first to combine a biological and a chemical treatment. They used *Fusarium lateritium* and a reduced dose of a benzimidazole fungicide against *Eutypa armeniacae* on apricot trees. In such experiments antagonistic isolates with stable resistance to fungicides are needed. Resistant isolates have been obtained in three ways, viz. after UV irradiation (Papavizas *et al.* 1982; Papavizas & Lewis 1983; Gullino *et al.* 1985), by gamma irradiation (Troutman & Matejka 1978), and by selection of isolates able to grow on agar containing fungicides (Abd-El Moity *et al.* 1982; Tronsmo 1985 a). It has been shown that many mutants are as antagonistic as the wild-type in laboratory tests. In some trials they are even more antagonistic than the wild strain, even without application of fungicides (Papavizas *et al.* 1982; Papavizas & Lewis 1983; Tronsmo 1985 a). In combination with a reduced dose of fungicides, disease control has been as good as that obtained with the recommended dosage of the fungicide alone (Tronsmo 1985 a).

Integrated control with *Trichoderma* spp. and fungicides seems to have a promising future and deserves more attention. By this technique one is able to combine the immediate effect of the fungicide with the long-term effect of the biocontrol agent.

### *Selection of antagonistic isolates*

Selection of antagonistic isolates has traditionally been done after testing them under laboratory conditions on the basis of different antagonistic mechanisms. However, there are several reports which state that there is not necessarily any relationship between antagonistic mechanisms in the laboratory and the ability to control diseases under field conditions (e.g. Corke & Hunter 1979; Tronsmo 1986). This indicates that mechanisms other than those which can be measured in petri dishes, are of importance under field conditions. Among these are probably the ability to live and grow on the plant surface under different nutritional and climatic conditions, and the ability to colonize the plant in such a way that establishment of the necrotrophic pathogen is prevented.

This makes it very difficult to select antagonists for field trials, and the problems are further complicated by the fact that an isolate, although being effective against a pathogen on one crop, may not be effective against the same pathogen on another crop (Tronsmo 1983; Gullino *et al.* 1985). The lack of good selection methods is probably one of the main obstacles in the progress of biological control.

## *DISEASES CAUSED BY TRICHODERMA SPP.*

*Trichoderma* spp. have frequently been isolated from dead or diseased plants and in most cases they are present as a saprophyte or hyperparasite on the primary colonizer (Domsch & Gams 1970). However, *Trichoderma* spp. have also been described as parasites on plants. Årsvoll (1975) occasionally isolated *T. viride* from winter-damaged grassland in Norway, and in pathogenicity tests on cold-stressed grasses the fungus was able to attack the plant. Foot rot of *Cicer arietinum* caused by *T. harzianum* (Kaiser & Sen Gupta 1975) and first internode lesion on maize caused by *T. koningii*, *T. harzianum* and *T. hamatum* (Sutton 1972; McFadden & Sutton 1975) have also been described. Conway (1983) showed that *T. harzianum* could cause rot in artificially infected apple. This means that under some conditions isolates of *Trichoderma* spp. may act as plant pathogens, and this aspect must be taken into consideration when selecting isolates for biocontrol purposes.

## *CONCLUSION*

*Trichoderma* spp. have the advantage as biocontrol agents that they can use many different antagonistic mechanisms, but have the disadvantage that they are not organisms native to the phyllosphere. However, *Trichoderma* spp. are occasionally found in the phyllosphere and reisolation experiments have shown that they are able to survive in the phyllosphere for more than one year.

Success or failure of *Trichoderma* spp. as biocontrol agents is, however, probably dependent on the ability of the isolate used to establish itself in such a way that it can act as an active antagonist in the phyllosphere. The differing ability to do so is probably the reason that there is usually no correlation between the antagonistic mechanism tested in the laboratory and the effect as a biocontrol agent in the field.

Better methods for selection of active isolates either from the natural population or from mutants made by biotechnological methods are needed for further progress in this field. However, I believe that the results obtained so far indicate that further work with *Trichoderma* spp. in the biocontrol of necrotrophic pathogens in the phyllosphere should be continued, and that, if the toxicological aspects of biocontrol can be clarified, I think that *Trichoderma* spp. will be used alone or in an integrated program against several diseases in the future.

## *REFERENCES*

Abd-El Moity, T.H., Papavizas, G.C. & Shatla, M.N. (1982). Induction of new isolates of *Trichoderma harzianum* tolerant to fungicides and their experimental use for control of white rot of onion. Phytopathology, *72*, 396-400.

Ale-Agha, N., Dubos, B., Grosclaude, C. & Ricard, J.L. (1974). Antagonism between nongerminated spores of *Trichoderma viride* and *Botrytis cinerea, Monilia laxa, M. fructigena* and *Phomopsis viticola*. Plant Disease Reporter, *58*, 915-7.

Årsvoll, K. (1975). Fungi causing winter damage on cultivated grasses in Norway. Medlinger fra Norges Landbrukshøgskole, *54* (9), 49 pp.

Austin, R.A. (1657). A Treatise of Fruit Trees. Oxford: Henry Hall.

Aytoun, R.S.C. (1953). The genus *Trichoderma*: its relationship with *Armillaria mellea* (Vahl ex Fries) Quél. and *Polyporus schweinitzii* Fr., together with preliminary observations on its ecology in woodland soils. Transactions of the Botanical Society of Edinburgh, II, *36*, 99-114.

Barak, R., Elad, Y., Mirelman, D. & Chet, I. (1985). Lectins: a possible basis for specific recognition in the interaction of *Trichoderma* and *Sclerotium rolfsii*. Phytopathology, *75*, 458-62.

Bhatt, D.D. & Vaughan, E.K. (1962). Preliminary investigations on biological control of gray mold (*Botrytis cinerea*) of strawberries. Plant Disease Reporter, *46*, 342-5.

Bilaǐ, V.I. (1956). [Volatile antibiotics in fungi of the genus *Trichoderma*.] Microbiology (Moscow), *25*, 458-65 (English transl.).

Bisiach, M., Minervini, G., Vercesi, A. & Zerbetto, F. (1985). Six years of experimental trials on biological control against grapevine grey mould. Quaderni della Scuola di Specializzazione in Viticoltura ed Enologia, Università di Torino, *9*, 285-97.

Blakeman, J.P. (1985). Ecological succession of leaf surface micro-organisms in relation to biological control. *In* Biological Control on the Phylloplane, eds C.E. Windels & S.E. Lindow, 6-30. St. Paul: Am. Phytopath. Soc.

Blakeman, J.P. & Fokkema, N.J. (1982). Potential for biological control of plant diseases on the phylloplane. Annual Review of Phytopathology, *20*, 167-92.

Brian, P.W., Curtis, P.J., Hemming, H.G. & McGowan, J.C. (1946). The production of viridin by pigment-forming strains of *Trichoderma viride*. Annals of Applied Biology, *33*, 190-200.

Brian, P.W. & Hemming, H.G. (1945). Gliotoxin, a fungistatic metabolic product of *Trichoderma viride*. Annals of Applied Biology *32*, 214-20.

Bruce, A., Austin, W.J. & King, B. (1984). Control of growth of *Lentinus lepideus* by volatiles from *Trichoderma*. Transactions of the British Mycological Society, *82*, 423-8.

Bugaret, Y., Dubos, B. & Bulit, J. (1983). L'utilisation du *Trichoderma harzianum* Rifaï dans la pratique viticole pour lutter contre l'excoriose (*Phomopsis viticola* Sacc.). Les Colloques de l'INRA, *18*, 297-302.

Carter, M.V. & Mullett, L.F. (1978). Biological control of *Eutypa armeniacae*. 4. Design and performance of an applicator for metered delivery of protective aerosols during pruning. Australian Journal of Experimental Agriculture and Animal Husbandry, *18*, 287-93.

Carter, M.V. & Price, T.V. (1974). Biological control of *Eutypa armeniacae*. II. Studies of the interaction between *E. armeniacae* and *Fusarium lateritium*, and their relative sensitivities of benzimidazole chemicals. Australian Journal of Agricultural Research, *25*, 105-19.

Chet, I. & Elad, Y. (1983). Mechanism of mycoparasitism. Les Colloques de l'INRA, *18*, 35-40.

Coley-Smith, J.R., Verhoeff, K. & Jarvis, W.R., eds (1980). The Biology of *Botrytis*. London: Academic Press.

Conway, W.S. (1983). *Trichoderma harzianum*: a possible cause of apple decay in storage. Plant Disease, *67*, 916-7.

Cook, R.J. & Baker, K.F. (1983). The Nature and Practice of Biological Control of Plant Pathogens. St. Paul: Am. Phytopath. Soc.

Corke, A.T.K. (1974). The prospect for biotherapy in trees infected by silver leaf. Journal of Horticultural Science, *49*, 391-4.

Corke, A.T.K. (1978). Microbial antagonisms affecting tree diseases. Annals of Applied Biology, *89*, 89-93.

Corke, A.T.K. & Hunter, T. (1979). Biocontrol of *Nectria galligena* infection of pruning wounds on apple shoots. Journal of Horticultural Science, *54*, 47-55.

Corke, A.T.K. & Rishbeth, J. (1981). Use of microorganisms to control plant diseases. *In* Microbial Control of Insects, Mites and Plant Diseases 1970-1980, ed. H.D. Burges, pp. 717-36. London: Academic Press.

De Matos, A.P. (1983). Chemical and microbiological factors influencing the infection of lemons by *Geotrichum candidum* and *Penicillium digitatum*. Ph.D. thesis, University of California, Riverside.

Dennis, C. & Webster, J. (1971 a). Antagonistic properties of species-groups of *Trichoderma* I. Production of non-volatile antibiotics. Transactions of the British Mycological Society, *57*, 25-39.

Dennis, C. & Webster, J. (1971 b). Antagonistic properties of species-groups of *Trichoderma* II. Production of volatile antibiotics. Transactions of the British Mycological Society, *57*, 41-8.

Dennis, C. & Webster, J. (1971 c). Antagonistic properties of species-groups of *Trichoderma* III. Hyphal interaction. Transactions of the British Mycological Society, *57*, 363-9.

D'Ercole, N. (1985). Lotta biologica alla muffa grigia (*Botrytis cinerea*) della fragola con applicazioni di *Trichoderma viride*. Informatore Fitopatologico, *35*, 35-8.

Domsch, K.H. & Gams, W. (1970). Pilze aus Agrarböden. Stuttgart: Gustav Fischer Verlag.

Dubos, B. (1985). L'utilisation des *Trichoderma* comme agent de lutte biologique à l'égard de deux parasites aériens: *Chondrostereum purpureum* (Pers. ex Fr.) Pouzar (plomb des arbres fruitiers) et *Botrytis cinerea* Pers. (pourriture grise de la vigne). Les Colloques de l'INRA (in press).

Dubos, B. & Bulit, J. (1981). Filamentous fungi as biocontrol agents on aerial plant surfaces. *In* Microbial Ecology of the Phylloplane, ed. J.P. Blakeman, pp. 353-67. London: Academic Press.

Dubos, B., Bulit, J., Bugaret, Y. & Verdu, D. (1978). Possibilités d'utilisation du

*Trichoderma viride* Pers. comme moyen biologique de lutte contre la pourriture grise (*Botrytis cinerea* Pers.) et l'excoriose (*Phomopsis viticola* Sacc.) de la vigne. Comptes Rendus des Séances de l'Académie d'Agriculture de France, *64*, 1159-68.

Dubos, B., Jailloux, F. & Bulit, J. (1982 a). Protection du vignoble contre la pourriture grise: les propriétés antagonistes du *Trichoderma* à l'égard du *Botrytis cinerea*. Les Colloques de l'INRA, *11*, 205-19.

Dubos, B., Jailloux, F. & Bulit, J. (1982 b). L'antagonisme microbien dans la lutte contre la pourriture grise de la vigne. EPPO Bulletin, *12*, 171-5.

Dubos, B. & Ricard, J.L. (1974). Curative treatment of peach trees against silver leaf disease (*Stereum purpureum*) with *Trichoderma viride* preparations. Plant Disease Reporter, *58*, 147-50.

Dubos, B., Roudet, J., Bulit, J. & Bugaret, Y. (1983). L'utilisation du *Trichoderma harzianum* Rifaî dans la pratique viticole pour lutter contre la pourriture grise (*Botrytis cinerea* Pers.). Les Colloques de l'INRA, *18*, 289-96.

Durand, A. (1983). Les potentialités de la culture à l'état solide en vue de la production de microorganismes filamenteux. Les Colloques de l'INRA, *18*, 263-77.

Elad, Y., Barak, R. & Chet, I. (1983 a). Possible role of lectins in mycoparasitism. Journal of Bacteriology, *154*, 1431-5.

Elad, Y., Barak, R., Chet, I. & Henis, Y. (1983 b). Ultrastructural studies of the interaction between *Trichoderma* spp. and plant pathogenic fungi. Phytopathologische Zeitschrift, *107*, 168-75.

Elad, Y., Chet, I., Boyle, P. & Henis, Y. (1983 c). Parsitism of *Trichoderma* spp. on *Rhizoctonia solani* and *Sclerotium rolfsii* - Scanning electron microscopy and fluorescence microscopy. Phytopathology, *73*, 85-8.

Elad, Y., Chet, I. & Henis, Y. (1982). Degradation of plant pathogenic fungi by *Trichoderma harzianum*. Canadian Journal of Microbiology, *28*, 719-25.

Elad, Y., Chet, I. & Katan, J. (1980). *Trichoderma harzianum*: a biocontrol agent effective against *Sclerotium rolfsii* and *Rhizoctonia solani*. Phytopathology, *70*, 119-21.

Gear, A. (1985). *Trichoderma* Newsletter. Braintree, Essex: Henry Doubleday Research Ass., 15 pp.

Greig, B.J.W. (1976). Biological control of *Fomes annosus* by *Peniophora gigantea*. European Journal of Forest Pathology, *6*, 65-71.

Grosclaude, C. (1970). Premiers essais de protection biologique des blessures de taille vis-à-vis du *Stereum purpureum* Pers. Annales de Phytopathologie, *2*, 507-16.

Grosclaude, C., Ricard, J. & Dubos, B. (1973). Inoculation of *Trichoderma viride* spores via pruning shears for biological control of *Stereum purpureum* on plum tree wounds. Plant Disease Reporter, *57*, 25-8.

Gullino, M.L. & Garibaldi, A. (1983). Situation actuelle et perspectives d'avenir de la lutte biologique et intégrée contre la pourriture grise de la vigne en Italie. Les Colloques de l'INRA, *18*, 91-7.

Gullino, M.L., Mezzalama, M. & Garibaldi, A. (1985). Biological and integrated control of *Botrytis cinerea* in Italy: experimental results and problems. Quaderni della Scuola di Specializzazione in Viticoltura ed Enologia, Università di Torino, *9*, 299-308.

Hadar, Y., Chet, I. & Henis, Y. (1979). Biological control of *Rhizoctonia solani* damping-off with wheat bran culture of *Trichoderma harzianum*. Phytopathology, *69*, 64-8.

Hulme, M.A. & Shields, J.K. (1970). Biological control of decay fungi in wood by competition for non-structural carbohydrates. Nature (London), *227*, 300-1.

Hulme, M.A. & Shields, J.K. (1972). Interaction between fungi in wood blocks. Canadian Journal of Botany, *50*, 1421-7.

Jones, D. & Watson, D. (1969). Parasitism and lysis by soil fungi of *Sclerotinia sclerotiorum* (Lib.) de Bary, a phytopathogenic fungus. Nature (Lon-

don), *224*, 287-8.
Jones, K.G., Morgan, N.G. & Corke, A.T.K. (1975). Experimental application equipment. Long Ashton Research Station, Bristol, Report 1974, 107.
Kaiser, S.A.K.M. & Sen Gupta, P.K. (1975). A pathogenic strain of *Trichoderma harzianum* causing foot rot of *Cicer arietinum*. Phytopathologische Zeitschrift, *83*, 185-7.
Kallio, T. & Hallaksela, A.-M. (1979). Biological control of *Heterobasidion annosum* (Fr.) Bref. (*Fomes annosus*) in Finland. European Journal of Forest Pathology, *9*, 298-308.
Kuč, J. (1982). Induced immunity to plant disease. BioScience, *32*, 854-60.
Lantero, E., Bazzano, V. & Gullino, M.L. (1982). Tentativi di impiego della lotta integrata nei confronti della *Botrytis cinerea* della vite. La Difesa delle Piante, *5*, 11-20.
McFadden, A.G. & Sutton, J.C. (1975). Relationships of populations of *Trichoderma* spp. in soil to disease in maize. Canadian Journal of Plant Science, *55*, 579-86.
Mercer, P.C., Kirk, S.A., Gendle, P. & Clifford, D.R. (1983). Chemical treatments for control of decay in pruning wounds. Annals of Applied Biology, *102*, 435-53.
Papavizas, G.C. (1985). *Trichoderma* and *Gliocladium*: biology, ecology, and potential for biological control. Annual Review of Phytopathology, *23*, 23-54.
Papavizas, G.C., Dunn, M.T., Lewis, J.A. & Beagle-Ristaino, J. (1984). Liquid fermentation technology for experimental production of biocontrol fungi. Phytopathology, *74*, 1171-5.
Papavizas, G.C. & Lewis, J.A. (1983). Physiological and biocontrol characteristics of stable mutants of *Trichoderma viride* resistant to MBC fungicides. Phytopathology, *73*, 407-11.
Papavizas, G.C., Lewis, J.A. & Abd-El Moity, T.H. (1982). Evaluation of new biotypes of *Trichoderma harzianum* for tolerance to benomyl and enhanced biocontrol capabilities. Phytopathology, *72*, 126-32.
Pottle, H.W., Shigo, A.L. & Blanchard, R.O. (1977). Biological control of wound hymenomycetes by *Trichoderma harzianum*. Plant Disease Reporter, *61*, 687-90.
Ricard, J.L., Wilson, M.M. & Bollen, W.B. (1969). Biological control of decay in Douglas-fir poles. Forest Products Journal, *19*, 41-5.
Rifai, M.A. (1969). A revision of the genus *Trichoderma*. Mycological Papers, No. 116.
Rishbeth, J. (1963). Stump protection against *Fomes annosus*, III. Inoculation with *Peniophora gigantea*. Annals of Applied Biology, *52*, 63-77.
Seaby, D.A. (1977). The possibility of using *Trichoderma viride* for the control of *Heterobasidion annosum* on conifer stumps. *In* Proceedings of Seminar on Biological Control, ed. J.J. Duggan, pp. 57-72. Dublin: Royal Irish Academy.
Seaby, D.A. & Swinburne, T.R. (1976). Protection of pruning wounds on apple trees from *Nectria galligena* Bres. using modified pruning shears. Plant Pathology, *25*, 50-4.
Shigo, A.L. & Wilson, C.L. (1971). Are tree wound dressings beneficial? Arborists News, *36*, 85-8.
Smith, K.T., Blanchard, R.O. & Shortle, W.C. (1979). Effect of spore load of *Trichoderma harzianum* on wood-invading fungi and volume of discolored wood associated with wounds in *Acer rubrum*. Plant Disease Reporter, *63*, 1070-1.
Smith, K.T., Blanchard, R.O. & Shortle, W.C. (1981). Postulated mechanism of biological control of decay fungi in red maple wounds treated with *Trichoderma harzianum*. Phytopathology, *71*, 496-8.
Stople, E. (1982). Tradisjonell og alternativ bekjempelse av gråskimmel i jordbaer. Hovedoppgave NLH, 91 pp.
Strizyk, S. (1982). Modélisation et évaluation du risque parasitaire. Phytoma-

Défense des Cultures, *335*, 6-7.

Sutton, J.C. (1972). *Trichoderma koningii* as a parasite of maize seedlings. Canadian Journal of Plant Science, *52*, 1037-42.

Tamimi, K.M. & Hutchinson, S.A. (1975). Differences between the biological effects of culture gases from several species of *Trichoderma*. Transactions of the British Mycological Society, *64*, 455-63.

Tronsmo, A. (1983). *Trichoderma harzianum* used as a biocotrol agent against *Botrytis cinerea* on apple. Les Colleques de l'INRA, *18*, 109-13.

Tronsmo, A. (1985 a). Muligheter for integrert bekjempelse av soppsykdommer. Informasjonsmøte i Plantevern, Aktuelt fra Statens Fagtjeneste for Landbruket, 1985 (2), 107-13.

Tronsmo, A. (1985 b). *Botrytis cinerea* on stored carrots. Effect of biological control and storage conditions on rot development during long term storage. Quaderni della Scuola di Specializzazione in Viticoltura ed Enologia, Università di Torino, *9*, 310 (Abstr.).

Tronsmo, A. (1986). *Trichoderma* used as a biocontrol agent against *Botrytis cinerea* rots on strawberry and apple. Meldinger fra Norges Landbrukshøgskole, *65* (in press).

Tronsmo, A. & Dennis, C. (1977). The use of *Trichoderma* species to control strawberry fruit rots. Netherlands Journal of Plant Pathology, *83* (suppl. 1), 449-55.

Tronsmo, A. & Dennis, C. (1978). Effect of temperature on antagonistic properties of *Trichoderma* species. Transactions of the British Mycological Society, *71*, 469-74.

Tronsmo, A. & Hoftun, H. (1984). Storage and distribution of carrots. Effect on quality of long term storage in ice bank cooler and cold room, and of different packing materials during distribution. Acta Horticulturae, *163*, 143-50.

Tronsmo, A. & Raa, J. (1977 a). Life cycle of the dry eye rot pathogen *Botrytis cinerea* Pers. on apple. Phytopathologische Zeitschrift, *89*, 203-7.

Tronsmo, A. & Raa, J. (1977 b). Antagonistic action of *Trichoderma pseudokoningii* against the apple pathogen *Botrytis cinerea*. Phytopathologische Zeitschrift, *89*, 216-20.

Tronsmo, A. & Raa , J. (1978). Morphological/cytological description of dry eye rot in apple fruit caused by *Botrytis cinerea* Pers. Acta Agriculturae Scandinavica, *28*, 218-20.

Tronsmo, A., Tronsmo, A.M. & Raa, J. (1977). Cytology and biochemistry of pathogenic growth of *Botrytis cinerea* Pers. in apple fruit. Phytopathologische Zeitschrift, *89*, 208-15.

Tronsmo, A. & Ystaas, J. (1980). Biological control of *Botrytis cinerea* on apple. Plant Disease, *64*, 1009.

Troutman, J.L. & Matejka, J.C. (1978). Induced tolerance of *Trichoderma viride* to benomyl. Phytopathology News, *12*, 131 (Abstr.).

Webster, J. & Lomas, N. (1964). Does *Trichoderma viride* produce gliotoxin and viridin? Transactions of the British Mycological Society, *47*, 535-40.

Weindling, R. (1934). Studies on a lethal principle effective in the parasitic action of *Trichoderma lignorum* on *Rhizoctonia solani* and other soil fungi. Phytopathology, *24*, 1153-79.

Wilson, C.L. & Pusey, P.L. (1985). Potential for biological control of postharvest plant diseases. Plant Disease, *69*, 375-8.

Wood, R.K.S. & Tveit, M. (1955). Control of plant diseases by use of antagonistic organisms. Botanical Review, *21*, 441-92.

Woodgate-Jones, P. & Hunter, T. (1983). Integrated control of *Chondrostereum purpureum* in plum by treatment of pruning wounds. Journal of Horticultural Science, *58*, 491-5.

# USE OF MICROBIAL METABOLITES INDUCING RESISTANCE AGAINST PLANT PATHOGENS

F. Schönbeck and H.-W. Dehne

*Institut für Pflanzenkrankheiten und Pflanzenschutz, University of Hannover, Herrenhäuser Strasse 2, D-3000 Hannover 21, Fed. Rep. Germany.*

## *INTRODUCTION*

The resistance of plants to disease is not a static condition, but has to be regarded as a dynamic process, due to a continuous interaction between host, pathogen and the environment. Whereas complete resistance of plants to diseases is rather rare, within a genetically defined potential, the resistance actually expressed can be mediated by the plant's environment. In practical plant production the influence of climatic conditions can only be controlled in glasshouse-cultivated crops, but not in the field. The use of other cultural measures, e.g. fertilization, irrigation etc., is generally dominated by economic requirements; they are applied to provide for a maximum yield regardless of their influence on the disposition of plants to diseases. No attempt has been made to activate the natural ability of plants susceptible to diseases to withstand plant pathogens. The application of highly efficient chemicals to control the development of pathogens directly on plant surfaces or inside the host made the search for the activation of resistance potentials in plants unnecessary. The problems of pathogen resistance to certain pesticides after repeated application, as well as the increasing economic pressure to keep the expenses in plant production as low as possible, may stimulate interest in alternative ways of disease control.

Besides physical factors, e.g. climate, soil and water, which can influence the actual disposition of plants to disease, micro-organisms are also an important part of the plant's environment. There are indications that phyllosphere organisms, potential plant pathogens as well as saprophytes, can not only reduce the development of parasites by direct antagonism, but are also able to stimulate host plant resistance to diseases. The stimulation of resistance by moderate pre-infection has already been described early in the history of plant pathology (Chester, 1933). Generally, acquired or induced resistance in plants to various diseases could be achieved by different infectious agents and in various plant parts. Many examples of acquired resistance have been shown against leaf pathogens, but also the development of symbiotic fungi in the roots could efficiently induce non-specific resistance in host plants against soil-borne pathogens (Dehne, 1982). Most of the reports demonstrated a high specificity for the inducing agent, but especially for the challenging pathogen or even races of the pathogen (Kuč, 1982). There have been very few attempts to stimulate the non-specific resistance to certain diseases or to improve plant resistance in general.

Although the phenomenon of induced resistance has been investigated for several decades, this basic research did not lead to a new measure in biological plant protection, which could widely be applied in practical agri- and horticulture. The problems of transferring successful experiments into practical plant production are in part due to the use of living organisms in most of the models of induced resistance. Under field conditions the intensity and efficiency of pre-inoculations with pathogens or the application of phyllosphere micro-organisms cannot be controlled and a successful stimulation of disease resistance cannot be guaranteed.

There have been very few attempts to imitate the biological activity of phyllosphere micro-organisms by using living bacteria and fungi or microbial metabolites. For example the experiments of Sequeira (1979) indicated that the application of Gram-negative bacteria could induce resistance in tobacco against bacterial diseases. This activity could be attributed to metabolites formed by the micro-organisms. Also culture filtrates of saprophytic bacteria and fungi could induce resistance against the bean rust fungus, a biotrophic leaf pathogen, without any direct antagonistic effect on the pathogen (Schönbeck *et al.*, 1980). These and further studies on the effects of bacterial and fungal culture filtrates will be discussed.

## *PRINCIPLES OF DISEASE RESISTANCE INDUCED BY MICROBIAL METABOLITES*

As there had been no systematic research on the potential of nonpathogenic micro-organisms to stimulate disease resistance in higher plants, we developed a screening system. Various bacteria and fungi were isolated from plant surfaces as well as from different soils. These organisms were propagated in liquid culture and the filtrates were applied to leaves of bean or wheat plants prior to inoculation with rust or powdery mildew (Fig. 1). Microscopic examination of pathogen development on the plant surface and during initial stages of infection assured that the only filtrates used for further studies, were those which altered the disposition of the plant to disease, but did not directly affect the pathogens during germination and growth before penetration (Schönbeck *et al.* 1980). Among hundreds of bacterial and fungal species or isolates tested some proved to be able to stimulate resistance. In particular, those saprophytes which were capable of inducing resistance against bean rust or wheat powdery mildew, had only a weak or non-detectable direct antagonistic activity against other micro-organisms. When culture filtrates of these saprophytes were applied to wheat or beet plants under field conditions, they had no effect on the numbers of phyllosphere micro-organisms. No significant changes could be observed in the composition of the populations of bacteria, yeasts and filamentous fungi under the influence of the microbial metabolites.

For further experiments, the only bacterial and fungal filtrates used, were those which had no direct effect on the pre-penetration development of the pathogens on the host plant. This lack of antagonistic activity was continuously checked by microscopic examinations.

The induction of resistance by application of microbial culture filtrates was rather non-specific for the host plant. The examination of various host-parasite relationships demonstrated that this kind of resistance was effective only against biotrophic leaf pathogens forming haustoria, whereas perthotrophic organisms were not affected. In a broad screening the culture filtrates proved to be effective in a number of host plants of different families against diseases caused by fungal species from the Peronosporales, Erysiphales and Uredinales (Schönbeck *et al.* 1980; Table 1).

Although living micro-organisms could induce resistance in host plants as well, for further studies only the culture filtrates were used to provide a chemical basis of the inducing agent. The identification of active microbial metabolites is problematic due to the need for a bio-assay to evaluate the active fractions. In order to minimize the interference of accessory metabolites in crude culture filtrates, only those constituents which were necessary to produce an active inducing filtrate were left in the liquid medium. In this way, the complexity of the media could be drastically reduced. The microbial metabolites used for further investigations were produced by a special strain of *Bacillus subtilis*, which had no detectable capacity to produce antifungal compounds (Schönbeck *et al.* 1980; Beicht 1981).

Microscopic examinations of the pathogens' development on the leaf surface of susceptible and induced resistant plants showed that the formation of germ tubes, superficial hyphae and appressoria remained

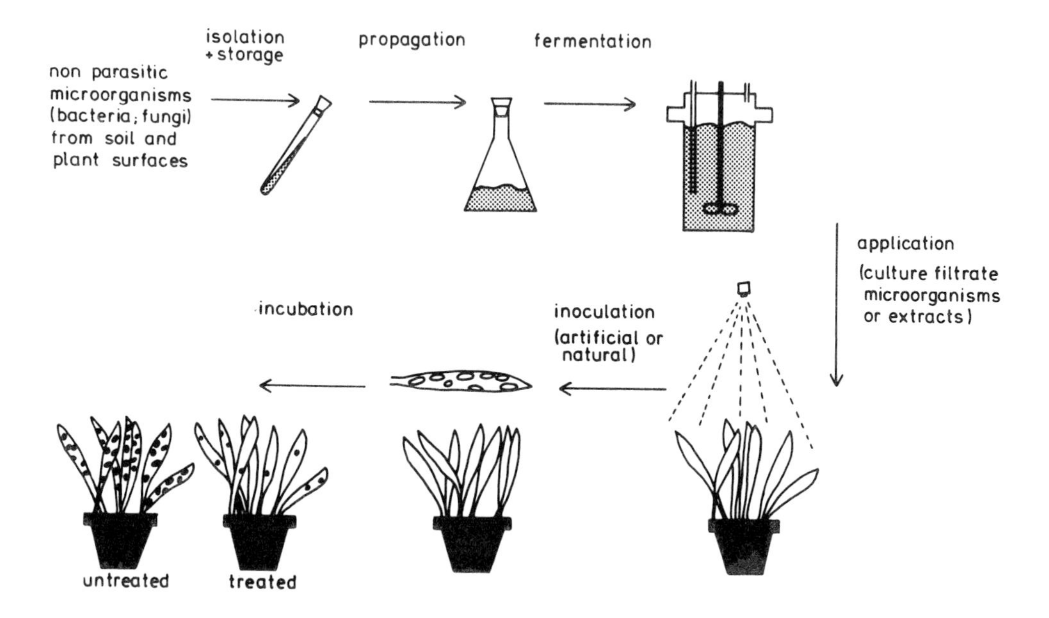


Fig. 1. Scheme for the production and application of microbial metabolites inducing resistance against biotrophic leaf pathogens.

unchanged (Fig. 2). No differences could be observed between the growth of rust or powdery mildew fungi on the leaf surface of treated or untreated plants.

### *CHARACTERISTICS OF RESISTANCE INDUCED BY MICROBIAL METABOLITES*

In induced resistant plants, the pathogen's development was affected, when the pathogen came into contact with the host: the number of successful penetrations was reduced as well as later stages such as the formation of haustoria and secondary mycelium or sporulation (Fig. 2). In general, resistance could be induced successfully in every host plant variety, but the degree of resistance varied with the genotype. The genetic potential of the host has to be regarded as a frame for the induction of resistance. Obviously, it also determines the phase of pathogenesis in which the altered resistance is manifested in the host plant. A most effective reduction of the penetration rate of the powdery mildew fungus could be seen, for example, in one wheat variety, but a lower degree of haustoria formation in another (Fig. 2). In both varieties the induction of resistance led to a reduced colony density and a decrease in sporulation. As this example illustrates, not only the intensity of infection but also the complete development of a disease is influenced by this kind of induced resistance.

The differential reaction of host plant varieties could also be demonstrated for other characteristic parameters such as size and sporulating area of each colony remaining on induced resistant leaves. All colonies formed on the leaves of different wheat varieties were smaller and, in particular, the area in which conidiophores had formed was reduced (Table 2). This effect could also be seen on a variety where the application of the inducing agents did not lead to a reduction in the numeric

Table 1. Hosts in which resistance against a biotrophic pathogen could be induced by treatment with microbial culture filtrates.

| Biotrophic pathogen | | |
|---|---|---|
| Powdery mildew | Downy mildew | Rust |
| apple | cabbage | barley |
| barley | cucumber | bean |
| beet | grape-vine | carnation |
| carrot | lettuce | chrysanthemum |
| grape-vine | rape | maize |
| oat | | wheat |
| wheat | | |

colony density. This characteristic influence on the disease development in general could also be found under practical conditions. As described previously, the efficiency of induced resistance could be confirmed for various host-parasite relationships even in the field (Schönbeck *et al.* 1982). In all experiments, the pathogens were not eliminated completely by the application of microbial metabolites, but the disease intensity could always be kept under an economic threshold. In large-scale experiments, especially the reduced reproduction rate of the pathogen on induced resistant leaves proved to be an epidemiologically interesting phenomenon. For wheat powdery mildew it could be demonstrated that under field conditions not only the number of colonies per leaf was reduced, but also the number of conidia produced in the remaining colonies on induced resistant leaves (Fig. 3). As these colonies were also smaller than those on untreated plants, the sporulation rate per leaf was reduced drastically. When all individual effects are considered together, which means fewer colonies as well as specifically decreased sporulation

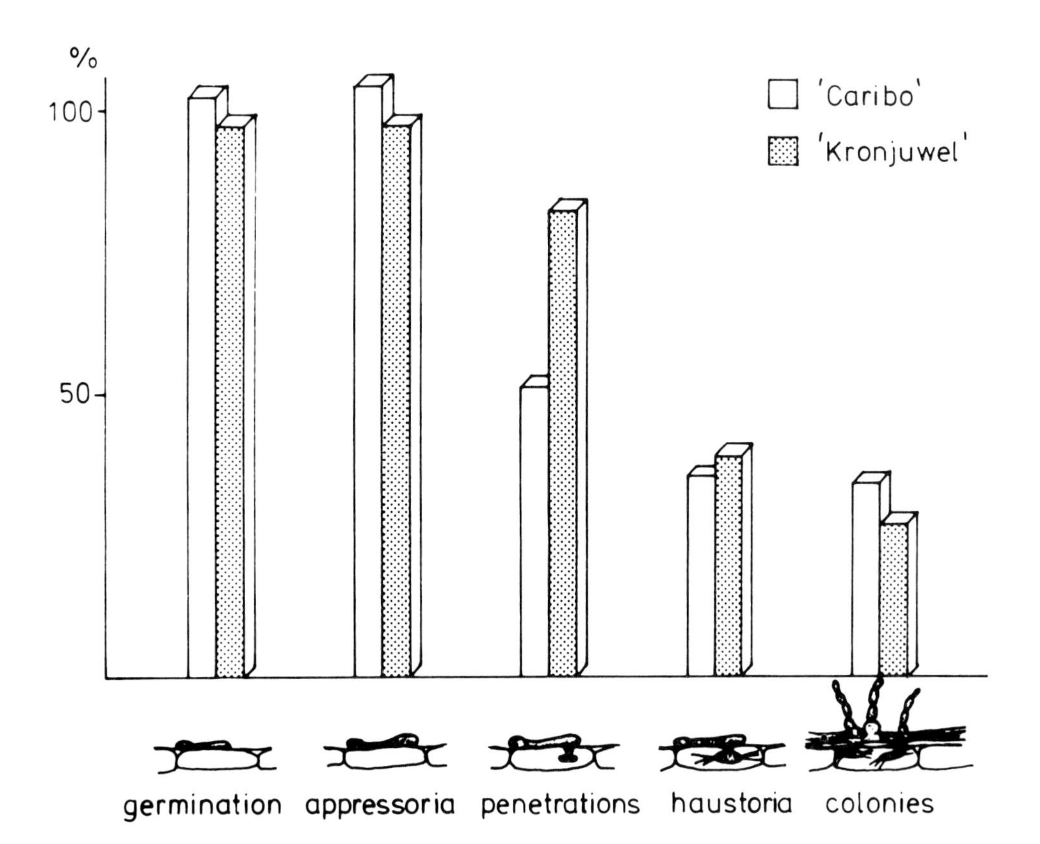


Fig. 2. Effect of microbial culture filtrates on different stages of powdery mildew infection of two wheat varieties, expressed as the percentage of corresponding values found on the untreated variety (Schönbeck & Dehne, unpublished results).

Table 2. Influence of host genotype on the reduction of colony number, colony size and sporulating colony area of wheat powdery mildew on leaves of induced resistant plants, 8 days after inoculation (Dehne & Schönbeck, unpublished results).

| Variety | Number of colonies leaf$^{-1}$ | | Colony size ($mm^{-2}$) | | Sporulating area ($mm^{-2}$) colony$^{-1}$ | |
|---|---|---|---|---|---|---|
| | Un-treated | Induced resistant | Un-treated | Induced resistant | Un-treated | Induced resistant |
| Caribo | 80.2 | 32.7*[1] | 1.08 | 1.00 | 0.45 | 0.25* |
| Kanzler | 67.5 | 15.8* | 1.10 | 0.82* | 0.58 | 0.25* |
| Okapi | 52.3 | 16.9* | 1.06 | 0.82 | 0.52 | 0.23* |
| Kronjuwel | 29.5 | 20.8* | 1.24 | 0.78* | 0.47 | 0.21* |
| Wattiness | 16.7 | 15.9 | 1.21 | 0.56* | 0.46 | 0.16* |

[1] *: $P \leqq 0.05$ (induced resistant versus untreated).

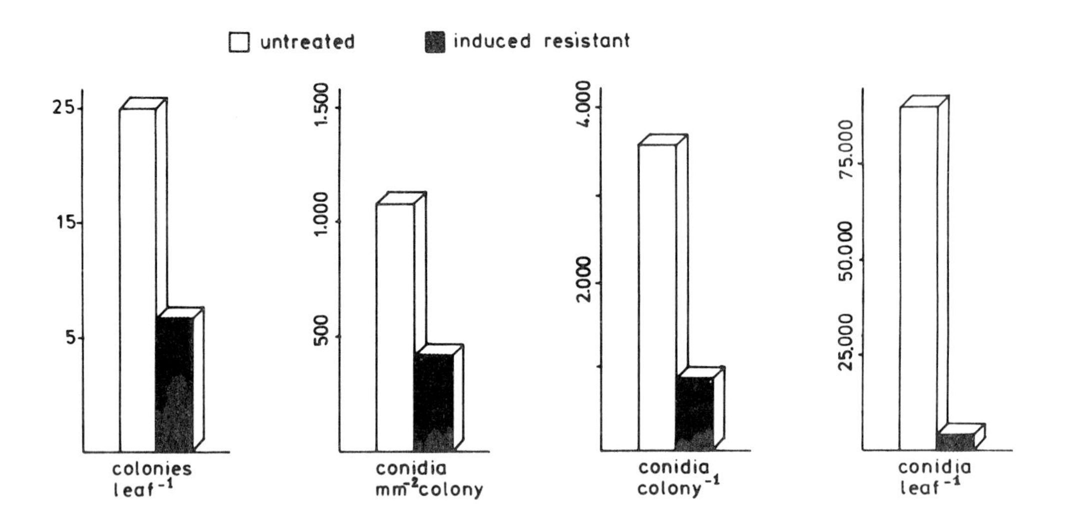


Fig. 3. Influence of induced resistance on wheat powdery mildew development and sporulation under practical field conditions (cv. Caribo; from Dehne *et al.* 1984, modified).

rate, they led to a reduction in disease intensity, which exceeded the initial reduction of colony density considerably. It has to be recognized as a general characteristic of induced resistance that its effectiveness is always less than 100%. A disease will never be completely inhibited by such a system, but it can be kept at a very low level.

Under practical conditions it could also be demonstrated, that repeated applications of microbial metabolites were able to reduce wheat powdery mildew and could prevent yield loss (Dehne *et al.* 1984). The development of powdery mildew could be efficiently reduced by the application of microbial metabolites while the control of the pathogen was almost complete if a commercial fungicide was used (Fig. 4). The reduction of grain yield could be prevented with the application of microbial metabolites as well as with the fungicide spray.

In general, experiments for biological control of plant pathogens and diseases can be very effective under laboratory conditions, but tend to fail, if they are transferred to practical conditions (Leben *et al.* 1965). In the case of resistance induced by microbial metabolites almost the opposite situation could be found: the reduction of powdery mildew of wheat on induced resistant plants was higher in field experiments compared to glasshouse experiments (Table 3). Increasing variation of temperatures stimulated the effectiveness of induced resistance, whereas nearly constant day and night temperatures led only to a slight reduc-

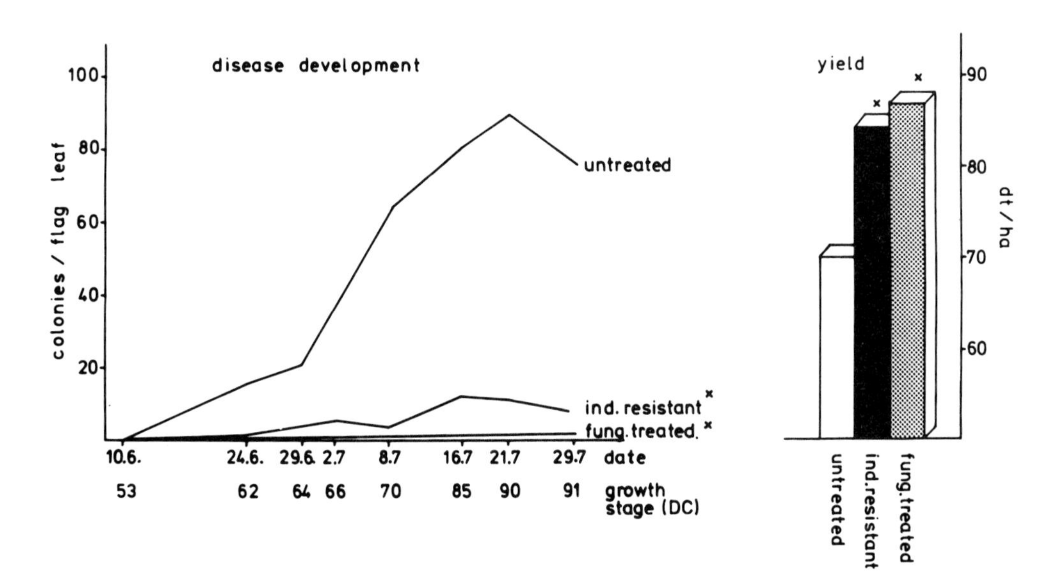


Fig. 4. Reduction of powdery mildew development on wheat (cv. Caribo) and improvement of yield under practical conditions after repeated applications of inducer (5x between stages DC 53 and 85) or one application of a systemic fungicide (triadimefon, at stage DC 52). ×: P≦0.05 (untreated versus treated). From Dehne *et al.* (1984), modified.

tion of powdery mildew density in experiments carried out under controlled conditions. These and other results suggest that environmental stresses, e.g. more extreme temperatures, sensitize plants to induced resistance rather than interfere negatively with it. This fact may in part explain the high efficiency of microbial metabolites in stimulating resistance under field conditions.

The induction of resistance in plants against biotrophic leaf pathogens could also be characterized by a reduction of haustoria formation and a decreased functioning of these pathogenic infection structures. As it could be seen in glasshouse and field experiments, the number of epidermal cells under each colony, containing many haustoria, was much smaller in induced resistant leaves (Table 3). Additionally, the efficiency of these haustoria was reduced under the influence of the microbial metabolites. There were fewer and smaller colonies (Table 3) with a smaller number of haustoria on induced resistant leaves, and the production of conidia per conidiophore and especially that of conidia per haustorium per day was decreased (Table 4). This reduced efficiency of single haustoria has to be regarded as a further parameter for this type of induced resistance.

## *MECHANISMS OF RESISTANCE INDUCED BY MICROBIAL METABOLITES*

The induction of non-specific resistance in plants to disease has been correlated with many possible mechanisms. The stimulation

Table 3. Effect of inducer treatment on powdery mildew development on wheat (cv. Caribo) grown in a glasshouse or in the field. Disease intensity on untreated plants was similar in both conditions. Ten days after the inducer treatment the 2nd leaves and flag leaves were examined in the glasshouse and the field experiment, respectively (Schönbeck & Dehne, unpublished results).

| Treatment of leaves | Number of colonies $leaf^{-1}$ | Colony size ($mm^2$) | Number of infected epidermal cells $colony^{-1}$ with | |
|---|---|---|---|---|
| | | | ≦5 haustoria | >5 haustoria |
| *Glasshouse experiment* | | | | |
| None | 112.3 | 1.9 | 33.9 | 58.5 |
| Inducer treatment | 64.0*[1] | 1.4 | 50.6 | 19.9* |
| *Field experiment* | | | | |
| None | 82.4 | 5.6 | 77.8 | 313.0 |
| Inducer treatment | 11.2* | 2.7* | 106.2 | 87.9* |

[1] *: P≦0.05 (inducer treatment versus untreated).

of plants to form inhibitors of fungal growth, especially phytoalexins, as well as an enhanced lignification have been described as a cause of increased resistance (Kuč & Caruso 1977; Hammerschmidt & Kuč 1982; Kuč 1982). Furthermore, induced resistance has been related to altered receptor proteins (Kerr 1969; Lippincott & Lippincott 1975; Sequeira 1979). The type of resistance against biotrophic leaf pathogens induced by microbial metabolites could not be related to these defense mechanisms. Furthermore, the formation of papillae-like cellular depositions has been described as a potential defense mechanism (Aist 1976). Under the influence of induced resistance these cellular reactions could be stimulated as well as decreased (Table 5). No correlation could be found between the efficiency of the microbial metabolites to induce resistance in leaves of different wheat varieties against powdery mildew and the formation of cell wall depositions.

The reduction of fungal development in induced resistant leaves was not correlated with any deformations of the fungus such as can be found after fungicide treatment or accumulation of antifungal compounds. Additionally no morphological alterations of host cells, such as in cells undergoing necrosis or a hypersensitive response, were observed. Induction of resistance by microbial metabolites did not change host plant physiology substantially. No antifungal compounds such as phytoalexins could be detected (Table 6). But those physiological reactions in a susceptible plant, which are found as a response to powdery mildew infection, could be seen in induced resistant plants to a lesser extent. For example, the inhibition of carbohydrate translocation from a susceptible leaf due to powdery mildew development normally leads to an accumulation of starch in the chloroplasts under the colony. This situation, typical for compatible host-parasite relationships, could not be found under the influence of the microbial metabolites. The stimulation of enzymatic activities, which occurs in diseased tissues, was less pronounced in induced resistant leaves. In part this may only be a consequence of a reduced disease intensity, but the lack of those reactions,

Table 4. Effect of inducer treatment on haustoria formation and efficiency of powdery mildew haustoria in leaves of glasshouse-grown wheat plants (cv. Caribo, 8 days after inoculation) (from Stenzel *et al.* 1985, modified).

| Treatment of leaves | Number of haustoria colony$^{-1}$ | Number of haustoria mm$^{-2}$ colony | Number of conidia conidiophore$^{-1}$ d$^{-1}$ | Number of conidia haustorium$^{-1}$ d$^{-1}$ |
|---|---|---|---|---|
| None | 1908 | 310 | 7.64 | 2.75 |
| Inducer treatment | 594*[1] | 176* | 4.13* | 1.50* |

[1] *: P≦0.05 (inducer treatment versus untreated).

Table 5. Effect of inducer treatment on host reaction and development of powdery mildew in leaves of different glasshouse-grown wheat varieties (Schönbeck & Dehne, unpublished results).

| Variety | Relative occurrence (%)[1] of | | | |
|---|---|---|---|---|
| | Cellular depositions appressorium$^{-1}$ | Successful penetrations appressorium$^{-1}$ | Primary haustoria appressorium$^{-1}$ | Colonies leaf$^{-1}$ |
| Kronjuwel | -39*[2] | -29 | -53* | -71* |
| Caribo | +46* | -47* | -53* | -66* |
| Okapi | -35 | -45* | -64* | -64* |
| Kanzler | -20 | -33 | -50* | -56* |
| Jubilar | -2 | -5 | -23 | -40* |
| Wattiness | +26* | -11 | -12 | -12 |

[1] Values indicate differences (%) from corresponding values on untreated plants (100%).
[2] *: $P \leq 0.05$ (inducer treatment versus untreated).

Table 6. Effect of inducer treatment on host plant metabolism after inoculation of wheat leaves with powdery mildew (K. Stenzel, unpublished results).

| Compounds or enzymes | Change in content or activity[1] | |
|---|---|---|
| | Untreated plants | Induced resistant plants |
| Inhibitory compounds | 0 | 0 |
| Carbohydrates | ++ | + |
| Peroxidase | ++ | + |
| Glucose-6-P dehydrogenase | ++ | + |
| Malate dehydrogenase | 0 | 0 |
| Isocitrate dehydrogenase | ++ | ++ |
| Invertase | ++ | + |
| Alkaline phosphatase | ++ | + |

[1] 0: unchanged content or activity; +, ++: increased content or activity.

which are necessary to establish a successful host-parasite relationship, may also be regarded as the reason for the reduced infection in induced resistant plants.

The described type of resistance against biotrophic leaf pathogens induced by microbial metabolites could not be correlated with often quoted typical defense mechanisms. But it shows similarities to the characteristics of partial resistance. It appears to be based on an impairment of fungal nutrition as a result of the reduction in haustorial efficiency. This may explain that this type of resistance is limited to those diseases caused by biotrophic fungi forming haustoria (Stenzel *et al.* 1985). Reduced nutrient supply and lower efficiency of haustoria are correlated with partial resistance of plants (Shaner 1973; Carver & Carr 1978; Asher & Thomas 1983; Royer *et al.* 1984). Both systems, induced resistance as well as partial resistance, lead to a similar restricted development of colonies. It is likely that the induction of resistance by microbial metabolites is not based on the activation of new resistance mechanisms, but on the stimulation of those mechanisms which are described for partial resistant plants and are present in susceptible plants to a much lower degree. The identification of active, inducing agents can offer not only access to the actual mechanisms responsible for this kind of resistance, but may also offer a potential key to alter the compatibility of host and pathogen.

## *CONCLUSIONS*

The use of microbial metabolites to induce resistance in plants against biotrophic leaf pathogens provides for a new measure in biological plant protection. The efficiency of this system to control a disease especially under field conditions stimulates the hope that this can be used in practical horti- and agriculture in the future. The application of living micro-organisms in practical disease control has been mostly disappointing. Bacterial antagonists, for example, were effective in inhibiting the development of leaf spot in maize, but the micro-organisms died immediately when the leaf surface dried out (Sleesman & Leben 1976). The successful stimulation of disease resistance in plants by pre-inoculation with various infectious agents has been reported frequently (Yarwood 1956; Ouchi *et al.* 1976; Randall & Helton 1976; Kuč 1982). The establishment of induced resistance by pre-inoculation with pathogens is also difficult to obtain under practical conditions. Additionally some of the described systems proved to be rather specific for the inducing agent or a certain disease. The specificity of microbial metabolites to induce resistance was rather low. The whole range of diseases caused by biotrophic leaf pathogens could be controlled in various host plants and even under field conditions. The application of effective microbial metabolites can give further access to the general principles of induced resistance, once the mechanisms of this phenomenon are elucidated. Because these metabolites are not applied against a particular pathogen like conventional pesticides, but are used against an entire disease, this also offers new aspects for plant protection in the future. The pathogen does not need to be eliminated, which subsequently leads to problems of resistance against the controlling agent, but the disease must be retarded or kept under an economic threshold. Resistance induced by microbial metabolites or chemical substances in general will not displace highly efficient pesticides, but can become an additional tool to protect plants from diseases and to prevent yield losses.

*ACKNOWLEDGEMENTS*

This research was supported by the Deutsche Forschungsgemeinschaft.

*REFERENCES*

Aist, J.R. (1976). Papillae and related wound plugs of plant cells. Annual Review of Phytopathology, *14*, 145-63.

Asher, M.J.C. & Thomas, L.E. (1983). The expression of partial resistance to *Erysiphe graminis* in spring barley. Plant Pathology, *32*, 79-89.

Beicht, W. (1981). Untersuchungen zur Induktion von Resistenzmechanismen in Pflanzen durch mikrobielle Stoffwechselprodukte. Dissertation, University of Hannover.

Carver, T.L.W. & Carr, A.J.H. (1978). Effects of host resistance on the development of haustoria and colonies of oat mildew. Annals of Applied Biology, *88*, 171-8.

Chester, K.S. (1933). The problem of acquired physiological immunity in plants. Quarterly Review of Biology, *8*, 129-54, 275-324.

Dehne, H.-W. (1982). Interaction between vesicular-arbuscular mycorrhizal fungi and plant pathogens. Phytopathology, *72*, 1115-9.

Dehne, H.-W., Stenzel, K. & Schönbeck, F. (1984). Zur Wirksamkeit induzierter Resistenz unter praktischen Anbaubedingungen. III. Reproduktion Echter Mehltaupilze auf induziert resistenten Pflanzen. Zeitschrift für Pflanzenkrankheiten und Pflanzenschutz, *91*, 258-65.

Hammerschmidt, R. & Kúc, J. (1982). Lignification as a mechanism for induced systemic resistance in cucumber. Physiological Plant Pathology, *20*, 61-71.

Kerr, A. (1969). Crown gall of stone fruit. I. Isolation of *Agrobacterium tumefaciens* and related species. Australian Journal of Biological Sciences, *22*, 111-6.

Kuč, J. (1982). Induced immunity to plant disease. BioScience, *32*, 854-60.

Kuč, J. & Caruso, F. (1977). Activated coordinated chemical defense against disease in plants. *In* Host Plant Resistance to Pests, ed. P. Hedin, pp. 78-89, American Chemical Society Symposium Series 63. Washington, DC: Am. Chem. Soc. Press.

Leben, C., Daft, G.C., Wilson, J.D. & Winter, H.F. (1965). Field tests for disease control by an epiphytic bacterium. Phytopathology, *55*, 1375-6.

Lippincott, J.A. & Lippincott, B.B. (1975). The genus *Agrobacterium* and plant tumorigenesis. Annual Review of Microbiology, *29*, 377-405.

Ouchi, S., Hibino, C. & Oku, H. (1976). Effect of earlier inoculation on the establishment of a subsequent fungus as demonstrated in powdery mildew of barley by a triple inoculation procedure. Physiological Plant Pathology, *9*, 25-32.

Randall, H. & Helton, A.W. (1976). Effect of inoculation date on induction of resistance to *Cytospora* in Italian prune trees by *Cytospora cincta*. Phytopathology, *66*, 206-7.

Royer, M.H., Nelson, R.R., MacKenzie, D.R. & Diehle, D.A. (1984). Partial resistance of near-isogenic wheat lines compatible with *Erysiphe graminis* f. sp. *tritici*. Phytopathology, *74*, 1001-6.

Schönbeck, F., Dehne, H.-W. & Balder, H. (1982). Zur Wirksamkeit induzierter Resistenz unter praktischen Anbaubedingungen. I. Echter Mehltau an Reben, Gurken und Weizen. Zeitschrift für Pflanzenkrankheiten und Pflanzenschutz, *89*, 177-84.

Schönbeck, F., Dehne, H.-W. & Beicht, W. (1980). Untersuchungen zur Aktivierung unspezifischer Resistenzmechanismen in Pflanzen. Zeitschrift für Pflanzenkrankheiten und Pflanzenschutz, *87*, 654-66.

Sequeira, L. (1979). The acquisition of systemic resistance by prior inoculation. *In* Recognition and Specificity in Plant Host-Parasite Interactions, eds J.M. Daly & I. Uritani, pp. 231-51. Tokyo: Japan Scientific Societies Press; Baltimore: University Park Press.

Shaner, G. (1973). Reduced infectability and inoculum production as factors of slow mildewing in Knox wheat. Phytopathology, *63*, 1307-11.

Sleesman, J.P. & Leben, C. (1976). Microbial antagonists of *Bipolaris maydis*. Phytopathology, *66*, 1214-8.

Stenzel, K., Steiner, U. & Schönbeck, F. (1985). Effect of induced resistance on the efficiency of powdery mildew haustoria in wheat and barley. Physiological Plant Pathology, *27*, 357-67.

Yarwood, C.E. (1956). Cross protection with two rust fungi. Phytopathology, *46*, 540-4.

# INDEX